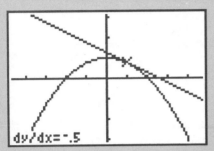

A function is drawn with a tangent line at $x = 1$. (Section 2.6)

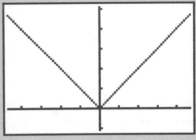

$y = |x|$ is a function that is everywhere continuous but not differentiable at $x = 0$. (Section 2.7)

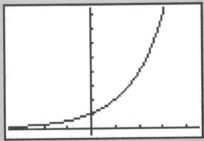

$y = b^x$ displays exponential growth if $b > 1$. (Section 5.1)

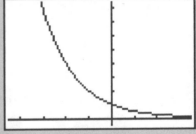

$y = b^x$ exhibits exponential decay if $0 < b < 1$. (Section 5.1)

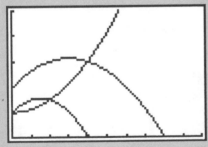

Marginal analysis: revenue, cost, and profit functions. (Section 3.5)

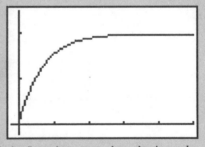

Logistics/learning curves have horizontal asymptotes. (Sections 5.1, 8.3)

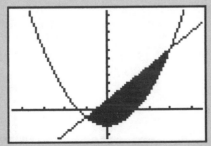

The shaded area between two curves. (Section 6.4)

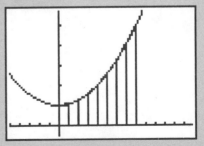

The area under a curve can be estimated using trapezoids. (Section 7.3)

APPLIED CALCULUS

Fourth Edition

APPLIED CALCULUS

Fourth Edition

Claudia Dunham Taylor
University of Cincinnati

Lawrence Gilligan
University of Cincinnati

Brooks/Cole Publishing Company

I(T)P™ An International Thomson Publishing Company

Pacific Grove • Albany • Bonn • Boston • Cincinnati • Detroit • London • Madrid • Melbourne
Mexico City • New York • Paris • San Francisco • Singapore • Tokyo • Toronto • Washington

Sponsoring Editor: *Robert W. Pirtle*
Marketing Team: *Patrick Farrant and Jean Vevers Thompson*
Editorial Assistant: *Linda Row*
Production Editor: *Laurel Jackson*
Manuscript Editor: *Carol Dondrea*
Interior Design: *Detta Penna*
Interior Illustration: *John Foster, Precision Graphics, and ST Associates, Inc.*

Cover Design: *Roy R. Neuhaus*
Cover Photo: *Doris De Witt / Tony Stone Images*
Art Editor: *Lisa Torri*
Typesetting: *Weimer Graphics*
Cover Printing: *Color Dot Graphics, Inc.*
Printing and Binding: *R.R. Donnelley & Sons Company, Crawfordsville Mfg. Division*

About the cover: This photograph is of the water cannon on the Chicago River. It shoots every hour on the hour. The path the water takes is a parabolic arc, one of the fundamental function shapes discussed in this book. See Section 2.6 for more about the calculus of parabolic arcs.

For more information, contact:

BROOKS/COLE PUBLISHING COMPANY
511 Forest Lodge Road
Pacific Grove, CA 93950
USA

International Thomson Publishing Europe
Berkshire House 168-173
High Holborn
London WC1V 7AA
England

Thomas Nelson Australia
102 Dodds Street
South Melbourne, 3205
Victoria, Australia

Nelson Canada
1120 Birchmount Road
Scarborough, Ontario
Canada M1K 5G4

International Thomson Editores
Campos Eliseos 385, Piso 7
Col. Polanco
11560 México D.F. México

International Thomson Publishing GmbH
Königswinterer Strasse 418
53227 Bonn
Germany

International Thomson Publishing Asia
221 Henderson Road
#05-10 Henderson Building
Singapore 0315

International Thomson Publishing Japan
Hirakawacho Kyowa Building, 3F
2-2-1 Hirakawacho
Chiyoda-ku, Tokyo 102
Japan

Printed in the United States of America

10 9 8 7 6 5 4 3 2 1

Library of Congress Cataloging-in-Publication Data

Taylor, Claudia [date]
    Applied calculus / Claudia Dunham Taylor, Lawrence Gilligan. —
4th ed.
        p.   cm.
    Includes index.
    ISBN 0-534-33971-9 (hardcover : alk. paper)
    1. Calculus. I. Gilligan, Lawrence G. II. Title.
QA303.T204 1996
515—dc20
                                    95-36624
                                    CIP

This book is dedicated to
Richard and Susan

PREFACE

This book represents the fourth generation of a text that was written expressly for students majoring in business, management science, economics, life science, social science, and psychology. As such, it recognizes both the particular needs of that student audience and the unique set of constraints placed on instructors teaching the calculus in a three- or four-hour, one-semester course.

Our main purpose in writing *Applied Calculus* was to make the calculus *meaningful* to students who would be going directly into applied disciplines. To achieve our goal, we have presented differential and integral calculus within an applied, problem-solving setting. If students leave this course with a strong sense of the usefulness of calculus as a tool for solving real-world problems, then we as educators have succeeded in meeting the most fundamental goal of the course.

We are proud of the fact that, as the manuscript for this book was reviewed by our colleagues across the country, the most often-repeated comment was made about our writing style. From using the text in class, we know that students *do* read it and don't just use it as a collection of homework exercises. To enhance the book's "coefficient of readability," we have incorporated the following features:

- New concepts and techniques are presented directly and concisely. Prose is enhanced by numerous step-by-step examples. In our applied calculus classes, each of us has found the technique of teaching by carefully selecting examples to be highly effective.
- We have included enough theory to substantiate the mathematical development of concepts for the purposes of this course and this student audience. As a result, the instructor can decide how much theory to include in the course because proofs and verifications generally appear at the end of their sections. (See, for example, the treatment of the product and quotient rules in Section 3.2.)
- The design and format of the book have been especially crafted with its audience in mind. Important rules, theorems, and properties are boxed for easy reference. Graphics and diagrams are used liberally throughout the text to help students visualize ideas.
- We have found that students need to master vocabulary in order to read any mathematics text. Consequently, for easy reference, new vocabulary terms are listed in vocabulary/formula lists at the end of each chapter. We have also included a glossary in this edition, immediately following Appendix C.
- Exercise sets are carefully constructed. There are four types of exercises: Set A, Set B, Writing and Critical Thinking, and Using Technology in Calculus. Set A exercises contain groups of exercises keyed to specific examples in that section. Each example serves as a direct problem-solving model for the student. Set B

includes application problems and the more challenging exercises. Writing and Critical Thinking exercises are optional and are designed to encourage students to think about concepts and express their reasoning in writing. Instructors may also use these exercises as group activities for their students. Using Technology in Calculus exercises are also optional and are designed for students who have access to computer programs or graphics calculators. These exercises enhance the concepts by using the power of the technology. As with the Writing and Critical Thinking exercises, instructors may want to consider assigning Using Technology in Calculus exercises as group activities.

ADDITIONAL FEATURES

New visual examples and exercises have been added to this edition. For instance, see Section 2.1, Example 9; Section 2.7, Exercises 17–22; Section 4.1, Example 2; and Section 4.2, Exercise 30. We have also added numerous graphs to help students visualize particular applications.

Because the use of technology varies from instructor to instructor, we have built in a great deal of flexibility. This edition contains eight Alternative Sections that focus on using the graphics calculator. They include the following:

1.6A Lines on a Graphics Calculator
2.5A Asymptotes on a Graphics Calculator
2.6A Slopes of Tangents/Rates of Change on a Graphics Calculator
4.1A Increasing and Decreasing Functions on a Graphics Calculator
4.4A Absolute Extrema on a Graphics Calculator
5.1A Exponential Functions on a Graphics Calculator
5.2A Logarithmic Functions on a Graphics Calculator
7.3A Simpson's Rule on a Graphics Calculator

These sections are designed to replace the traditional sections on these topics. Although the actual graphics screens from a TI-82 calculator are displayed, the sections are not calculator-specific—any graphing technology can be used.

The Using Technology in Calculus exercises offer a challenging look at the concepts of calculus using either computer software or graphics calculators.*

A new section, "Problem Solving," appears between Chapters 1 and 2 and is referred to several times throughout the text. This section is intended to help students increase their problem-solving ability.

Chapter Reviews include vocabulary/formula lists, review exercises (keyed to specific sections), and chapter tests. Each chapter test has two parts: Part 1 is a

---

*We have chosen to concentrate on three technology components: *DERIVE*®, a computer algebra system; *CONVERGE*®, a general-purpose mathematics software package; and the *TI-82 Graphics Calculator*®. The authors would like to thank the following three companies for their cooperation in and enthusiasm toward this project:

    DERIVE is a registered trademark of Soft Warehouse, Inc. (3660 Waialae Ave., Suite 304, Honolulu, HI 96816; (808) 734-5801).
    CONVERGE is a registered trademark of JEMware (The Kawaiahao Plaza Executive Center, 567 South King Street, Suite 178, Honolulu, HI 96813; (808) 947-1853).
    TI-82 Graphics Calculator is a trademark of Texas Instruments, Inc. (7800 Banner Drive, Dallas, TX 75265; (214) 917-1539).

multiple-choice test; Part 2 is a more standard test composed of questions that require complete answers.

ANCILLARIES

The *Instructor's Manual* contains the answers to even-numbered exercises, suggestions on how to use the Writing and Critical Thinking exercises, answers to and suggestions regarding the Writing and Critical Thinking exercises, and answers to the Using Technology in Calculus exercises.

The *Test Items* offer two forms of tests for each chapter.

An *Electronic Test Bank,* using Brooks/Cole's *EXP-Test®*, contains a wide variety of questions (ranked easy, medium, and difficult) and can be used to create many different forms of tests of equivalent difficulty. *EXP-Test* is available free to adopters of Brooks/Cole texts.

A *Student Solutions Manual* is available for sale in your college bookstore. It contains the completely worked-out solutions to all odd-numbered exercises.

A manual, *Calculus and the DERIVE® Program: Experiments with the Computer, Third Edition,* by L. Gilligan and J. Marquardt, is available as a supplement for instructors or students interested in using the computer algebra system *DERIVE* as part of the optional technology component of the text. Ordering information for this manual can be obtained by writing Gilmar Publishing, P.O. Box 6376, Cincinnati, OH 45206 or by calling (513) 751-8688.

OTHER FOURTH EDITION CHANGES

- Section 1.5 from the third edition has been split into two sections, 1.5 ("Rectangular Coordinate System") and 1.6 ("Lines").
- Section 2.1 from the third edition has been split into two sections to incorporate additional visual examples and exercises.
- Section 3.3, "The Chain Rule," and Section 5.5, "Application to Economics: Elasticity of Demand," have been rewritten.
- Appendix A has been rewritten for using the TI-82 Graphics Calculator.

ACKNOWLEDGMENTS

This book has benefited from the contributions of many student users—from early versions of the original manuscript to users of the first three editions of this work. We would like to thank the excellent reviewers of this new edition for their many valuable suggestions. These reviewers include Christopher D. Barone, Bowling Green State University; Linda Becerra, University of Houston–Downtown; Dennis Brewer, University of Arkansas; Ellen Church, Golden West College; Darrell Clevidence, Carl Sandburg College; Lily Crowley, Lexington Community College; Yvette Hester, Texas A & M University; Vuryl Klassen, California State University at Fullerton; Phillip McGill, Illinois Central College; Beverly K. Michael; R. L. Richardson, Appalachian State University; Charles E. Schwarz; Henry Smith; Dan Streeter, Portland State University; Lenore Vest, Lower Columbia College; and Charles Votaw, Ft. Hays State University. Jeanne Bowman has worked with us for the last two editions of this text and has been a valuable contributor. We appreciate her talent and her input.

We would also like to thank editors Gary Ostedt and Bob Pirtle for their support on this project. Also, a special thanks to the production crew at Brooks/Cole for their hard work. They include Dorothy Bell, Patrick Farrant, Laurie Jackson, Roy Neuhaus, Elizabeth Rammel, Linda Row, Jean Thompson, and Lisa Torri. In addition, we want to thank manuscript editor Carol Dondrea and designer Detta Penna.

Claudia Taylor
Lawrence Gilligan

CONTENTS

APPENDIX B   TABLES 611

APPENDIX C   ANSWERS TO SELECTED EXERCISES 618

APPLIED CALCULUS

Fourth Edition

ALGEBRA REVIEW

1.1   SIMPLIFYING EXPRESSIONS

There are many situations in solving mathematical problems in which it is important to perform some kind of *simplification* process. One of the most basic types of simplification is called **combining like terms. Terms** are parts of an expression joined by addition (or subtraction) and **like terms** are terms that differ from each other *only* by the numerical coefficients. To *combine* like terms, add (or subtract) their coefficients. The following examples illustrate this process.

Example 1   Simplify each of the following by combining like terms:
a. $2x + 3 + 5x - 2 - y + 1$
b. $7x^2 - 4y + x - 2x^2 + 3x + y$

Solution   a.   Rearranging the terms, we have

$$\underbrace{2x + 5x}_{7x} - \underbrace{y}_{-y} + \underbrace{3 - 2 + 1}_{2}$$

(Notice that the sign preceding a term is part of it.) Therefore, the simplified form is

$$7x - y + 2$$

b. $\underbrace{7x^2 - 2x^2}_{5x^2} + \underbrace{x + 3x}_{4x} - \underbrace{4y + y}_{3y}$

The result is $5x^2 + 4x - 3y$.   ∎

The next example involves expressions containing *grouping* symbols of parentheses and brackets. The procedure is to start on the inside and work your way out.

Example 2   Simplify the following:
a. $(2x + 3) - (4x - 1)$ b. $2[5x - 3(x + 2)]$

Solution   a.   When removing the parentheses, change the sign of the terms in the second "piece," because the operation is subtraction.

$$(2x + 3) - (4x - 1)$$
$$2x + 3 - 4x + 1 \qquad \text{Combine like terms.}$$
$$\text{or } -2x + 4$$

b. Start inside with the parentheses.

$$2[5x - 3(x + 2)]$$
$$2[5x - 3x - 6] \qquad \text{Use the distributive property.}$$
$$2[2x - 6] \qquad \text{Combine like terms.}$$
$$4x - 12 \qquad \text{Distributive property.} \qquad \blacksquare$$

In addition to *simplifying* an algebraic expression, we also, under certain circumstances, want to *find the value of* an expression. To *evaluate* $3x - 4y$ when $x = 2$ and $y = 7$, for example, means to substitute 2 for x and 7 for y in $3x - 4y$.

$$\begin{array}{cc} 3x - & 4y \\ \downarrow & \downarrow \end{array}$$
$$3 \cdot 2 - 4 \cdot 7 \qquad \text{Multiplication is performed before subtraction.}$$
$$6 - 28$$
$$-22$$

Example 3 Evaluate the following expression for $x = -2$:

$$4x^3 - 2x + 3$$

Solution We need to replace x with -2 everywhere x appears in the expression.

$$4(-2)^3 - 2(-2) + 3$$
$$4(-8) + \quad 4 + 3$$
$$-32 + \quad 7$$
$$-25 \qquad \blacksquare$$

Before we simplify any expressions that involve multiplying variables, we review some definitions and properties for exponents and radicals in Table 1.1.

Table 1.1

EXPONENTS AND RADICALS

Property	Example
1. $x^m \cdot x^n = x^{m+n}$	$x^3 \cdot x^5 = x^8$
2. $\dfrac{x^m}{x^n} = x^{m-n}, x \neq 0$	$\dfrac{x^7}{x^4} = x^3$
3. $(x^m)^n = x^{mn}$	$(x^3)^2 = x^6$
4. $(xy)^n = x^n y^n$	$(xy)^4 = x^4 y^4$
5. $\left(\dfrac{x}{y}\right)^n = \dfrac{x^n}{y^n}, y \neq 0$	$\left(\dfrac{x}{y}\right)^3 = \dfrac{x^3}{y^3}$
6. $x^{-n} = \dfrac{1}{x^n}, x \neq 0$	$x^{-3} = \dfrac{1}{x^3}$
7. $x^0 = 1, x \neq 0$	$3^0 = 1$
8. $x^{1/n} = \sqrt[n]{x}$	$x^{1/3} = \sqrt[3]{x}$
9. $x^{m/n} = \sqrt[n]{x^m}$ or $(\sqrt[n]{x})^m$	$x^{2/3} = \sqrt[3]{x^2}$ or $(\sqrt[3]{x})^2$

Property 1 is used in conjunction with the distributive property in the next two examples.

Example 4 Simplify each of the following:
  a. $2x(3x^2 + 1)$ **b.** $(x^4 + 3)(2x^4 - 4)$

Solution **a.** Using the distributive property, we get

$$2x(3x^2 + 1)$$
$$2x \cdot 3x^2 + 2x \cdot 1$$
or $\qquad 6x^3 + 2x$

b. Using the distributive property twice produces four products, as illustrated:

$$(x^4 + 3)(2x^4 - 4)$$
$$x^4 \cdot 2x^4 + 3 \cdot 2x^4 - 4 \cdot x^4 - 4 \cdot 3$$
$$2x^8 \;+\; 6x^4 \;-\; 4x^4 \;-\; 12 \qquad \text{Combine like terms.}$$
or $\qquad 2x^8 + 2x^4 - 12$

Example 5 Simplify the following:
  $x(2x - 3) - 4x(x^2 - 1)$

Solution
$$x(2x - 3) - 4x(x^2 - 1)$$
$$2x^2 - 3x - 4x^3 + 4x$$
or $\qquad -4x^3 + 2x^2 + x$

The process of multiplying two expressions by applying the distributive property sometimes results in expressions that have a specific form. It will be important, in Section 1.2, to recognize these forms, called **special products,** and we summarize them in Table 1.2.

The presence of a *fractional* exponent means that an expression may be rewritten using a radical symbol, as properties 8 and 9 in Table 1.1 suggest. For example, $8^{1/3}$ is another way of writing $\sqrt[3]{8}$ and $9^{3/2}$ means $(\sqrt{9})^3$. Examples 6 and 7 show how you can convert from fractional exponent form to radical form and vice versa.

Table 1.2

Example	Name	Form
$(x + 4)(x - 4) = x^2 + 4x - 4x - 16$ $= x^2 - 16$	Difference of two squares	$(x + a)(x - a)$ $= x^2 - a^2$
$(x + 5)(x - 6) = x^2 + 5x - 6x - 30$ $= x^2 - x - 30$	Trinomial (x^2 coefficient is 1)	$(x + a)(x + b)$ $= x^2 + (a + b)x + ab$
$(2x - 5)(x - 7) = 2x^2 - 5x - 14x + 35$ $= 2x^2 - 19x + 35$	Trinomial (x^2 coefficient $\neq$ 1)	$(ax + b)(cx + d)$ $= acx^2 + (ad + bc)x + bd$
$(x - 2)(x^2 + 2x + 4) = x^3 + 2x^2 + 4x - 2x^2$ $- 4x - 8 = x^3 - 8$	Difference of two cubes	$(x - a)(x^2 + ax + a^2)$ $= x^3 - a^3$
$(x + 3)(x^2 - 3x + 9) = x^3 - 3x^2 + 9x + 3x^2$ $- 9x + 27 = x^3 + 27$	Sum of two cubes	$(x + a)(x^2 - ax + a^2)$ $= x^3 + a^3$

Example 6   Change to radical form:

     a. $2x^{2/3}$ **b.** $x^{-1/2}$ **c.** $(x - 1)^{1/3}$ **d.** $x^{-4/3}$

     Solution   **a.** $2\sqrt[3]{x^2}$ **b.** $\dfrac{1}{\sqrt{x}}$ **c.** $\sqrt[3]{x - 1}$ **d.** $\dfrac{1}{\sqrt[3]{x^4}}$

             (not $\sqrt[3]{2x^2}$)   ∎

Example 7   Change to exponential form:

     a. $-5\sqrt{x}$ **b.** $\dfrac{x}{\sqrt[4]{2 - x}}$ **c.** $\sqrt[5]{x^4}$

     Solution   **a.** $-5x^{1/2}$ **b.** $x(2 - x)^{-1/4}$ **c.** $x^{4/5}$ ∎

Example 8   Evaluate each of the following for the given value:

     a. $2x^{2/3}$ for $x = -8$

     b. $\dfrac{1}{\sqrt[4]{1 - x}}$ for $x = 0$

c. $-5x^{-1/2}$ for $x = 9$

Solution a. $2(-8)^{2/3} = 2[(-8)^{1/3}]^2 = 2[-2]^2 = 2[4] = 8$

b. $\dfrac{1}{\sqrt[4]{1-0}} = \dfrac{1}{\sqrt[4]{1}} = 1$

c. $-5(9)^{-1/2} = \dfrac{-5}{\sqrt{9}} = \dfrac{-5}{3}$ ∎

It is sometimes desirable in a rational expression involving radicals to have no radicals in the denominator or in some cases no radicals in the numerator. The process of removing radicals from the denominator (or numerator) is called **rationalizing** the denominator (or numerator).

Example 9   Rationalize the denominator in the following:

a. $\dfrac{3}{\sqrt{2}}$ b. $\dfrac{1}{1 - \sqrt{x}}$

Solution a. Multiply by $\sqrt{2}$ in both the numerator and denominator.

$$\frac{3}{\sqrt{2}} \cdot \frac{\sqrt{2}}{\sqrt{2}} = \frac{3\sqrt{2}}{\sqrt{4}} = \frac{3\sqrt{2}}{2}$$

b. In this case, we need to multiply by $1 + \sqrt{x}$ in the numerator and denominator.

$$\frac{1}{1 - \sqrt{x}} \cdot \frac{1 + \sqrt{x}}{1 + \sqrt{x}} = \frac{1 + \sqrt{x}}{1 - \sqrt{x} + \sqrt{x} - \sqrt{x^2}}$$
$$= \frac{1 + \sqrt{x}}{1 - x}$$ ∎

Example 10   Rationalize the numerator for $\dfrac{\sqrt{x} + \sqrt{x + 3}}{3}$.

Solution   Multiply by $\sqrt{x} - \sqrt{x + 3}$ in the numerator and denominator.

$$\frac{\sqrt{x}+\sqrt{x+3}}{3}\cdot\frac{\sqrt{x}-\sqrt{x+3}}{\sqrt{x}-\sqrt{x+3}}$$

$$=\frac{\sqrt{x^2}+\sqrt{x}\sqrt{x+3}-\sqrt{x}\sqrt{x+3}-\sqrt{(x+3)^2}}{3(\sqrt{x}-\sqrt{x+3})}$$

$$=\frac{x-(x+3)}{3(\sqrt{x}-\sqrt{x+3})}$$

$$=\frac{-3}{3(\sqrt{x}-\sqrt{x+3})}$$

$$=\frac{-1}{\sqrt{x}-\sqrt{x+3}}$$ ∎

We conclude this section by multiplying and simplifying expressions involving fractional exponents.

Example 11   Simplify by multiplying and combining like terms:

$$2x^{1/2}(x^{-1/2}+1)-x^{-1/2}(x^{1/2}-x)$$

Solution $2x^{1/2}(x^{-1/2}+1)-x^{-1/2}(x^{1/2}-x)$

$2x^{1/2}\cdot x^{-1/2}+2x^{1/2}-x^{-1/2}\cdot x^{1/2}+x^{-1/2}\cdot x^1$

$\qquad 2x^0\quad+2x^{1/2}-\quad x^0\quad+\quad x^{1/2}\qquad (x^0=1)$

or $\qquad\qquad 3x^{1/2}+1$ ∎

1.1 EXERCISES

SET A

In exercises 1 through 14, perform the indicated operations and simplify. (See Examples 1, 2, 4, and 5.)

1. $4x^2-2x-x^2-5x+3$
2. $5a-4b+2a-3+b$
3. $(2x-x^2)-(5+3x^2)$
4. $(x^2+1)-(x^2+2x)-(x+4)$
5. $4(x-3)-2(4x+1)$
6. $2x(x+1)-4(2x+5)$
7. $-2[5-3(x-2)]$
8. $6[2(1-x)-3(2-x)]$
9. $x^2(2x^2-x+4)$
10. $-2x^3(x^2+x-1)$
11. $(2x+3)(x-7)$
12. $x(x+3)(x-4)$
13. $5x-x[2x-(x-3)]$
14. $2x[4-2(x^2-1)]+3x$

In exercises 15 through 22, let $x=-2$ and $y=3$. Evaluate each expression. (See Example 3.)

15. $2x^3-3x+5$
16. $4y^2-5y+1$
17. $3x^4-x+6$
18. y^3-y+7
19. $5x^2y-3xy^2$
20. $8x^3y-5xy^3$
21. x^2+y^2
22. $(x+y)^2$

In exercises 23 through 28, change each expression to a radical form. (See Example 6.)

23. $x^{3/4}$

24. $-x^{4/3}$

25. $4x^{1/2}$

26. $-4x^{-1/2}$

27. $3(x + 2)^{-1/2}$

28. $-x(1 - 2x)^{2/3}$

In exercises 29 through 34, change each radical expression to an exponential form. (See Example 7.)

29. $\sqrt[3]{x}$

30. $x\sqrt{x}$

31. $-2x\sqrt[4]{x}$

32. $\dfrac{1}{\sqrt[3]{x^2 + 1}}$

33. $\dfrac{2x}{\sqrt[3]{x}}$

34. $\dfrac{x - 2}{\sqrt{x + 1}}$

In exercises 35 through 44, evaluate each expression for the given value. (See Example 8.)

35. $x^{1/2}$ for $x = 16$

36. $x^{1/3}$ for $x = 27$

37. $x^{1/3}$ for $x = -27$

38. $x^{2/3}$ for $x = 64$

39. $x^{2/3}$ for $x = -64$

40. $x^{-2/3}$ for $x = 64$

41. $5x^{3/4}$ for $x = 16$

42. $-2x^{-1/3}$ for $x = 27$

43. $\dfrac{x}{\sqrt[3]{7 - x}}$ for $x = -1$

44. $(3x - 1)\sqrt{x + 4}$ for $x = 0$

In exercises 45 through 50, rationalize the denominator. (See Example 9.)

45. $\dfrac{5}{\sqrt{6}}$

46. $\dfrac{2}{\sqrt{8}}$

47. $\dfrac{\sqrt{x}}{1 + \sqrt{x}}$

48. $\dfrac{\sqrt{x}}{1 - \sqrt{x}}$

49. $\dfrac{6}{\sqrt{x} - 4}$

50. $\dfrac{6}{\sqrt{x} - \sqrt{y}}$

In exercises 51 through 56, rationalize the numerator. (See Example 10.)

51. $\dfrac{\sqrt{x}}{x}$

52. $\dfrac{\sqrt{x}}{5}$

53. $\dfrac{\sqrt{x} + 1}{2}$

54. $\dfrac{1 - \sqrt{x}}{x}$

55. $\dfrac{\sqrt{x} + \sqrt{x + 1}}{2}$

56. $\dfrac{\sqrt{x + h} - \sqrt{x}}{h}$

In exercises 57 through 60, simplify each expression. (See Example 11.)

57. $x^{-2/3}(x + 2x^{2/3})$

58. $x^{-1/4}(x - x^{1/2})$

59. $x^{1/2}(x - 1) + 2x^{-1/2}(3x)$

60. $(x^{3/2} + 1)(x^{3/2} - 1)$

SET B

In exercises 61 through 64, perform the indicated operations and simplify.

61. $-3[4 - 2(2x - 1)] + 4x$

62. $-2[4(-1 - x) + 4(2 - 3x)]$

63. $(2x + 3)(x^2 + x - 1)$

64. $(x + 1)(x + 2)(x + 3)$

In exercises 65 and 66, change each radical expression to an exponential form.

65. $\dfrac{2x}{\sqrt[3]{(x^2 + 1)^2}}$

66. $\dfrac{2x + 1}{\sqrt[4]{x^2 + x + 3}}$

In exercises 67 and 68, evaluate each expression assuming that x = 16 and y = −27.

67. $x^{1/2} + y^{1/3}$

68. $\dfrac{-x^{3/2}}{y^{2/3}}$

In exercises 69 and 70, rationalize the denominator.

69. $\dfrac{2}{\sqrt{3} - \sqrt{5}}$

70. $\dfrac{\sqrt{x}}{\sqrt{x} - \sqrt{y}}$

71. The relationship between algebra and geometry can be seen in multiplying two binomials. Algebra-ically finding $(x + 5)(x + 3)$, for example, is geometrically equivalent to finding the area of a rectangle with sides $(x + 5)$ and $(x + 3)$:

Algebra: $(x + 5)(x + 3) = x^2 + 5x + 3x + 15 = x^2 + 8x + 15$

Geometry:

Total area:

$x^2 + 5x + 3x + 15$

or

$x^2 + 8x + 15$

	x	3
x	x^2	$3x$
5	$5x$	15

Use a similar diagram to find $(x + 9)(x + 2)$.

72. **a.** Expand $(x + y)^2$.

b. The area, A, of a square with side s is: $A = s^2$. What expression represents the area of the $(x + y)$ by $(x + y)$ square shown here?

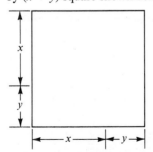

c. Find the area of the following figure by summing the four distinct areas shown and compare your answer with parts a and b.

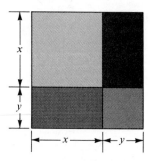

1.2 FACTORING

To **factor** an expression is to rewrite it as the product of other expressions. The process of factoring is an important skill we use often in later chapters. In this section, we emphasize the distributive "multiplying-out" process but *in reverse*. For example, in Section 1.1 we multiplied $2x$ by $3x^2 + 1$ and obtained

$2x(3x^2 + 1)$ The "$2x$" distributes over $3x^2$ and 1.
$6x^3 + 2x$

In this section, we begin with an expression like $6x^3 + 2x$, and our goal is to write it as a product, $2x(3x^2 + 1)$.

The first category of factoring that we examine involves searching for a **common factor.** An expression is called a common factor if it is a factor in *every* term of the expression. In $3x^2y^2 - 6xy^2 + 12x$, for example, notice that 3 and x are factors present in each term:

$3x^2y^2 \rightarrow 3x \cdot xy^2$
$-6xy^2 \rightarrow 3x(-2y^2)$
$+12x \rightarrow 3x \cdot 4$

Thus, we write

$3x^2y^2 - 6xy^2 + 12x = 3x(xy^2 - 2y^2 + 4)$

An example of finding the common factor follows.

Example 1   Factor $2x^3 - 4x^2$.

Solution
$2x^3 \quad - \quad 4x^2$
$\underline{2 \cdot x \cdot x \cdot x} - 2 \cdot \underline{2 \cdot x \cdot x}$ $2 \cdot x \cdot x$ or $2x^2$ is common to each term.
$2x^2(x - 2)$ Notice we are "left" with $x - 2$
    after $2x^2$ is extracted.   ∎

Once the common factor is found, the remaining factor(s) may be further factorable. If they are, they may be one of the *special products* summarized in Table 1.2.

The strategy for *completely* factoring an expression is summarized in Table 1.3.

Table 1.3

Step 1. Always check for the *common factor* first.

Step 2. Is the expression a ⎡*two*⎤-termed expression?
If yes, then try one of these three forms:
  1. Difference of two squares:

$$x^2 - a^2 = (x - a)(x + a)$$

  2. Difference of two cubes:

$$x^3 - a^3 = (x - a)(x^2 + ax + a^2)$$

  3. Sum of two cubes:

$$x^3 + a^3 = (x + a)(x^2 - ax + a^2)$$

Step 3. If it is a ⎡*three*⎤-termed expression (or trinomial), it may fall into one of these groups (see also Table 1.2).
  1. The coefficient of x^2 is 1. Example: $x^2 + 7x + 12$. Find two numbers whose sum is 7 and whose product is 12. They are 3 and 4:

$$x^2 + 7x + 12 = (x + 3)(x + 4)$$

  2. The coefficient of x^2 is not 1. Example: $6x^2 + 19x + 15$.
    a. Find the product of first and last coefficients: $6 \cdot 15 = 90$.
    b. Look for two numbers whose product is 90 and whose sum is 19: 9 and 10.
    c. Write the expression as *four* terms:

$$6x^2 + 9x + 10x + 15$$

    d. Proceed to use Step 4 as follows:

$$6x^2 + 9x + 10x + 15$$
$$3x(2x + 3) + 5(2x + 3)$$
$$(2x + 3)(3x + 5)$$

Step 4. If it is a ⎡*four*⎤-termed expression, try factoring by grouping.

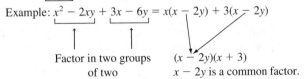

Example: $x^2 - 2xy + 3x - 6y = x(x - 2y) + 3(x - 2y)$

Factor in two groups $(x - 2y)(x + 3)$
of two $x - 2y$ is a common factor.

Examples of completely factoring various expressions follow.

Example 2 Completely factor $2x^3 - 8x$.

Solution There is a common factor of $2x$

$$2x^3 - 8x = 2x(x^2 - 4)$$

Now, $x^2 - 4$ is factorable—it is the difference of two squares. We have

$$2x^3 - 8x = 2x(x^2 - 4)$$
$$= 2x(x + 2)(x - 2)$$ ∎

Example 3   Factor completely:

a. $2x^2 - 5x - 12$ **b.** $2x^3 - 2x^2 - 12x$ **c.** $x^4 - 3x^2 - 4$

Solution   **a.**   Using Table 1.3, Step 3, we find the product of the first and last coefficients, $2(-12) = -24$, and look for two numbers whose product is -24 and whose sum is -5. They are 3 and -8, and we write

$$2x^2 - 5x - 12 = 2x^2 + 3x - 8x - 12$$
$$= x(2x + 3) - 4(2x + 3)$$
$$= (2x + 3)(x - 4)$$

b.   First factor out $2x$ from each term.

$$2x^3 - 2x^2 - 12x = 2x\underbrace{(x^2 - x - 6)}_{\text{factors}}$$
$$= 2x(x - 3)(x + 2)$$

c. $x^4 - 3x^2 - 4 = \underbrace{(x^2 - 4)}_{\text{factors}}(x^2 + 1)$

$$= (x - 2)(x + 2)(x^2 + 1) \qquad \text{\small\textit{Note: Sum of squares is not factorable.}}$$ ∎

To get us ready for the next examples, let us examine the expression $2x^3 + 4x^{-2}$ and see what happens if we factor out $2x^{-2}$. We make this choice because 2 is common to both terms and x^{-2} is the smallest power of x.

$$2x^3 + 4x^{-2} = 2x^{-2}(x^5 + 2)$$

Check this result by using the distributive property to multiply. Writing the result without negative exponents gives

$$\frac{2(x^5 + 2)}{x^2}$$

In the expression

$$2x(x + 3) - 5(x + 3)$$

we have two main terms:

$$2x(x + 3) \qquad \text{and} \qquad -5(x + 3)$$

We can factor out the common factor, which is $x + 3$ (see Table 1.3, Step 4):

$$2x(x + 3) - 5(x + 3) = (x + 3)(2x - 5)$$

Both of the previous ideas will be used in the following examples.

Example 4   In the expression

$$-3x^2(x + 4)^{-4} + 2x(x + 4)^{-3}$$

factor out x and $(x + 4)^{-4}$ and rewrite the result without negative exponents.

Solution $-3x^2(x + 4)^{-4} + 2x(x + 4)^{-3} = x(x + 4)^{-4}[-3x + 2(x + 4)^1]$
$$= x(x + 4)^{-4}[-3x + 2x + 8]$$
$$= x(x + 4)^{-4}(-x + 8)$$
$$= \frac{x(-x + 8)}{(x + 4)^4}$$ ∎

Example 5   Factor and leave all exponents positive:

   a. $x^2(2x + 1)^{-1/2} + 2x(2x + 1)^{1/2}$ **b.** $-\frac{1}{3}x(1 - x)^{-4/3} + (1 - x)^{-1/3}$

Solution   **a.**   Factor out x and $(2x + 1)^{-1/2}$.

$$x^2(2x + 1)^{-1/2} + 2x(2x + 1)^{1/2} = x(2x + 1)^{-1/2}[x + 2(2x + 1)^1]$$
$$= x(2x + 1)^{-1/2}(5x + 2)$$
$$= \frac{x(5x + 2)}{(2x + 1)^{1/2}} \quad \text{or} \quad \frac{x(5x + 2)}{\sqrt{2x + 1}}$$

   b.   Factor out $(1 - x)^{-4/3}$. Factoring out $\frac{1}{3}$ will also help.

$$-\frac{1}{3}x(1 - x)^{-4/3} + (1 - x)^{-1/3} = \frac{1}{3}(1 - x)^{-4/3}[-x + 3(1 - x)^1]$$
$$= \frac{1}{3}(1 - x)^{-4/3}(3 - 4x)$$
$$= \frac{3 - 4x}{3(1 - x)^{4/3}} \quad \text{or} \quad \frac{3 - 4x}{3\sqrt[3]{(1 - x)^4}}$$ ∎

Applications of factoring include simplifying algebraic fractions and adding and subtracting fractions. These are explored in Examples 6 and 7.

Example 6   Simplify by factoring and reducing to lowest terms:

   a. $\dfrac{2x^2 - 8}{x^2 + x - 6}$ **b.** $\dfrac{x^2 + 2x + 1}{\sqrt[3]{(x + 1)^2}}$

Solution   **a.** $\dfrac{2x^2 - 8}{x^2 + x - 6} = \dfrac{2(x^2 - 4)}{(x + 3)(x - 2)}$ Factor.

$$= \frac{2(x + 2)(x - 2)}{(x + 3)(x - 2)}$$ Divide numerator and denominator by the common factor, $x - 2$.

$$= \frac{2(x + 2)}{x + 3}$$

b.   Factor the numerator and write the denominator in exponential form.

$$\frac{x^2 + 2x + 1}{\sqrt[3]{(x + 1)^2}} = \frac{(x + 1)^2}{(x + 1)^{2/3}}$$

$$= (x + 1)^{4/3} \qquad \text{Subtract exponents.} \qquad \blacksquare$$

In Example 7, the process of factoring is used to help find the lowest common denominator when adding fractions.

Example 7   Add:

$$\frac{2x}{x - 2} + \frac{3x}{x^2 - 5x + 6}$$

Solution   To add two fractions, we find the lowest common denominator of $(x - 2)$ and $x^2 - 5x + 6$. Writing $x^2 - 5x + 6$ in factored form as $(x - 2)(x - 3)$, however, helps us to see that the common denominator is $(x - 2)(x - 3)$.

$$\frac{2x}{x - 2} + \frac{3x}{(x - 2)(x - 3)} = \frac{2x(x - 3)}{(x - 2)(x - 3)} + \frac{3x}{(x - 2)(x - 3)}$$

$$= \frac{2x^2 - 6x + 3x}{(x - 2)(x - 3)}$$

$$= \frac{2x^2 - 3x}{(x - 2)(x - 3)}$$

$$= \frac{x(2x - 3)}{(x - 2)(x - 3)} \qquad \blacksquare$$

1.2   EXERCISES

SET A

In exercises 1 through 16, factor each expression completely. (See Examples 1, 2, and 3.)

1. $6x^4 + 2x^3$
2. $-4x^3 - 6x$
3. $x^2 + 7x + 12$
4. $x^2 + 5x - 36$
5. $3x^2 + 4x - 4$
6. $6x^2 + 5x - 4$
7. $8 + 2x - x^2$
8. $3 + x - 2x^2$
9. $x^4 - 9x^2$
10. $x^4 - 2x^2 + 1$
11. $x^3 - 27$
12. $8 - 27x^3$
13. $6x^3 + 3x^2 - 3x$
14. $-x^4 - x^3 + 2x^2$
15. $5x(x - 3) + 2(x - 3)$
16. $x^2(2x + 1) - (2x + 1)$

In exercises 17 through 24, factor each expression and leave all exponents positive in the answer. (See Examples 4 and 5.)

17. $-8x^3(2x + 1)^{-3} + 6x^2(2x + 1)^{-2}$

18. $-x^3(2x^2 + 3)^{-2} + 3x^2(2x^2 + 3)^{-1}$

19. $-2(x + 1)^2(x - 4)^{-3} + 3(x + 1)(x - 4)^{-2}$

20. $-8(x + 2)^2(2x - 1)^{-5} + 2(x + 2)(2x - 1)^{-4}$

21. $3x^2(2x + 1)^{1/3} - x^3(2x + 1)^{-2/3}$

22. $2x(x + 1)^{1/3} + \frac{1}{3}(x + 1)^{-2/3}(x^2 + 1)$

23. $x^{-1/2}(1 - x)^2 - 4x^{1/2}(1 - x)$

24. $3(2x)^{1/2}(x + 1)^2 + (2x)^{-1/2}(x + 1)^3$

In exercises 25 through 34, factor and reduce to lowest terms. (See Example 6.)

25. $\dfrac{x^2 - 9}{x - 3}$

26. $\dfrac{x^2 + x - 2}{x^2 - 2x + 1}$

27. $\dfrac{x^2 - 9}{3 - x}$

28. $\dfrac{9x^2 - 1}{3x^2 + 11x - 4}$

29. $\dfrac{x^6 - 16x^2}{x^4 + 5x^2 + 4}$

30. $\dfrac{1 - y^2}{y^2 - 1}$

31. $\dfrac{x^2 - 2x - 3}{\sqrt{x - 3}}$

32. $\dfrac{x^2 - 2x + 1}{\sqrt[3]{1 - x}}$

33. $\dfrac{x^2 + x}{\sqrt{x}}$

34. $\dfrac{2x(x + 3) - (x + 3)}{\sqrt{x + 3}}$

In exercises 35 through 42, add or subtract as indicated and simplify. (See Example 7.)

35. $\dfrac{2}{x} + \dfrac{1}{x^2}$

36. $\dfrac{3}{2x} - \dfrac{2}{x^2}$

37. $\dfrac{x}{x + 1} - \dfrac{2x}{x - 3}$

38. $\dfrac{x}{x + 1} - \dfrac{2x}{x^2 - 1}$

39. $5y - \dfrac{1}{y^2}$

40. $y + 3 + \dfrac{1}{y}$

41. $\dfrac{2}{x - 1} + \dfrac{3}{x - 2} + \dfrac{4}{x - 3}$

42. $\dfrac{2x + 1}{x - 1} + \dfrac{3x + 4}{x^2 + 2}$

SET B

In exercises 43 through 48, factor each expression and leave all exponents positive in the answer.

43. $6x^4(3x - 1)^{-3} + 12x^3(3x - 1)^{-2}$

44. $2x(x + 1)^{-1/2} - 4(x + 1)^{1/2}$

45. $4x^2(2 - x)^{1/3} - x^3(2 - x)^{-2/3}$

46. $x^{-1/2}(4 - x) - 4x^{1/2}(4 - x)^2$

47. $(x + 2)^2(x + 4)^5 + (x + 2)^3(x + 4)^4$

48. $(x + 2)^{-2}(x + 4)^{-5} + (x + 2)^{-3}(x + 4)^{-4}$

BUSINESS
COST

49. The Firmware Corporation has determined that its cost for producing x items is given by

$$0.02x^2 - 4x^{-3}$$

Factor this expression, leaving all exponents positive in the answer.

PHYSICAL SCIENCE
FREE-FALLING BODIES

50. When a ball is thrown upward with an initial velocity of 112 feet per second from the top of a 480-foot-tall building, its height in feet is given by the expression

$$-16t^2 + 112t + 480$$

Factor the expression.

WRITING AND CRITICAL THINKING

Because you had algebra before taking calculus, you have already used factoring for various kinds of problems. Give an example of a type of problem where factoring is useful and explain in words how, in your example, the factoring facilitates your work.

USING TECHNOLOGY IN CALCULUS

In this text, when we give the directions to "factor" an expression, we really mean "factor the expression over the rational numbers." Many symbolic algebra software packages and some high-end calculators can perform factoring over the rational numbers as well as over the real (including irrational numbers) and even over the complex numbers.

Consider $x^2 - 7$. Although you may answer "not factorable," note how *DERIVE*® factors it here. See also how *DERIVE*® factors $x^2 - 3x - 1$.

```
        2
#1:    x  - 7

#2:    (x + √7)·(x - √7)

        2
#3:    x  - 3·x - 1

           √13   3      √13   3
#4:    [x + --- - -]·[x - --- - -]
            2    2        2    2

COMMAND: Author Build Calculus Declare Expand Factor Help Jump soLve Manage
         Options Plot Quit Remove Simplify Transfer Unremove moVe Window approX
Enter option                                                    Derive XM
Factor using RADICALS                    Free:100% Ins              Algebra
```

1. Explain how to find the expressions in lines 2 and 4 on the screen.
2. Of course, one of the advantages of symbolic software packages is their ability to perform operations on very complicated expressions. Consider the expression $x^2 + 60x - 864$. If you have access to such software, factor this trinomial over the rational numbers. If you do not, explain how you would go about factoring this expression.
3. Repeat question 2 for the expression $6x^3 - 101x^2 + 420x$.

1.3 SOLVING EQUATIONS

In this section, we concentrate on solving equations that fall into the following four categories:

1. Linear equations (for example, $3x - 7 = 18$)
2. Quadratic equations (for example, $x^2 + 7x + 12 = 0$)
3. Equations involving radicals (or fractional exponents) (for example, $\sqrt[3]{x + 1} = 2$)
4. Rational equations $\left(\text{for example, } \dfrac{1 - 2x}{x^2 + 1} = 0 \right)$

Starting with linear equations, recall that you may add (or subtract) any real number (or expression representing a real number) to both sides of the equation and the equality will be preserved. You may also multiply (or divide) by any nonzero real number. With this information, we proceed to some examples.

Example 1 Solve for x: $5x - 3 = 2x + 9$.

Solution
$$5x - 3 = 2x + 9$$
$$3x - 3 = 9$$
$$3x = 12$$
$$x = 4$$ ∎

Example 2 Solve for x: $5[x - 3(2x + 1)] = 0$.

Solution
$$5[x - 3(2x + 1)] = 0$$
$$5(x - 6x - 3) = 0$$
$$5(-5x - 3) = 0$$
$$-25x - 15 = 0$$
$$-25x = 15$$
$$x = \frac{-15}{25} = \frac{-3}{5}$$ ∎

We use one of three methods in solving quadratic equations. The first technique involves factoring and is illustrated by the next example. We use a basic property of real numbers that if $a \cdot b = 0$, then $a = 0$ or $b = 0$, or both.

Example 3 Solve the following quadratic equations for x:

  a. $x^2 - 2x - 8 = 0$ **b.** $2x^2 - 6x = 0$ **c.** $2x^2 - 3x = 5$

Solution **a.** First factor and then set each factor equal to zero to solve.

$$x^2 - 2x - 8 = 0$$
$$(x - 4)(x + 2) = 0$$
$$x - 4 = 0 \qquad x + 2 = 0$$
$$x = 4 \qquad\quad x = -2$$

b. $2x^2 - 6x = 0$
$2x(x - 3) = 0$
$2x = 0 \qquad x - 3 = 0$
$x = 0 \qquad\quad x = 3$

c. First subtract 5 from both sides so that the right side is zero.
$$2x^2 - 3x = 5$$
$$2x^2 - 3x - 5 = 0$$
$$(2x - 5)(x + 1) = 0$$
$$2x - 5 = 0 \qquad x + 1 = 0$$
$$x = \frac{5}{2} \qquad\quad x = -1$$

The next technique can be referred to as the *extraction of roots* method, and one of the following two properties will be used (see exercises 41 and 42):

If $x^2 = p$, then $x = \pm\sqrt{p}$.
If $(x - m)^2 = p$, then $x = m \pm \sqrt{p}$.

Examples 4 and 5 illustrate how this works.

Example 4   Solve for x in each of the following:
  a. $x^2 = 9$ **b.** $2x^2 - 12 = 0$

Solution   **a.** $x^2 = 9$ **b.** $2x^2 - 12 = 0$
      $x = \pm\sqrt{9}$ $2x^2 = 12$
      $x = \pm 3$ $x^2 = 6$
               $x = \pm\sqrt{6}$

Example 5   Solve for x: $(x - 3)^2 = 8$.

Solution $(x - 3)^2 = 8$
    $x - 3 = \pm\sqrt{8}$
      $x = 3 \pm \sqrt{8} = 3 \pm 2\sqrt{2}$

The final method we use to solve quadratic equations is the **quadratic formula.** Note that this method works for *any* quadratic equation, including those solved in Examples 3, 4, and 5.

Quadratic formula

If $ax^2 + bx + c = 0$ and $a \neq 0$, then

$$x = \frac{-b \pm \sqrt{b^2 - 4ac}}{2a}$$

Example 6   Solve for x using the quadratic formula:

    a. $x^2 - 3x - 10 = 0$ **b.** $2x^2 + x - 2 = 0$

Solution   **a.** We have $a = 1$, $b = -3$, and $c = -10$. Substituting into the formula gives

$$x = \frac{-(-3) \pm \sqrt{(-3)^2 - 4(1)(-10)}}{2(1)}$$

$$= \frac{3 \pm \sqrt{49}}{2}$$

$$= \frac{3 \pm 7}{2}$$

$$x = \frac{3 + 7}{2} \qquad x = \frac{3 - 7}{2}$$

$$x = 5 \qquad\quad x = -2$$

b. Since $a = 2$, $b = 1$, $c = -2$,

$$x = \frac{-1 \pm \sqrt{1^2 - 4(2)(-2)}}{2(2)}$$

$$= \frac{-1 \pm \sqrt{17}}{4}$$

    ■

Before we leave quadratic equations, we present an example that combines the factoring technique and extraction of roots on a fourth-degree equation.

Example 7   Solve for x: $x^4 - 6x^2 + 5 = 0$.

Solution   Factoring, we have

$$x^4 - 6x^2 + 5 = 0$$
$$(x^2 - 5)(x^2 - 1) = 0$$
$$x^2 - 5 = 0 \qquad x^2 - 1 = 0$$
$$x^2 = 5 \qquad\quad x^2 = 1$$

Now extracting roots, we get

$$x = \pm\sqrt{5} \qquad x = \pm 1$$

    ■

The third type of equation under consideration requires the use of the following idea. Let L represent the left side of an equation and R represent the right side. We raise both sides of the equation to the same power, n, producing

$$L^n = R^n$$

The solution set for this new equation contains all the solutions to the original equation, $L = R$, but may contain extra answers, called *extraneous solutions*. Extraneous solutions may be eliminated by *checking* your solutions in the original equation. For example, the simple equation $x = 1$ has one solution: 1. Squaring both sides yields $x^2 = 1$, which has two solutions: 1 and -1. The -1 is extraneous to the original equation.

Example 8   Solve for x:

  a. $\sqrt{x} = 4$ **b.** $\sqrt[3]{x + 1} = 2$

Solution   **a.**   Squaring both sides, we have

$$(\sqrt{x})^2 = 4^2$$
or $x = 16$

Checking by substitution gives

$$\sqrt{16} = 4 \quad \checkmark$$

  b.   Cubing both sides, we get

$$(\sqrt[3]{x + 1})^3 = 2^3$$
$$x + 1 = 8$$
$$x = 7$$

Checking gives

$$\sqrt[3]{7 + 1} = 2$$
$$\sqrt[3]{8} = 2 \quad \checkmark$$

■

Example 9   Solve for x: $\sqrt{x + 1} - \sqrt{2x} = 1$.

Solution $\sqrt{x + 1} - \sqrt{2x} = 1$

$\sqrt{x + 1} = 1 + \sqrt{2x}$

$(\sqrt{x + 1})^2 = (1 + \sqrt{2x})^2$ Squaring both sides

$x + 1 = 1 + 2\sqrt{2x} + 2x$

$-x = 2\sqrt{2x}$

$(-x)^2 = (2\sqrt{2x})^2$ Squaring both sides again

$x^2 = 8x$

$x^2 - 8x = 0$

$x(x - 8) = 0$

$x = 0 \quad \text{or} \quad x = 8$

Checking both solutions, we have

$$\sqrt{0 + 1} - \sqrt{2 \cdot 0} = 1 \qquad \sqrt{8 + 1} - \sqrt{2 \cdot 8} = 1$$
$$1 - 0 = 1 \qquad\qquad \sqrt{9} - \sqrt{16} = 1$$
$$1 = 1 \quad \checkmark \qquad\qquad 3 - 4 = 1$$
$$-1 \neq 1$$

Notice that $x = 8$ is an extraneous solution. The only solution is $x = 0$. ∎

Example 10 Solve for x:

    a. $2x^{1/2} = 6$ **b.** $(1 - x)^{2/3} = 2$

Solution **a.**
$$2x^{1/2} = 6$$
$$x^{1/2} = 3$$
$$(x^{1/2})^2 = 3^2$$
$$x = 9$$

b.
$$(1 - x)^{2/3} = 2$$
$$[(1 - x)^{2/3}]^3 = 2^3$$
$$(1 - x)^2 = 8$$
$$1 - x = \pm\sqrt{8}$$
$$1 - x = \pm 2\sqrt{2}$$
$$-x = -1 \pm 2\sqrt{2}$$
$$x = 1 \pm 2\sqrt{2}$$

∎

The final type of equations to be considered in this section is called *rational equations,* equations involving variables in denominators.

Example 11 Solve the following equation for x:
$$\frac{2x}{x - 3} = 4$$

Solution We begin by noticing that x cannot be 3 ($x \neq 3$) because division by zero is not defined. Now, multiplying both sides by $x - 3$ gives

$$(x - 3)\left(\frac{2x}{x - 3}\right) = 4(x - 3)$$
$$2x = 4x - 12$$
$$-2x = -12$$
$$x = 6$$

∎

In most situations when we encounter rational type equations in this book, the right side will be zero. When this occurs, we can immediately assume that the *numerator* is zero based on the property that if $\frac{a}{b} = 0$, since $b \neq 0$, then $a = 0$.

Example 12 Solve for x:

    a. $\dfrac{1 - 2x}{x^2 + 1} = 0$ **b.** $\dfrac{x^2 - 3x}{\sqrt{x + 1}} = 0$

Solution   **a.** $\dfrac{1 - 2x}{x^2 + 1} = 0$ **b.** $\dfrac{x^2 - 3x}{\sqrt{x + 1}} = 0$

$1 - 2x = 0$ $x^2 - 3x = 0$

$-2x = -1$ $x(x - 3) = 0$

$x = \dfrac{1}{2}$ $x = 0$ $x = 3$ ∎

1.3   EXERCISES

SET A

In exercises 1 through 8, solve the equations. (See Examples 1 and 2.)

1. $4x + 1 = 3x - 2$

2. $x - 5 = 7 - 4x$

3. $5(x + 1) - 3 = 7$

4. $7 - 2(2t + 1) = 5$

5. $\frac{1}{3}t - 2 = 1$

6. $\frac{2}{3}S - 1 = \frac{1}{2}$

7. $3(1 - y) + 2 = 0$

8. $2x + 3(x - 4) = 0$

In exercises 9 through 26, solve the equations. (See Examples 3, 4, 5, 6, and 7.)

9. $x^2 - 6x = 0$

10. $-4x^2 - 12x = 0$

11. $S^2 - S - 12 = 0$

12. $10 + 3x - x^2 = 0$

13. $2x^2 + 4x + 1 = 0$

14. $x^3 - 9x = 0$

15. $x^2 = 5$

16. $t^2 = 12$

17. $3x^2 - 12 = 0$

18. $4x^2 - 7 = 0$

19. $(x + 5)^2 = 7$

20. $(p - 4)^2 = 24$

21. $t^2 + 2t - 2 = 0$

22. $3y^2 + 7y + 3 = 0$

23. $4x^2 - x + 2 = 0$

24. $t^2 + 8t + 8 = 0$

25. $x^4 - 5x^2 + 4 = 0$

26. $2t^4 - 11t^2 + 5 = 0$

In exercises 27 through 34, solve the equations. (See Examples 8, 9, and 10.)

27. $\sqrt[3]{x} = 4$

28. $2\sqrt{x} = 3$

29. $\sqrt[4]{1 - x} = 0$

30. $\sqrt[3]{2x + 1} = 1$

31. $x^{3/2} - 8 = 0$

32. $x^{-1/3} = 2$

33. $\dfrac{2}{\sqrt{x}} = 3$

34. $-2x^{-1/2} = 1$

In exercises 35 through 40, solve the equations. (See Examples 11 and 12.)

35. $\dfrac{x + 1}{x - 3} = 2$

36. $\dfrac{-x}{1 - x} = 3$

37. $\dfrac{x^2 - 4}{2x + 1} = 0$

38. $\dfrac{x^2 - x - 2}{(x^2 + 1)^2} = 0$

39. $\dfrac{x - x^2}{\sqrt{x + 3}} = 0$

40. $\dfrac{x^2 + 2x}{\sqrt[3]{x + 1}} = 0$

SET B

41.   Solve $x^2 = p$, or equivalently $x^2 - p = 0$, by factoring. (*Hint:* Factor as the difference of squares using the fact that $p = (\sqrt{p})^2$.)

42.   Solve $(x - m)^2 = p$, or equivalently $(x - m)^2 - p = 0$, by factoring. (*Hint:* Factor as the difference of two squares. See exercise 41.)

BUSINESS
PROFIT

43. The Electrolyte Corporation's research department has found that its profit, in thousands of dollars, can be determined by the expression

$$6x - x^2 - 5, \qquad 0 \le x \le 4$$

where x represents the number of units sold (in hundreds). Find the "break-even" point; that is, find the value of x for which the profit is zero.

REVENUE

44. Acme Paints can represent its revenue by the expression

$$1000x - 2x^2, \qquad x \ge 50$$

where x represents the number of units sold. If its revenue is $19,200, how many units were sold?

MANUFACTURING

45. Container design is an important aspect of the manufacturing process. In particular, consider open-box construction. An open box is constructed from a square piece of cardboard by cutting 2-centimeter squares out of each corner and then folding up the sides, as indicated in the figures.

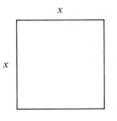 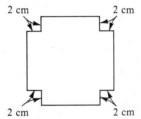 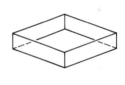

If the volume of the box is to be 968 cubic centimeters, what should be the dimension of the original square piece of cardboard?

**PHYSICAL
SCIENCE**
VIBRATING CRYSTALS

46. Solid-state digital watches use a vibrating quartz crystal to regulate the recording of time. The formula

$$t = \frac{v}{32,768}$$

is an algebraic model that gives the time t (in seconds) after v vibrations of the quartz crystal.
  a. After the crystal vibrates 1,048,576 times, how many seconds have elapsed?
  b. How many vibrations does the quartz crystal make in 15 seconds?

BIOLOGY
GENETICS

47. A biologist is counting fruit flies in a sterilization experiment. The number of flies present after t minutes can be approximated by the following expression:

$$-4t^2 + 76t + 80$$

After how many minutes is the fruit fly population equal to 360?

WRITING AND CRITICAL THINKING

In this section, we reviewed the quadratic formula and used it to solve *quadratic* equations. Do you think this formula can be useful in solving other types of equations? Explain why or why not.

USING TECHNOLOGY IN CALCULUS

It is easy to see that the equation $5 - x^3 = 3\sqrt{x^2 + 1}$ is equivalent to the equation $5 - x^3 - \sqrt{x^2 + 1} = 0$. It is *not* easy, however, to **solve** this equation. In fact, it is not possible to find an exact solution. One option is to realize that the value of x that solves $5 - x^3 - \sqrt{x^2 + 1} = 0$ is the same value of x that solves $5 - x^3 - \sqrt{x^2 + 1} = y$ when $y = 0$. We use a graphics calculator to approximate a solution to $5 - x^3 - \sqrt{x^2 + 1} = 0$ by graphing $y = 5 - x^3 - \sqrt{x^2 + 1}$.

First, we graph $y = 5 - x^3 - \sqrt{x^2 + 1}$ in a suitable window:

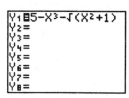 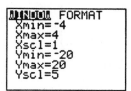 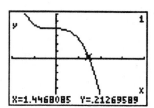

Notice that we are interested in finding the x value that corresponds to the point where y is zero. (This is called the x intercept of the graph.) We zoom in on a point close to that, where y is approximately 0.21269589, and obtain the following screens:

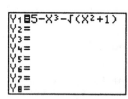 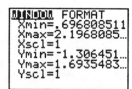 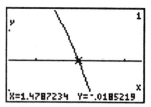

From the graph, it appears that the value of x that solves $5 - x^3 - \sqrt{x^2 + 1} = 0$ is approximately 1.48.

Use this technique to find two solutions to the equation $x - x^4 = \sqrt[3]{2x^2 - 1}$.

1.4 INEQUALITIES

In many instances, we are not interested in finding exact values for quantities, as in the last section, but merely *ranges* or *intervals* of values for a quantity. We say that the temperature is between 70°F and 75°F, the price of a certain car is between $8000 and $9000, the time it takes to drive to work is between 15 and 20 minutes, the profits for a company will exceed $2 million this year, or the tolerance level for an error in the diameter of a machine part is less than 0.003 inch. To be represented symbolically, all these examples require a particular type of notation. The notation used involves the symbols $>$, $<$, $\geq$, and $\leq$, which are referred to as inequality (or order) symbols and are used to express sets or intervals of real numbers. For example,

Symbolism	Represents Real Numbers
$x < 3$	"less than 3"
$t \geq 0$	"greater than or equal to 0"
$3 < x < 5$	"between 3 and 5"

Many inequalities can be solved using the same techniques that were used in the previous section for equations. The main difference between them is the result or solution.

To see the difference, we compare an equation and an inequality that look very similar by solving both and interpreting the results.

Equation	Inequality
$2x + 1 = 5$	$2x + 1 < 5$
$2x = 4$	$2x < 4$
$x = 2$	$x < 2$

The solution to the equation gives x as being exactly equal to 2, and the solution to the inequality gives x as being any real number less than 2.

Before solving any more inequalities, let us review some notation and symbolism. There are two main techniques for representing solutions to inequalities. Both of these are shown in Table 1.4, together with a graphical representation on a real number line. Notice that in the notation on the left, the variable appears in the representation. In the one on the right, referred to as **interval notation,** the variable does not appear. The symbol $+\infty$ is used to represent positive infinity, and the symbol $-\infty$ is used to represent negative infinity.

In Table 1.4, notice that when the endpoint of an interval is not included, parentheses are used; on the real number line, we use an open circle. When the endpoint is included, brackets and a closed circle are used.

Table 1.4

1. $x < b$	$(-\infty, b)$	
2. $x > a$	$(a, +\infty)$	
3. $x \leq b$	$(-\infty, b]$	
4. $x \geq a$	$[a, +\infty)$	
5. $a < x < b$	(a, b)	
6. $a < x \leq b$	$(a, b]$	
7. $a \leq x < b$	$[a, b)$	
8. $a \leq x \leq b$	$[a, b]$	

Note: (a, b) is called an *open interval*.
      $[a, b]$ is called a *closed interval*.
      $(a, b]$ and $[a, b)$ are called *half-open intervals*.

It is also beneficial to review some general rules for solving inequalities.

Rule 1. Any real number may be added to (or subtracted from) both sides of an inequality.

Rule 2. Both sides of an inequality may be multiplied (or divided) by the same positive real number.

Rule 3. If both sides of an inequality are multiplied or divided by the same negative real number, then the *sense* (or direction) of the inequality is reversed (that is, "<" becomes ">" and vice versa).

Example 1   Solve the inequalities for x and interpret the results:
      a. $3x - 5 \geq 7$ **b.** $2x + 6 < 5$

Solution   **a.** $3x - 5 \geq 7$ **b.** $2x + 6 < 5$
            $3x \geq 12$ $\leftarrow$ Rule 1 $\rightarrow$ $2x < -1$
            $x \geq 4$ $\leftarrow$ Rule 2 $\rightarrow$ $x < -\dfrac{1}{2}$

The solution to the first inequality is all real numbers greater than or equal to 4. The second solution is all real numbers less than $-\frac{1}{2}$. ∎

Example 2   Solve for x: $4 - 5x \le 7$.

Solution $4 - 5x \le 7$
$-5x \le 3$
$x \ge \dfrac{-3}{5}$ Using Rule 3

The solution in interval notation is $[-\frac{3}{5}, +\infty)$.   ■

Example 3   Solve for x and write the answer in interval notation:

$$5 + 3(x - 1) > 2x$$

Solution $5 + 3(x - 1) > 2x$
$5 + 3x - 3 > 2x$
$3x + 2 > 2x$
$x + 2 > 0$ Subtracted 2x from both sides
$x > -2$

The answer is $(-2, +\infty)$.   ■

In this book, we are interested in being able to solve four types of inequalities. The type studied thus far is referred to as a **linear inequality.** In addition to linear inequalities we examine the following types:

Type	Example
quadratic inequalities	$x^2 + 9x + 20 \le 0$
rational inequalities	$\dfrac{x - 1}{x + 2} > 0$
inequalities involving radicals	$\sqrt[3]{1 - x} < 0$

To solve quadratic inequalities, we use a graphing technique on a real number line. We outline the graphing technique step by step in Example 4.

Example 4   Solve for x: $x^2 - x - 6 > 0$.

Solution   **Step 1.**   Solve the associated equality for x to determine **key points.**

$x^2 - x - 6 = 0$ Replaced inequality symbol with equal sign
$(x - 3)(x + 2) = 0$ Solve resulting quadratic equation.
$x - 3 = 0 \qquad x + 2 = 0$
$x = 3 \qquad\quad x = -2$

The key points are -2 and 3.

Step 2.   Mark key points on a number line with open or closed circles.

Open circles are used here because the inequality symbol in the original is >, not ≥; that is, the equality is not included here.

Step 3.   Pick a test point in each region, substitute into each factor, and determine whether the factor is positive or negative.

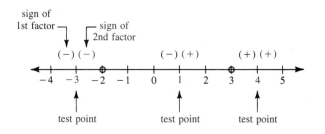

Step 4.   Examine the result in each region.

Step 5.   Refer back to the inequality to determine the desired result. Because $x^2 - x - 6 > 0$ means we want $x^2 - x - 6$ to be positive (greater than zero), the correct result is Regions 1 and 3, which gives

$$x < -2 \quad \text{or} \quad x > 3$$

(*Note:* These may *not* be combined into one inequality.)   ∎

Example 5   Solve for x: $2x^2 + 3x \le 2$.

Solution   The inequality should always be rewritten (if necessary) so you have zero on one side.

$$2x^2 + 3x - 2 \le 0 \qquad \text{Means we want } 2x^2 + 3x - 2 \text{ to be negative or zero}$$
$$2x^2 + 3x - 2 = 0 \qquad \text{Associated equality}$$
$$(2x - 1)(x + 2) = 0$$
$$x = \frac{1}{2} \qquad x = -2 \quad \text{Key points}$$

Now mark key points with closed circles (≤) and check regions.

$$(-)(-) = + \qquad (-)(+) = - \qquad (+)(+) = +$$

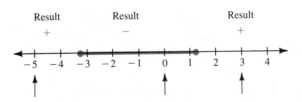

The result is shaded on a number line. The solution is $[-2, \frac{1}{2}]$.   ■

A question probably now arises as to what to do if the quadratic expression does not factor readily. The quadratic formula is used to solve the associated equality, a quadratic equation, and a slight variation in the method is used, as illustrated by Example 6.

Example 6   Solve the following inequality for x: $x^2 + 2x - 4 \le 0$.

Solution   Determine key points by solving the associated equality using the quadratic formula.

$$x^2 + 2x - 4 = 0$$
$$x = \frac{-2 \pm \sqrt{(2)^2 - 4(1)(-4)}}{2(1)}$$
$$x = \frac{-2 \pm \sqrt{20}}{2}$$
$$x = \frac{-2 \pm 2\sqrt{5}}{2} = -1 \pm \sqrt{5}$$
$$x \approx -3.236 \quad \text{or} \quad 1.236$$

The decimal approximations are used to help mark the key points on the number line. Because we did not factor this time, substitute the test points into the left side of the original inequality and determine whether the result is positive or negative.

```
  Result          Result              Result
    +               -                   +

◄──┼──┼──●──┼──┼──┼──┼──●──┼──┼──┼──►
  -5  -4  -3  -2  -1   0   1   2   3   4
        ↑               ↑       ↑
```

The original inequality required a negative or zero result, so the solution is $-1 - \sqrt{5} \le x \le -1 + \sqrt{5}$ ■

Occasionally you may encounter a quadratic inequality whose associated equation has no real solutions (a negative number results under the radical). In such a case, the inequality is either always true (solution is all real numbers) or never true (solution is an empty set). A single test point (arbitrary) will determine the result. For example, consider the inequality

$$x^2 + 3x + 3 < 0$$

and use $x = 1$ as a test point:

$$(1)^2 + 3(1) + 3 < 0$$
$$7 < 0$$

This is false, and therefore the solution to the inequality is the empty set.

To solve rational inequalities, the graphing technique can again be used. First, make sure that one side of the inequality is zero and if it is not, take the necessary algebraic steps to produce a zero. To find the key points, set the numerator and the denominator equal to zero and solve. Remember that division by zero is undefined so that key points determined by the denominator are always represented by open circles and are not included in the solution.

Example 7   Solve for x: $\dfrac{2x}{x - 3} \geq 0$.

Solution   Setting the numerator and the denominator equal to zero for key points gives

$$2x = 0 \qquad x - 3 = 0$$
$$x = 0 \qquad x = 3$$

Mark $x = 0$ with a closed circle (≥ 0) and $x = 3$ with an open circle because the denominator can never be zero. Pick test points and substitute them into the numerator and the denominator.

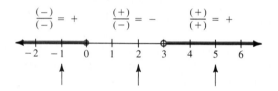

The solution is $x \leq 0$ or $x > 3$, which in interval notation is $(-\infty, 0] \cup (3, +\infty)$. Note that the union symbol, $\cup$, replaces the word "or."   ■

Example 8   Solve for x: $\dfrac{5}{x + 2} > 1$.

Solution   Subtract 1 from both sides and rewrite the left side as a single expression.

$$\frac{5}{x + 2} - 1 > 0$$
$$\frac{5 - (x + 2)}{x + 2} > 0$$
$$\frac{3 - x}{x + 2} > 0$$

Key points are $x = -2$ and $x = 3$.

$$\frac{(+)}{(-)} = - \qquad \frac{(+)}{(+)} = + \qquad \frac{(-)}{(+)} = -$$

The solution is $(-2, 3)$. ∎

Some inequalities are best solved by inspection. The techniques used in the previous examples could still be used but require more work.

Example 9 Solve the following inequalities for x:

   a. $x^2 + 4 > 0$ **b.** $x^2 - 4x + 4 < 0$ **c.** $\dfrac{x-4}{x^2+1} \geq 0$

Solution **a.** The left side is always positive (greater than zero) for any x value, so the solution is all real numbers. In interval notation, this solution is expressed as

$$(-\infty, +\infty)$$

   b. Factoring gives

$$(x-2)^2 < 0$$

The left side is always positive or zero (when $x = 2$), so there is no solution to this inequality.

   c. Since the denominator is always positive and we want the left side to be positive or zero, this implies that the numerator must be positive or zero.

$$x - 4 \geq 0$$

or $x \geq 4$ is the solution. ∎

In Chapter 3, we may need to determine when an expression involving radicals is positive (greater than zero) or negative (less than zero). This leads us to our fourth type of inequality.

Example 10 Solve for x: $\sqrt[3]{1-x} < 0$.

Solution In working with the cube root (or any odd root) of a number, the sign of the number does not change, so we can solve by merely removing the root sign, in this case giving

$$1 - x < 0$$
$$1 < x$$

or $x > 1$ is the solution. ∎

Example 11 Solve for x: $\dfrac{x-1}{\sqrt{x+2}} > 0$.

Solution   Using a method similar to that for rational inequalities, we find that the key points are $x = 1$ and $x = -2$. Both points have open circles on the number line.

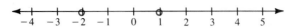

In choosing test points, it is not necessary to choose a point from the region to the left of -2, because $x + 2$ under the radical sign in the denominator must be positive.

If $x + 2 > 0$

then $x > -2$

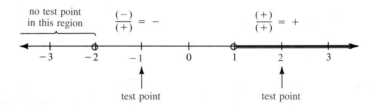

The solution is $x > 1$. Note that this example could also have been solved similarly to the method used in part c of Example 9. ∎

1.4   EXERCISES

SET A

In exercises 1 through 14, solve the given linear inequalities. (See Examples 1, 2, and 3.)

1. $x + 2 \leq 0$ **2.** $3 - x < 0$ **3.** $2x - 1 > 0$

4. $3x + 5 \leq 0$ **5.** $7 + 2x > 3$ **6.** $\frac{1}{2}x - 4 \geq 6$

7. $x - 4 \leq 3x + 4$ **8.** $5x < 2x - 3$ **9.** $2x + 1 < x + \frac{1}{2}$

10. $\frac{5}{2} - x > x + \frac{1}{2}$ **11.** $6x - 3(x + 1) > 0$ **12.** $6 - 3(x + 1) > 0$

13. $2(x - 3) \leq 3x$ **14.** $2(x + 4) - x \leq 3x$

In exercises 15 through 30, solve the given quadratic inequalities. (See Examples 4, 5, 6, and 9.)

15. $x^2 - 9 < 0$ **16.** $x^2 - 9 \geq 0$ **17.** $x^2 + 7x + 12 \geq 0$

18. $x^2 - 7x + 12 < 0$ **19.** $x^2 + 2x < 8$ **20.** $x^2 > 3x$

21. $3x^2 - 10x - 8 \leq 0$ **22.** $3x^2 + 10x - 8 \geq 0$ **23.** $4x^2 + 4x + 1 > 0$

24. $x^2 + 6x + 9 > 0$ **25.** $x^2 - 7 \geq 0$ **26.** $x^2 + 7 < 0$

27. $x^2 + x + 1 < 0$ **28.** $x^2 + x - 1 \geq 0$ **29.** $2x^2 + 1 < 0$

30. $2x^2 > 0$

In exercises 31 through 42, solve the given rational inequalities. (See Examples 7, 8, and 9.)

31. $\dfrac{2}{x} > 0$ **32.** $\dfrac{1}{x + 2} < 0$ **33.** $\dfrac{3x}{x - 1} \leq 0$

34. $\dfrac{-3x}{x+1} \geq 0$ **35.** $\dfrac{x+4}{x+2} \geq 0$ **36.** $\dfrac{x+2}{x+4} \leq 0$

37. $\dfrac{2x-1}{3-x} > 0$ **38.** $\dfrac{2x-1}{3-x} < 0$ **39.** $\dfrac{x}{x^2-1} < 0$

40. $\dfrac{x}{x^2+1} < 0$ **41.** $\dfrac{x}{x-3} < 2$ **42.** $\dfrac{4}{x+1} \geq 1$

SET B

In exercises 43 through 66, solve the given inequalities. (See Examples 10 and 11.)

43. $\sqrt{x} > 0$ **44.** $\sqrt{-x} > 0$ **45.** $\sqrt{x-2} \leq 0$

46. $\sqrt{1-x} < 0$ **47.** $\sqrt[3]{x+4} < 0$ **48.** $\sqrt[3]{x+4} > 0$

49. $\sqrt{2x-1} > 0$ **50.** $\sqrt[3]{2x-1} > 0$ **51.** $\dfrac{2}{\sqrt[3]{x}} > 0$

52. $\dfrac{-2}{\sqrt[3]{x}} > 0$ **53.** $\dfrac{-1}{\sqrt[3]{(x-1)^2}} < 0$ **54.** $\dfrac{-1}{\sqrt[3]{x-1}} < 0$

55. $\dfrac{5-x}{\sqrt{x}} \geq 0$ **56.** $\dfrac{2x}{\sqrt{x-1}} \geq 0$ **57.** $\dfrac{x+3}{\sqrt{x-3}} < 0$

58. $\dfrac{2x+3}{\sqrt{5-x}} < 0$ **59.** $\dfrac{1}{\sqrt[4]{2x}} > 0$ **60.** $\dfrac{2-x}{\sqrt[4]{x+1}} > 0$

61. $(x+1)(x-2)(x+3) > 0$ **62.** $2x(x-1)(x+4) \leq 0$

63. $x(x-5)^2 \leq 0$ **64.** $(x+2)^2(x-1) > 0$

65. $x^3 - x^2 - 6x \geq 0$ **66.** $x^3 - 1 \geq 0$

BUSINESS
PROFIT

67. The Averbley Company, which manufactures paper products, has determined that its profit from selling x hundred boxes of envelopes can be approximated by the following expression:

$$-5x^2 + 55x - 50, \qquad 0 \leq x < 10$$

a. How many boxes must the company sell to break even? (*Hint:* Its profit is zero.)

b. How many boxes must the company sell to make money? (*Hint:* Its profit is greater than zero.)

PROFIT

68. At Shirley's Chrysanthemum Farm, the revenue, R, is given by the equation $R = 2x$, where x is the number of chrysanthemums sold. The cost equation is given by $C = 100 + 1.5x$. How many chrysanthemums must be sold to make a profit ($R > C$)?

WRITING AND CRITICAL THINKING

Think of a "real-world" situation that would be appropriately described or *modeled* by an inequality. Write a paragraph describing it.

USING TECHNOLOGY IN CALCULUS

A graphical way to solve an inequality such as $x^2 + 2x - 4 < 0$ involves graphing the two-variable related equality, $y = x^2 + 2x - 4$. Once we graph $y = x^2 + 2x - 4$, to find the region(s) where $x^2 + 2x - 4 < 0$ means to find where $y < 0$. Equivalently, we look at the graph to see where the curve has negative y values—that is, where points are *below* the x axis. Let us elaborate.

First, we graph $y = x^2 + 2x - 4$.

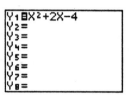  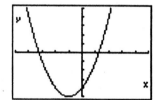

Next, because y equals $x^2 + 2x - 4$ and we are searching for x values where $x^2 + 2x - 4$ is negative (that is, less than zero), we examine the graph for the region where y is negative. Notice that that is the collection of points below the x axis.

The values of x that solve $x^2 + 2x - 4 = 0$ (the x intercepts) are found, using the quadratic formula, to be $x = -1 - \sqrt{5} \approx -3.236$ and $x = -1 + \sqrt{5} \approx 1.236$.

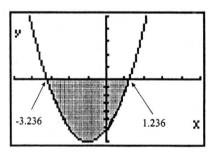

Thus, the solution to $x^2 + 2x - 4 < 0$ is the set of x values between $-1 - \sqrt{5}$ and $-1 + \sqrt{5}$:

$$-1 - \sqrt{5} < x < -1 + \sqrt{5}$$

Use this technique to solve each of the following inequalities:

1. $x^2 + 3x - 1 < 0$ **2.** $x^2 - 5x - 1 \geq 0$

1.5 RECTANGULAR COORDINATE SYSTEM

In this section, we work with points, distances, lines, and geometric concepts in two dimensions. Before moving into the realm of two dimensions, however, we give the definition of the absolute value of a real number, a, denoted $|a|$, and then show how it can be used to describe the distance between two numbers on a real number line.

Absolute value of a
real number

For any real number a,

$$|a| = \begin{cases} a, & \text{if } a \geq 0 \\ -a, & \text{if } a < 0 \end{cases}$$

Note: An alternative definition is $|a| = \sqrt{a^2}$.

Geometrically, the absolute value of a number gives its distance from zero on a real number line, as illustrated here. Both 3 and -3 are 3 units from zero.

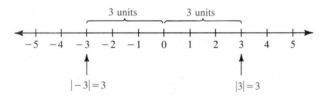

This idea can be extended to describe the distance between any two numbers on a real number line.

Distance between
real numbers

The distance between two real numbers a and b on a real number line is given by

$$|a - b| \quad \text{or} \quad |b - a|$$

Example 1 Find the distance between the following pairs of numbers:
   a. -2 and 4   **b.** -2 and -7

Solution **a.** Letting $a = -2$ and $b = 4$ in $|b - a|$ gives

$$|4 - (-2)| = |6| = 6$$

By letting $a = 4$ and $b = -2$, we still have

$$|-2 - 4| = |-6| = 6 \text{ units}$$

b. $|-2 - (-7)| = |5| = 5 \text{ units}$ ∎

We now move from the one-dimensional number line to a two-dimensional configuration called the **rectangular** (or **Cartesian**) **coordinate system,** which is composed of two number lines placed perpendicular to each other, as shown in Figure 1.1. The point of intersection of the number lines is called the **origin,** and the two number lines are referred to as **axes.** The horizontal axis is usually called the *x* **axis** and the vertical axis is usually called the *y* **axis.**

Figure 1.1

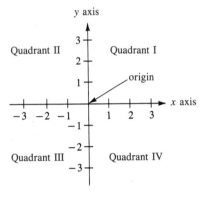

There are four distinct regions, which have been labeled **quadrants** and numbered counterclockwise using Roman numerals, as in Figure 1.1. We describe points in each of these four quadrants and on the axes using **ordered pairs** of real numbers. Symbolically, ordered pairs are represented by (x, y); geometrically, they are represented by points located "x units" horizontally from the y axis and "y units" vertically from the x axis (see Figure 1.2).

Figure 1.2   *Figure 1.3*

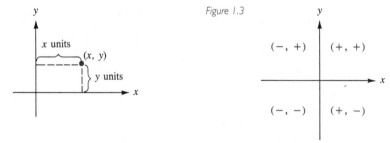

In the ordered pair (x, y), we refer to the first number as the *x* **coordinate** (or **abscissa**) and the second as the *y* **coordinate** (or **ordinate**). Figure 1.3 shows the signs for the x and y coordinates in each quadrant. If a point lies on the x axis, then the y coordinate is 0. A point on the y axis has x coordinate 0.

Example 2   Locate (or plot) the following points:

$(1, 3), (-2, 2), (-3, -1), (4, -3),$ and $(-5, 0)$

Solution   See Figure 1.4.

Figure 1.4

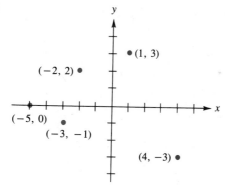

Example 3   Determine the coordinates of each of the points labeled in Figure 1.5.

Figure 1.5

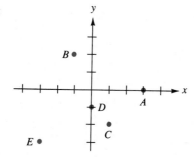

Solution   Point: *A*   *B*   *C*   *D*   *E*
Coordinates: $(3, 0)$ $(-1, 2)$ $(1, -2)$ $(0, -1)$ $(-3, -3)$ ■

Two formulas are associated with any two points in a plane. One is the **distance formula** and the other is the **slope formula.** We use a point P with coordinates (x_1, y_1) and a point Q with coordinates (x_2, y_2) in the formulas.

Distance formula

> The distance d between P and Q is given by
> $$d = \sqrt{(x_2 - x_1)^2 + (y_2 - y_1)^2}$$

By studying Figure 1.6, we can see that the distance formula is merely a special application of the **Pythagorean theorem.** In the diagram (Figure 1.6), $a = |x_2 - x_1|$, $b = |y_2 - y_1|$, and $c = d$. Notice that $a^2 = |x_2 - x_1|^2$, but because $|x_2 - x_1|^2$ is equivalent to $(x_2 - x_1)^2$ for any real number, the absolute value symbols are not necessary.

Figure 1.6

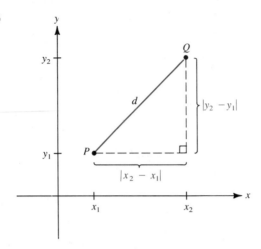

$$a^2 + b^2 = c^2$$

Example 4   Find the distance between $(1, -3)$ and $(4, 1)$.

Solution $(1, -3)$ and $(4, 1)$

        ↑ ↑   ↑ ↑

      x_1 y_1 x_2 y_2

$$d = \sqrt{(4 - 1)^2 + [1 - (-3)]^2}$$
$$= \sqrt{3^2 + 4^2} = \sqrt{25} = 5$$

■

Example 5   Find the distance between $(-2, 5)$ and $(0, 9)$.

Solution $d = \sqrt{[0 - (-2)]^2 + (9 - 5)^2}$
$$= \sqrt{2^2 + 4^2} = \sqrt{20} = 2\sqrt{5}$$

or, using a calculator, $d \approx 4.472$.

■

We close this section with a visualization example. Your ability to do mathematics can be enhanced by learning to "think with pictures." Sketches, diagrams, and graphs help to make concepts more vivid and concrete in your mind.

Example 6   Working in a rectangular coordinate system, what does $x > 0$ imply?

Solution   First, we need to interpret the symbolism $x > 0$. It can be read as "x is greater than zero"; however, a more meaningful way to read it might be "x is positive." Now we visualize a rectangular coordinate system—or we can actually sketch one. Our original question becomes, Where is x positive? We can answer by thinking in terms of quadrants. For what quadrant(s) is x positive? The answer is quadrants I and IV. We now know that if we are working with a graph under the condition that $x > 0$, our attention will be focused in quadrants I and IV. Something abstract has been replaced with something visual.

■

1.5 EXERCISES

SET A

In exercises 1 through 6, find the distance between the given pair of numbers. (See Example 1.)

1. 0 and -4 **2.** 4 and -4 **3.** 2 and -5

4. -2 and 9   **5.** -3 and -8 **6.** -3.5 and -6

In exercises 7 through 10, plot the list of ordered pairs given on the same coordinate system. (See Examples 2 and 3.)

7. $(0, 2), (-3, -1), (1, 6), (-4, 0), (3, -1)$

8. $(-4, 0), (2, 1), (3, -3), (5, 0), (-6, -2)$

9. $(1, 10), (0, \frac{5}{2}), (\frac{-1}{3}, 0), (-1, -5)$

10. $(1, 1), (0, \frac{5}{2}), (\frac{5}{3}, 0), (-1, 4)$

In exercises 11 through 18, find the distance between the given pairs of points. (See Examples 4 and 5.)

11. $(2, 3)$ and $(-2, 6)$ **12.** $(0, 4)$ and $(-3, 0)$

13. $(-2, 1)$ and $(-5, -2)$ **14.** $(1, 3)$ and $(-3, -1)$

15. $(0, 1)$ and $(-1, 3)$ **16.** $(-4, 2)$ and $(-4, 5)$

17. $(-3.2, -1)$ and $(1.2, -1)$ **18.** $(5\sqrt{2}, 1)$ and $(\sqrt{2}, 3)$

19. Working in a rectangular coordinate system, what does $x < 0$ imply? (*Hint:* See Example 6.)

20. Working in a rectangular coordinate system, what does $y > 0$ imply?

21. For what quadrant(s) would each of the following be true?
a. $xy > 0$ **b.** $xy < 0$ **c.** $x/y > 0$

SET B

22. The formula for finding the point that is halfway between the two points (x_1, y_1) and (x_2, y_2) is given by

$$\left(\frac{x_1 + x_2}{2}, \frac{y_1 + y_2}{2} \right)$$

This can be referred to as the **midpoint formula**. Use this formula to find the point halfway between each given pair of points.
a. $(-1, 4)$ and $(5, 2)$
b. $(4, -3)$ and $(3, -2)$
c. $(1.4, -0.4)$ and $(2.6, 1.3)$

23. Use the distance formula and the Pythagorean theorem to show that the points $(1, 3), (-2, 4)$, and $(2, 6)$ form a right triangle.

24. Find all points with x coordinate 3 that are 13 units from the point $(-2, 1)$.

25. Find all points with y coordinate -2 that are 5 units from the point $(3, -5)$.

BUSINESS
PLANNING

26. A fast-food chain has decided to open a third restaurant in a particular sector of the city. A grid has been placed over a map of that portion of the city.

Thinking of the grid as the first quadrant of a rectangular coordinate system, we see that the two current restaurant locations correspond to the points (1, 2) and (8, 7), as shown in the figure.

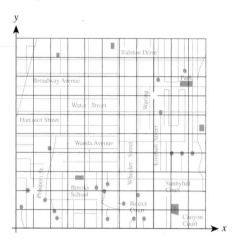

a. As the "crow flies," how far apart are the current restaurants, assuming that on both axes 1 unit = 0.5 mile?

b. If the owners would like the new restaurant to be halfway between the other two, what would its coordinates be?

c. How far will the new restaurant be from the other two?

WRITING AND CRITICAL THINKING

The standard form for the equation of a circle with center (h, k) and radius r is given by

$$(x - h)^2 + (y - k)^2 = r^2$$

Explain in words the relationship between this equation and the distance formula.

1.6   LINES

We begin our discussion with the concept of slope. The **slope** is a numerical value used to describe the "steepness" of a line or line segment. Analytically, for any two points on the line, it gives the ratio of the change in the y coordinates to the change in the x coordinates as we move from one point to the other. Using m to represent the slope, we get

$$m = \frac{\text{change in } y}{\text{change in } x}$$

or symbolically $m = \Delta y/\Delta x$. The Greek letter Δ represents *change*.

If P with coordinates (x_1, y_1) and Q with coordinates (x_2, y_2) are two distinct points on a line, we have the following formula:

Slope formula

The slope m of the line joining P and Q is given by

$$m = \frac{y_2 - y_1}{x_2 - x_1}$$

Example 1   Find the slope of the line joining the points (5, 3) and (2, −4). Sketch a graph of the line.

Solution $m = \dfrac{3 - (-4)}{5 - 2}$

$ = \dfrac{7}{3}$

Plotting the two points and connecting them, we have the line shown in Figure 1.7.

Figure 1.7

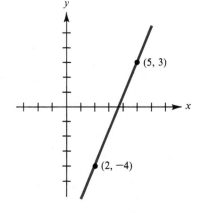

Example 2   Find the slope of the line through (−2, 4) and (3, −6).

Solution $m = \dfrac{-6 - 4}{3 - (-2)}$

$= \dfrac{-10}{5} = -2$ ∎

From the preceding two examples, we see that it is possible for the slope of a line to be positive or negative. Figure 1.8 shows graphically the difference between a positive slope ($m > 0$) and a negative slope ($m < 0$).

Figure 1.8

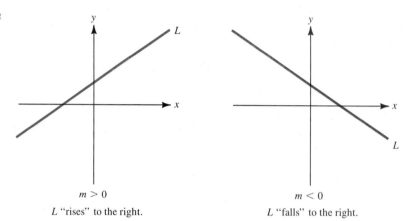

$m > 0$ $m < 0$

L "rises" to the right. L "falls" to the right.

It is also possible for a line to have zero slope ($m = 0$) or no slope (m is undefined). Example 3 shows how these events can occur.

Example 3   Find the slope of the line through the given points and sketch a graph of the lines:
a. $(2, 4)$ and $(-3, 4)$ **b.** $(4, -1)$ and $(4, 2)$

Solution   **a.** $m = \dfrac{4 - 4}{-3 - 2} = \dfrac{0}{-5} = 0$

By plotting the points, we see that the two points lie on a horizontal line (see Figure 1.9).

Figure 1.9

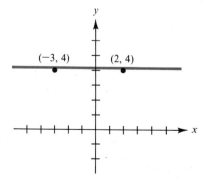

b. $m = \dfrac{2 - (-1)}{4 - 4} = \dfrac{3}{0}$, which is undefined

Notice that these two points lie on a vertical line (see Figure 1.10).

Figure 1.10

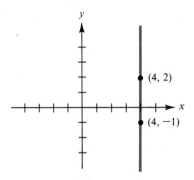

(4, 2)

(4, −1)

The slope of a horizontal line will always be zero because the difference of the y coordinates in the numerator will always be zero. The slope of a vertical line is always undefined.

The slope formula can be used to generate the equation of a line with a given slope and passing through a specified point. If we let m be the given slope, (x_1, y_1) the given point, and (x, y) any other point on the line, then

$$m = \frac{y - y_1}{x - x_1}$$

Multiplying both sides by $x - x_1$ gives the desired equation:

Point-slope form

> The **point-slope form** for the equation of a line is
>
> $$y - y_1 = m(x - x_1)$$

Example 4   Write the equation of the line through $(1, -3)$ with slope 2.

Solution   With $x_1 = 1$, $y_1 = -3$, and $m = 2$, we have

$$\begin{aligned}
y - y_1 &= m(x - x_1) \\
y - (-3) &= 2(x - 1) \\
y + 3 &= 2x - 2 \\
y &= 2x - 5
\end{aligned}$$

In the previous example, the equation was left in a form in which y is expressed explicitly in terms of x (solved for y). This form is referred to as the **slope-intercept form.**

Slope-intercept form

> The **slope-intercept form** for the equation of a line is
>
> $$y = mx + b$$
>
> where m is the slope and b is the **y intercept** (y coordinate of the point where the line crosses the y axis).

Example 5   Find the equation of the line that crosses the y axis at the point $(0, -4)$ and has slope $\frac{2}{3}$.

Solution   Using the slope-intercept form with $m = \frac{2}{3}$ and $b = -4$, we have

$$y = \tfrac{2}{3}x - 4$$

Note that the point-slope form could also be used.   ∎

We will also use a third form for the equation of a line.

Standard form

> The **standard form** for the equation of a line is
>
> $$Ax + By + C = 0$$
>
> where A and B are not both zero.

We could take the answer for the last example, $y = \frac{2}{3}x - 4$, and put it in the standard form as follows:

$$y = \tfrac{2}{3}x - 4$$
$$0 = \tfrac{2}{3}x - y - 4 \qquad \text{Subtract } y \text{ from both sides.}$$
$$0 = 2x - 3y - 12 \qquad \text{Multiply both sides by 3.}$$

or $2x - 3y - 12 = 0$

Consider what happens in the standard form if $A = 0$. The equation reduces to

$$By + C = 0 \quad \text{or} \quad y = \frac{-C}{B}$$

which means y is a constant.

Horizontal line

> The equation of a horizontal line is
>
> $$y = K$$
>
> where K is a constant.

If $B = 0$, we get a vertical line.

Vertical line	The equation of a vertical line is $$x = K$$ where K is a constant.

Finally, we would like to see how to sketch the graph of a line from the equation of a line or from information about the line. If the equation is given, we can construct a table of values, as in the next example.

Example 6   Sketch the graph of the line whose equation is $y = 3x - 1$.

Solution   To make a table of values, select values for x and compute the corresponding y values by substituting into the equation and arrange in table form.

x	y	
-2	-7	since $3(-2) - 1 = -7$
-1	-4	since $3(-1) - 1 = -4$
0	-1	since $3(0) - 1 = -1$
1	2	since $3(2) - 1 = 5$

Now, plotting these points and connecting them give the graph. It was really necessary to plot only two points in this case because two points determine a line (see Figure 1.11).

Figure 1.11

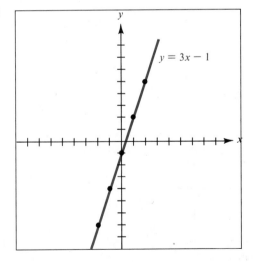

$y = 3x - 1$

Because only two points are necessary, a second method using the x and y intercepts will be used. To find the **x intercept** (point where the line crosses the x axis), let $y = 0$ and solve for x. To find the **y intercept,** let $x = 0$ and solve for y.

Example 7   Sketch the graph of the line $2x - 3y + 6 = 0$ using the x and y intercepts.

Solution

x intercept	*y intercept*
let $y = 0$	let $x = 0$
$2x - 3(0) + 6 = 0$	$2(0) - 3y + 6 = 0$
$2x = -6$	$-3y = -6$
$x = -3$	$y = 2$

This gives the ordered pairs $(-3, 0)$ and $(0, 2)$, which are points on the x and y axes, respectively (see Figure 1.12).

Figure 1.12

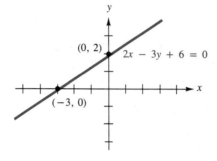

Example 8   Sketch the graph of the line with slope $\frac{-3}{2}$ that passes through the point $(2, 1)$.

Solution   We could write the equation of the line and use either a table of values or the x and y intercepts. Instead, we will use the fact that the slope gives

$$m = \frac{\text{change in } y \text{ coordinate}}{\text{change in } x \text{ coordinate}} = \frac{-3}{2}$$

Figure 1.13

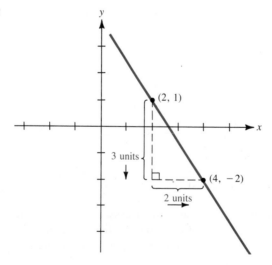

Starting at the given point, (2, 1), move 3 units in the negative y direction (down) and 2 units in the positive x direction (right). The new point is (4, -2) (see Figure 1.13). ∎

Before leaving the topic of lines, we give the relationship between the slope of lines that are parallel and that of lines that are perpendicular.

Parallel and perpendicular lines

Let line L_1 have slope m_1 and line L_2 have slope m_2. If L_1 and L_2 are **parallel,** then

$$m_1 = m_2$$

If L_1 and L_2 are **perpendicular,** then

$$m_1 m_2 = -1 \quad \text{or} \quad m_1 = \frac{-1}{m_2}$$

Example 9

Find the equation of the line that passes through the point $(-4, 2)$ and is parallel to the line $y = 5x - 2$.

Solution

From the slope-intercept form, the slope of the given line is 5 and therefore the slope m for the desired parallel line is 5. Using the point-slope form with $m = 5$ and the point $(-4, 2)$, we have

$$y - y_1 = m(x - x_1)$$
$$y - 2 = 5[x - (-4)]$$
$$y - 2 = 5x + 20$$
$$\text{or} \qquad y = 5x + 22$$ ∎

Example 10

Determine whether or not the lines $2x - 3y + 12 = 0$ and $3x + 2y + 6 = 0$ are perpendicular. Sketch both lines on the same set of axes.

Solution

Putting both equations in the slope-intercept form by solving for y gives

L_1	L_2
$2x - 3y + 12 = 0$	$3x + 2y + 6 = 0$
$-3y = -2x - 12$	$2y = -3x - 6$
$y = \frac{2}{3}x + 4$	$y = \frac{-3}{2}x - 3$

The slopes are $m_1 = \frac{2}{3}$ and $m_2 = \frac{-3}{2}$.

$$m_1 m_2 = \left(\frac{2}{3}\right)\left(\frac{-3}{2}\right) = -1$$

and so the lines are perpendicular (see Figure 1.14).

Figure 1.14

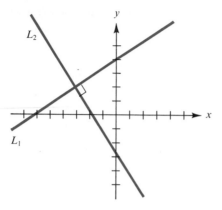

In Example 10, we found that the two lines given were indeed perpendicular, and from Figure 1.14, we see that the two lines intersect. We will use the two lines from this example for our next question. What are the coordinates of the point of intersection of these two lines?

From Figure 1.14, it appears that the point of intersection is approximately $(-3, 2)$. However, in this case the actual point of intersection is $(\frac{-42}{13}, \frac{24}{13})$, as will be determined in Example 11. It is not always possible to get the exact point by looking at the graph, but we can find the exact point using an algebraic process. We can find the point of intersection by solving the two equations simultaneously, as shown in Example 11.

Example 11   Find the point of intersection for $2x - 3y + 12 = 0$ and $3x + 2y + 6 = 0$.

Solution   Using a technique called *substitution*, we solve the first equation for y and substitute this into the second equation.

$$2x - 3y + 12 = 0$$
$$-3y = -2x - 12$$
$$y = \frac{2}{3}x + 4$$

Now substituting into the second equation gives

$$3x + 2y + 6 = 0$$
$$3x + 2\left(\frac{2}{3}x + 4\right) + 6 = 0$$
$$3x + \frac{4}{3}x + 8 + 6 = 0$$
$$\frac{13}{3}x + 14 = 0$$
$$\frac{13}{3}x = -14$$
$$x = \frac{-42}{13}$$

To get the corresponding y coordinate, we need to use substitution again.

$$y = \frac{2}{3}x + 4$$

$$= \frac{2}{3}\left(\frac{-42}{13}\right) + 4$$

$$= \frac{-28}{13} + 4 = \frac{24}{13}$$

The point of intersection is $(\frac{-42}{13}, \frac{24}{13})$.   ■

In many business applications we use what are called **supply and demand equations,** which can often be represented by linear equations, as we see in the next example.

Example 12   The supply equation for a certain product relates the number of units x supplied and the price per unit p (measured in dollars) and is given by

$$p = x + 200$$

The demand equation relates the number of units x demanded and the price per unit p. For this product the demand equation is

$$p = 500 - 2x$$

Find the point at which the supply and demand are equal, called the **equilibrium point.**

Solution   To find the equilibrium point, we need to find the point of intersection of the two lines using the substitution method. Because both are solved for p, we can write

$$x + 200 = 500 - 2x$$
$$3x = 300$$
$$x = 100$$

Substituting gives

$$p = 100 + 200$$
$$= 300$$

The equilibrium point occurs for 100 units (x) at a price p of $300. Figure 1.15 is a graphical representation of the facts in this example.

Figure 1.15

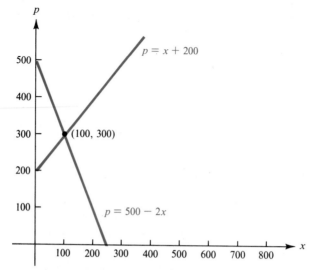

1.6 EXERCISES

SET A

In exercises 1 through 12, determine the slope of the line joining the given pair of points. (See Examples 1, 2, and 3.)

1. (4, 2) and (2, 8)

2. (2, 4) and (8, 2)

3. (−3, −1) and (2, −6)

4. (5, −5) and (3, −11)

5. (0, 3) and (−2, 0)

6. (−6, 0) and (0, −6)

7. (5, −2) and (1, −1)

8. (7, −3) and (5, 2)

9. (0, 3) and (−5, 3)

10. (1, 1) and (−1, 1)

11. (−2, −1) and (−2, 7)

12. $(\sqrt{2}, \sqrt{3})$ and $(\sqrt{2}, 5\sqrt{3})$

In exercises 13 through 24, write the equation of the line through the given point with the given slope. Put answers in slope-intercept form, if possible. (See Examples 4 and 5.)

13. (3, 4); $m = 6$

14. (5, 3); $m = 4$

15. (−1, 4); $m = −3$

16. (2, −6); $m = −1$

17. (5, 4); $m = \frac{1}{2}$

18. (3, 2); $m = \frac{5}{2}$

19. $(\frac{1}{2}, \frac{5}{4})$; $m = \frac{-2}{3}$

20. $(\frac{1}{2}, \frac{-1}{2})$; $m = \frac{-5}{3}$

21. (3, 5); $m = 0$

22. (−2, 7); $m = 0$

23. (−2, −3); m is undefined

24. (0, 3); m is undefined

In exercises 25 through 30, find the equation for the line through the given pairs of points. Write answers in standard form. (Hint: First use the slope formula to compute m.)

25. (1, 3) and (−2, 6)

26. (2, 6) and (−2, −6)

27. (5, 5) and (0, 5)

28. (0, 3) and (−2, 0)

29. (−2, 3) and (−4, 8)

30. (−1, 6) and (−1, 4)

In exercises 31 through 38, determine the slope and y intercept of the given line. (Hint: Put the equation in slope-intercept form.)

31. $y = 6 − 4x$

32. $y = \frac{1}{3}x + 7$

33. $x − 2y + 6 = 0$

34. $3x − 3y + 1 = 0$

35. $5x + 4y = 0$

36. $2x + 4y − 5 = 0$

37. $y = 6$

38. $x = −2$

In exercises 39 through 52, sketch the graph of the line. (See Examples 6, 7, and 8.)

39. $y = x$

40. $y = -2x$

41. $y = 4x + 3$

42. $y = 3 - 2x$

43. $y = \frac{1}{2}x - \frac{3}{2}$

44. $y = \frac{3}{2}x + 1$

45. $x + y = 0$

46. $x - y = 0$

47. $2x + y - 3 = 0$

48. $x + 2y - 5 = 0$

49. $2x - 4y - 8 = 0$

50. $4x - 3y - 12 = 0$

51. $x = 4$

52. $y + 2 = 0$

In exercises 53 through 62, find the equation of the line as described. (See Examples 9 and 10.)

53. through $(-2, -3)$ and parallel to $y = 4x$

54. through $(1, -3)$ and parallel to $y = 1 - 3x$

55. through $(0, 5)$ and parallel to $x + 2y = 0$

56. through $(-3, 0)$ and parallel to $x + 2y + 1 = 0$

57. through $(5, 2)$ and parallel to $x + 3 = 0$

58. through $(3, 4)$ and parallel to $y = 2$

59. through $(1, 1)$ and perpendicular to $y = 5x - 2$

60. through $(3, -2)$ and perpendicular to $y = -4x$

61. through $(0, -1)$ and perpendicular to $x + 3y + 4 = 0$

62. through $(0, 3)$ and perpendicular to $2x - 5y - 7 = 0$

In exercises 63 through 67, find the point of intersection of the given lines. (See Example 11.)

63. $y = 2x - 1$ and $x - y + 4 = 0$

64. $x + y - 3 = 0$ and $x + 2y = 0$

65. $y = 2x - 5$ and $y = 3x - 3$

66. $y = -4x$ and $y = 2x + 2$

67. $2x - y + 5 = 0$ and $x + 2y - 5 = 0$

SET B

In exercises 68 through 71, a demand equation is given. Determine the number of units x demanded for the price given.

68. $p = 200 - x; p = 5$

69. $p = 2000 - 2x; p = 20$

70. $p = 500 - 0.1x^2; p = 10$

71. $p = 740 - 0.2x^2; p = 20$

In exercises 72 through 75, a supply equation is given. Determine the number of units x supplied for the price given.

72. $p = x + 200; p = 250$

73. $p = 4x + 1200; p = 1500$

74. $p = 0.2x + 110; p = 125$

75. $p = 0.02x^2 + 350; p = 400$

In exercises 76 through 79, solve the given supply and demand equations simultaneously to find the equilibrium point. Give the number of units x and the price p. (See Example 12.)

76. supply: $p = x + 100$
demand: $p = 500 - x$

77. supply: $p = 0.2x + 40$
demand: $p = 50 - 0.3x$

78. supply: $p = 0.5x^2 + 100$
demand: $p = 500 - 10x$

79. supply: $p = 0.1x^2 + 200$
demand: $p = 400 - x$

80.   Sketch the supply and demand equations from exercise 76 on the same set of axes.

BUSINESS
SALES

81.   The Desmond Development Company has determined that it can sell 500 units of its new product at a price of $7. In an attempt to increase sales, the price was lowered and it was found that with each $1 reduction in price, sales increased by 200 units. Express the linear relationship between the price p and the number of units x. (*Hint:* Use the information to generate two ordered pairs.)

REAL ESTATE

82.   In a certain apartment complex, when the rent was $350 per month, 52 units were rented. When the rent was raised to $375 a month, the number of units rented decreased to 45. Express the linear relationship between the rent, r, and the number of units rented, n.

WRITING AND CRITICAL THINKING

In the diagram you see the point $(1, 4)$ and the line $y = x + 1$. Outline a procedure you might use to find the (shortest) distance between the point and the line. Explain your procedure in words. (*Hint:* We have a formula for finding the distance between two points. Can you find the coordinates of an appropriate point on the line you can use together with the given point?) Now try your procedure. Does your answer seem reasonable based on the diagram?

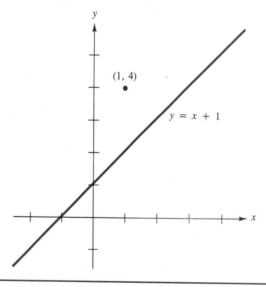

1.6A   LINES ON A GRAPHICS CALCULATOR

In this section we use a graphics calculator to observe some properties of straight lines. We begin with the concept of *slope*. The **slope** of the line segment joining the points P and Q is a numerical value used to describe the "steepness" of the line segment. Analytically, it gives the ratio of the change in the y coordinates to the change in the x coordinates as we move from one point to the other. Using m to represent the slope, we get

$$\text{slope} = m = \frac{\text{change in } y}{\text{change in } x} = \frac{\Delta y}{\Delta x}$$

Using the coordinates (x_1, y_1) for P and (x_2, y_2) for Q, we have the following formula:

Slope formula

The slope m of the line joining $P(x_1, y_1)$ and $Q(x_2, y_2)$ is given by

$$m = \frac{y_2 - y_1}{x_2 - x_1}$$

We can use technology, in this case a TI-82 graphics calculator, to draw the line segment connecting two points. In particular, we choose to plot the line segment connecting the two points $(2, -4)$ and $(5, 3)$ (see Figure 1.6A.1).

Figure 1.6A.1

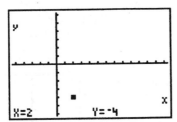

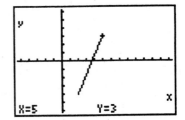

From the graphics screen, choose LINE from the DRAW menu. Move the cursor to the first point, $(2, -4)$, and press ENTER.

Move the cursor to the second point, $(5, 3)$, and press ENTER. The TI-82 will draw the line segment connecting the two points.

A right triangle can be formed with that line segment as a hypotenuse. In Figure 1.6A.2, we see that the coordinates of the right angle are $(5, -4)$.

Figure 1.6A.2

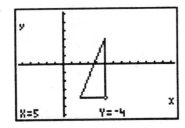

The slope connecting these two points is the length of the vertical leg of the right triangle divided by the length of the horizontal leg of the triangle. That is, $m = \frac{-4-3}{2-5} = \frac{7}{3}$.

The slope of a slanted line can be positive (as in Figure 1.6A.2) or negative. Lines with negative slopes fall from upper left to lower right, as Figure 1.6A.3 demonstrates.

Figure 1.6A.3

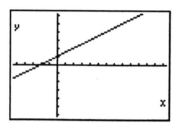 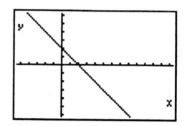

$m > 0$
The line "rises" to the right.

$m < 0$
The line "falls" to the right.

For an example of a horizontal line, consider the line $y = 4$ (see Figure 1.6A.4). Choose any two points on that line, say $(2, 4)$ and $(3, 4)$, and the calculation of the slope is $\frac{4-4}{2-3} = 0$. If a line is vertical, its slope is undefined (see Figure 1.6A.5). For example, the vertical line $x = 4$ contains the points $(4, -5)$ and $(4, 7)$. The calculation of the slope is: $\frac{-5-7}{4-4} = \frac{-12}{0}$, which is undefined.

Figure 1.6A.4

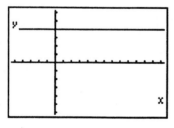

Figure 1.6A.5

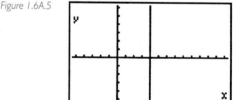

The horizontal line $y = 4$ has a slope of 0. Every horizontal line has zero for the value of its slope.

The vertical line $x = 4$ has a slope that is undefined. Every vertical line has an undefined slope.

The slope formula can be used to generate the equation of a line with a given slope and passing through a specified point. If we let m be the given slope, (x_1, y_1) the given point, and (x, y) any other point on the line, then

$$m = \frac{y - y_1}{x - x_1}$$

Multiplying both sides by $x - x_1$ gives the desired equation:

The point-slope form

> The **point-slope form** for the equation of a line with slope m passing through (x_1, y_1) is given by
>
> $$y - y_1 = m(x - x_1)$$

Unfortunately, many graphics calculators do not use lowercase letters or subscripts. We will adapt the preceding formula for those calculators. Now, given the slope M and a given point (S, T), we have the following:

The point-slope form
(calculator version)

> The point-slope form for the equation of a line with slope M and passing through (S, T) is given by
>
> $$Y = M(X - S) + T$$

For an example, consider the problem of finding the equation of the line through $(1, -3)$ with slope 2. The point-slope form, with $S = 1$, $T = -3$, and $M = 2$, gives us:

$$
\begin{aligned}
Y &= M(X - S) + T \\
&= 2(X - 1) + (-3) \\
&= 2X - 2 - 3 \\
Y &= 2X - 5
\end{aligned}
$$

The advantage of entering the constants and general equation, as in Figure 1.6A.6, is that if many different equations have to be graphed, we need only enter the general equation (right-hand screen) once:

Figure 1.6A.6 The screen on the left assigns the appropriate values to S, T, and M. The Y= key allows us to enter the general point-slope formula as can be seen in the screen on the right.

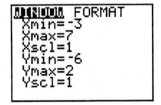

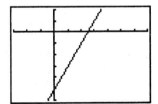

The graph of $Y = 2X - 5$ is shown in Figure 1.6A.7.

Figure 1.6A.7

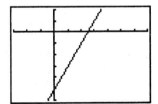

Two points determine a line. So, given two distinct points, we can find the equation of the line passing through them. Actually, this is almost identical to the situation of knowing a point and the line's slope because, if two points are known, the slope is easily calculable. We can alter the point-slope form and make it into the **two-point form,** as follows:

The two-point form

The **two-point form** for the equation of a line passing through the points (x_1, y_1) and (x_2, y_2) is given by

$$y - y_1 = \frac{y_2 - y_1}{x_2 - x_1}(x - x_1)$$

or $$y = \frac{y_2 - y_1}{x_2 - x_1}(x - x_1) + y_1$$

On a graphics calculator, there are several ways to graph a line knowing two points. One way is to use the LINE option under the DRAW menu. We discussed this earlier in this section. Another way is to use the two-point form. For example, if a line passes through the two points $(1, 2)$ and $(0, -1)$, then we can calculate the line's equation using the two-point form:

$$y = \frac{y_2 - y_1}{x_2 - x_1}(x - x_1) + y_1$$

$$= \frac{2 - (-1)}{1 - 0}(x - 1) + 2$$

$$= 3x - 1$$

The TI-82 graph of $y = 3x - 1$ is shown in Figure 1.6A.8.

Figure 1.6A.8

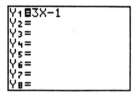
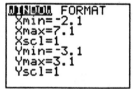
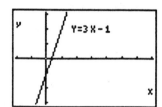

In addition to the point-slope form and the two-point form, there are two other forms for the equation of a straight line. They are the slope-intercept form and the standard form; these are discussed in the previous section.

Two special points on a line are its x intercept and its y intercept. For the graph of $y = 3x - 1$ (shown in Figure 1.6A.8), the y intercept is $(0, -1)$—the point where the graph crosses the y axis. (*Note:* Every point on the **y** axis has an **x** value of 0.) To find a y intercept, substitute 0 for x and find the value of y. If your calculator is equipped with a TABLE feature, this value can be obtained there (see Figure 1.6A.9).

Figure 1.6A.9

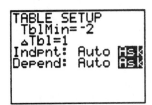
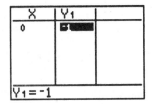

Many calculators are equipped with a SOLVE feature that will approximate a graph's x intercept. We use that feature to determine that the x intercept for $y = 3x - 1$ is $(\frac{1}{3}, 0)$ (see Figure 1.6A.10).

Figure 1.6A.10

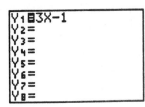
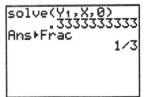

The 0 (zero) in the command **solve(Y₁,X,0)** represents our initial guess for the x intercept. Although we used 0, many other numbers would have given the same result.

Of course, graphics calculators can be used to graph a collection of different lines. In particular, they can be used to find—or at least approximate—the simultaneous solution to a system of two equations. In the following example, we see how a graphics calculator can be used to *approximate* a point of intersection of two straight lines and then how algebra is used to find the *exact* solution.

Example 1   Find the point of intersection for $2x - 3y + 12 = 0$ and $3x + 2y + 6 = 0$.

Solution   We can use a graph as a check or estimate in finding the solution to a system of equations. Notice that we had to solve each of the two equations for y before we graphed them on a graphics calculator (see Figure 1.6A.11).

Figure 1.6A.11

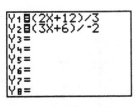
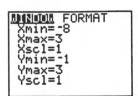
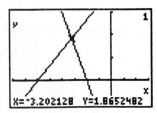

Although the calculator *approximates* the solution to be $x \approx -3.2$ and $y \approx 1.87$, we need to rely on our algebraic skills to find the *exact* solution to the system. Using a technique called *substitution*, we solve the first equation for y and substitute this into the second equation.

$$2x - 3y + 12 = 0$$
$$-3y = -2x - 12$$
$$y = \frac{2}{3}x + 4$$

Now, substituting $\frac{2}{3}x + 4$ for y into the second equation gives

$$3x + 2y + 6 = 0$$
$$3x + 2\left(\frac{2}{3}x + 4\right) + 6 = 0$$
$$3x + \frac{4}{3}x + 8 + 6 = 0$$
$$\frac{13}{3}x + 14 = 0$$
$$\frac{13}{3}x = -14$$
$$x = \frac{-42}{13}$$

To get the corresponding y coordinate, we need to use substitution again.

$$y = \frac{2}{3}x + 4$$
$$= \frac{2}{3}\left(\frac{-42}{13}\right) + 4$$
$$= \frac{-28}{13} + 4 = \frac{24}{13}$$

The point of intersection is *exactly* $(\frac{-42}{13}, \frac{24}{13})$. ■

We conclude this section with a study of parallel and perpendicular lines. Parallel lines have the same slope. This should not be too surprising because we have seen that slope is a measure of a line's "steepness." A quick examination of $y = 2x - 1$, $y = 2x + 1$, and $y = 2x + 2$ shows this nicely (see Figure 1.6A.12).

Figure 1.6A.12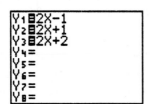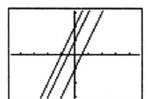

It can also be shown (see exercise 40) that if two lines are perpendicular, their slopes are negative reciprocals of one another. The slope relationships of parallel and perpendicular lines are as follows:

Parallel and perpendicular lines

Let line L_1 have slope m_1 and line L_2 have slope m_2.
If L_1 and L_2 are **parallel,** then

$$m_1 = m_2$$

If L_1 and L_2 are **perpendicular,** then

$$m_1 m_2 = -1 \quad \text{or} \quad m_1 = \frac{-1}{m_2}$$

The concept of perpendicularity comes into play when we want to find the distance a point is from a line. That is, the distance of a point from a line is the perpendicular (or shortest) distance between the point and the line. For example, we want to find the distance between the point $(-3, 2)$ and the line $y = 2x + 3$. The point and line are pictured in Figure 1.6A.13, a graphics calculator screen.

Figure 1.6A.13

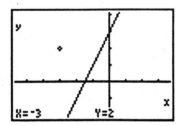

The first step is to find the point on $y = 2x + 3$ that is also on the line containing $(-3, 2)$ and perpendicular to $y = 2x + 3$. That line has slope $-\frac{1}{2}$ and contains $(-3, 2)$ so we use the point-slope form to find its equation:

$$y = m(x - x_1) + y_1$$
$$y = -\frac{1}{2}(x - (-3)) + 2$$
$$y = -\frac{1}{2}x + \frac{1}{2}$$

Now, because the point falls on both $y = -\frac{1}{2}x + \frac{1}{2}$ and $y = 2x + 3$, we can find it by solving the simultaneous system:

$$y = 2x + 3$$
$$-\frac{1}{2}x + \frac{1}{2} = 2x + 3$$
$$\frac{5}{2}x = -\frac{5}{2}$$
$$x = -1$$

If $x = -1$, then $y = 1$, and now we must find the distance between $(-3, 2)$ and $(-1, 1)$ (see Figure 1.6A.14).

$$Distance = \sqrt{(-3-(-1))^2 + (2-1)^2} = \sqrt{5} \text{ units.}$$

Figure 1.6A.14

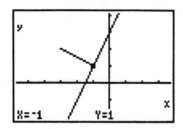

1.6A   EXERCISES

In exercises 1 through 10, (a) find the slope of the line connecting the two given points; (b) find the equation of the line connecting the two points; and (c) using a suitable collection of WINDOW values, graph each line.

1. (4, 2) and (2, 8)
2. (2, 4) and (8, 2)
3. (−3, −1) and (2, −6)
4. (5, −5) and (3, −11)
5. (0, 3) and (−2, 0)
6. (−6, 0) and (0, −6)
7. (5, −2) and (1, −1)
8. (7, −3) and (5, 2)
9. (0, 3) and (−5, 3)
10. (−2, −1) and (−2, 7)

In exercises 11 through 16, read directly from the graphics calculator screen and approximate the slope and y intercept of the drawn lines.

11.

WINDOW FORMAT
Xmin=-5
Xmax=5
Xscl=1
Ymin=-4
Ymax=4
Yscl=1

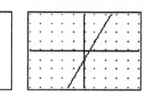

12.

WINDOW FORMAT
Xmin=-5
Xmax=5
Xscl=1
Ymin=-4
Ymax=4
Yscl=1

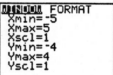

13.

WINDOW FORMAT
Xmin=-4
Xmax=4
Xscl=1
Ymin=-3
Ymax=3
Yscl=1

14.

WINDOW FORMAT
Xmin=-4
Xmax=4
Xscl=1
Ymin=-3
Ymax=3
Yscl=1

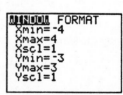

15.

WINDOW FORMAT
Xmin=-4
Xmax=4
Xscl=1
Ymin=-25
Ymax=-18
Yscl=1

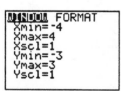

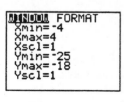

16.

WINDOW FORMAT
Xmin=-10
Xmax=10
Xscl=2
Ymin=-3
Ymax=3
Yscl=1

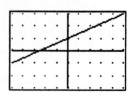

In exercises 17 through 30, find the equation of the line with the given information. Also, using a suitable collection of WINDOW values, graph each line.

17.   through (3, 4) with slope 6

18.   through (−1, 4) with slope −3

19.   through (2, −6) with slope −1

20.   through (5, 4) with slope $\frac{1}{2}$

21.   through (−2, 7) with slope 0

22.   through (3, 5) with slope 0

23.   through (−2, −3) with undefined slope

24.   through (0, 3) with undefined slope

25.   *x* intercept (2, 0) and *y* intercept (0, 4)

26.   *x* intercept (3, 0) and *y* intercept (0, −4)

27.   through (−2, −3) and parallel to $y = 4x$

28.   through (1, −3) and parallel to $y = 1 - 3x$

29.   through (1, 1) and perpendicular to $y = 5x - 2$

30.   through (3, −2) and perpendicular to $y = -4x$

In exerc.   1 through 34, estimate the point of intersection of the lines, given the graphics calculator screen and W₁NDOW information.

31.

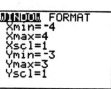

32.

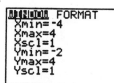

33.

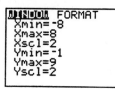

34.

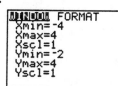

In exercises 35 through 39, find the point of intersection of the given lines.

35. $y = 2x - 1$ and $x - y + 4 = 0$

36. $x + y - 3 = 0$ and $x + 2y = 0$

37. $y = 2x - 5$ and $y = 3x - 3$

38. $y = -4x$ and $y = 2x + 2$

39. $2x - y + 5 = 0$ and $x + 2y - 5 = 0$

40.   Complete this analytic geometry proof that if two lines have slopes that are negative reciprocals of one another, then the two lines are perpendicular:

   a.   Suppose the slope of one line is *m*. How would you represent the slope of the other line? Draw a picture.

   b.   Suppose the line with slope *m* has *y* intercept (0, y_1) and the other line has *y* intercept (0, y_2). How would you represent the two lines algebraically?

   c.   Find the point of intersection of the two lines. Call this point *P*.

   d.   Find the distance between point *P* and each of the *y* intercepts. Also find the distance between the two *y* intercepts.

   e.   Use the Pythagorean theorem to show that the angle formed at *P* must be a right angle.

CHAPTER I REVIEW

VOCABULARY/FORMULA LIST

term

factor

laws of exponents

linear equation

quadratic equation

quadratic formula

$$x = \frac{-b \pm \sqrt{b^2 - 4ac}}{2a}$$

radical equation

rational equation

linear inequality

quadratic inequality

rational inequality

radical inequality

absolute value

ordered pair

distance formula

$$d = \sqrt{(x_2 - x_1)^2 + (y_2 - y_1)^2}$$

slope formula

$$m = \frac{y_2 - y_1}{x_2 - x_1}$$

point-slope form

$$y - y_1 = m(x - x_1)$$

perpendicular

slope-intercept form

$$y = mx + b$$

standard form

$$Ax + By + C = 0$$

x intercept

y intercept

parallel

CHAPTER 1 REVIEW EXERCISES

[1.1]   *In exercises 1 through 3, simplify each expression.*

1. $3x^2(x + 2) - 5(x^2 - 4x)$

2. $\dfrac{2x}{x - 1} + \dfrac{3}{x + 2}$

3. $x^{-1/2}(x + x^{1/2})$

[1.2]   *In exercises 4 through 8, factor. Leave all exponents positive in answers.*

4. $2x^2 + 2x - 40$

5. $10x(x + 4)^4 + 2(x + 4)^5$

6. $x^8 - 1$

7. $8y^3 - 27$

8. $x^{-3} - 4x^{-2}$

[1.3]   *In exercises 9 through 15, solve each equation for x.*

9. $4x + 3(2x - 4) = 0$

10. $x^2 - 8x = 0$

11. $(x - 3)^2 = 16$

12. $x^2 = 5x + 1$

13. $\sqrt[3]{3x - 1} = 1$

14. $\dfrac{2x + 1}{2x - 3} = 2$

15. $\dfrac{x^2 - x}{\sqrt{x + 1}} = 0$

[1.4]   *In exercises 16 through 20, solve the given inequality for x.*

16. $1 - 3x < -2x + 7$

17. $x^2 - 4 \geq 0$

18. $x^2 \leq 2x$

19. $\dfrac{3x + 1}{x} < 0$

     20. $\sqrt{4-x} \geq 0$

[1.5] **21.** Find the distance between $(-2, 5)$ and $(4, 13)$.

[1.6] **22.** Find the equation of the line through the points $(0, 6)$ and $(-1, 8)$.

[1.6] **23.** Find the slope and y intercept of the line whose equation is

$$2x + 4y + 7 = 0$$

BUSINESS   [1.6] **24.** The demand equation for a certain product is given by
DEMAND

$$p = 250 - 0.2x$$

where x is the number of units demanded and p is the corresponding price per unit. If the price per unit is \$75, determine the number of units demanded.

PROFIT   [1.3] **25.** The Creative Card Company has determined that its monthly profit can be determined by the following equation:

$$P = -125 + 30x - x^2, \qquad 0 \leq x \leq 20$$

where P is the profit (in dollars) and x is the number of cards produced and sold (in hundreds). How many cards must the company sell per month to break even? (*Hint:* Let $P = 0$.)

CHAPTER 1 TEST

PART ONE

Multiple choice. In exercises 1 through 10, put the letter of your correct response in the blank provided.

_____ **1.** Rationalize the denominator: $\dfrac{2}{\sqrt{2x}}$.

    a. $\sqrt{2x}$ **b.** $\dfrac{\sqrt{2x}}{x}$

    c. $2\sqrt{2x}$ **d.** $\dfrac{\sqrt{4x}}{2x}$

_____ **2.** Factor completely: $x^4 - 4x^2$.

    a. $(x^2 + 2x)(x^2 - 2x)$ **b.** $x^2(x - 2)(x + 2)$

    c. $x^2(x - 4)(x + 4)$ **d.** $x^2(x^2 - 4)$

_____ **3.** Solve for x: $\sqrt[3]{x^2 + 1} = 2$.

    a. $x = \pm 1$ **b.** $x = \pm 2$

    c. $x = \pm\sqrt{3}$ **d.** $x = \pm\sqrt{7}$

_____ **4.** Multiply and simplify: $(x^2 - 5)(x + 7)$.

    a. $x^2 - 35$ **b.** $x^3 + 7x^2 - 5x - 35$

    c. $x^3 + 7x^2 + 5x + 35$ **d.** $x^3 - 35$

_____ 5. Solve for x: $x^2 + 4x - 2 = 0$.

    a. $-2 \pm \sqrt{24}$ **b.** $2 \pm \sqrt{6}$

    c. $-2 \pm \sqrt{2}$ **d.** $-2 \pm \sqrt{6}$

_____ 6. Find the complete solution for x: $\dfrac{x}{x + 3} > 0$.

    a. $-3 < x < 0$ **b.** $x < -3$ or $x > 0$

    c. $x > -3$ **d.** $x > 0$

_____ 7. Find the distance between the points $(5, -1)$ and $(3, 2)$.

    a. 6   **b.** $\sqrt{5}$

    c. $2\sqrt{5}$ **d.** $\sqrt{13}$

_____ 8. Find the equation of the line passing through the point $(5, -4)$ and having slope $\frac{1}{2}$.

    a. $x - 2y - 9 = 0$ **b.** $x - 2y - 13 = 0$

    c. $x + 2y + 13 = 0$ **d.** $x - 2y + 14 = 0$

_____ 9. The graph of $y = 1 - 2x$ is represented by which of the following?

    a.

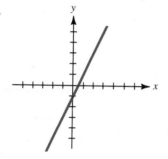

    b.

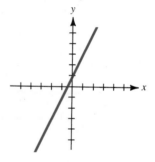

    c.

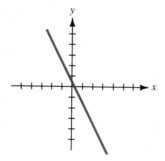

    d.

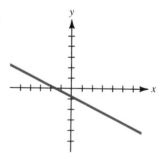

_____ 10. The equation of the line through $(-6, -7)$ and perpendicular to the x axis is

    a. $x + 6 = 0$ **b.** $y + 7 = 0$

    c. $x - 6 = 0$ **d.** $x + y + 13 = 0$

PART TWO

Solve and show all work.

11. Solve for x: $\sqrt{3 - 2x} = 0$.

12. Evaluate for $x = 16$: $-2x^{-3/4}$.

13. Solve for x: $x^2 \geq 4x$.

14. Factor completely and leave all exponents positive in the answer:

$$2x(x-1)^{-2} + 4x^2(x-1)^{-3}$$

15. Find the equation of the line that passes through $(2, -4)$ and $(0, -3)$.

16. Solve for x: $\dfrac{x^2 - 2x}{x + 3} = 0$.

BUSINESS
DEMAND

17. The Handyman Hobby Company has determined that the demand x for its model airplane kits is related to the price per kit p by the equation

$$p = 12.5 - 0.01x$$

Determine the number of kits demanded when the price is $3.50.

PROFIT

18. The profit equation for the Acme Company is given by

$$P = 0.1x^2 - 250, \qquad x > 0$$

Determine the values of x for which the company loses money ($P < 0$) and the values of x for which the company makes money ($P > 0$).

PROBLEM SOLVING

Problem solving is an important skill to develop if you want to be successful in mathematics. In fact, problem solving is an important "life" skill, and the ideas we present here can be used in other disciplines as well as in your everyday life. We feel there are three basic stages to problem solving, and we list them next.

Three basic stages to problem solving

> 1. Thinking, planning, research
> 2. Gathering tools, executing plan
> 3. Thinking about and testing result

In this section, we give you an overview of each of the three stages, discuss the importance of diagrams, and then lead you through some examples using these ideas.

The first stage is probably the most important and the one that is often overlooked by students. When you are presented with a problem you need to solve, you do not need to start writing something down immediately. You need time to **think** first and to **develop a plan.** Once you have a plan in mind, even if it is sketchy or incomplete, begin to write your ideas down. You can always rearrange and add things as you go. It is also helpful, if possible, to arrange your plan into steps. This will help you in the execution stage. As you develop your plan, you may find that some words, ideas, or concepts are not clear to you. In this case, you may need to do some **research,** which may be something as simple as looking up a word in the dictionary.

Many students neglect the power that writing can provide in learning mathematics. Writing gives you a way to "spill" your ideas onto paper so you can examine them more carefully, sort them, edit (change) them, and give them structure. Once you have done this, you are ready to translate your ideas into mathematical symbols.

One last word on the first stage. If at all possible, make an **estimate** of what you think the result will be. It can help you in two ways. First, it can help you find errors in your plan as you work, and second, it is valuable for comparing with your final result.

When you move to the next stage, you are ready to **gather your tools** so you can **execute your plan.** The tools might be things like equations, formulas (algebraic or geometric), ruler, cardboard, scissors, calculator, computer, and so on. As you now begin to execute your plan, it is extremely important that you document what you are doing. Describe in words what you are doing for each step in your plan. Doing so will enable you to come back to your work at a later date and easily follow what occurred. You may also find it useful for solving other related problems.

Stage three is also sometimes skipped. When we arrive at a result to a problem, our tendency in many cases is to feel satisfied and ready to move on. However, we need to spend some time thinking about our result and, if possible, compare it with our plan to decide if it "fits." Also, if we were able to make an estimate of the result in the beginning, we should at this time compare our result with the estimate to see if the result seems reasonable.

We have not finished with stage three. We should now explore other possibilities or techniques for solving a problem. Figure PS.1 is a visual example that illustrates the importance of further exploration.

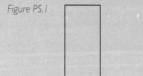

Using three lines, see if you can divide this figure into four pieces that are the same size and shape. Is there more than one way to do this?

Figure PS.2 shows three possible results:

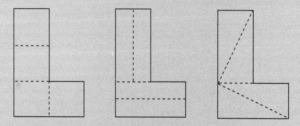

Most of us would probably come up with the diagram on the left and feel finished. We don't usually "stretch" ourselves to look for other possible results. But it is that stretching or further exploration that leads to important ideas and discoveries. It also prepares us to tackle more difficult problems when they arise.

This problem also shows that you can carefully follow instructions, work something through, and have a result that does not look exactly like the result your instructor or another classmate may have. There may be many ways to approach the same problem.

We did not mention diagrams or sketches in any particular stage of the problem-solving process because it is important to incorporate sketches or diagrams throughout the process where possible. A common myth is that people who are good at mathematics do everything "in their head." Quite the opposite is true. It is now and always has been important not only to write ideas down but, whenever possible, to enhance or clarify those ideas by using some sort of diagram. The diagram may be something as simple as "doodling" or something with a more structured look, such as a graph. Think about these ideas for your diagrams:

1. Diagrams can be two-dimensional or three-dimensional. Colors and patterns may also help.
2. If you have access to computer facilities or a graphing calculator, you may want to experiment with those.
3. Some diagrams need to have more accuracy than others, depending on their use. Graph paper may be useful.
4. Some of the pictures you use to help with your study of mathematics may be visible only in your "mind's eye."

Remember that diagrams and graphs are powerful, and using them may enable you to display information in a manner that is clearer or more concise than using just words.

One final note before we present some examples. Keep in mind that we cannot always clearly distinguish one stage from another in the problem-solving process— sometimes stages overlap. The important thing is to use the process as a "package." Resist the temptation to use only "pieces" of it.

Example I   Find the area of the shaded region between the two squares in Figure PS.3. The larger square is 5 cm on a side and the smaller square is 3 cm on a side.

Figure PS.3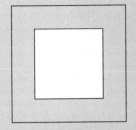

Solution   *Stage 1*

We begin by noticing that there are two squares, a large square with a smaller square inside. To find the area of the shaded region we could subtract the area of the smaller square from the area of the larger square. Here's what we might write down for ourselves to get started:

> Shaded area **is** area of large square **minus** area of small square.

or

> shaded area = area of large square − area of small square

Notice in the second case the use of some symbols (the equal sign and the subtraction symbol). Feel free to use all words or a mixture of words and symbols to get started. The important thing is not to be afraid to write your ideas down in whatever form is comfortable for you. Next, we organize our plan into steps:

Step 1.   Determine area of large square.
Step 2.   Determine area of small square.
Step 3.   Determine area of shaded region by subtracting the area of the small square from the area of the large square.

If you have forgotten the formula for the area of a square, you may need to do some research at this point.

Stage 2

The only "tool" we seem to need here is the formula for the area of a square. The area of a square with side of length s is given by $A = s^2$. Now we are ready to execute the plan.

Step 1.   Determine area of large square.

The large square is 5 cm on a side so

$$A = 5^2 \quad \text{or} \quad 25 \text{ cm}^2$$

Step 2. Determine area of small square.

$$A = 3^2 \quad \text{or} \quad 9 \text{ cm}^2$$

Step 3. Determine area of shaded region by subtracting the area of the small square from the area of the large square.

$$\text{area of shaded region} = 25 - 9$$
$$= 16 \text{ cm}^2$$

Stage 3

Does our result seem reasonable? Can you think of another way to solve this problem? Think about the following. Move the small square into one corner of the large square as shown in Figure PS.4. Now compute the area of this new shaded region. Is it the same? Should it be?

Figure PS.4

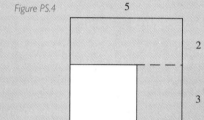

This new shaded region can be divided into two rectangles, one 2 cm by 5 cm and the other 2 cm by 3 cm. If we compute the area of each rectangle (length times width) and add them together, we get a result of 16 cm².

$$\text{area} = (2)(5) + (2)(3) = 16$$

∎

The next example shows that if you spend the time to think and develop a plan, you may arrive at the result without much extra work. In this case, it is difficult to distinguish each stage, so we will not label them. However, you will still be able to see how this process works.

Example 2   Find the equation of a line that is perpendicular to the *x* axis and that passes through the point (3, 5).

Solution   We want a line perpendicular to the *x* axis. A diagram will be helpful (see Figure PS.5).

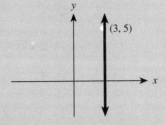

Figure PS.5

From the diagram, we notice that the line we want will be a vertical line, so we need the equation for a vertical line. Researching this, we find that it is

$x = k$　where k is a constant

We are told that the line passes through the point (3, 5). Because the x coordinate of this point is 3, and all points on a vertical line have the same x coordinate, then x must always be 3.

We conclude that the equation for the line will be

$x = 3$

Does this result seem reasonable? How would we alter this plan if the line is to be perpendicular to the y axis instead?　　■

Example 3

The points $(-2, -1)$ and $(4, 5)$ are endpoints of the diameter of a circle. Find the equation for the circle.

Solution

First we draw a picture to help us with our thinking and with our plan (see Figure PS.6). It does not need to be an accurate diagram on a coordinate system.

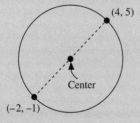

Figure PS.6

After doing some research, we find that one form for the equation of a circle is given by

$$(x - h)^2 + (y - k)^2 = r^2$$

where (h, k) is the center of the circle and r is the radius. This form is called the *standard form*. To use this form for the equation of a circle we need to find the center and the radius. Looking at the picture, we can see that the center will be halfway between the two points that are given. We are now ready to write a plan.

Step 1.　Calculate the center of the circle using the midpoint formula.

Problem Solving

Step 2. Find the radius by using the distance formula to calculate the distance from the center to one of the given points.

Step 3. Use the standard form for the equation of a circle, substituting the center point from Step 1 and the radius from Step 2.

We may need to do further research to find the other formulas we need (our tools):

midpoint formula: $\left(\dfrac{x_1 + x_2}{2}, \dfrac{y_1 + y_2}{2}\right)$

distance formula: $d = \sqrt{(x_2 - x_1)^2 + (y_2 - y_1)^2}$

Step 1. Use the midpoint formula to find the center:

$$\left(\dfrac{-2 + 4}{2}, \dfrac{-1 + 5}{2}\right)$$

Simplifying gives the center as $(1, 2)$.

Step 2. Use the distance formula to find the radius. Use the center and either one of the other points. We use the point $(4, 5)$:

$$r = \sqrt{(4 - 1)^2 + (5 - 2)^2}$$
$$= \sqrt{3^2 + 3^2} = \sqrt{18}$$

Step 3. Use the standard form for the equation of a circle, substituting the center point from Step 1 and the radius from Step 2:

$$(x - h)^2 + (y - k)^2 = r^2$$
$$(x - 1)^2 + (y - 2)^2 = (\sqrt{18})^2$$
$$(x - 1)^2 + (y - 2)^2 = 18$$

Can we check the result? Try substituting the point we did not use in Step 2 into the equation for x and y—that is, the point $(-2, -1)$. Does it make a true statement?

What if the two points given were just any two points on the circle, not endpoints of a diameter. Will our plan still work? Draw a series of diagrams to convince yourself that given any two points that are not endpoints of a diameter, a unique circle is not determined. ∎

The next example illustrates a case in which the original plan does not get us all the way to the desired result, and we need to devise a subplan.

Example 4 A closed box with square base is to hold 12 cubic feet. Let x be the length of one side of the square base. Express the amount of material A used to construct the box as a function of x.

Solution As in the last example, we draw a picture first (see Figure PS.7).

Figure PS.7

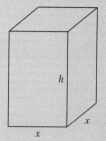

Notice that we have labeled the height of the box as h.

We need to express the amount of material A as a function of x. This means that when we are finished, we want the following:

$A = $ "x stuff"

The preceding may not be very mathematical, but it helps us to remember our goal.

We need to think about what "amount of material" translates into geometrically. It is the surface area of the box. If we think of taking the box apart, we know it consists of a top, a bottom, and four sides. Because the base or bottom is square, so is the top. The four sides are rectangles. We can then conclude the following:

surface area = area of top + area of bottom + area of 4 sides

We can think of this statement as our plan (instead of using steps). Think of it as an outline.

As tools, we need the formulas for the area of a square and the area of a rectangle. Actually, we can just use the formula for the area of a rectangle because a square is just a special kind of rectangle.

area of rectangle = (length)(width)

Using the diagram as a guide and our "outline" from above, we have

$$\begin{array}{cccc} \text{amount of material} & \text{top} & \text{bottom} & \text{4 sides} \\ \text{surface area } (A) = & (x)(x) + & (x)(x) + & 4(x)(h) \end{array}$$

$$A = x^2 + x^2 + 4xh$$

$$A = 2x^2 + 4xh$$

We now have A as a function of x **and** h. To eliminate h, we need a subplan. Notice that we have not used the information that the boxes should hold 12 cubic feet. Looking at the units "cubic feet" gives us the hint that this is the volume. The tool we need this time is the formula for the volume of a box.

volume of a box = (length)(width)(height)

Using this formula with x as the length, x also as the width (square base), and h as the height gives us:

$$
\begin{array}{cccc}
\text{volume} & \text{length} & \text{width} & \text{height} \\
12 \;\; = & x \;\; \cdot & x \;\; \cdot & h \\
12 \;\; = & x^2 h & &
\end{array}
$$

Now we have another equation with x and h. If we put the two equations together, though, we should be able to eliminate h. Remember that our earlier equation represented the amount of material, so it is our main equation or function. The $12 = x^2 h$ is our secondary equation. Because h is the variable we want to eliminate, we need to solve our secondary equation for h. After dividing both sides by x^2, this gives:

$$
h = \frac{12}{x^2}
$$

We can now get our desired result by merely substituting this expression for h into our main function:

$$
A = 2x^2 + 4x\left(\frac{12}{x^2}\right)
$$

Simplifying gives

$$
A = 2x^2 + \frac{48}{x}
$$

Here are some questions we might want to think about: What if the box is open? How would the function change? What if the base is not square? What if the base is twice as long as it is wide? How would the function change? ▪

PROBLEM-SOLVING EXERCISES

Use the problem-solving process to find the result for each exercise.

1. A circle of radius 7 cm is inscribed in a square. Find the area of the region between the circle and the square (outside the circle and inside the square).

2. A square is inscribed in a circle of radius 3 in. Find the area of the region between the square and the circle.

3. A square is inscribed in a circle of radius r. Express the area of the square as a function of r.

4. Find the equation of the line that is parallel to the x axis and that passes through the point $(-2, 6)$.

5. Find the distance from the point $(4, -3)$ to the line $y = 5$.

6. Find the distance from the point $(1, 3)$ to the line $y = 2x + 5$.

7. A window is in the shape of a square surmounted by a semicircle of radius r, as pictured. Express the total area of the window as a function of r.

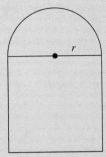

8. Find the equation of the circle that passes through the three points $(2, 5)$, $(-1, 0)$, and $(-3, 4)$.

9. An open box with a square base is to hold 16 cubic feet. Let x be the length of one side of the square base. Express the amount of material A used to construct the box as a function of x.

10. A closed box is to hold 12 cubic feet. The box is twice as long as it is wide. Let w represent the width of the box. Express the amount of material A used to construct the box as a function of w.

11. A company is designing new boxes for one of its products. The boxes have a square base and are closed (have a top). The boxes are to hold 24 cubic inches. The cost of the material for the top and the bottom is 4¢ per square inch and the material for the sides is 3¢ per square inch. Let x represent one side of the base. Express the total cost of a box as a function of x.

12. A travel agency is running a special on cruises to the Bahamas. The price per person is $425; however, the minimum number of people that must sign up to get this special rate is 30. Each time an additional person signs up (after the initial 30), the price of the cruise goes down $10 for each person. The maximum number of people the agency can accommodate is 50. Let x be the number of additional people above 30 that sign up for the cruise. Express the total revenue for the travel agency as a function of x.

13. Using the information from exercise 12 and the fact that the cost per person for the travel agency is $100, express the profit for the travel agency as a function of x.

2

FUNCTIONS, LIMITS, AND THE DERIVATIVE

2.1 FUNCTIONS AND THEIR GRAPHS

Calculus begins with the study of **functions**, special relationships between numbers. For example, the area of a square is related to the length of a side, the price of an item is a function of the demand for that item, and profit is a function of sales.

In Section 1.5 of Chapter 1, we discussed the idea of an ordered pair of real numbers and saw how it can be represented graphically on a rectangular coordinate system. In this section, we deal with sets of ordered pairs, starting with the following definition.

Relation

> A **relation** is a set of ordered pairs.

There are basically three ways to represent a relation: a list of the ordered pairs, an equation describing the ordered pairs, and a graph representing the ordered pairs. It is the last two that interest us. If the ordered pairs are represented by (x, y), then the set of all x values for the relation is called the **domain** and the set of all y values is called the **range.**

Example 1

For each of the relations give the domain and range:

a. $\{(1, 2), (3, 5), (-2, 7), (3, 6)\}$

b. $y = x + 2$

c.

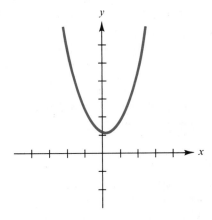

Solution **a.** The domain is $\{-2, 3, 1\}$.
The range is $\{2, 5, 6, 7\}$.

b. We recognize $y = x + 2$ as the equation of a line from Section 1.6. Notice that any real number may be substituted for x, and when 2 is added, we again get a real number. The domain is all real numbers or, symbolically, $x \in R$, and the range is all real numbers or $y \in R$.

c. From the graph, the domain is all real numbers, but the y values are 1 or larger, which means the range is $y \geq 1$. ∎

It is often useful to find the domain and range to help graph a relation. When finding the domain from an equation, we look for *restrictions* on *x*—values of *x* for which *y* is not a real number. Such restrictions, for the work in this book, may be of two types: even roots of negative numbers and zero denominators. Example 2 examines possible restrictions on the domain and the range.

Example 2   Find the domain and range for $y = \sqrt{x + 2}$.

Solution   Because we are taking the square root of a quantity—that is, $x + 2$—it must be nonnegative:

$$x + 2 \geq 0$$
$$x \geq -2$$

The domain is $x \geq -2$.

For the range, we observe that y is always the principal square root of a nonnegative number; thus, the range is $y \geq 0$. The graph of $y = \sqrt{x + 2}$ is shown in Figure 2.1.

Figure 2.1

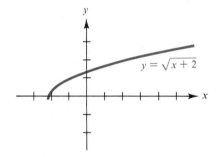

A particular type of relation, called a **function,** is an extremely important concept throughout this text; we present its definition as follows:

Function   A *function* is a relation in which no two ordered pairs have the same first component, *x*, and different second component, *y*. More generally, we say that a function is a *correspondence* between two sets such that each element of the first set corresponds to exactly one element of the second set.

Example 3   Determine whether the following relation is a function:

$$\{(1, 2), (2, 5), (3, 7), (4, 5), (3, 8)\}$$

Solution   It is *not* a function, because the ordered pairs (3, 7) and (3, 8) have the same first component and different second components. ■

Example 4   Determine whether the relation represented by $y = 2x - 1$ is a function.

Solution   The question we need to ask is, when we choose a number for x, will we get more than one y value corresponding to that particular x value? Because the answer is no, we conclude that it is a function.   ∎

In Example 4, the function $y = 2x - 1$ is written so that y is expressed *explicitly* in terms of x, which means our equation is solved for y. If our function is written this way, we can use a new notation reserved exclusively for functions. We will call our function the function f and replace y with the symbol $f(x)$, which is read "f of x" (**not** "f times x"). This is referred to as **functional notation.** For the function above, we have, symbolically,

$$f(x) = 2x - 1$$

This notation can be used to evaluate the function for a particular x value. For example, $f(3)$ represents the number that should be paired with an x value of 3. To compute this number, we merely substitute $x = 3$ into the right side, giving

$$f(3) = 2(3) - 1 = 5$$

So $f(3) = 5$. Graphically, this would correspond to the ordered pair $(3, 5)$. (Remember that $f(x)$ is just a notational replacement for y.)

Example 5   For $f(x) = x^2 + 3x$, compute each of the following:

$$f(-2), \quad f(0), \quad f(3), \quad f(h), \quad f(-x), \quad f(x + h), \quad \frac{f(x + h) - f(x)}{h}$$

Solution

$$f(-2) = (-2)^2 + 3(-2) = 4 - 6 = -2$$
$$f(0) = 0^2 + 3(0) = 0 + 0 = 0$$
$$f(3) = 3^2 + 3(3) = 9 + 9 = 18$$
$$f(h) = h^2 + 3h$$
$$f(-x) = (-x)^2 + 3(-x) = x^2 - 3x$$
$$f(x + h) = (x + h)^2 + 3(x + h) = x^2 + 2hx + h^2 + 3x + 3h$$

$$\frac{f(x + h) - f(x)}{h} = \frac{x^2 + 2hx + h^2 + 3x + 3h - (x^2 + 3x)}{h}$$
$$= \frac{2hx + h^2 + 3h}{h}$$
$$= 2x + h + 3$$ ∎

Although curve sketching is the major focus of a later chapter, in the next example we graph functions by constructing tables of values and finding intercepts (the points on the graph that intersect the axes).

Example 6   Make a table of values and sketch each of the following:
   a. $f(x) = x^2$ **b.** $f(x) = x^2 + 1$ **c.** $f(x) = x^2 - 4$

Solution   **a.**

x	$y = f(x)$
-3	9
-2	4
-1	1
0	0
1	1
2	4
3	9

See Figure 2.2.

Figure 2.2

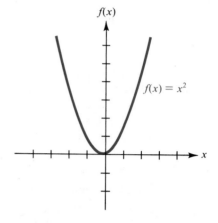

$f(x) = x^2$

b.

x	$f(x)$
-3	10
-2	5
-1	2
0	1
1	2
2	5
3	10

See Figure 2.3.

Figure 2.3

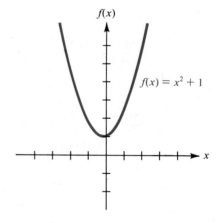

$f(x) = x^2 + 1$

c.

x	$f(x)$
-3	5
-2	0
-1	-3
0	-4
1	-3
2	0
3	5

See Figure 2.4.

Figure 2.4

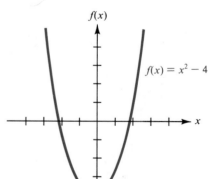

$f(x) = x^2 - 4$

Notice that some key points for this graph are the x and y intercepts. Recall that to calculate the y intercept, set $x = 0$ and solve for y.

$$y = 0^2 - 4 = -4$$

For the x intercept, set $y = 0$.

$$0 = x^2 - 4$$
$$0 = (x + 2)(x - 2)$$
$$x = \pm 2$$

■

All of the graphs in Example 6 have the same shape but are placed differently relative to the axes. The curve $f(x) = x^2$ or $y = x^2$ is called a **parabola.** The other two, $f(x) = x^2 + 1$ and $f(x) = x^2 - 4$, are also parabolas because they have the same shape. The graph of $f(x) = x^2$ is *translated* or moved up 1 unit (+1) to produce the graph of $f(x) = x^2 + 1$ and translated down 4 units (−4) to produce $f(x) = x^2 - 4$.

Before leaving this example, another observation is in order. Notice that each parabola in Example 6 is **symmetric** with respect to the y axis. Three kinds of symmetry can be very useful in sketching graphs.

Symmetry tests

1. **Symmetry with respect to the y axis.**

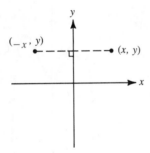

To test it: Substitute $-x$ for x. If the equation remains unchanged, then the graph is symmetric with respect to the y axis.

2. **Symmetry with respect to the x axis.**

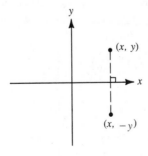

To test it: Substitute $-y$ for y. If the equation remains unchanged, then the graph is symmetric with respect to the x axis.

(continued)

(*continued*)

3. Symmetry with respect to the origin.

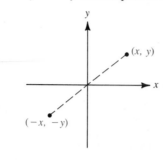

To test it: Substitute $-x$ for x and $-y$ for y. If the equation remains unchanged, then the graph is symmetric with respect to the origin.

(*Note:* Any relation whose graph is symmetric with respect to the x axis is not a function.)

If the graph of a function is symmetric with respect to the y axis, the function is called an **even function** (see Figure 2.5).

Figure 2.5

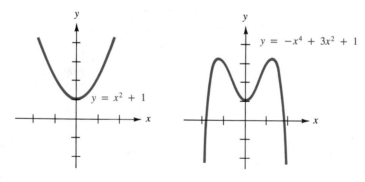

If the graph of a function is symmetric with respect to the origin, the function is called an **odd function** (see Figure 2.6).

Figure 2.6

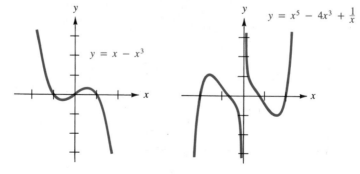

Example 7   Check for symmetry and sketch the graph of $y = 4 - x^2$.

Solution   Substituting $-x$ for x, we have $y = 4 - (-x)^2$ or $y = 4 - x^2$, which is the same as the original equation, so the graph is symmetric with respect to the y axis. Next, we have a table of values.

x	y
0	4
1	3
2	0
3	-5

Plotting these points gives Figure 2.7, and using the symmetry, we have Figure 2.8.

Figure 2.7   *Figure 2.8*

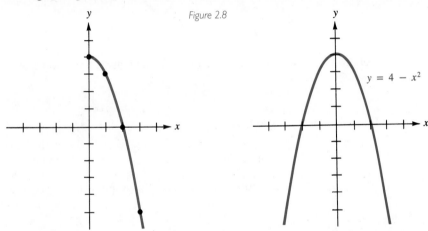

$y = 4 - x^2$

Notice that we need to plot fewer points than we did for the parabolas in Example 6 because we used the symmetry. ■

Example 8   Make a table of values and sketch the graph of $f(x) = \sqrt{x + 1}$.

Solution   The domain for this function is $x \geq -1$, so we start our table of values with $x = -1$.

x	y
-1	0
0	1
1	$\sqrt{2} \approx 1.4$
2	$\sqrt{3} \approx 1.7$
3	2
4	$\sqrt{5} \approx 2.3$
5	$\sqrt{6} \approx 2.4$

Figure 2.9

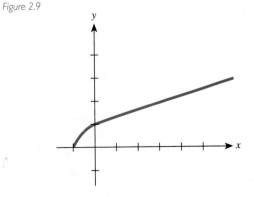

See Figure 2.9.

■

Example 9   Using the graph in Figure 2.10, answer the following questions:

    1. For what values of x is $f(x) < 0$?
    2. For what values of x is $f(x) > 0$?

Figure 2.10

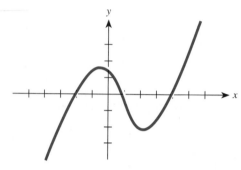

Solution   This can best be done by asking and answering a series of questions.

Question	Answer
What is $f(x)$?	y
What does "< 0" mean?	Negative
Where is y negative?	Below the x axis
What quadrants?	III and IV

Now, color-code the sections of the graph in quadrants III and IV (see Figure 2.11).

Figure 2.11

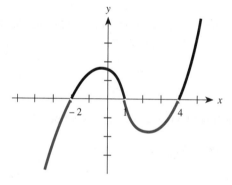

What values of x correspond to these portions of the graph?

    $x < -2$ and $1 < x < 4$ ∎

Example 10   A company has determined that the cost of producing x units of its product can be modeled by the function $C(x) = 800 + x + 0.002x^2$, where C is measured in dollars. Find the cost of producing 500 units.

Solution   The cost of producing 500 units is

$$C(500) = 800 + 500 + 0.002(500)^2$$
$$= \$6300$$ ∎

2.1 EXERCISES

SET A

In exercises 1 through 10, determine if the given relation is a function. (See Examples 3 and 4.)

1. $\{(1, 2), (-1, 3), (3, 4), (5, 4), (10, 5)\}$

2. $\{(-2, -2), (-1, -1), (0, 0), (1, 1), (2, 2)\}$

3. $\{(-1, 3), (0, 3), (1, 3), (5, 3), (7, 3), (10, 3)\}$

4. $\{(1, 6), (1, -3), (1, 4), (1, 5)\}$

5. $y = 1 - 5x$

6. $y = -3x$

7. $y = x^2$

8. $y = x^2 + 3$

9. $y = \dfrac{1}{x + 3}$

10. $y^2 = x$

In exercises 11 through 18, give the domain and range. (See Examples 1 and 2.)

11. $\{(-3, 2), (-2, 1), (-1, 3)\}$

12. $\{(0, 1), (2, 3), (4, 3)\}$

13. $y = 1 - x$

14. $f(x) = 2x + 3$

15. $f(x) = \sqrt{x}$

16. $y = x^2$

17.

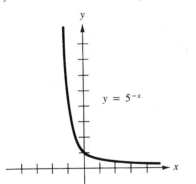

$y = 5^{-x}$

18.

$y = 2^{x^2}$

In exercises 19 through 24, compute the functional values as described for each function. (See Example 5.)

19. $f(x) = 5$
  a. $f(2)$
  b. $f(-3)$

20. $g(x) = -4$
  a. $g(0)$
  b. $g(a)$

21. $f(x) = 7 - 4x$
  a. $f(-4)$
  b. $f(1)$
  c. $f(0)$
  d. $f(h)$

22. $f(x) = 5x + 1$
  a. $f(-1)$
  b. $f(0)$
  c. $f(3)$
  d. $f(h)$

23. $g(x) = 3x^2 + x - 4$
  a. $g(-1)$
  b. $g(0)$
  c. $g(-x)$
  d. $g(x + h)$

24. $f(x) = 2x - 2x^2$
  a. $f(-2)$
  b. $f(2)$
  c. $f(-x)$
  d. $f(x + h)$

For each of the functions in exercises 25 through 30, compute the following: (a) $f(x + h)$; (b) $f(x + h) - f(x)$;
(c) $\dfrac{f(x + h) - f(x)}{h}$. (See Example 5.)

25. $f(x) = 4x + 6$

26. $f(x) = 3 - 7x$

27. $f(x) = x^2 + x$

28. $f(x) = 2x^2 + x$

29. $f(x) = \dfrac{1}{x}$

30. $f(x) = \dfrac{1}{x + 2}$

In exercises 31 through 46, make a table of values and sketch the graph. (See Examples 6, 7, and 8.)

31. $y = 4$

32. $f(x) = -2$

33. $f(x) = 2x$

34. $f(x) = -3x$

35. $y = 5 - 3x$

36. $y = 2x + 5$

37. $y = -x^2$

38. $f(x) = 9 - x^2$

39. $f(x) = (x + 3)^2$

40. $y = (x - 2)^2$

41. $y = \sqrt{x}$

42. $y = -\sqrt{x}$

43. $y = \dfrac{-1}{x}$

44. $f(x) = \dfrac{1}{x}$

45. $y = x^3$

46. $y = x^3 + 1$

In exercises 47 through 49, refer to Example 9.

47. Explain why, in Example 9, $x = -2$, $x = 1$, and $x = 4$ are not included as part of the result.

48. Use the following graph. For what values of x is $f(x) < 0$?

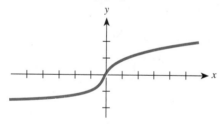

49. Use the following graph. For what values of x is $f(x) \geq 0$?

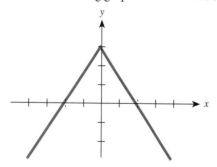

SET B

*A function is called a **one-to-one function** if, for every y value, there corresponds exactly one x value.*
In exercises 50 through 53, determine whether or not the given function is one-to-one.

50. $f(x) = 2x - 1$

51. $y = x^2$

52. $y = |x - 1|$

53. $g(x) = \sqrt{x}$

In exercises 54 through 57, a cost function C(x) is given. Determine the cost of producing the given number of units x. (See Example 10.)

54. $C(x) = 900 + x^2, x = 250$

55. $C(x) = 800 + 0.1x^2, x = 100$

56. $C(x) = 500 + 3x + x^2, x = 50$

57. $C(x) = 1000 + 0.1x^3, x = 200$

*In business, a **revenue function** gives the total income from producing and selling x units of a product. In exercises 58 through 61, a revenue function R(x) is given. Determine the revenue received from selling the given number of units x.*

58. $R(x) = 30x^2 - x^3, x = 20$

59. $R(x) = 70x - 0.5x^2, x = 100$

60. $R(x) = 50x - 0.1x^2, x = 50$

61. $R(x) = 3000x - 0.2x^3, x = 100$

BUSINESS
COST FUNCTION

62. If the cost C, measured in thousands of dollars, for producing x hundred units of a product is given by $C(x) = x^2 - 4x + 7$, then $C(2) = 3$ means that the cost of producing 200 units is $3000. Find $C(5)$ and interpret the result.

REVENUE

63. A small company that produces sports watches has determined that its monthly revenue can be modeled by the function

$$R(x) = -x^2 + 400x - 37,000, \qquad x \geq 0$$

Sketch a graph of this function and, from the graph, approximate the number of watches that must be produced and sold to maximize the monthly revenue. (*Hint:* The result corresponds to the highest point on the graph.)

BIOLOGY
BOTANY

64. An experiment involving a new variety of heat-tolerant ground cover was conducted that measured the total "spread" in square centimeters per week at a temperature of 95°F. The data for the first 6 weeks are given in the following table:

Week	1	2	3	4	5	6
Spread (sq cm)	6	12	20	28	34	40

Plot the points on a rectangular coordinate system, letting w represent the week number on the horizontal axis and $s(w)$ the spread on the vertical axis. Using the table and the sketch, decide which of the following functions would give the "best fit" for these data.

$$s(w) = 7w - 1 \qquad \text{or} \qquad s(w) = w^2 + 5$$

Explain your choice.

WRITING AND CRITICAL THINKING

Write a definition for function as you understand it. In addition, explain what is meant by the domain and range of a function.

USING TECHNOLOGY IN CALCULUS

It is important to understand the relationship between the graphs of $y = f(x)$ and $y = f(x) + k$. When a constant, k, is added to a function, the y value is increased k units and the graph is shifted up (if k is positive) k units. Compare the *DERIVE*® graphs of $y = x^2 + 3x$ and $y = x^2 + 3x + 4$.

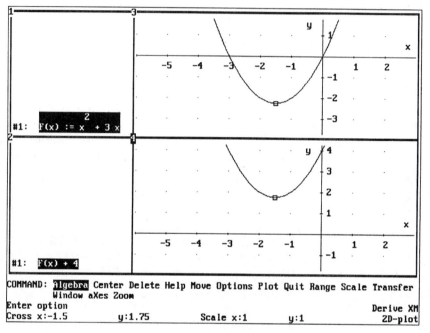

1. Use a graphics calculator or a computer graphing package to graph $y = x^2 - x^3$ and $y = x^2 - x^3 + 3$ on the same axes.
2. Use a graphics calculator or a computer graphing package to graph $y = \frac{1}{x}$ and $y = \frac{1}{x} - 3$ on the same axes. In general, describe the graph of $y = f(x) + k$ (in terms of the graph of $y = f(x)$) when k is negative.

It is also important to understand the relationship between the graphs of $y = f(x)$ and $y = f(x - c)$. The effect of the constant, c, is to shift the graph horizontally. Compare the *DERIVE*® graphs of $y = f(x) = x^2 + 3x$, $y = f(x - 2)$, and $y = f(x + 1)$.

(continued)

(*continued*)

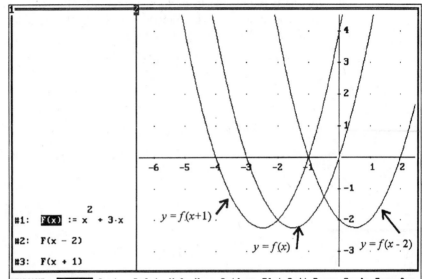

```
#1:  F(x) := x² + 3·x
#2:  F(x - 2)
#3:  F(x + 1)
```

$y = f(x+1)$

$y = f(x)$

$y = f(x - 2)$

```
COMMAND: Algebra Center Delete Help Move Options Plot Quit Range Scale Transfer
         Window aXes Zoom
Enter option                                              Derive XM
Cross x:0           y:0           Scale x:1      y:1       2D-plot
```

3. Use a graphics calculator or a computer graphing package to graph $y = f(x)$, where $f(x) = x^2 - x^3$ and $y = f(x - 2)$ on the same axes.

4. Use a graphics calculator or a computer graphing package to graph $y = \frac{1}{x}$ and $y = \frac{1}{x+1}$ on the same axes. In general, describe the graph of $y = f(x - c)$ (in terms of the graph of $y = f(x)$) if c is positive; if c is negative.

5. Following is a graphics calculator depiction of a cubic function, $y = f(x)$. Find the values of x for which $f(x) \geq 0$.

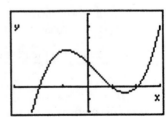

```
WINDOW FORMAT
Xmin=-3
Xmax=3
Xscl=1
Ymin=-2
Ymax=6
Yscl=1
```

2.2   MORE ON FUNCTIONS

Just as it is possible to perform operations with real numbers, so we can perform operations involving functions. Consider the functions $f(x) = 2x^2 + 4$ and $g(x) = 3x - 2$. We can add these functions.

$$f(x) + g(x) = (2x^2 + 4) + (3x - 2) = 2x^2 + 3x + 2$$

We can subtract them.

$$f(x) - g(x) = (2x^2 + 4) - (3x - 2) = 2x^2 - 3x + 6$$

It is also possible to multiply and divide them.

$$f(x) \cdot g(x) = (2x^2 + 4)(3x - 2) = 6x^3 - 4x^2 + 12x - 8$$
$$\frac{f(x)}{g(x)} = \frac{2x^2 + 4}{3x - 2}, \qquad x \neq \frac{2}{3}$$

Example 1   For $f(x) = 1 - x^2$ and $g(x) = 4 - x$, find:

$$f(x) + g(x), \quad f(x) - g(x), \quad f(x) \cdot g(x), \quad \frac{f(x)}{g(x)}$$

Solution $f(x) + g(x) = 1 - x^2 + 4 - x = 5 - x - x^2$
$f(x) - g(x) = (1 - x^2) - (4 - x) = -x^2 + x - 3$
$f(x) \cdot g(x) = (1 - x^2)(4 - x) = 4 - x - 4x^2 + x^3$
$\dfrac{f(x)}{g(x)} = \dfrac{1 - x^2}{4 - x}, \qquad x \neq 4$ ∎

In business, a profit function can be obtained by subtracting the cost function from the revenue function.

Profit = Revenue − Cost
$P(x) = \quad R(x) \quad - C(x)$

We use this idea in the next example.

Example 2   A small company that manufactures clipboards has determined its cost and revenue functions to be as follows:

$$C(x) = 800 + 5x \qquad \text{and} \qquad R(x) = 20x - 0.05x^2$$

where x represents the number of clipboards produced and sold, and C and R are measured in dollars.
a.   Find the profit function, $P(x)$.
b.   Evaluate $P(100)$ and explain what it represents.

Solution **a.** $P(x) =$ $R(x)$ $-$ $C(x)$
$= 20x - 0.05x^2 - (800 + 5x)$
$P(x) = -0.05x^2 + 15x - 800$

b. $P(100) = -0.05(100)^2 + 15(100) - 800 = 200$

The profit from producing and selling 100 clipboards is $200. ∎

Another method for combining functions is called **composition of functions.** For the functions f and g, there are two ways to symbolically represent the composition:

$(f \circ g)(x)$ read "f composition g of x"
or $f[g(x)]$ read "f of g of x"

We will use the latter notation. Suppose that $f(x) = 2x + 3$ and $g(x) = x^2$. To compute $f[g(2)]$ we work in two stages. First, calculate $g(2)$ by substituting 2 for x into $g(x)$.

$g(2) = 2^2 = 4$

Now take that result, 4, and substitute the 4 for x in $f(x)$.

$f(4) = 2(4) + 3 = 11$

Therefore, $f[g(2)] = 11$.

Example 3 For $f(x) = 1 - 2x^2$ and $g(x) = x + 4$, compute each of the following:
a. $f[g(0)]$ **b.** $g[f(1)]$

Solution **a.** $g(0) = 0 + 4 = 4$ **b.** $f(1) = 1 - 2(1)^2 = -1$
So $f[g(0)] = f(4)$ $g[f(1)] = g(-1)$
$= 1 - 2(4)^2$ $= -1 + 4$
$= 1 - 32$ $= 3$
$= -31$ ∎

In Example 3 the composition involved a particular value of x in each case. It is also important to be able to form a composition for any value of x. Example 4 illustrates how this is possible.

Example 4 For $f(x) = x^3$ and $g(x) = 2x - 1$, find $f[g(x)]$ and $g[f(x)]$.

Solution To compute $f[g(x)]$, think of putting $g(x)$ "into" the function f in place of x.

$f[\quad] = [\quad]^3$
   ↑   ↑
$g(x)$ $g(x)$
$f[g(x)] = (2x - 1)^3$

We put $f(x)$ in for x in g to get

$$g[\ \ ] = 2[\ \ ] - 1$$

$$\uparrow \qquad \uparrow$$

$$f(x) \qquad f(x)$$

$$g[\,f(x)] = 2(x^3) - 1$$

$$g[\,f(x)] = 2x^3 - 1 \qquad \blacksquare$$

There are certain pairs of functions f and g such that $f[g(x)] = x$ and $g[\,f(x)] = x$. This means that if you substitute an x value into one function and then substitute that result into the other function, you get your original x value back again. This property is true for functions called **inverse functions.** What one function "does" to a number the other function "undoes."

Let's look at the function $f(x) = x + 2$ and $g(x) = x - 2$. Notice that if we substitute a number into $f(x)$, then 2 will be added to it. If we take that result and substitute it into $g(x)$, then 2 will be subtracted and we will be back to our original number.

Example 5   Verify that $f(x) = x + 2$ and $g(x) = x - 2$ are inverse functions by using the properties $f[g(x)] = x$ and $g[\,f(x)] = x$.

Solution $f[\ \ ] = [\ \ ] + 2$ and $g[\ \ ] = [\ \ ] - 2$

$\quad\uparrow\qquad\uparrow \qquad\qquad\qquad \uparrow\qquad\uparrow$

$\quad g(x)\quad g(x) \qquad\qquad\qquad f(x)\quad f(x)$

$f[g(x)] = (x - 2) + 2 \qquad\qquad g[\,f(x)] = (x + 2) - 2$

$f[g(x)] = x \qquad\qquad\qquad\qquad g[\,f(x)] = x \qquad \blacksquare$

Notationally, the inverse of a function f is sometimes represented by f^{-1}, read "f inverse" (the -1 is *not* an exponent). In the previous example, we say that $f(x) = x + 2$ and $f^{-1}(x) = x - 2$ are inverses of each other.

Example 6   Verify that $f(x) = 2x - 3$ and $g(x) = \dfrac{x + 3}{2}$ are inverse functions.

Solution $f[g(x)] = 2\left(\dfrac{x + 3}{2}\right) - 3 \qquad g[\,f(x)] = \dfrac{(2x - 3) + 3}{2}$

$\qquad\qquad\quad = (x + 3) - 3 \qquad\qquad\qquad\quad = \dfrac{2x}{2}$

$\qquad\qquad\quad = x \qquad\qquad\qquad\qquad\qquad = x \qquad \blacksquare$

Before leaving the topic of inverse functions, we would like to see how inverse functions are related graphically. We will graph the inverse functions from Example 6 on the same set of axes (see Figure 2.12).

Figure 2.12

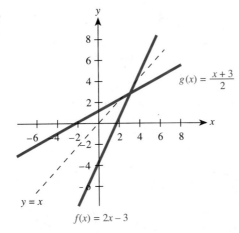

$y = x$

$f(x) = 2x - 3$

Notice that the graphs are symmetric with respect to the line $y = x$ because the x and y coordinates for each point have been interchanged. This will always be true of inverse functions. In fact, if you construct a table of values for one function, a table for the other function can easily be constructed by merely interchanging all the x and y coordinates.

We now consider some functions that require special attention. Example 7 involves a function that is defined in "pieces." We call such functions **piecewise functions.**

Example 7   Sketch the graph of the function

$$f(x) = \begin{cases} x^3, & x \le 1 \\ 2 - x, & x > 1 \end{cases}$$

Solution   Make two tables of values.

x	$f(x) = x^3$	x	$f(x) = 2 - x$	
-2	-8	1	1	← Not included for this piece
-1	-1	2	0	but shows starting point
0	0	3	-1	
1	1	4	-2	

Now graph both on the same set of axes (see Figure 2.13).

Figure 2.13

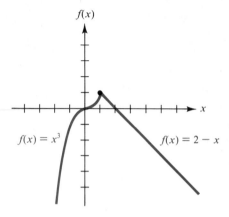

$f(x) = x^3$ $f(x) = 2 - x$

Example 8   Sketch the graph of the function

$$f(x) = \begin{cases} x^2 + 1, & x \le 0 \\ x, & x > 0 \end{cases}$$

Solution   Refer to Figure 2.3 for a sketch of $f(x) = x^2 + 1$. We need only the "left half," since $x \le 0$. Also, the graph of $f(x) = x$ is a line through the origin at a 45° angle. Notice the use of the open circle to indicate that $x = 0$ was not included for this portion of the graph (see Figure 2.14).

Figure 2.14

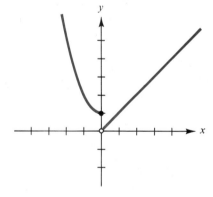

We now define and graph two functions that deserve special attention—the **absolute value function** and the **greatest integer function.**

Example 9   Sketch the graph of the absolute value function $f(x) = |x|$.

Solution   Using the definition of absolute value from Chapter 1, we can write $f(x) = |x|$ as a piecewise function. The graph is shown in Figure 2.15.

$$f(x) = |x| = \begin{cases} x, & x \ge 0 \\ -x, & x < 0 \end{cases}$$

Figure 2.15

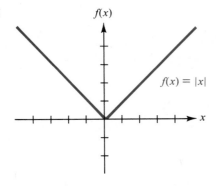

$f(x)$

$f(x) = |x|$

Example 10   Sketch the graph of the greatest integer function.

$$y = [\![x]\!]$$

Solution   By $y = [\![x]\!]$, we mean y is always an integer and that integer is the *greatest* integer less than or equal to x. Some examples of x and y values are

$[\![3.2]\!] = 3$ $[\![4.7]\!] = 4$ $[\![2]\!] = 2$

$[\![-1.2]\!] = -2$ $[\![-2]\!] = -2$ $[\![-2.7]\!] = -3$

The table of values representing the pairs is given here, and the graph of the greatest integer function is shown in Figure 2.16.

x	y
3.2	3
4.7	4
2	2
−1.2	−2
−2	−2
−2.7	−3

Figure 2.16

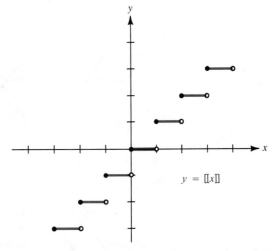

$y = [\![x]\!]$

Notice from the graph that the function forms "steps" and is sometimes referred to as the **step function.**   ∎

Example 11   The graph of a function $y = f(x)$ is given in Figure 2.17. Draw the graphs for $y = -f(x)$ and $y = |f(x)|$. Explain the visual effect in each case.

Figure 2.17

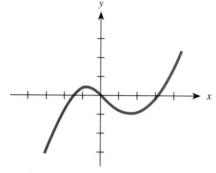

Solution   For $y = -f(x)$, the y coordinate for each point has been multiplied by -1. That is, each y coordinate has "changed sign," which causes the graph to appear to have "flip-flopped" across the x axis. Points in quadrants I and II went to quadrants III and IV and vice versa (see Figure 2.18).

Figure 2.18

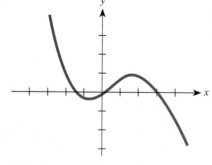

For $y = |f(x)|$, all y coordinates that were negative are now positive. Those points whose y coordinates are negative are below the x axis, and they appear to have "flipped up" across the x axis into quadrants I and II (see Figure 2.19).

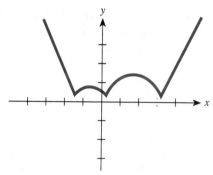

Figure 2.19

We conclude this section with two examples that involve functions in an applied setting. You may want to reread the section on problem solving on pages 66–75.

Example 12

A rectangular plot of land is to be fenced in using 800 feet of fencing. Express the area of the plot as a function of the width only.

Solution

Since the area of a rectangle is length times width, we can express the area of the plot as $A = lw$.

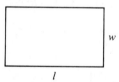

The instructions indicate that we would like the area as a function of the width only. To do this, we can determine the relationship between the length and width by using the fact that there is 800 feet of fencing available or, more directly, that the perimeter is 800 feet. Symbolically this is given by

$$2w + 2l = 800$$

Solving for l, we get

$$2l = 800 - 2w$$
$$l = 400 - w$$

Now, we substitute $400 - w$ for l in the area formula.

$$A = (400 - w)w$$
or $A = 400w - w^2$

Since A is a function of w, we can also write

$$A(w) = 400w - w^2$$

Example 13

The Encore Company has determined that the **total cost** of producing x mattresses per month depends on both its **fixed cost** (for example, rent, electricity, phone) and its **variable cost** (for example, labor, materials, advertising). If the fixed cost per month is $8000 and the variable cost per month is $2x + 0.3x^2$, express the total cost, $C(x)$, in terms of the fixed cost and variable cost. Using this cost function, determine the cost of producing 200 mattresses.

Solution   The total cost is the sum of the fixed cost and the variable cost, so that

$$C(x) = 8000 + 2x + 0.3x^2$$

and the cost of producing 200 mattresses will be

$$C(200) = 8000 + 2(200) + 0.3(200)^2$$
$$= 8000 + 400 + 12{,}000$$
$$= \$20{,}400$$

2.2 EXERCISES

SET A

In exercises 1 through 8, use the functions $f(x) = 2x^2 - x$ and $g(x) = 3x + 4$ to compute each result. (See Examples 1 and 3.)

1. $f(x) + g(x)$

2. $f(x) - g(x)$

3. $f(x) \cdot g(x)$

4. $\dfrac{f(x)}{g(x)}$

5. $\dfrac{f(1)}{g(0)}$

6. $f(1) \cdot g(1)$

7. $f[g(-2)]$

8. $f[g(-1)]$

In exercises 9 through 12, find $f[g(x)]$ and $g[f(x)]$. (See Example 4.)

9. $f(x) = 2 - x, g(x) = 3x$

10. $f(x) = -2x, g(x) = 3x + 1$

11. $f(x) = x^2 + x, g(x) = x + 1$

12. $f(x) = x^2 - x, g(x) = x + 1$

In exercises 13 through 18, verify that the given pairs of functions are inverses. Sketch both on the same set of axes. (See Examples 5 and 6.)

13. $f(x) = 4 - x, g(x) = 4 - x$

14. $f(x) = \dfrac{1}{2x}, g(x) = \dfrac{1}{2x}$

15. $f(x) = \dfrac{x + 1}{5}, g(x) = 5x - 1$

16. $f(x) = \frac{1}{2}x - 2, g(x) = 2x + 4$

17. $f(x) = x^3, g(x) = \sqrt[3]{x}$

18. $f(x) = x^3 - 1, g(x) = \sqrt[3]{x + 1}$

In exercises 19 through 26, sketch the graph of each function. (See Examples 7, 8, 9, and 10.)

19. $y = -[\![x]\!]$

20. $y = -|x|$

21. $f(x) = |x| + 2$

22. $f(x) = [\![x]\!] + 2$

23. $f(x) = \begin{cases} -x, & x \le 1 \\ x, & x > 1 \end{cases}$

24. $y = \begin{cases} -x, & x \le 1 \\ x + 1, & x > 1 \end{cases}$

25. $y = \begin{cases} 1 - x^2, & x < 0 \\ x + 1, & x \ge 0 \end{cases}$

26. $f(x) = \begin{cases} 2, & x < -2 \\ x^2 - 4, & x \ge -2 \end{cases}$

27. Sketch the graph of $y = x^2 - 1$ and then, using Example 11 as a guide, sketch the graph of $y = -(x^2 - 1)$ and $y = |x^2 - 1|$.

28. Sketch the graph of $y = x^3$ and then, using Example 11 as a guide, sketch the graph of $y = -x^3$ and $y = |x^3|$.

29. Explain why the graphs of $y = x^2$ and $y = |x^2|$ look the same.

30. The perimeter of a rectangle is 120 cm. Express the area as a function of the length.

31. The area of a rectangle is 108 cm^2. Express the perimeter as a function of the width.

SET B

In exercises 32 through 36, sketch the graph of each function using the graph of $y = f(x)$ given here.

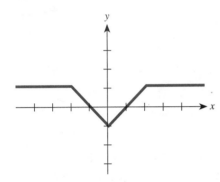

32. $y = f(x) + 1$

33. $y = f(x) - 2$

34. $y = -f(x)$

35. $y = |f(x)|$

36. $y = \begin{cases} f(x), & x \ge 0 \\ -f(x), & x < 0 \end{cases}$

37. Graph all ordered pairs (x, y) such that $x - 1 < y \le x$, where x is a real number and y is an integer. Show that the function is equivalent to $y = [\![x]\!]$.

*In business, the **profit function**, $P(x)$, is defined to be the revenue minus the cost, or $P(x) = R(x) - C(x)$. In exercises 38 through 41, find the profit function for the given revenue and cost functions.*

38. $R(x) = 30x - x^2$ and $C(x) = 50 + 5x$

39. $R(x) = 40x - x^2$ and $C(x) = 100 + 12x$

40. $R(x) = 6x^2 - x^3$ and $C(x) = x^3 + 100$

41. $R(x) = 1.75x^2 - 0.5x^3$ and $C(x) = 100 + 0.25x^2$

BUSINESS
COST FUNCTION

42. The Taylor Button Manufacturing Company has determined that its fixed cost per month is $5000 and the variable cost is given by $0.02x + \sqrt{x}$, where x represents the number of buttons produced per month. Find the total cost

function, $C(x)$, and determine the cost of producing 10,000 buttons. (*Hint:* See Example 13.)

REVENUE

43. The Spike and Pepper Spice Company has determined that the revenue per month, $R(x)$, from selling x pounds of its new seasoning salt is given by the function

$$R(x) = 13x - 0.01x^2$$

where $R(x)$ is measured in dollars. Determine the income (revenue) from selling 500 pounds of its new salt.

MANUFACTURING

44. A box is to be constructed for the shipping department of a manufacturing plant. Because of the size and shape of the item to be shipped, the height of the box is to be three times the length of one side of its square base. Express the volume of the box as a function of the length of the base.

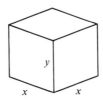

MANUFACTURING

45. The volume of a box with a square base is to be 64 in.^{3}. The box has both a top and a bottom. Express the surface area of the box as a function of the length of the base.

BIOLOGY

46. In a laboratory experiment involving mice, there were 300 mice at the beginning of the experiment. If the number of mice that have died after t days is given by $3t^2$, determine the function, $N(t)$, which gives the number of mice still alive after t days. Determine how many days will pass before all the mice are dead.

BACTERIAL GROWTH

47. If the number of bacteria in a culture is given by

$$B(t) = 500 + 4t^3$$

where t is measured in days, determine the number of bacteria present after three days. How many bacteria were there initially ($t = 0$)? How many days will it take for the number of bacteria to double?

SOCIAL SCIENCE
INCOME LEVEL

48. A survey was conducted to determine the number of families in certain regions of the country whose income is below the poverty level. From the sample, it appears that the function shown here will closely approximate the number of families, $N(x)$, with incomes below the poverty level for a total population of x people.

$$N(x) = 0.01x + \sqrt{0.5x}$$

In a population of 20,000 people, approximately how many families have incomes below the poverty level?

In exercises 49 through 52, match the graph of the function in column I with its inverse's graph in column II. The WINDOW values for each graphic are

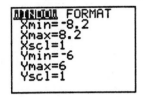

```
WINDOW FORMAT
 Xmin=-8.2
 Xmax=8.2
 Xscl=1
 Ymin=-6
 Ymax=6
 Yscl=1
```

	I		II

49.

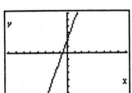

a.

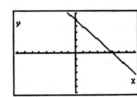

50.

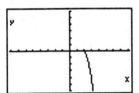

b.

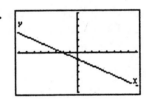

51.

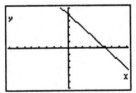

c.

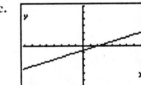

52.

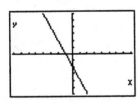

d.

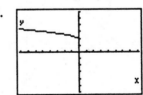

53. Pictured here is a graphics calculator screen for the graph of $y = f(x)$. Name one point on the graph of $y = f^{-1}(x)$.

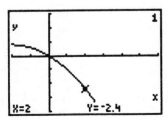

54. Pictured here is a graphics calculator screen for the graph of $y = g(x)$. Name one point on the graph of $y = g^{-1}(x)$.

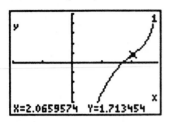

X=2.0659574 Y=1.713454

WRITING AND CRITICAL THINKING

For the piecewise function

$$f(x) = \begin{cases} x + 1, & x \le 0 \\ 1 - x^2, & x > 0 \end{cases}$$

explain the role of the "$x \le 0$" and the "$x > 0$" part of the definition of this function. For instance, if we want to find $f(-2)$, how do we decide which "piece" to use?

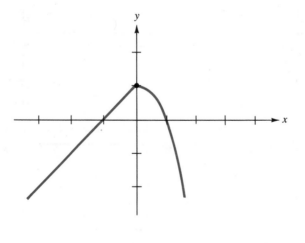

USING TECHNOLOGY IN CALCULUS

1. If you interpret the absolute value function as "never negative," then the graph of $y = |f(x)|$ will never be below the x axis. For example, consider $y = |9 - x^2|$; its graph is as follows:

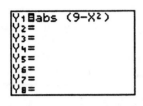 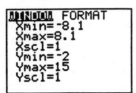 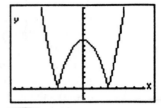

Graph $y = |f(x)|$ if the graph of $y = f(x)$ is as follows:

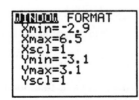 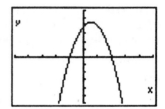

2. *A Technology Tip:* Graphing piecewise functions takes a bit of ingenuity on some graphics calculators. To graph the function

$$f(x) = \begin{cases} x^3, & x \le 1 \\ 2 - x, & x > 1 \end{cases}, \text{ for example, we use the following settings:}$$

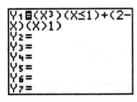 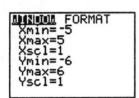 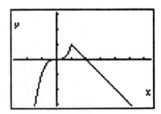

2.3   LIMITS

Let's assume that you're planning a trip and will be traveling 500 miles. Suppose on the first day you travel half the distance, or 250 miles. If you continue each day covering half the remaining distance from your present position to the final destination, how many days will pass before you reach your final destination?

Theoretically, of course, you will never reach your final destination, but you will certainly be very close after, say, two weeks (a little over 160 feet to go). We don't advise this method of travel if you're in a hurry, but this example does serve to introduce an important concept in the study of calculus called the **limit**—in particular, the **limit of a function.**

This concept is best explained using an example. Consider the function $f(x) = x^2 + 1$. We will determine what happens to the functional values (the y's or $f(x)$'s) as x "gets close to" or approaches a value of 2. We are not concerned with the value of the function *at* 2. In the tables in Figure 2.20, we let x get close to 2 (our final destination) by starting at $x = 1$ and moving toward 2 from the left or below (Table A) and also by starting at $x = 3$ and moving toward 2 from the right or above (Table B). We use the "half of the remaining distance" idea in choosing the x values. By using decimals, it is very easy to verify our calculations using a calculator.

Figure 2.20   Table A

x	$f(x) = x^2 + 1$
1.00000000	2.00000000
1.50000000	3.25000000
1.75000000	4.06250000
1.87500000	4.51562500
1.93750000	4.75390625
1.96875000	4.87597656
1.98437500	4.93774414
1.99218750	4.96881104
1.99609375	4.98439026
1.99804688	4.99219131

Table B

x	$f(x) = x^2 + 1$
3.00000000	10.00000000
2.50000000	7.25000000
2.25000000	6.06250000
2.12500000	5.51562500
2.06250000	5.25390625
2.03125000	5.12597656
2.01562500	5.06274414
2.00781250	5.03131104
2.00390625	5.01564026
2.00195313	5.00781631

(*Note*: All values are rounded to eight decimal places.)

Notice that as x gets closer to 2 in each table, the value of $f(x)$ seems to be getting close to 5.

It may be advantageous at this point to introduce some preliminary notation before going to a new example. Symbolically, we can write $x \to 2$ to mean that x gets close to 2 or x *approaches* 2. Our conclusion from the tables symbolically becomes

if $x \to 2$,   then $f(x) \to 5$

Taking this a step further, we say that "the **limit** as x approaches 2 of $f(x)$ is 5." Symbolically, we write

$$\lim_{x \to 2} f(x) = 5 \quad \text{or} \quad \lim_{x \to 2} (x^2 + 1) = 5$$

If we look at a graph of $f(x) = x^2 + 1$, we can see that for $x = 2$, the $f(x)$ or y coordinate is *exactly* 5 (see Figure 2.21). In introducing the idea of a limit, it is important to emphasize that we are not concerned with the value of the function at $x = 2$, but merely the values when x is *close to* 2.

Figure 2.21

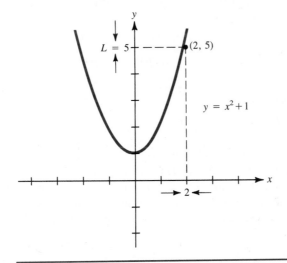

Definition of limit

If $f(x)$ approaches some finite number L as x approaches c (see Figure 2.22), then we say that the **limit** of $f(x)$ as x approaches c is L and symbolically write

$$\lim_{x \to c} f(x) = L$$

Figure 2.22

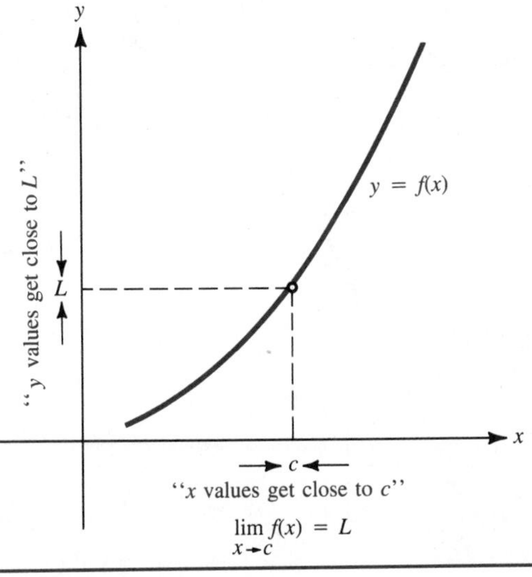

"y values get close to L"

$y = f(x)$

L

"x values get close to c"

c

$$\lim_{x \to c} f(x) = L$$

Example 1   Approximate $\lim_{x \to 3} (x^2 + 1)$ by completing the following table:

x	2	2.5	2.9	2.99	2.999	3.001	3.01	3.1	3.5	4
$f(x)$										

Solution   A calculator gives

| | | | | | close to 3 | | | | | |
					↓	↓				
x	2	2.5	2.9	2.99	2.999	3.001	3.01	3.1	3.5	4
$f(x)$	5	7.25	9.41	9.9401	9.994001	10.006001	10.0601	10.61	13.25	17
					↑	↑				
					close to 10					

Conclusion: $\lim\limits_{x \to 3} (x^2 + 1) = 10$.   ∎

Notice that in Example 1 we could get the result, 10, by substituting 3 for x in the function $f(x) = x^2 + 1$. Many of the limits we need to calculate can be done using direct substitution. Whether or not we can use this method depends on the behavior of the particular function around the number we are approaching. In the next four examples we use direct substitution to calculate each limit.

Example 2   Evaluate $\lim\limits_{x \to 1} (2x + 3)$.

Solution   By substitution we have

$$\lim\limits_{x \to 1} (2x + 3) = 2(1) + 3 = 5$$

See Figure 2.23.

Figure 2.23

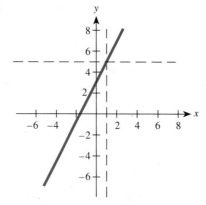

      ∎

Example 3   Evaluate the following limits:

    a. $\lim\limits_{x \to 0} (x^3 + 4x - 7)$ **b.** $\lim\limits_{x \to -2} \dfrac{2x}{x + 3}$

Solution   **a.** $\lim\limits_{x \to 0} (x^3 + 4x - 7) = 0^3 + 4 \cdot 0 - 7 = -7$

b. $\lim\limits_{x \to -2} \dfrac{2x}{x+3} = \dfrac{2(-2)}{-2+3} = -4$

See Figure 2.24.

Figure 2.24

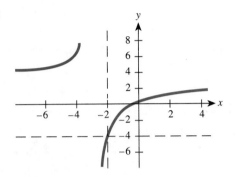

Example 4   Write symbolically and evaluate the limit of $f(x) = x - \dfrac{1}{x}$ as x approaches -4.

Solution $\lim\limits_{x \to -4} \left(x - \dfrac{1}{x} \right) = -4 - \dfrac{1}{-4} = \dfrac{-15}{4}$ ∎

The next example deals with limits that involve radicals. The substitution technique still applies here.

Example 5   Evaluate each of the following:

a. $\lim\limits_{x \to 2} \sqrt{x + 7}$ **b.** $\lim\limits_{x \to 0} \dfrac{x}{\sqrt[3]{x+1}}$ **c.** $\lim\limits_{x \to 9} (5 - \sqrt{x})$

Solution   **a.** $\lim\limits_{x \to 2} \sqrt{x + 7} = \sqrt{2 + 7} = \sqrt{9} = 3$

See Figure 2.25.

Figure 2.25

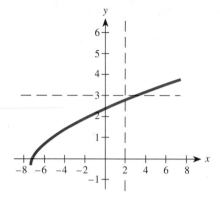

b. $\displaystyle\lim_{x \to 0} \frac{x}{\sqrt[3]{x+1}} = \frac{0}{\sqrt[3]{0+1}} = 0$

c. $\displaystyle\lim_{x \to 9} (5 - \sqrt{x}) = 5 - \sqrt{9} = 2$

See Figure 2.26.

Figure 2.26

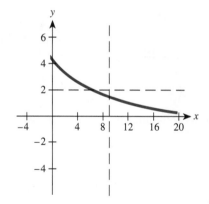

We are now ready to consider some functions for which a limit does not exist. To help us prepare for one of these situations, let's return to the original function of this section, $f(x) = x^2 + 1$. In Figure 2.20, we constructed two tables of values: one for x approaching 2 from the left (Table A) and one for x approaching 2 from the right (Table B). We can discuss these two situations separately by addressing what are called **one-sided limits.**

Left-hand limit

"x approaches c from below" (or from the left)

$$\lim_{x \to c^-}$$

Right-hand limit

"x approaches c from above" (or from the right)

$$\lim_{x \to c^+}$$

This means that Figure 2.20, Table A, represents the left-hand limit or $x \to 2^-$ and Table B represents the right-hand limit or $x \to 2^+$. For the limit of a function to exist as x approaches c, we must have the two one-sided limits equal or, symbolically,

$$\lim_{x \to c} f(x) = L \quad \text{if and only if} \quad \lim_{x \to c^-} f(x) = L \quad \text{and} \quad \lim_{x \to c^+} f(x) = L$$

Example 6   Evaluate the following one-sided limits:

a. $\lim\limits_{x \to 3^-} (4x^2 - 1)$ b. $\lim\limits_{x \to 2^+} \sqrt{x - 2}$

Solution   a.   We can still use the substitution technique that gives

$$\lim_{x \to 3^-} (4x^2 - 1) = 4(3)^2 - 1 = 36 - 1 = 35$$

(Be careful: 3^- does not mean that the 3 is negative; it means that we approach positive 3 from the left.)

b. $\lim\limits_{x \to 2^+} \sqrt{x - 2} = \sqrt{2 - 2} = 0$

For this function, it makes sense to approach 2 only from the positive side because the function is defined only for $x \geq 2$.   ∎

Example 7   Sketch the graph of the greatest integer function, $f(x) = [\![x]\!]$, and evaluate the following limits by examining the graph in Figure 2.27:

a. $\lim\limits_{x \to 1^-} f(x)$ b. $\lim\limits_{x \to 2^+} f(x)$ c. $\lim\limits_{x \to -1/2} f(x)$

Figure 2.27

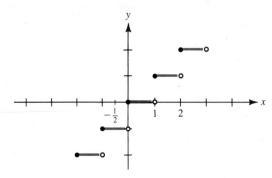

Solution   **a.** $x \to 1^-$ means x approaches 1 from the left.

$$\lim_{x \to 1^-} f(x) = 0$$

b. $x \to 2^+$ means x approaches 2 from the right.

$$\lim_{x \to 2^+} f(x) = 2$$

c. $x \to \dfrac{-1}{2}$ means x approaches $\dfrac{-1}{2}$ (from both sides).

$$\lim_{x \to -1/2} f(x) = -1$$ ■

Example 8   Sketch the function and evaluate the limits:

$$f(x) = \begin{cases} -x^2, & x \le 1 \\ x + 1, & x > 1 \end{cases}$$

a. $\displaystyle\lim_{x \to 0} f(x)$ **b.** $\displaystyle\lim_{x \to 3} f(x)$ **c.** $\displaystyle\lim_{x \to 1^-} f(x)$

d. $\displaystyle\lim_{x \to 1^+} f(x)$ **e.** $\displaystyle\lim_{x \to 1} f(x)$

Solution   A sketch of this piecewise function can be made using the technique described in Section 2.2 (see Figure 2.28).

Figure 2.28

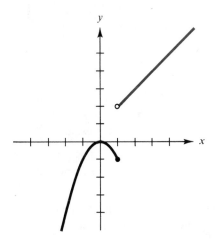

a. Because x is approaching 0, which is less than 1, the top portion of the function gives

$$\lim_{x \to 0} f(x) = \lim_{x \to 0} (-x^2) = -0^2 = 0$$

b. $\lim_{x \to 3} f(x) = \lim_{x \to 3} (x + 1) = 3 + 1 = 4$

We use $x + 1$ because $3 > 1$.

c. $\lim_{x \to 1^-} f(x) = \lim_{x \to 1^-} (-x^2) = -1^2 = -1$

(1^- means "below" or less than 1)

d. $\lim_{x \to 1^+} f(x) = \lim_{x \to 1^+} (x + 1) = 1 + 1 = 2$

(1^+ means "above" or greater than 1)

e. Because parts c and d are not equal, this means that $\lim_{x \to 1} f(x)$ **does not exist.**

From Figure 2.28 you can see that there is a "jump" in the graph at $x = 1$. To the left of $x = 1$, the y values are getting close to -1, and on the right side of $x = 1$, the y values are getting close to 2. ∎

In Example 8, we saw a limit that does not exist because of a "jump" in the graph. A second type of limit that does not exist is illustrated in Example 9, parts a and b.

Example 9   Evaluate the following:

a. $\lim_{x \to 3} \dfrac{5}{x - 3}$ **b.** $\lim_{x \to -2} \dfrac{x - 1}{x^2 + 2x}$ **c.** $\lim_{x \to 1} \dfrac{x - 1}{2x - 3}$

Solution   **a.**   If we substitute 3 for x in $\dfrac{5}{x-3}$, we obtain $\dfrac{5}{3-3} = \dfrac{5}{0}$, which is undefined. We

say that $\lim\limits_{x \to 3} \dfrac{5}{x-3}$ does not exist.*

b.   If we substitute -2 for x in $\dfrac{x-1}{x^2+2x}$, we obtain $\dfrac{-2-1}{(-2)^2+2(-2)} = \dfrac{-3}{0}$, which is

undefined. Thus, $\lim\limits_{x \to -2} \dfrac{x-1}{x^2+2x}$ does not exist.

c. $\lim\limits_{x \to 1} \dfrac{x-1}{2x-3} = \dfrac{1-1}{2(1)-3} = \dfrac{0}{-1} = 0$ Notice that zero is in the numerator this time.   ∎

In Example 10, we start with direct substitution but find we need to perform an algebraic manipulation before substitution will work.

Example 10   Evaluate the following limit:

$$\lim_{x \to 1} \frac{x-1}{x^2+x-2}$$

Solution   Here, substituting 1 for x in $\dfrac{x-1}{x^2+x-2}$ yields

$$\frac{1-1}{1^2+1-2} = \frac{0}{0}$$

which is called an *indeterminate form* (it is not equal to zero and it is not undefined). When this occurs, we need to see if the function can be transformed into an equivalent form. We proceed as follows:

$$\lim_{x \to 1} \frac{x-1}{x^2+x-2} = \lim_{x \to 1} \frac{x-1}{(x-1)(x+2)} \qquad \text{Factor the denominator.}$$

$$= \lim_{x \to 1} \frac{1}{x+2} \qquad \text{Reduce.}$$

$$= \frac{1}{1+2} \qquad \text{Substitute for } x.$$

So $\lim\limits_{x \to 1} \dfrac{x-1}{x^2+x-2} = \dfrac{1}{3}$

We are permitted to reduce the function because we are interested only in the limit as x gets close to 1 and the functions

$$f(x) = \frac{x-1}{x^2+x-2} \qquad \text{and} \qquad f(x) = \frac{1}{x+2}$$

---

*See Section 2.5 for a graphical interpretation.

behave exactly the same except for $x = 1$. Notice that the function

$$f(x) = \frac{x-1}{x^2 + x - 2}$$

is not defined for $x = 1$, but it *need not be for the limit to exist.*

The graph of $y = \dfrac{x-1}{x^2 + x - 2}$ appears in Figure 2.29.

Figure 2.29

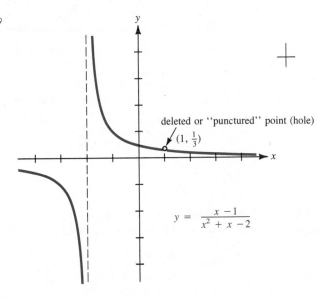

deleted or "punctured" point (hole)

$(1, \frac{1}{3})$

$$y = \frac{x-1}{x^2 + x - 2}$$

Example 11   Evaluate $\displaystyle\lim_{x \to 0} \frac{x^2 - 3x}{x}$.

Solution   Strict substitution yields the indeterminate form $\frac{0}{0}$. We factor the numerator and rewrite an equivalent form for $\dfrac{x^2 - 3x}{x}$ as follows:

$$\lim_{x \to 0} \frac{x^2 - 3x}{x} = \lim_{x \to 0} \frac{x(x-3)}{x}$$
$$= \lim_{x \to 0} (x - 3)$$
$$= 0 - 3$$

So $\displaystyle\lim_{x \to 0} \frac{x^2 - 3x}{x} = -3$

The graph of $y = \dfrac{x^2 - 3x}{x}$ appears in Figure 2.30.

Figure 2.30

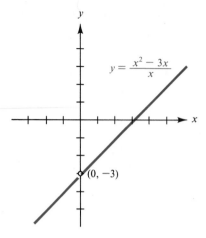

$$y = \frac{x^2 - 3x}{x}$$

$(0, -3)$

Example 12 shows that a different technique of transformation may be needed for the indeterminate form.

Example 12   Evaluate $\lim\limits_{x \to 1} \dfrac{1 - x}{1 - \sqrt{x}}$.

Solution   We obtain the indeterminate form $\frac{0}{0}$ by substituting 1 for x. We can remedy the situation here by rationalizing the denominator (see Figure 2.31).

$$\lim_{x \to 1} \frac{1 - x}{1 - \sqrt{x}} = \lim_{x \to 1} \frac{1 - x}{1 - \sqrt{x}} \cdot \frac{1 + \sqrt{x}}{1 + \sqrt{x}}$$

$$= \lim_{x \to 1} \frac{(1 - x)(1 + \sqrt{x})}{1 - x} \qquad \text{Reduce to lowest terms.}$$

$$= \lim_{x \to 1} (1 + \sqrt{x})$$

$$= 1 + \sqrt{1}$$

$$\lim_{x \to 1} \frac{1 - x}{1 - \sqrt{x}} = 2$$

Figure 2.31

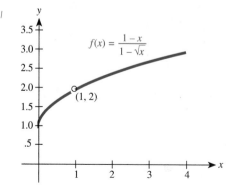

$$f(x) = \frac{1 - x}{1 - \sqrt{x}}$$

$(1, 2)$

We conclude this section with some properties of limits and a summary of some of our results. We have already used these properties in the examples.

Properties of limits

Given that $\lim\limits_{x \to c} f(x) = L$ and $\lim\limits_{x \to c} g(x) = M,$

1. $\lim\limits_{x \to c} K = K$ (K is a constant)

2. $\lim\limits_{x \to c} x^n = c^n$

3. $\lim\limits_{x \to c} [f(x) + g(x)] = \lim\limits_{x \to c} f(x) + \lim\limits_{x \to c} g(x) = L + M$

4. $\lim\limits_{x \to c} [f(x)g(x)] = \left[\lim\limits_{x \to c} f(x) \right]\left[\lim\limits_{x \to c} g(x) \right] = LM$

5. $\lim\limits_{x \to c} \dfrac{f(x)}{g(x)} = \dfrac{\lim\limits_{x \to c} f(x)}{\lim\limits_{x \to c} g(x)} = \dfrac{L}{M},$ provided $M \neq 0$

In evaluating limits, we produced four different results that we summarize in Table 2.1. Assume that L represents some nonzero real number whenever L appears in the denominator.

Table 2.1

Limit evaluation result	Conclusion
L	limit is L
$\dfrac{0}{L}$	limit is 0
$\dfrac{L}{0}$	limit does not exist
$\dfrac{0}{0}$	transform function to produce one of the forms above

2.3 EXERCISES

SET A

In exercises 1 through 4, approximate each limit by completing the table. (See Example 1.) A calculator may be helpful for these exercises.

1. $\lim\limits_{x \to 3} (2x - 6)$

x	2.5	2.9	2.99	2.999	3.001	3.01	3.1	3.5
$f(x)$								

2. $\lim\limits_{x \to 0} \sqrt{x + 4}$

x	-0.5	-0.1	-0.01	-0.001	0.001	0.01	0.1	0.5
$f(x)$								

3. $\lim\limits_{x \to -2} \dfrac{1}{x + 3}$

x	-2.5	-2.1	-2.01	-2.001	-1.999	-1.99	-1.9	-1.5
$f(x)$								

4. $\lim\limits_{x \to -1} (x^2 - 4)$

x	-1.5	-1.1	-1.01	-1.001	-0.999	-0.99	-0.9	-0.5
$f(x)$								

In exercises 5 through 22, evaluate the limits that exist. If the limit does not exist, state that it does not exist. (See Examples 2–5 and 9.)

5. $\lim\limits_{x \to 3} 5$

6. $\lim\limits_{x \to 2} (-4)$

7. $\lim\limits_{x \to -1} (2x + 1)$

8. $\lim\limits_{x \to -3} (1 - 2x)$

9. $\lim\limits_{x \to 0} (x^2 + 2x)$

10. $\lim\limits_{x \to 5} (x^2 - x)$

11. $\lim\limits_{x \to 2} \dfrac{x + 1}{x - 4}$

12. $\lim\limits_{x \to -4} \dfrac{-x}{x - 4}$

13. $\lim\limits_{x \to -2} \dfrac{5x}{x + 2}$

14. $\lim\limits_{x \to 3} \dfrac{x + 1}{x^2 - 3x}$

15. $\lim\limits_{x \to 7} \sqrt{2x - 5}$

16. $\lim\limits_{x \to -1} \sqrt[3]{7 - x}$

17. $\lim\limits_{x \to 9} \dfrac{1}{3 - \sqrt{x}}$

18. $\lim\limits_{x \to 4} \dfrac{2}{\sqrt{x} - 2}$

19. $\lim\limits_{x \to -4} \left(x^2 + \dfrac{1}{x} \right)$

20. $\lim\limits_{x \to 0} \left(3 - \dfrac{1}{x - 1} \right)$

21. $\lim\limits_{x \to 1/2} \dfrac{x - 1}{2x + 3}$

22. $\lim\limits_{x \to -3/4} \dfrac{x - 1}{4x + 3}$

In exercises 23 through 30, evaluate the one-sided limits. (See Example 6.)

23. $\lim\limits_{x \to 1^+} (x + 3)$

24. $\lim\limits_{x \to 1^-} (x + 3)$

25. $\lim\limits_{x \to 2^-} (3x^2 + x - 1)$

26. $\lim\limits_{x \to 2^+} (2x - 5x^2)$

27. $\lim\limits_{x \to -4^+} \dfrac{x + 4}{x + 1}$

28. $\lim\limits_{x \to -4^-} \dfrac{x + 4}{x + 1}$

29. $\lim\limits_{x \to 0^+} (5 - \sqrt{x})$

30. $\lim\limits_{x \to 0^+} (2\sqrt{x} + 1)$

In exercises 31 through 36, sketch the graph of the given function and use the graph to evaluate the limits that exist. (See Examples 7 and 8.)

31. $f(x) = [\![x]\!]$

    a. $\lim\limits_{x \to 0^+} f(x)$

    b. $\lim\limits_{x \to 0^-} f(x)$

    c. $\lim\limits_{x \to 0} f(x)$

    d. $\lim\limits_{x \to 1} f(x)$

33. $g(x) = \begin{cases} x^2 + 1, & x < 0 \\ x, & x \geq 0 \end{cases}$

    a. $\lim\limits_{x \to -2} g(x)$

    b. $\lim\limits_{x \to 3} g(x)$

    c. $\lim\limits_{x \to 0^-} g(x)$

    d. $\lim\limits_{x \to 0^+} g(x)$

35. $f(x) = \begin{cases} 1, & x \leq 2 \\ -1, & x > 2 \end{cases}$

    a. $\lim\limits_{x \to 2^-} f(x)$

    b. $\lim\limits_{x \to 2^+} f(x)$

    c. $\lim\limits_{x \to 2} f(x)$

    d. $\lim\limits_{x \to 0} f(x)$

32. $f(x) = -[\![x]\!]$

    a. $\lim\limits_{x \to 0^+} f(x)$

    b. $\lim\limits_{x \to 0^-} f(x)$

    c. $\lim\limits_{x \to 0} f(x)$

    d. $\lim\limits_{x \to -1} f(x)$

34. $g(x) = \begin{cases} x^2 - 1, & x < 0 \\ 1 - x, & x \geq 0 \end{cases}$

    a. $\lim\limits_{x \to 1} g(x)$

    b. $\lim\limits_{x \to -3} g(x)$

    c. $\lim\limits_{x \to 0^-} g(x)$

    d. $\lim\limits_{x \to 0^+} g(x)$

36. $f(x) = \begin{cases} 3, & x \leq 1 \\ -3, & x > 1 \end{cases}$

    a. $\lim\limits_{x \to 1^-} f(x)$

    b. $\lim\limits_{x \to 1^+} f(x)$

    c. $\lim\limits_{x \to 1} f(x)$

    d. $\lim\limits_{x \to 0} f(x)$

SET B

In exercises 37 through 50, evaluate the limits that exist. If the limit does not exist, state that it does not exist. (See Examples 10, 11, and 12.)

37. $\lim\limits_{x \to 3} \dfrac{x^2 - 9}{x - 3}$

38. $\lim\limits_{x \to 2} \dfrac{x^2 - 4}{x - 2}$

39. $\lim\limits_{x \to -2} \dfrac{x + 2}{x^2 + 2x}$

40. $\lim\limits_{x \to -3} \dfrac{2x + 6}{x + 3}$

41. $\lim\limits_{x \to 0} \dfrac{x^4 - x^2}{x^2}$

42. $\lim\limits_{x \to 0} \dfrac{2x^3}{x^6 - x^3}$

43. $\lim\limits_{x \to 9} \dfrac{\sqrt{x} - 3}{x - 9}$

44. $\lim\limits_{x \to 4} \dfrac{x - 4}{2 - \sqrt{x}}$

45. $\lim\limits_{x \to -5^+} \dfrac{x + 5}{\sqrt{x + 5}}$

46. $\lim\limits_{x \to 3^-} \dfrac{\sqrt{3 - x}}{x - 3}$

47. $\lim\limits_{x \to 0} \dfrac{1}{x}(x^2 + x)$

48. $\lim\limits_{x \to 0} x\left(1 + \dfrac{1}{x}\right)$

49. $\lim\limits_{x \to 1/2} \dfrac{2x^2 + x - 1}{2x - 1}$

50. $\lim\limits_{x \to -2/3} \dfrac{3x + 2}{6x^2 + 4x}$

51. **Calculator Exercise**

a. Complete the following table of values for the function

$$f(x) = \frac{x^2 - 3x}{x}$$

x	-0.5	-0.1	-0.01	-0.001	-0.0001	0.0001	0.001	0.01	0.1	0.5
$f(x)$										

b. From the table, estimate $\lim\limits_{x \to 0} f(x)$.

c. Compare the estimate from part b with the actual limit calculated in Example 11.

52. **Calculator Exercise**

a. Complete the following table of values for the function

$$f(x) = \frac{1 - x}{1 - \sqrt{x}}$$

x	0.9	0.99	0.999	0.9999	0.99999	1.00001	1.0001	1.001	1.01	1.1
$f(x)$										

b. From the table, estimate $\lim\limits_{x \to 1} \dfrac{1 - x}{1 - \sqrt{x}}$.

c. Compare the estimate from part b with the limit calculated in Example 12.

BUSINESS
PRODUCTION LEVEL

53. The following graph gives the production level, $P(t)$, for a factory during the first ten days of the month before the new equipment was installed, and for the ten-day period immediately after the new equipment was installed. From the graph, evaluate the following limits:

a. $\lim\limits_{t \to 10^-} P(t)$ **b.** $\lim\limits_{t \to 10^+} P(t)$

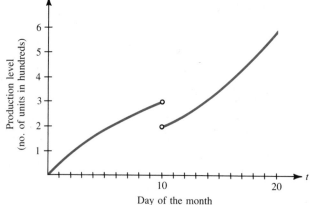

DEPRECIATION

54. The value of a piece of farm equipment is depreciating at a rate

$$V(t) = \frac{5000}{t + 1}$$

where t is measured in years.
a. What is the value of the piece of equipment initially ($t = 0$)?
b. Find $\lim_{t \to 9} V(t)$ and interpret the result.

INCOME TAX

55. Tax tables issued by the Internal Revenue Service can be described as piece-wise functions. For example, consider the 1993 Tax Rate Schedule for Single Taxpayers (Schedule X):

1993 TAX RATE SCHEDULE

| If the amount on Form 1040, line 37, is: | | Then your tax is: | |
over	but not over	This amount + this %	of the amount over
$ 0	$ 22,100	. . . . 15%	$ 0
22,100	53,500	$ 3,315.00 + 28%	22,100
53,500	115,000	12,107.00 + 31%	53,500
115,000	250,000	31,172.00 + 36%	115,000
250,000	. . . .	79,772.00 + 39.6%	250,000

If x represents your income, then your tax, $T(x)$, as a function of x, is obtained from the schedule. For example, if $x = 25,000$, then using the second line from the table,

$$T(x) = 3315 + 0.28(25,000 - 22,100) = \$4127$$

Find the following:
a. $\lim_{x \to 30,000} T(x)$ **b.** $\lim_{x \to 100,000} T(x)$
c. $\lim_{x \to 53,500^-} T(x)$ **d.** $\lim_{x \to 53,500^+} T(x)$

56. Using the information in exercise 55, graph $y = T(x)$.

BIOLOGY
BOTANY

57. A nursery is conducting experiments with a new variety of ferns. The experimenters started with 1600 young ferns and have found that the ferns are very sensitive to changes in temperature. The number of ferns that survive is given in terms of the temperature by

$$N(t) = 1600 - (t - 75)^2$$

where t is measured in degrees Fahrenheit. Determine the limit of $N(t)$ as the temperature approaches 35°F.

SOCIOLOGY
POPULATION

58. The population for a rural region is increasing due to the construction of a new industrial plant. The population is given by

$$P(t) = 100\sqrt{t} + 280$$

where t is measured in months, with $t = 0$ corresponding to March 1991.
a. What was the population initially?
b. What is the population after two months?
c. Calculate the limit of the population as the end of 1991 approached.

WRITING AND CRITICAL THINKING

    1. Consider the following three true statements:

$$\lim_{x \to 1} \sqrt{x - 1} \text{ does not exist}$$

$$\lim_{x \to 1^-} \sqrt{x - 1} \text{ does not exist}$$

$$\lim_{x \to 1^+} \sqrt{x - 1} = 0$$

Explain in complete sentences why the answer to the first two limits is not also 0.

    2. Let $f(x) = \dfrac{1}{x - 2}$ and $g(x) = \begin{cases} x^2, & x \le 2 \\ 3 - x, & x > 2 \end{cases}$

Notice that $\lim\limits_{x \to 2} f(x)$ does not exist and $\lim\limits_{x \to 2} g(x)$ does not exist. These limits do not exist for different reasons. For each function, explain why the limit as x approaches 2 does not exist.

2.4   CONTINUITY

It is probably not an exaggeration to say that all of us have at least once been tempted to try our luck at doing a maze in some book or magazine. You remember that the object was to begin with your pencil at some starting point and to trace out the correct path to the end without lifting your pencil.

We would like to use this same idea for describing a characteristic of functions. That characteristic is called **continuity** (both at a point and on an interval). Very informally, we can say that a function is *continuous* on an interval if we can trace out the graph of the function over that interval without lifting our pencil. If, during the course of tracing out the function, we encounter a break, jump, or hole, then we say that the function is discontinuous and give the location of the discontinuity by giving its *x* coordinate.

As an example of a discontinuous function, assume that you are the manager of a large company and are not satisfied with the present advertising campaign for your product. The advertising department goes to work to come up with a new campaign that is more expensive but that causes sales to jump significantly (see Figure 2.32).

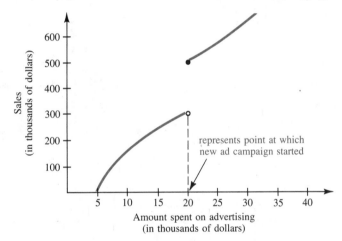

This is an example of what can be referred to as a **jump** discontinuity. For more examples of jump discontinuities, other types of discontinuities, and also some continuous functions, study Figure 2.33.

Figure 2.33

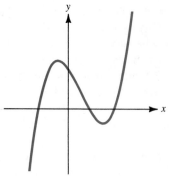

(a) continuous on $(-\infty, \infty)$

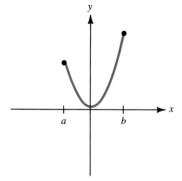

(b) continuous on $[a, b]$

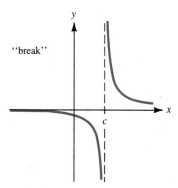

"break"

(c) discontinuous at $x = c$

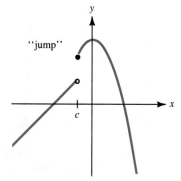

"jump"

(d) discontinuous at $x = c$

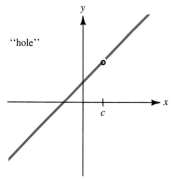

"hole"

(e) discontinuous at $x = c$

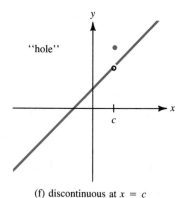

"hole"

(f) discontinuous at $x = c$

Example 1   Graph the following functions and determine from the graphs if the functions are continuous. If not, give the location of any discontinuities.

a. $f(x) = 1 - x^2$ **b.** $f(x) = \begin{cases} -x, & x \le 0 \\ x^2, & x > 0 \end{cases}$

c. $f(x) = \begin{cases} 1 - x, & x \le 0 \\ x^2, & x > 0 \end{cases}$

Solution   **a.**

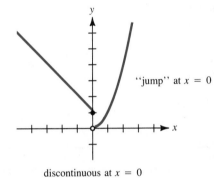

continuous for all x

b.

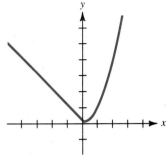

continuous for all x

c.

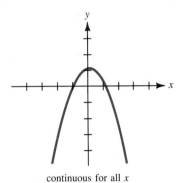

"jump" at $x = 0$

discontinuous at $x = 0$ ■

To give a definition for continuity, we need to return to the idea of limit presented in Section 2.3. Three conditions can occur to cause a function to have a discontinuity at a point. One, the function may not be defined for some $x = c$, as in the functions $f(x) = \dfrac{1}{(x - 2)^2}$ and $f(x) = \dfrac{x^2 - 1}{x - 1}$. The graphs are shown in Figure 2.34.

Figure 2.34

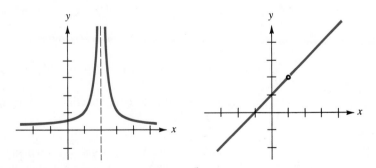

Two, the limit as x approaches some number c may not exist, as in the piecewise function

$$f(x) = \begin{cases} -x, & x \le 0 \\ x + 1, & x > 0 \end{cases}$$

The reason the limit does not exist is that

$$\lim_{x \to 0^-} f(x) \ne \lim_{x \to 0^+} f(x)$$

That is, the right-hand and left-hand limits are not the same.

Finally, the third reason for a discontinuity at some $x = c$ is that $f(c)$, the value of the function at $x = c$, is not equal to the limit of the function as x approaches c. This third situation occurs for a function such as

$$f(x) = \begin{cases} 2x + 1, & x \ne 1 \\ 5, & x = 1 \end{cases}$$

We can now use these three criteria to make a formal definition for continuity of a function at a point.

Continuous function

> A function is **continuous** at $x = c$ if and only if
>
> 1. f is defined for $x = c$
> 2. $\lim_{x \to c} f(x)$ exists
> 3. $\lim_{x \to c} f(x) = f(c)$

To say that a function is continuous on an interval, it must be true that the function is continuous at each point on that interval.

Example 2   Use the definition of continuity to determine whether or not the function $f(x) = 3x^2 + x - 4$ is continuous for $x = 2$.

Solution   We need to check the three parts of the definition.

1. $f(2) = 3(2)^2 + 2 - 4 = 10$, so we may conclude that the function is defined for $x = 2$.
2. $\lim_{x \to 2} (3x^2 + x - 4) = 3(2)^2 + 2 - 4 = 10$, so the limit exists.
3. Since $f(2) = 10$ and $\lim_{x \to 2} f(x) = 10$, we have $\lim_{x \to 2} f(x) = f(2)$.

All three parts have been verified, so the function is continuous at $x = 2$. ∎

Example 3   Psychologists are studying the effect of negative advertising on a group of subjects. Subjects are shown a commercial advertising a certain product, and the level of effect in terms of buying the product is measured at time intervals following the commercial. At some point, a conflicting message is introduced, producing a sudden drop in the effect of the original commercial, as shown by the graph in Figure 2.35. Using the definition, explain why the function is not continuous at $x = t_0$. Let $E(x)$ represent the function.

Figure 2.35

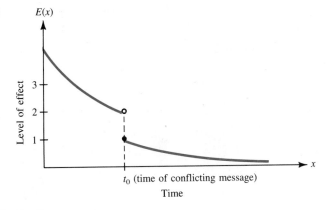

Solution First, the function is defined for $x = t_0$,

$$E(t_0) = 1$$

Second, checking the one-sided limits, we get

$$\lim_{x \to t_0^-} E(x) = 2 \qquad \text{and} \qquad \lim_{x \to t_0^+} E(x) = 1$$

Because these are not equal, $\lim_{x \to t_0} E(x)$ does not exist and the function is not continuous at $x = t_0$. ∎

2.4 EXERCISES

SET A

In exercises 1 through 8, examine the graphs and give the location (x coordinate) of any discontinuities, or state that the function is continuous. (See Example 1 and Figure 2.33.)

1.

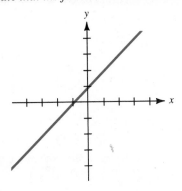

2.

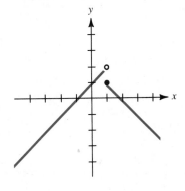

3.

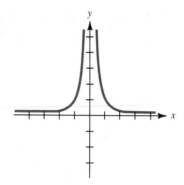

4.

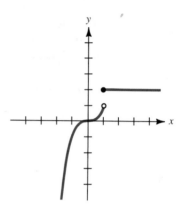

5.

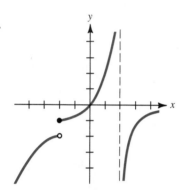

6.

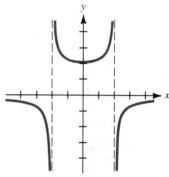

7.

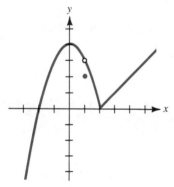

8.

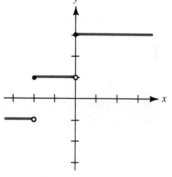

In exercises 9 through 18, sketch the graph of each function and give the location of any discontinuities.
(See Example 1.)

9. $f(x) = [\![x]\!]$

10. $f(x) = |x|$

11. $f(x) = \dfrac{1}{x}$

12. $f(x) = \dfrac{1}{x - 3}$

13. $f(x) = \begin{cases} -2x, & x \le -1 \\ x, & x > -1 \end{cases}$

14. $f(x) = \begin{cases} -1, & x \le 1 \\ x - 2, & x > 1 \end{cases}$

15. $f(x) = \begin{cases} x^2 + 1, & x < 0 \\ 1 - x^2, & x \ge 0 \end{cases}$

16. $f(x) = \begin{cases} x^2 + 1, & x < 0 \\ x^2 - 1, & x \ge 0 \end{cases}$

17. $f(x) = \dfrac{x^2 - 1}{x - 1}$

18. $f(x) = \dfrac{x^2 - 2x}{x}$

(*Hint:* Factor the numerator
and reduce.)

SET B

In exercises 19 through 24, use the definition of continuity to determine whether or not the function is continuous for the given x value. (See Example 2.)

19. $f(x) = 1 - 2x, x = 4$

20. $f(x) = 1 - 2x^2, x = 4$

21. $f(x) = \dfrac{3}{x + 2}, x = -2$

22. $f(x) = \dfrac{3}{x - 2}, x = -2$

23. $f(x) = \begin{cases} x^2 + 2, & x < 1 \\ 4 - x, & x \geq 1 \end{cases}; x = 1$

24. $f(x) = \begin{cases} x^2 + 2, & x < 1 \\ 4 - x, & x \geq 1 \end{cases}; x = 2$

25. Using the definition of continuity, find the value of a so that the given function will be continuous at $x = 2$.

$$f(x) = \begin{cases} 3x - 1, & x \leq 2 \\ 2x + a, & x > 2 \end{cases}$$

26. Use the definition of continuity to find the value of m so that the given function will be continuous for $x = 1$.

$$f(x) = \begin{cases} x^2, & x \leq 1 \\ mx + 3, & x > 1 \end{cases}$$

27. Determine whether or not the function h is continuous at $x = -2$:

$$h(x) = \begin{cases} 2 - |x|, & |x| < 2 \\ x^2 - 4, & |x| \geq 2 \end{cases}$$

BUSINESS
COST FUNCTION

28. The weekly cost, $C(m)$, of renting a car from the Autorent Company depends on the number of miles driven. The cost function is given by

$$C(m) = \begin{cases} 100, & 0 < m \leq 200 \\ 100 + 0.2m, & 200 < m \leq 500 \\ 150 + 0.1m, & m > 500 \end{cases}$$

Is this function discontinuous at any value of m where $m > 0$? If so, give the value for m.

POSTAL RATES

29. The cost of first-class postage is 29 cents for the first ounce and 23 cents an ounce for each additional ounce or fraction thereof. If the weight in ounces is represented by x, determine the function, $P(x)$, that describes the cost of mailing a first-class letter weighing x ounces. Is the function $P(x)$ continuous at $x = 1, x = 2, x = 3$, and $x = 6.5$? Why?

PERSONAL INCOME TAX

30. The amount of federal income taxes owed by a single taxpayer can be considered a function, where $T(x)$ is the amount of taxes owed and x is the taxable income. A partial definition of this function is given as follows:

$$T(x) = \begin{cases} \vdots \\ 4216 & \text{for} & 23{,}000 \le x < 23{,}050 \\ 4230 & \text{for} & 23{,}050 \le x < 23{,}100 \\ 4244 & \text{for} & 23{,}100 \le x < 23{,}150 \\ 4258 & \text{for} & 23{,}150 \le x < 23{,}200 \\ 4272 & \text{for} & 23{,}200 \le x < 23{,}250 \\ \vdots \end{cases}$$

Is $T(x)$ a continuous function? Justify your answer.

BIOLOGY
IMMUNOLOGY

31. For a given concentration of antibody molecules, the greater the fraction of ligand bound by them, the greater their affinity for the ligand. This can be expressed using the equation

$$K = \frac{r}{(n - r)c}$$

where r = ratio of the concentration of bound ligand to the concentration of antibody molecules
n = number of ligand-binding sites
c = concentration of unbound ligand
K = association value

What happens to this function if we take the limit as $r \to n$?

GENETICS

32. Some mutant or defective genes are temperature sensitive. They may behave normally at a temperature below 34°C but as a mutant above 39°C. Temperature-sensitive genes are important in laboratory research because they enable one to turn off the activity simply by raising the temperature. The activity level of a certain gene is given by

$$A = \frac{45 - t}{(t - 39)^2}$$

where t is measured in degrees Celsius.
a. For what temperature is the activity level zero?
b. Find $\lim\limits_{t \to 40} A$.

WRITING AND CRITICAL THINKING

1. Explain in concise sentences how limits are used in the concept of continuity for functions.

2. For the function $f(x) = \begin{cases} x + 3, & x < 0 \\ 3 - x, & x > 0 \end{cases}$ pictured here, what value should be assigned to $f(0)$ so that f is continuous at $x = 0$? Explain your answer.

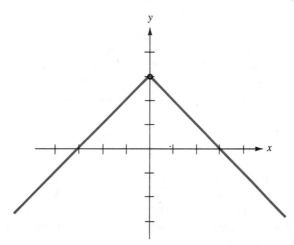

2.5 ASYMPTOTES

A dictionary defines infinity as "an indefinitely large number or amount." In this section, we deal with indefinitely large numbers and use the infinity symbol, ∞.

In Section 2.3, we discussed what happens to a function such as $f(x) = \dfrac{1}{x - 2}$ as x approaches 2 by trying to calculate

$$\lim_{x \to 2} \frac{1}{x - 2}$$

We concluded that the limit does not exist. We now wish to go back and take a closer look at the behavior of this function as x approaches 2.

By making tables of values for x close to 2, consider what happens to the y values for the function $f(x) = \dfrac{1}{x - 2}$. We will make tables for the case of x approaching 2 from the left ($x \to 2^-$) and for x approaching 2 from the right ($x \to 2^+$) (see Figure 2.36).

Figure 2.36

Table A $x \to 2^{-}$

x	y
1.5	-2
1.9	-10
1.99	-100
1.999	-1000
1.9999	$-10{,}000$

Table B $x \to 2^{+}$

x	y
2.5	2
2.1	10
2.01	100
2.001	1000
2.0001	10,000

From Table B, it appears that as $x \to 2^{+}$, y is getting larger and larger, and it fits our dictionary definition for infinity. Symbolically we write as $x \to 2^{+}, y \to +\infty$ (positive infinity, because all y numbers in Table B are positive). From Table A, we can write as $x \to 2^{-}, y \to -\infty$ (negative infinity, because y numbers in Table A are negative). Our conclusions can be written as follows:

$$\lim_{x \to 2^{-}} \frac{1}{x-2} = -\infty \qquad \text{and} \qquad \lim_{x \to 2^{+}} \frac{1}{x-2} = +\infty$$

It is still true, however, that the limit as $x \to 2$ does not exist because the two one-sided limits do not produce the same result.

The type of limit we have been dealing with is called an **infinite limit** (result is $+\infty$ or $-\infty$). One interpretation of what we just encountered is that graphically we have a **vertical asymptote** for the function $f(x) = \dfrac{1}{x-2}$. Recall that a vertical asymptote occurs for a rational function whenever the denominator is zero (and the numerator is nonzero). In this case, the vertical asymptote occurs when x is 2, and we say that the equation for the asymptote is $x = 2$, represented graphically by a dashed line in Figure 2.37.

Figure 2.37

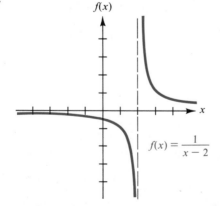

$$f(x) = \frac{1}{x-2}$$

Example 1   Discuss the behavior of the function $f(x) = \dfrac{1}{2-x}$ close to its vertical asymptote by completing the tables. Sketch the graph.

x	f(x)	x	f(x)
1.9		2.1	
1.99		2.01	
1.999		2.001	
1.9999		2.0001	

Solution   The vertical asymptote for the function is $x = 2$. The table on the left represents $x \to 2^-$ and the table on the right represents $x \to 2^+$.

x	f(x)	x	f(x)
1.9	10	2.1	−10
1.99	100	2.01	−100
1.999	1000	2.001	−1000
1.9999	10,000	2.0001	−10,000

From the tables

$$\text{as} \quad x \to 2^-, \quad y \to +\infty$$
$$\text{and as} \quad x \to 2^+, \quad y \to -\infty$$

The graph is shown in Figure 2.38.

Figure 2.38

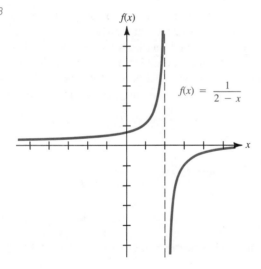

$$f(x) = \frac{1}{2 - x}$$

Example 2   Discuss the behavior of $f(x) = \dfrac{1}{(x - 2)^2}$ close to its vertical asymptote and sketch the graph.

Solution   Again, $x = 2$ is a vertical asymptote. No matter what value we substitute for x close to 2, the result is positive. Therefore, we can say

$$\lim_{x \to 2^-} \frac{1}{(x - 2)^2} = +\infty \quad \text{and} \quad \lim_{x \to 2^+} \frac{1}{(x - 2)^2} = +\infty$$

For simplicity, we use the symbolism "= $+\infty$" to indicate that the limit approaches positive infinity. The graph appears in Figure 2.39.

Figure 2.39

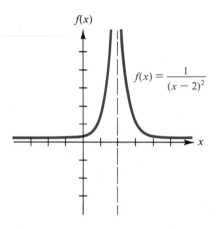

$$f(x) = \frac{1}{(x - 2)^2}$$

For our original function and both of our examples, $x = 2$ was a vertical asymptote, but the behavior of the function as $x \to 2^-$ and as $x \to 2^+$ was different for each one. Another observation that is interesting to us is that the x axis ($y = 0$) was acting as a horizontal asymptote in each case. Can we use our limit concept to help us determine the location of a horizontal asymptote if it exists for a rational function? The answer is yes. To do this, we use limits called **limits at infinity**. We wish to calculate, for $f(x) = \dfrac{1}{x - 2}$,

$$\lim_{x \to +\infty} \frac{1}{x - 2} \qquad \text{and} \qquad \lim_{x \to -\infty} \frac{1}{x - 2}$$

Notice that rather than the conclusion being infinite, it is the x value that is increasing (or decreasing) to an indefinitely large positive (or negative) number.

Again, by making tables of values, we can make some conclusion about the behavior of the function $f(x) = \dfrac{1}{x - 2}$ (see Figure 2.40).

From the tables, we have

as $x \to +\infty, f(x) \to 0^+$ $\qquad$ and $\qquad$ as $x \to -\infty, f(x) \to 0^-$

or $\displaystyle\lim_{x \to +\infty} \frac{1}{x - 2} = 0^+$ $\qquad$ and $\qquad$ $\displaystyle\lim_{x \to -\infty} \frac{1}{x - 2} = 0^-$

Figure 2.40

x	$f(x)$
10	$\dfrac{1}{8}$
100	$\dfrac{1}{98}$
1000	$\dfrac{1}{998}$
10,000	$\dfrac{1}{9998}$
100,000	$\dfrac{1}{99,998}$

x	$f(x)$
-10	$\dfrac{-1}{12}$
-100	$\dfrac{-1}{102}$
-1000	$\dfrac{-1}{1002}$
$-10,000$	$\dfrac{-1}{10,002}$
$-100,000$	$\dfrac{-1}{100,002}$

Example 3 Determine the behavior of $y = \dfrac{1}{(x-2)^2}$ as $x \to -\infty$ and $x \to +\infty$.

Solution $\displaystyle\lim_{x \to -\infty} \frac{1}{(x-2)^2} = 0^+$

$\displaystyle\lim_{x \to +\infty} \frac{1}{(x-2)^2} = 0^+$

Again, since $(x-2)^2$ will always be positive for $x \neq 2$

We can conclude that $y = 0$ is a horizontal asymptote (see Figure 2.39). ∎

We summarize some of our work with infinite limits and limits at infinity in Table 2.2.

Table 2.2

Infinite limits	Limits at infinity
$\displaystyle\lim_{x \to c^-} \frac{1}{x-c} = -\infty$	$\displaystyle\lim_{x \to -\infty} \frac{1}{x} = 0^-$
$\displaystyle\lim_{x \to c^+} \frac{1}{x-c} = +\infty$	$\displaystyle\lim_{x \to +\infty} \frac{1}{x} = 0^+$

As an example of where a *horizontal* asymptote may be important, consider the following situation. A company is spending a certain amount of money to promote a new product, and the money spent affects sales of the product directly. As the company spends more money on promotion, the sales continue to increase. At some point, however, increased promotion no longer leads to a significant increase in sales; the company has reached something called the "absolute ceiling" for sales. Graphically, the absolute ceiling is a horizontal asymptote or the upper bound for the function (see Figure 2.41).

Figure 2.41

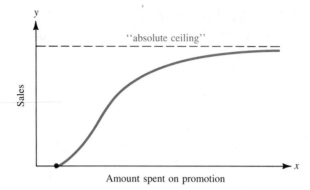

Amount spent on promotion

Example 4   The Krispee Cereal Company is spending x dollars (measured in millions) and yielding sales of $S(x)$ dollars (measured in millions), represented by the function

$$S(x) = \frac{5x^2 - 10x + 5}{x^2 - 2x + 2}, \quad x \geq 1$$

Determine the absolute ceiling for their sales.

Solution   We need to find the horizontal asymptote by calculating

$$\lim_{x \to +\infty} \frac{5x^2 - 10x + 5}{x^2 - 2x + 2}$$

As $x \to +\infty$, both the numerator and denominator increase without bound, resulting in the indeterminate form $+\infty/+\infty$. We can "transform" the function using some algebra to produce a form that can be determined. Note that we will use the fact that

$$\lim_{x \to +\infty} \frac{1}{x^n} = 0^+, \text{ where } n > 0.$$

$$
\begin{aligned}
\lim_{x \to +\infty} \frac{5x^2 - 10x + 5}{x^2 - 2x + 2} &= \lim_{x \to +\infty} \frac{5x^2 - 10x + 5}{x^2 - 2x + 2} \cdot \frac{1/x^2}{1/x^2} \\
&= \lim_{x \to +\infty} \frac{5 - (10/x) + (5/x^2)}{1 - (2/x) + (2/x^2)} \\
&= \frac{5 - 0 + 0}{1 - 0 + 0} \\
&= 5
\end{aligned}
$$

The absolute ceiling for sales is \$5 million. A graph of $y = \dfrac{5x^2 - 10x + 5}{x^2 - 2x + 2}$, where $x \geq 1$, appears in Figure 2.42.

Figure 2.42

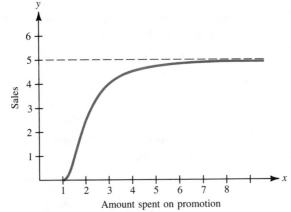

Amount spent on promotion

Example 5   Determine any horizontal asymptotes for $y = \dfrac{2x + 3}{x - 4}$.

Solution   Calculating limits at infinity,

$$\lim_{x \to -\infty} \frac{2x + 3}{x - 4} = \lim_{x \to -\infty} \frac{2x + 3}{x - 4} \cdot \frac{1/x}{1/x}$$

$$= \lim_{x \to -\infty} \frac{2 + (3/x)}{1 - (4/x)} = \frac{2}{1} = 2$$

(More specifically, as $x \to -\infty$, $y \to 2^-$.)

$$\lim_{x \to +\infty} \frac{2x + 3}{x - 4} = \lim_{x \to +\infty} \frac{2x + 3}{x - 4} \cdot \frac{1/x}{1/x}$$

$$= \lim_{x \to +\infty} \frac{2 + (3/x)}{1 - (4/x)} = 2$$

(That is, as $x \to +\infty$, $y \to 2^+$.) See Figure 2.43.

Figure 2.43

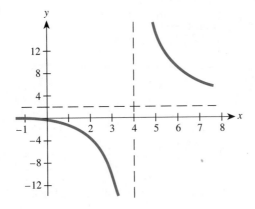

Example 6   Determine the behavior of the function at the vertical asymptotes and determine any horizontal asymptotes using limits and sketch:

$$y = \frac{2x}{x^2 - 1}$$

Solution   Because the denominator is zero for $x = \pm 1$, these are the vertical asymptotes. Calculate limits to determine behavior.

For $x = 1$
$$\begin{cases} \lim\limits_{x \to 1^-} \dfrac{2x}{x^2 - 1} = \dfrac{2}{0^-} = -\infty \\[2ex] \lim\limits_{x \to 1^+} \dfrac{2x}{x^2 - 1} = \dfrac{2}{0^+} = +\infty \end{cases}$$

(*Negative infinity:* Since $x \to 1^-$, the denominator is negative and the numerator is positive.)

For $x = -1$
$$\begin{cases} \lim\limits_{x \to -1^-} \dfrac{2x}{x^2 - 1} = \dfrac{-2}{0^+} = -\infty \\[2ex] \lim\limits_{x \to -1^+} \dfrac{2x}{x^2 - 1} = \dfrac{-2}{0^-} = +\infty \end{cases}$$

Now for the horizontal asymptote we need to calculate

$$\lim_{x \to -\infty} \frac{2x}{x^2 - 1} \quad \text{and} \quad \lim_{x \to +\infty} \frac{2x}{x^2 - 1}$$

First,

$$\lim_{x \to -\infty} \frac{2x}{x^2 - 1} = \lim_{x \to -\infty} \frac{2x}{x^2 - 1} \cdot \frac{1/x^2}{1/x^2} \qquad \leftarrow \text{Use highest power of } x.$$

$$= \lim_{x \to -\infty} \frac{2/x}{1 - (1/x^2)}$$

$$= 0^-$$

and

$$\lim_{x \to +\infty} \frac{2x}{x^2 - 1} = \lim_{x \to +\infty} \frac{2x}{x^2 - 1} \cdot \frac{1/x^2}{1/x^2}$$

$$= \lim_{x \to +\infty} \frac{2/x}{1 - (1/x^2)}$$

$$= 0^+.$$

This means that $y = 0$ is a horizontal asymptote (see Figure 2.44).

Figure 2.44

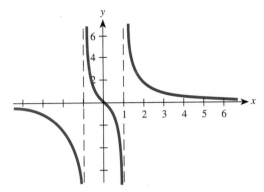

2.5 EXERCISES

SET A

In exercises 1 through 8, find the equations for the vertical asymptotes and determine the behavior of the functions. (See Examples 2 and 6.)

1. $f(x) = \dfrac{1}{x}$ **2.** $y = \dfrac{-1}{x}$ **3.** $y = \dfrac{1}{x+3}$ **4.** $f(x) = \dfrac{x}{x+2}$

5. $y = \dfrac{-2}{x+1}$ **6.** $f(x) = \dfrac{x-3}{x-4}$ **7.** $f(x) = \dfrac{1}{x^2+3x}$ **8.** $y = \dfrac{2}{x^2-4}$

In exercises 9 through 18, determine any horizontal asymptotes by calculating $\lim\limits_{x \to -\infty} f(x)$ and $\lim\limits_{x \to +\infty} f(x)$. (See Examples 3, 5, and 6.)

9. $f(x) = \dfrac{1}{x}$ **10.** $y = \dfrac{-1}{x}$ **11.** $y = \dfrac{x}{x+3}$ **12.** $f(x) = \dfrac{2x}{x-4}$

13. $y = \dfrac{x+1}{2x-3}$ **14.** $f(x) = \dfrac{2-x}{3x+4}$ **15.** $y = \dfrac{x-2}{x^2+1}$ **16.** $y = \dfrac{2}{2x^2+x}$

17. $f(x) = \dfrac{x^2-x+3}{4x^2+5}$ **18.** $f(x) = \dfrac{2x^2-1}{x^2+2}$

SET B

19. Calculator Exercise

Evaluate $\lim\limits_{x \to 1^-} \dfrac{2x}{x^2-1}$ *by completing the following table and compare to the result in Example 6.*

x	0	0.1	0.5	0.9	0.99	0.999	0.9999
y							

20.   Calculator Exercise

Evaluate $\lim\limits_{x \to -\infty} \dfrac{2x}{x^2 - 1}$ by completing the following table and compare to the result in Example 6.

x	-10	-100	-1000	$-10{,}000$	$-100{,}000$
y					

BUSINESS
SALES FUNCTION

21. The Statler Company manufactures shower heads and is spending x dollars (measured in thousands) on advertising. The income from sales, $S(x)$ (measured in thousands), is given by

$$S(x) = \frac{120x^2 - 600x + 3}{2x^2 - 10x + 1}, \qquad x \geq 5$$

Determine the absolute ceiling for their sales. (See Example 4.)

PRODUCTION LEVEL

22. A worker on an assembly line at the Electrak Company has been improving his production level each day at an amazing rate. His production level, $L(t)$, in number of units, is

$$L(t) = \frac{150t^2 + 20t}{2t^2 + 5t + 1}$$

where t represents the number of days the worker has been employed. What is the maximum production level that management can expect from this employee? (*Hint:* Calculate $\lim\limits_{t \to \infty} L(t)$.)

AVERAGE COST

23. A manufacturer has determined the total cost function for the production of a product to be $C(x) = 10x + 1000$, where x is the number of units produced. The average cost per item is given by the average cost function, which is found by dividing the total cost function by x and is denoted by $\overline{C}(x)$. Therefore, $\overline{C}(x) = C(x)/x = 10 + 1000/x$. If the manufacturer continually increases output, what value does the average cost approach?

SKILL TRAINING

24. When workers are trained for new skills, the proportion of workers passing a minimum skills test increases as the number of hours of training t increases. If the proportion p of workers passing the test is given by

$$p = 1 - \frac{70}{70 + t}$$

what is the largest value that p can be if t can be extended indefinitely?

BIOLOGY
ECOLOGY

25. State forest rangers are attempting to establish a herd of antelopes as part of a forest restoration project. They start by "seeding" the area with a herd of 100 antelopes and expect that the population of the antelopes, $N(t)$, can be represented by

$$N(t) = \frac{100 + 45t}{1 + 0.05t}$$

where t is time in years from now.

a.   Find $N(10)$.   b.   Find $N(25)$.

26. Using the model in exercise 25, find the limiting size of the population (that is, find $\lim_{t \to \infty} N(t)$).

SOCIAL SCIENCE
POPULATION

27. The population of a small town has been increasing over the years according to the function

$$P(t) = \frac{9200t^2 + 800}{t^2 + 5}$$

where t is measured in years. Find $\lim_{t \to \infty} P(t)$.

POPULATION GROWTH

28. A retailer in a small city is trying to decide on future expansion of her store. Sociologists at the local university have predicted that the population of the city t years from now will be given by the function

$$N(t) = 40{,}000 + \frac{8000}{(t + 1)^2}$$

Based on this prediction, what is the lower limit for the population?

Calculator Exercises

*In addition to horizontal and vertical asymptotes, graphs of certain functions may also approach **oblique asymptotes**. In exercises 29 through 32,*

a.   *Graph $y = f(x)$.*   b.   *Graph $y = g(x)$.*

c.   *Complete the table.*   d.   *Show that $\lim_{x \to +\infty} f(x) = \lim_{x \to +\infty} g(x)$.*

x	5	10	100	200	1000
$f(x)$					
$g(x)$					

29. $f(x) = \dfrac{2x^2 + 1}{x}$, $g(x) = 2x$

30. $f(x) = \dfrac{x^2}{x - 1}$, $g(x) = x + 1$

31. $f(x) = \dfrac{2x^2}{x - 1}$, $g(x) = 2x + 2$

32. $f(x) = \dfrac{2x^2 - 7x + 3}{x - 1}$, $g(x) = 2x - 5$

WRITING AND CRITICAL THINKING

For the function $f(x) = \dfrac{x^2 - 2x}{x^2 - 4}$, explain (in complete sentences) why there is a vertical asymptote at $x = -2$ but no asymptote at $x = 2$. It may be helpful to look at the limit of this function as x approaches -2 and as x approaches 2.

USING TECHNOLOGY IN CALCULUS

Consider the rational function $f(x) = \dfrac{2x^3 - 3x^2 - 23x + 12}{x^4 + 3x^2 + 2}$. When we

consider the expression $\lim\limits_{x \to \infty} f(x)$, we are interested in the *extreme behavior*
of the function. That is, we are interested in how f behaves for very large
x values. In general, as x gets very large, the rational function

$$\frac{a_n x^m + a_{n-1} x^{m-1} + \cdots + a_0}{b_n x^n + b_{n-1} x^{n-1} + \cdots + b_0}$$

approaches the function $\dfrac{a_n}{b_n} x^{m-n}$ for these "end values" (that is, values of x that
approach either $+\infty$ or $-\infty$).

Thus, $f(x) = \dfrac{2x^3 - 3x^2 - 23x + 12}{x^4 + 3x^2 + 2}$ should approach the function

$g(x) = \dfrac{2}{x}$ for large values of x. We use *DERIVE®* to graph each of these

functions. Notice that as x gets larger, the two curves approach one another.

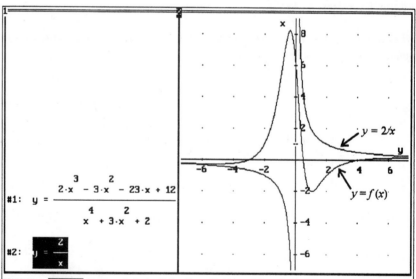

For a second example, consider $f(x) = \dfrac{3x^4 + 5x^3 - 20x^2 - x + 2}{2x^2 + 5x + 6}$. The

results of plotting $y = f(x)$ and its "approaching function" on a graphics
calculator are shown here:

(continued)

(*continued*)

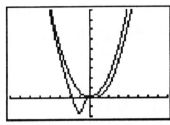

a. What function is *f* approaching?

b. Evaluate your answer to part a when $x = 100$. How does this compare to $f(100)$?

2.5A ASYMPTOTES ON A GRAPHICS CALCULATOR

We already know that when $x = 3$, the function $f(x) = \dfrac{2}{x-3}$ is undefined. In this section, we use the graphing capability of a calculator to picture what we mean by "undefined." To put it in the language of calculus, we would like to examine the behavior of the function "as *x approaches* 3," and we write:

$$\lim_{x \to 3} \frac{2}{x-3}$$

We begin by graphing the function over a suitable interval using the calculator's WINDOW option. It is important, of course, to include the value of 3 in that interval. Figure 2.5A.1 shows the graph and its chosen range values, as well as a particular "traced" point.

Figure 2.5A.1

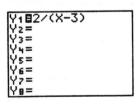

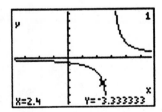

Either by using the tracing capability several times or by using a calculator's TABLE feature, we can develop the following two tables of values, one for values less than $x = 3$ and one for values greater than $x = 3$:

X	Y1
2.4	-3.333
2.5	-4
2.6	-5
2.7	-6.667
2.8	-10
2.9	-20
3	ERROR

X=3

X	Y1
3	ERROR
3.1	20
3.2	10
3.3	6.6667
3.4	5
3.5	4
3.6	3.3333

X=3

It appears from both the graph and the table that as x approaches 3 from the left $(x \to 3^-)$, $f(x)$ approaches negative infinity, and as x approaches 3 from the right $(x \to 3^+)$, $f(x)$ approaches positive infinity. Symbolically, we write:

$$\lim_{x \to 3^-} \frac{2}{x-3} = -\infty \quad \text{and} \quad \lim_{x \to 3^+} \frac{2}{x-3} = \infty$$

The limits from the left and right are not the same, and we say $\lim_{x \to 3} \dfrac{2}{x-3}$ does not exist.

The line $x = 3$ is a **vertical asymptote** for the function and, before we proceed with other examples of asymptotes, some things should be noted. First, notice that the mode for the graph produced in Figure 2.5A.1 is the CONNECTED option. The behavior of the function is often more easily observed in the CONNECTED mode, but the DOT (or discrete pixels) option is sometimes handy, too. We discuss more about these differences at the end of this section. Figure 2.5A.2 is the same graph as Figure 2.5A.1 but in DOT mode.

Figure 2.5A.2

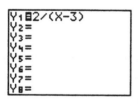
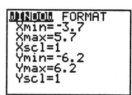
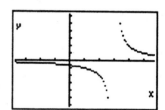

Keep in mind that the choice of WINDOW values for each variable is important when graphing rational functions. For the graph in Figure 2.5A.2, if we had chosen Xmin $= -2$ and Xmax $= 2$, we would not even see the region about the asymptote $x = 3$.

Next, we examine the behavior of the function $f(x) = \dfrac{x^2}{2(x-1)^2}$. The denominator is zero at $x = 1$ and the numerator is nonzero, so we expect a vertical asymptote of $x = 1$. Also, we observe that both the numerator and denominator are never negative (regardless of the value of x), so no portion of the graph can be below the x axis. We choose Ymin $= -0.5$ so the x axis is visible. See Figure 2.5A.3.

Figure 2.5A.3

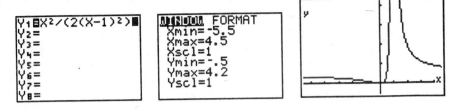

To be convinced of the asymptotic behavior around $x = 1$, try several other combinations of WINDOW values. Notice that

$$\lim_{x \to 1^-} \frac{x^2}{2(x-1)^2} = \infty \qquad \text{and} \qquad \lim_{x \to 1^+} \frac{x^2}{2(x-1)^2} = \infty$$

Now we can say that the two-sided limits are equal and write $\lim\limits_{x \to 1} \dfrac{x^2}{2(x-1)^2} = \infty$.

Next, we ask the question: What happens to this function for extreme values of x? Equivalently, we want to inquire about $\lim\limits_{x \to \infty} \dfrac{x^2}{2(x-1)^2}$ and $\lim\limits_{x \to -\infty} \dfrac{x^2}{2(x-1)^2}$, referred to as **limits at infinity**. Again, we choose "suitable" WINDOW values and have the graphics calculator redraw the graph. See Figure 2.5A.4 for the graph. Two points are traced: one for a large negative value of x and one for a large positive value of x. Notice that we have zoomed so far out that the behavior of the function around its vertical asymptote ($x = 1$) cannot be observed. (This is due to the resolution of the graphics calculator screen.)

Figure 2.5A.4

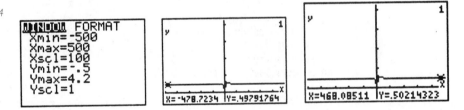

It appears that as x approaches $-\infty$, the function (that is, the y value) is approaching $\frac{1}{2}$ (from below) and as x approaches $+\infty$, the function is approaching $\frac{1}{2}$ (from above). In fact, $y = \frac{1}{2}$ is a **horizontal asymptote** for the graph of the function. This can be observed numerically by the values in the following TI-82 table:

X	Y1			X	Y1
-200	.49504			200	.50504
-250	.49602			250	.50402
-300	.49668			300	.50335
-350	.49716			350	.50287
-400	.49751			400	.50251
-450	.49779			450	.50223
-500				500	

Y1=.49800598404   Y1=.50200601604

This fact can be shown algebraically by expanding the denominator and then dividing each term by the highest power of x, x^2:

$$\lim_{x \to \infty} \frac{x^2}{2(x-1)^2} = \lim_{x \to \infty} \left(\frac{x^2}{2x^2 - 4x + 2} \cdot \frac{1/x^2}{1/x^2} \right) = \lim_{x \to \infty} \frac{1}{2 - 4/x + 2/x^2}$$

Now, as x gets infinitely large, the last two terms in the denominator approach zero and the value of the fraction is $\frac{1}{2}$.

We end this section by examining the graphs of several functions and concluding something about the asymptotes and limits from the graphs. We urge you to try to duplicate these graphs and experiment with other WINDOW values.

Consider the function $f(x) = \dfrac{5x^2 - 10x + 5}{x^2 - 2x + 2}$, $x \geq 1$, as graphed in Figure 2.5A.5. There is no vertical asymptote because it is impossible for the denominator to be zero. The line $y = 5$ is a horizontal asymptote. The value 5 is sometimes called an *absolute ceiling* because it is a horizontal asymptote and serves as an upper bound for the function. You may want to compare the graph in Figure 2.5A.5 with the graph drawn by an artist in Figure 2.42 of Section 2.5.

 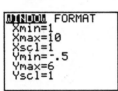 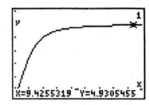 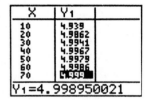

Figure 2.5A.5 $f(x) = \dfrac{5x^2 - 10x + 5}{x^2 - 2x + 2}$

To use a calculator to graph $y = \dfrac{2x}{x^2 - 1}$, we first observe that there will be vertical asymptotes when $x^2 - 1 = 0$, or simply $x = \pm 1$. So, we choose WINDOW values accordingly (see Figure 2.5A.6).

Figure 2.5A.6
$y = \dfrac{2x}{x^2 - 1}$

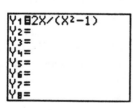 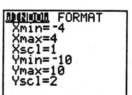 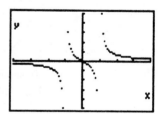

We have:

Vertical asymptotes occur at $x = 1$ and $x = -1$:

$$\lim_{x \to -1^-} \frac{2x}{x^2 - 1} = -\infty \qquad \lim_{x \to -1^+} \frac{2x}{x^2 - 1} = +\infty$$

$$\lim_{x \to 1^-} \frac{2x}{x^2 - 1} = -\infty \qquad \lim_{x \to 1^+} \frac{2x}{x^2 - 1} = +\infty$$

A horizontal asymptote occurs at $y = 0$:

$$\lim_{x \to -\infty} \frac{2x}{x^2 - 1} = \lim_{x \to -\infty} \left(\frac{2x}{x^2 - 1} \cdot \frac{1/x^2}{1/x^2}\right) = \lim_{x \to -\infty} \frac{2/x}{1 - 1/x^2} = 0$$

$$\lim_{x \to +\infty} \frac{2x}{x^2 - 1} = \lim_{x \to +\infty} \left(\frac{2x}{x^2 - 1} \cdot \frac{1/x^2}{1/x^2}\right) = \lim_{x \to +\infty} \frac{2/x}{1 - 1/x^2} = 0$$

In Figure 2.5A.6, notice that we chose the DOT and not the CONNECTED option. *Graphics calculators are not perfect!* For certain combinations of WINDOW values, graphics calculators will attempt to connect pixels (i.e., "points") across asymptotes and the (less than desirable) result is shown in Figure 2.5A.7. In Figure 2.5A.6 the graph appears correctly as three distinct pieces. In Figure 2.5A.7, it may appear that the calculator has drawn in the vertical asymptotes, but what really occurred is that the calculator has attempted to connect the three pieces.

Figure 2.5A.7
$y = \dfrac{2x}{x^2 - 1}$ in
CONNECTED mode.

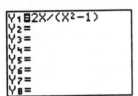
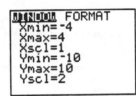
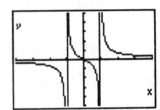

For our last example, we consider $y = \dfrac{3x + 2}{x - 4}$, which has a vertical asymptote $x = 4$ and a horizontal asymptote $y = 3$. (*Why?*) On the TI-82 calculator, we use the LINE option under the DRAW menu as one way of drawing the vertical asymptote on the graphics display.

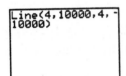

← LINE (4,10000, 4, −10000) draws a line between the point (4, 10000) and the point (4, −10000).

In Figure 2.5A.8, the curve's asymptotes are drawn on the graphics screen to emphasize their limiting relationship to the function. The LINE option was used for the vertical asymptote, but horizontal lines are functions and they can be entered in the same way a function is entered.

Figure 2.5A.8
$y = \dfrac{3x + 2}{x - 4}$ with vertical
asymptote $x = 4$ and
horizontal asymptote
$y = 3$.

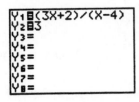
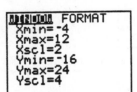
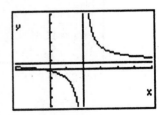

2.5A EXERCISES

In exercises 1 through 10, choose a suitable collection of WINDOW values and graph each function. State the horizontal and vertical asymptotes.

1. $f(x) = \dfrac{2}{x - 5}$

2. $f(x) = \dfrac{x}{4 - x}$

3. $f(x) = \dfrac{x}{x^2 - 1}$

4. $f(x) = \dfrac{x}{1 - x^2}$

5. $f(x) = \dfrac{x}{(1 - x)^2}$

6. $f(x) = \dfrac{x^2}{(1 - x)^2}$

7. $f(x) = \dfrac{x}{x^2 + 1}$

8. $f(x) = \dfrac{x + 5}{1 - x^2}$

9. $f(x) = \dfrac{3x^2 - 5x + 2}{x^2 + 1}$

10. $f(x) = \dfrac{3x^2 - 5x + 2}{x^2 - 1}$

11. Does $\lim\limits_{x \to 1} f(x)$ exist for the function in exercise 3 above? Explain your answer.

12. Does $\lim\limits_{x \to 1} f(x)$ exist for the function in exercise 5 above? Explain your answer.

13. Does $\lim\limits_{x \to 1} f(x)$ exist for the function in exercise 7 above? Explain your answer.

14. The Good Ol' Memory Company, which produces chips for calculators, is spending t thousands of dollars on advertising. Its income from sales, $S(t)$, is also measured in thousands of dollars and is given by

$$S(t) = \frac{30t^2 - 300t + 3}{0.5t^2 - 5t + 1}, \qquad t \geq 10$$

Find the absolute ceiling for sales.

15. The asymptotes discussed in this section are all straight lines, either vertical or horizontal. It can happen, however, that a curve approaches a straight line that is neither vertical nor horizontal. Such lines are called *oblique asymptotes*.

   a.   Consider the function $f(x) = \dfrac{2x^2 - 4x + 1}{x - 2}$. Graph it using a graphics calculator and use the following WINDOW values:

```
MINIMUM FORMAT
Xmin=-4
Xmax=12
Xscl=1
Ymin=-5
Ymax=15
Yscl=2
```

   b.   Now graph $g(x) = 2x$ on the same graphics screen with the same WINDOW values.

c. Use either the TRACE or TABLE options to complete the following table:

x	$f(x)$	$g(x)$	$f(x) - g(x)$
2.4			
5.6			
8.8			
12.0			

d. Are you convinced that f approaches g for extreme values of x?

16. Repeat exercise 15 for $f(x) = \dfrac{2x^3 - 3x^2 - 23x + 12}{x^4 + 3x^2 + 2}$ and $g(x) = \dfrac{2}{x}$ with suitable WINDOW values. The graph $y = g(x)$ is approached asymptotically by $y = f(x)$ but it is not a straight line!

17. *Don't become overly dependent on your calculator!* You have already seen that it can give erroneous results (that is, it can ignore asymptotes by connecting pixels). There is no substitute for thinking and knowledge; the calculator is merely a tool. For another example where the calculator fails, try using it to graph $y = \dfrac{x^2 - 7x + 12}{x^2 - 9}$. Keep in mind that there can be no point at $x = 3$. In fact, there is no asymptote there either. Does your calculator show a point? What should the graph look like?

18. Consider $y = \dfrac{5}{2x^5 - 3x^2 + 4x - 6}$. Use your graphics calculator to *approximate* (use the ZOOM and TRACE features) all the vertical asymptotes for this function.

19. **a.** A student attempted to graph the function $y = \dfrac{x - 3}{x^2 - 5}$ using the following WINDOW values:

```
WINDOW FORMAT
Xmin=2.6
Xmax=12
Xscl=1
Ymin=-1
Ymax=1
Yscl=.2
```

What is wrong?

b. Graph this function using a more appropriate collection of WINDOW values.

20. For large values of x, an added constant can have little effect. For example, consider the function $f(x) = \sqrt{x^2 + 1}$. The "+1" has virtually no effect on the function when x is "large." Thus, $f(x) = \sqrt{x^2 + 1}$ approaches $g(x) = \sqrt{x^2} = |x|$ for "large" values of x. In this case, for viewing purposes, "large" can be interpreted as x greater than 10 or less than -10. Graph $y = f(x)$ and $y = g(x)$ on the same graphics calculator screen. Be sure your WINDOW values include the interval $-10 \le x \le 10$.

2.6 SLOPES OF TANGENTS/RATES OF CHANGE

The concepts and skills learned in algebra, geometry, and trigonometry courses certainly have their applications. But the study of calculus—based on the limit concept examined in Section 2.3—opens up tremendous capabilities in applied settings. In particular, there are two concepts that will be the basis for the remainder of this text: differentiation and integration. When we differentiate a cost function, we are then able to *minimize* cost; if we differentiate a profit function, we may be able to *maximize* profit. In the remainder of this chapter and in Chapters 3 and 4, we study the *derivative*.

Consider the functions $y = 1 - \frac{1}{4}x$ and $y = 1 - \frac{1}{4}x^2$. We make a table of values for each (see Table 2.3) and watch how the y coordinates change as we change the x coordinates at a constant rate.

Table 2.3

x	$y = 1 - \frac{1}{4}x$	x	$y = 1 - \frac{1}{4}x^2$
0	1.000	0	1.000
	$) \rightarrow$ change is -0.125		$) \rightarrow$ change is -0.0625
0.5	0.875	0.5	0.9375
	$) \rightarrow$ change is -0.125		$) \rightarrow$ change is -0.1875
1.0	0.750	1.0	0.7500
	$) \rightarrow$ change is -0.125		$) \rightarrow$ change is -0.3125
1.5	0.625	1.5	0.4375
	$) \rightarrow$ change is -0.125		$) \rightarrow$ change is -0.4375
2.0	0.500	2.0	0.0000
	$) \rightarrow$ change is -0.125		$) \rightarrow$ change is -0.5625
2.5	0.375	2.5	-0.5625
	$) \rightarrow$ change is -0.125		$) \rightarrow$ change is -0.6875
3.0	0.250	3.0	-1.2500

Notice that as the x values are increased at a constant rate, the y coordinates for the function $y = 1 - \frac{1}{4}x$ are decreasing at a constant rate. If we compute this change in y and compare it to the change in x, we have

$$\frac{\text{change in } y}{\text{change in } x} = \frac{-0.125}{0.5} = \frac{-1}{4}$$

If we sketch a graph of $y = 1 - \frac{1}{4}x$, we see in Figure 2.45 that it is just a straight line, and the value we computed is what we defined in Chapter 1 as the slope of the line. The slope or "steepness" of the line will always be $-\frac{1}{4}$ (see Figure 2.45).

Figure 2.45

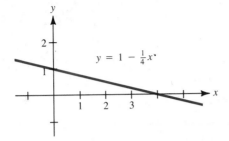

$$y = 1 - \tfrac{1}{4}x\cdot$$

Now examine Table 2.3 for the function $y = 1 - \frac{1}{4}x^2$ and notice that the rate at which the y coordinates are decreasing is *not* constant. A sketch of this function (Figure 2.46) shows that it is a parabola (opening down).

Figure 2.46

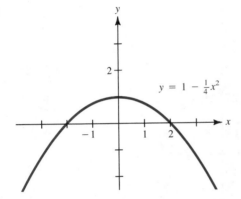

$$y = 1 - \tfrac{1}{4}x^2$$

We see from Table 2.3 and from Figure 2.46 that as the nonnegative x coordinate gets larger, the curve becomes "steeper" (in the negative direction). The question arises as to whether it is possible to find a formula to describe this change in "steepness." We are now ready to define what we mean geometrically by the "slope of a curve" at a point. Analytically, we will be describing the rate of change of the y coordinate compared to a change in the x coordinate. To do so, we choose a particular point on the curve and try to determine the slope of a line tangent to the curve at that point.

Suppose that $f(x) = 1 - \frac{1}{4}x^2$. We choose a point A with coordinates $(a, f(a))$ to be a fixed point on this curve. In particular, we choose $(1, 0.75)$ to be the point A. From our table, clearly the point $(1, 0.75)$ is on the graph of $y = f(x)$. We are interested in determining, if possible, the slope of a line tangent to this curve at point A. Figure 2.47 shows the function and the tangent line at point A, which is labeled L.

Figure 2.47

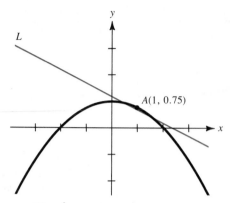

The slope of the tangent line is determined by considering the lines connecting point A with other points $(x, f(x))$ as we move closer to A along the curve—that is, as the x values approach 1 (see Figure 2.48).

Figure 2.48

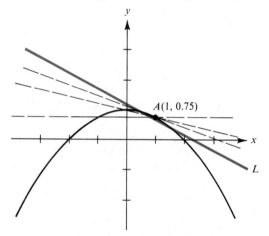

We calculate the slope of the lines connecting point A with various points on the curve that are close to A. These lines are called **secant lines** and their slopes are calculated using the formula from Chapter 1. Recall that the slope of a line can be represented as "rise over run" (see Figure 2.49).

Figure 2.49

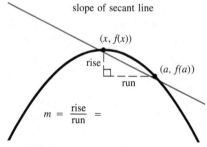

slope of secant line

With the coordinates of point A fixed at $(1, 0.75)$, we represent eight points close to A and the slopes of the corresponding secant lines joining the points to point A in Table 2.4.

Table 2.4

Points close to A $(x, f(x))$	Slope of the secant line $\dfrac{f(x) - f(a)}{x - a} = \dfrac{f(x) - 0.75}{x - 1}$
(0, 1)	$\dfrac{1 - 0.75}{0 - 1} = -0.25$
(0.5, 0.9375)	-0.375
(0.75, 0.859375)	-0.4375
(0.8, 0.84)	-0.45
(0.9, 0.7975)	-0.475
(0.99, 0.754975)	-0.4975
(0.999, 0.75049975)	-0.49975

As x approaches 1 (from the left),* it appears from Table 2.4 that the slope of the secant line is approaching -0.5. Using the concept of limit, we see that the slope of the *tangent* line at A can be expressed as the limit of the slopes of the *secant* lines. Symbolically, the slope of the tangent line at $(a, f(a))$ or A is

Slope of a tangent line

$$m_{\text{tan}} = \lim_{x \to a} \frac{f(x) - f(a)}{x - a}$$

More specifically for our function, since $a = 1$ we have

$$\begin{aligned} m_{\text{tan}} &= \lim_{x \to 1} \frac{f(x) - f(1)}{x - 1} \\ &= \lim_{x \to 1} \frac{[1 - \frac{1}{4}x^2] - [1 - \frac{1}{4}(1)^2]}{x - 1} \\ &= \lim_{x \to 1} \frac{\frac{-1}{4}(x^2 - 1)}{x - 1} \\ &= \lim_{x \to 1} \frac{\frac{-1}{4}(x - 1)(x + 1)}{x - 1} \\ &= \lim_{x \to 1} [\tfrac{-1}{4}(x + 1)] \end{aligned}$$

Now, as x approaches 1, we observe that:

1. Analytically, using the limit calculations from Section 2.3, we get

$$\lim_{x \to 1} [\tfrac{-1}{4}(x + 1)] = \tfrac{-1}{4}(1 + 1) = -0.5$$

---

*We have kept the table short by approaching x only from the left (numbers less than 1). The limit in the definition of the slope of a tangent line is, of course, a two-sided limit. The values obtained by approaching x from the right (numbers greater than 1) are calculated in exercise 50 at the end of this section.

2. Geometrically, the secant lines through $(x, f(x))$ and $(a, f(a))$ approach the tangent line at A. The slope of that tangent line is -0.5.

Example 1   Find the slope of the line tangent to the curve $f(x) = x^2 - 1$ at the point $(2, 3)$.

Solution   The formula for the slope of the tangent at $x = 2$ is

$$m_{tan} = \lim_{x \to 2} \frac{f(x) - f(2)}{x - 2}$$

$$= \lim_{x \to 2} \frac{(x^2 - 1) - (2^2 - 1)}{x - 2}$$

$$= \lim_{x \to 2} \frac{x^2 - 4}{x - 2} \qquad \text{Substituting } x = 2 \text{ gives } \tfrac{0}{0}, \text{ which is indeterminate.}$$

$$= \lim_{x \to 2} \frac{(x - 2)(x + 2)}{x - 2} \qquad \text{Factor the numerator and reduce.}$$

$$= \lim_{x \to 2} (x + 2)$$

$$= 4$$

The slope of the tangent line at the point $(2, 3)$ is 4 (see Figure 2.50).

Figure 2.50

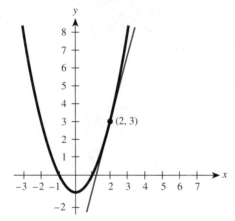

Example 2   For the function $f(x) = x^3$:

   a.   Find the slope of the tangent line at the point $(-1, -1)$.
   b.   Find the equation of the tangent line at the point $(-1, -1)$.
   c.   Sketch the graph of the function and the tangent line.

Solution   **a.** $m_{tan} = \lim_{x \to -1} \dfrac{f(x) - f(-1)}{x - (-1)}$

$$= \lim_{x \to -1} \frac{x^3 - (-1)^3}{x + 1}$$

$$= \lim_{x \to -1} \frac{x^3 + 1}{x + 1}$$

$$= \lim_{x \to -1} \frac{(x + 1)(x^2 - x + 1)}{x + 1} \qquad \text{Factoring the numerator as sum of cubes}$$

$$= \lim_{x \to -1} (x^2 - x + 1)$$

$$= (-1)^2 - (-1) + 1$$

$$= 3$$

b. Using the point-slope form for the equation of a line, we have

$$y - y_1 = m(x - x_1)$$
$$y - (-1) = 3[x - (-1)]$$
$$y + 1 = 3x + 3$$
$$y = 3x + 2$$

c. Figure 2.51 shows the graph of the function and the tangent line.

Figure 2.51

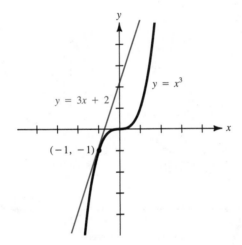

$y = x^3$

$y = 3x + 2$

$(-1, -1)$

As we saw in Chapter 1, the slope is the ratio of the change in y to the change in x. This ratio can also be referred to as the **average rate of change** for a function over some interval. Thus, each value $\dfrac{f(x) - f(a)}{x - a}$ we calculated in Table 2.4 represents the average rate of change for the function $y = 1 - \frac{1}{4}x^2$ from a point P to point A. For example, from Table 2.4, we see that the value -0.25 represents the average rate of change in the function from $x = 0$ to $x = 1$.

Average rate of change

$$\frac{f(x) - f(a)}{x - a}$$

Example 3   A company that manufactures typewriters has determined that the cost of producing x typewriters is given by the function

$$C(x) = 0.2x^2 + 800$$

where $C(x)$ is measured in dollars. Determine the average rate of change in the cost if the production level is increased from 100 to 150 typewriters.

Solution   We need to calculate the average rate of change for $C(x)$ from $x = 100$ to $x = 150$.

$$\text{average rate of change} = \frac{C(150) - C(100)}{150 - 100}$$
$$= \frac{5300 - 2800}{50}$$
$$= \$50 \text{ per typewriter}$$

See Figure 2.52.

Figure 2.52

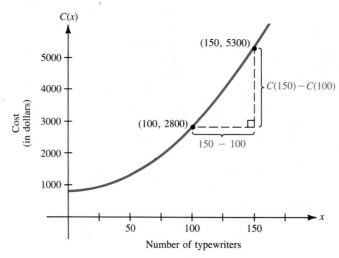

We are now ready to give another interpretation to what we calculated earlier as the slope of the line tangent to a curve $y = f(x)$ at the point $(a, f(a))$.

Instantaneous rate of change

The instantaneous rate of change of f at $x = a$ is given by

$$\lim_{x \to a} \frac{f(x) - f(a)}{x - a}$$

As you can see from the definition, the *instantaneous rate of change* (or slope of tangent) is merely the limit of the *average rate of change* (or slope of secant) as x approaches a.

Example 4   For $f(x) = 2x - x^2$, find the instantaneous rate of change of the function at $x = 3$.

Solution $\displaystyle \lim_{x \to 3} \frac{f(x) - f(3)}{x - 3} = \lim_{x \to 3} \frac{[2x - x^2] - [2(3) - 3^2]}{x - 3}$

$\displaystyle \qquad\qquad\qquad\quad = \lim_{x \to 3} \frac{-x^2 + 2x + 3}{x - 3}$

$\displaystyle \qquad\qquad\qquad\quad = \lim_{x \to 3} \frac{-(x - 3)(x + 1)}{x - 3}$

$\displaystyle \qquad\qquad\qquad\quad = \lim_{x \to 3} [-(x + 1)]$

$\displaystyle \qquad\qquad\qquad\quad = -4$ ■

In many applications for instantaneous rates of change, it is customary to simply ask for the *rate of change.*

2.6 EXERCISES

SET A

In exercises 1 through 12, find the slope of the line tangent to the given curve at the given point. (See Example 1.)

1. $f(x) = 3x^2$, $(1, 3)$
2. $f(x) = -3x^2$, $(1, -3)$
3. $f(x) = \frac{1}{2}x^2$, $(2, 2)$
4. $f(x) = \frac{1}{2}x^2 + 1$, $(2, 3)$
5. $f(x) = 1 - x^2$, $(2, -3)$
6. $f(x) = 4 - x^2$, $(-1, 3)$
7. $f(x) = x^2 + x$, $(-1, 0)$
8. $f(x) = x^2 + 3x$, $(0, 0)$
9. $f(x) = 2x^2 - x + 3$, $(2, 9)$
10. $f(x) = 6 - x - x^2$, $(1, 4)$
11. $f(x) = x^3 + 2$, $(1, 3)$
12. $f(x) = x^3 + x$, $(0, 0)$

In exercises 13 through 18, find the equation of the line tangent to the given curve at the given point. (See Example 2.)

13. $y = x^2 + 2$, $(1, 3)$
14. $y = x^2 - 2$, $(1, -1)$
15. $y = x^2 - 4x$, $(-1, 5)$
16. $f(x) = 4x - x^2$, $(-1, -5)$
17. $f(x) = 2x^3$, $(-1, -2)$
18. $y = 1 - x^3$, $(1, 0)$

In exercises 19 through 26, find the average rate of change. (See Example 3.)

19. $f(x) = 2x + 3$ from $x = 0$ to $x = 2$
20. $f(x) = 5 - 2x$ from $x = 1$ to $x = 4$
21. $f(x) = 2x^2 + 3$ from $x = 1$ to $x = 4$
22. $f(x) = 2x^2 + 4x$ from $x = 0$ to $x = 5$
23. $f(x) = 6 - x - x^2$ from $x = -2$ to $x = 2$
24. $f(x) = 1 - 2x^2$ from $x = -1$ to $x = 1$
25. $f(x) = x^3 + x$ from $x = 2$ to $x = 4$
26. $f(x) = x - x^3$ from $x = 1$ to $x = 3$

In exercises 27 through 36, find the rate of change (instantaneous) for the given function at the indicated value. (See Example 4.)

27. $f(x) = -x^2$, $x = -3$
28. $f(x) = -2x^2$, $x = -1$
29. $f(x) = 2x^2$, $x = \frac{1}{2}$
30. $f(x) = \frac{1}{3}x^2$, $x = 3$
31. $f(x) = x - x^2$, $x = 1$
32. $f(x) = x^2 + 2x$, $x = -1$

33. $f(x) = \frac{1}{3}x^3, x = -1$ **34.** $f(x) = 1 - x^3, x = 1$ **35.** $f(x) = \frac{1}{x}, x = 2$

36. $f(x) = \frac{-1}{x}, x = 2$

SET B

BUSINESS
DEMAND FUNCTION

37. The demand function for a certain product is given by

$$p = 500 - x^2$$

where x is the number of units and p is the price per unit.
a. Find the average rate of change from $x = 10$ to $x = 20$.
b. Find the rate of change for $x = 20$.

PROFIT FUNCTION

38. The Westfield Company, which produces stationary bicycles, has determined that its profit from selling x hundred bicycles is

$$P(x) = 2x^2 - x + 3$$

where $P(x)$ is measured in thousands of dollars.
a. Find the average rate of change in increasing sales from 200 to 300 bicycles. (*Hint:* Use $x = 2$ to $x = 3$ since x is measured in hundreds.)
b. Find the rate of change for $x = 3$—that is, for 300 bicycles.

COST FUNCTION

39. A company that manufactures cassette tapes has determined that the cost of producing x blank tapes a day is given by

$$C(x) = 0.04x^2 + 2000$$

Find the rate of change for $x = 200$ tapes.

REVENUE FUNCTION

40. A certain company has determined that the revenue from producing and selling x items is given by

$$R(x) = 30x - 0.1x^2$$

where $R(x)$ is measured in hundreds of dollars. Find the rate of change of this function for 50 items.

DEMAND FUNCTION

41. The price of a certain item is decreasing as the demand for the item increases. The price per item p is given in terms of the number of units x according to the function

$$p = 10,000 - 2x^2$$

a. Find the average rate at which the price changes as the number of units is increased from 40 to 50 units.
b. Find the rate at which the price is decreasing when the demand is 50 units.

UNEMPLOYMENT

42. The monthly unemployment rate last year in a certain city fluctuated due to the opening and closing of several industrial plants, as illustrated by the graph.

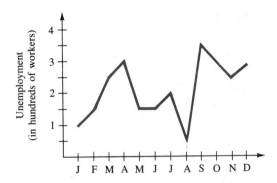

Find the average monthly rate of change in the unemployment figures from
a. March to June   **b.** August to December

SALES

43. Sales for a large corporation are given by the graph, where t is measured in years and S is measured in millions of dollars.

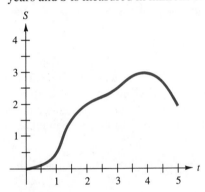

a. What is the sales figure for $t = 3$?
b. What is the average rate of change in sales for the first year to the fourth year?
c. Find the rate of change for sales at $t = 2$. (*Hint:* Draw a tangent line to the curve and estimate its slope.)
d. Over what time period were the sales increasing? decreasing?

BIOLOGY
BOTANY

44. In an experiment involving the growth rate of a certain type of plant exposed to varying amounts of light, it has been determined that the growth, $G(x)$ (measured in inches), is

$$G(x) = 0.001x^2$$

where x is the amount of light measured in watts per square meter. Find the rate at which the growth is changing when $x = 50$ watts per square meter.

BODY CONDITIONING

45. A patient recovering from an injury has been put on a weight-training program. The accompanying graph shows the patient's progress over a ten-week period.

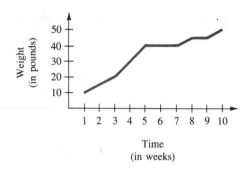

What is the average rate of change in the number of pounds from
a. week one to week three?
b. week three to week five?

SOCIAL SCIENCE
VOTER TURNOUT

46. The voter turnout of 12,560 in a recent election surprised county officials because, during the previous ten-year period, the turnout was given by

$$N = 13{,}900 + 10t - t^2$$

where N is the number of voters and t is measured in years.
a. How many people voted the fifth year that records were kept ($t = 5$)?
b. Find the number of voters expected for the current year ($t = 11$).
c. What was the rate of change in the number of voters for $t = 10$?

BIRTH RATE

47. In a certain rural region of the country, the birth rate is given by

$$N(t) = 85 + (t - 10)^2$$

where $N(t)$ is the number of births and t is measured in years. Find the rate at which the birth rate is changing at $t = 5$ years. Is this an increasing or decreasing rate of change?

PHYSICAL SCIENCE
METEOROLOGY

48. The amount of rainfall, $R(t)$ (measured in inches), for a certain city is

$$R(t) = 3 - t + \frac{t^2}{10}$$

where t is measured in days. Find the average rate of change in the rainfall from $t = 1$ to $t = 5$.

MOTION

49. The equation of motion for an object falling from the top of a 240-foot-tall building is $s = 240 - 16t^2$, where s (in feet) represents the altitude of the object at time t (in seconds).

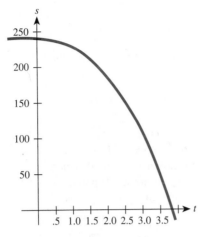

a. What is s when $t = 0$?
b. What is s when $t = 1$?
c. What is s when $t = 2$?
d. What is the *average* rate of change in altitude over the interval $t = 0$ to $t = 2$?
e. What is the *instantaneous* rate of change in altitude at $t = 1$?

Calculator Exercises

50. For the function $f(x) = 1 - \frac{1}{4}x^2$ illustrated in this section, we approached the point $(1, 0.75)$ only from the left (see Table 2.4). Complete the following table to help verify that the slope of the tangent line at $x = 1$ is -0.5.

$(x, f(x))$	$(a, f(a))$	$\dfrac{f(x) - f(a)}{x - a}$
$(2, 0)$	$(1, 0.75)$	-0.75
$(1.5, 0.4375)$	$(1, 0.75)$	-0.625
$(1.1, 0.6975)$	$(1, 0.75)$	
$(1.01, 0.744975)$	$(1, 0.75)$	
$(1.001, 0.74949975)$	$(1, 0.75)$	

In exercises 51 through 54, use the following TI-82 plots to find the requested information.

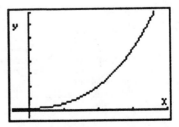

51. Find the slope of the secant line connecting the points on the curve where $x = 0$ to $x = 3$.

52. Find the slope of the secant line connecting the points on the curve where $x = 1$ to $x = 3$.

53. Find the slope of the secant line connecting the points on the curve where $x = 2$ to $x = 3$.

54. Use a ruler and draw a tangent line to the curve at $x = 3$. Estimate the slope of the tangent line to the curve at $x = 3$.

WRITING AND CRITICAL THINKING

1. Explain the difference between the slope of a secant line and the slope of a tangent line. Also, explain the difference between an average rate of change and an instantaneous rate of change.

2. Write a paragraph describing the procedure for finding the equation of a line tangent to a curve at a point. Use words, not symbols, for your explanation.

USING TECHNOLOGY IN CALCULUS

Graphics calculators and computer software packages have excellent graphing capabilities that can help you visualize the notion of a tangent line. Consider the graphs of $y = 1 - \frac{1}{4}x^2$ and $y = -\frac{1}{2}x + \frac{5}{4}$ introduced earlier in this section. We depict them here on the same axes using the software package called *DERIVE®*.

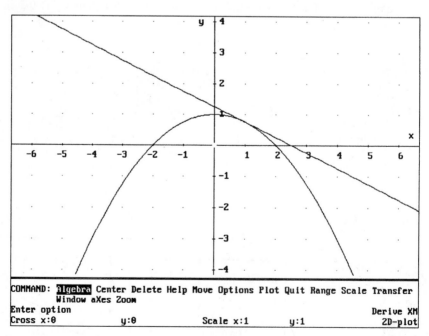

COMMAND: **Algebra** Center Delete Help Move Options Plot Quit Range Scale Transfer
Window aXes Zoom
Enter option   Derive XM
Cross x:0   y:0   Scale x:1   y:1   2D-plot

(continued)

(*continued*)

A big advantage of this technology is the capability to zoom in (or out) on a portion of the plane. The following figure shows the same functions after we have zoomed in on the critical point $(1, 0.75)$.

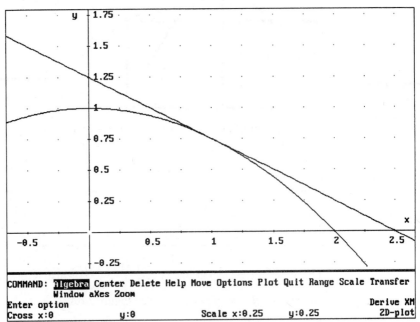

1. Calculate the equation of the line drawn tangent to the function $y = 4x - x^2$ at the point $(-1, -5)$.
2. Using a graphics calculator or a graphing software package, graph the curve and the tangent line on the same set of axes.
3. Now, zoom in on the point. Are you convinced of your answer to question 1?

2.6A   SLOPES OF TANGENTS/RATES OF CHANGE ON A GRAPHICS CALCULATOR

One of the two major processes of calculus, *differentiation*, enables us to maximize profit functions or minimize cost functions. The *derivative* is the subject of the rest of this chapter and Chapters 3 and 4. The derivative represents a *rate of change of a function*.

Suppose we are given a function $y = f(x)$ and a fixed point on its graph at $(a, f(a))$. If the point $(x, f(x))$ is chosen, the secant line between it and $(a, f(a))$ has slope $\dfrac{f(x) - f(a)}{x - a}$ and represents the *average rate of change of the function for the interval* [x, a]. See Figure 2.6A.1.

Figure 2.6A.1

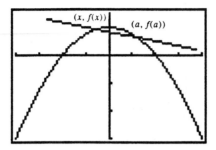

The slope of the secant line is given by

$$m_{\text{sec}} = \dfrac{f(x) - f(a)}{x - a}$$

and represents the *average* rate of change of the function.

As x approaches a (that is, as the point on the left approaches the fixed point), the secant line approaches a line tangent to the curve at (a, f(a)). The slope of the tangent line is

Slope of a tangent line

$$m_{\text{tan}} = \lim_{x \to a} \dfrac{f(x) - f(a)}{x - a}$$

The slope of the tangent line, $\lim\limits_{x \to a} \dfrac{f(x) - f(a)}{x - a}$, is also referred to as the *instantaneous rate of change of f at x = a*. See Figure 2.6A.2.

Figure 2.6A.2

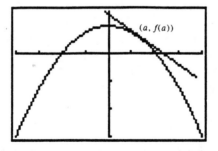

The slope of the tangent line is given by

$$m_{\text{tan}} = \lim_{x \to a} \dfrac{f(x) - f(a)}{x - a}$$

and represents the *instantaneous* rate of change of the function.

In this section, we use the programming feature of a graphics calculator to study the *slope of a function* because slope represents a function's rate of change. We begin by choosing a function and a fixed point [denoted in the program as (P, Q)]. We draw a *secant line* between (P, Q) and another point, (R, S), on the graph of the function. As the point (R, S) "moves" closer to (P, Q), the secant line looks more and more like a tangent line drawn at the point (P, Q). It is the slope of this tangent line that represents the slope of the curve at the point (P, Q). The program TANGENTS, which follows, helps the user see the approximation of secant lines to a tangent line as several lines are drawn that "close in" on the tangent line so it appears to be animated as (R, S) moves closer to the point (P, Q).

Frame 1

```
PROGRAM:TANGENTS
:ClrDraw
:ClrHome
:Disp "SECANT LI
NES"
:Disp "APPROACH"

:_
```

The first two lines clear the graphics and text screens.

Frame 2

```
PROGRAM:TANGENTS
:Disp "TANGENT L
INES"
:Disp " "
:Disp " "
:Disp " "
:Disp "(PRESS EN
TER)"
```

The seven "Disp" lines set up the title screen in Figure 2.6A.3.

Frame 3

```
PROGRAM:TANGENTS
:Pause
:ClrHome
:Disp "FIXED POI
NT IS"
:Disp "(P,Q)"
:Disp "ENTER P"
:Input P
```

The "Pause" command waits for the user to press the ENTER key.

The user inputs the x value of the fixed point. The program stores it in the location P.

Frame 4

```
PROGRAM:TANGENTS
:P→X
:Y₁→Q
:Disp "Q ="
:Disp Q
:Disp "(PRESS EN
TER)"
:Pause
```

To calculate the y value of the fixed point, we use the fact that Y1 is calculated only for x values. Then the Y1 value is stored in Q.

Frame 5

```
PROGRAM:TANGENTS
:ClrHome
:(Xmax-Xmin)/2→K

:Lbl 1
:P-K→R
:R→X
:Y₁→S
```

The "roving point" is (R,S). K represents half of the screen's horizontal distance [K=(Xmax–Xmin)/2]. That is, R is K units from P. It is best to choose the fixed point on the right side of the graphics screen.

"Lbl 1" is a label that marks the beginning of a loop. As K is recalculated, the values of R and S are recalculated in this loop.

Initially, R takes on the value P–K. Then S=f(R) and is found by calculating Y1 and then storing that value in S.

Frame 6

```
PROGRAM:TANGENTS
:DrawF ((Q-S)/(P
-R))*(X-P)+Q
:K/2→K
:If K>0.1
:Goto 1
:DispGraph
:Pause
```

The equation of the secant line connecting (P,Q) with (R,S) is drawn with "DrawF"; from the two-point form, it is the line Y=((Q–S)/(P–R))*(X–P)+Q.

The value of K is halved each time through the "Lbl 1" loop until it is less than or equal to 0.1.

(continued)

(*continued*)

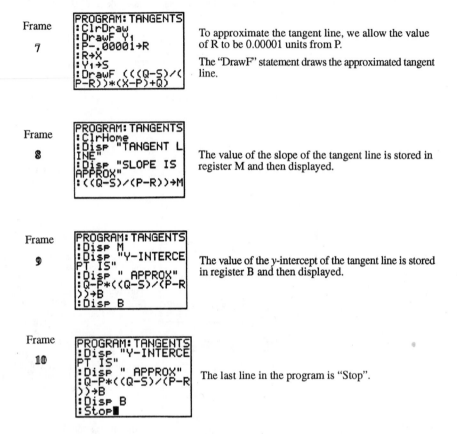

Frame 7

To approximate the tangent line, we allow the value of R to be 0.00001 units from P.

The "DrawF" statement draws the approximated tangent line.

Frame 8

The value of the slope of the tangent line is stored in register M and then displayed.

Frame 9

The value of the y-intercept of the tangent line is stored in register B and then displayed.

Frame 10

The last line in the program is "Stop".

Figure 2.6A.3 displays the first three screens encountered when the program TANGENTS is run using $y = 1 - 0.25x^2$ as the function and (1, 0.75) as the fixed point.

Figure 2.6A.3

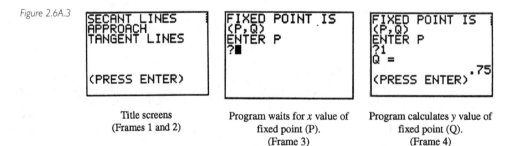

Title screens
(Frames 1 and 2)

Program waits for *x* value of fixed point (P).
(Frame 3)

Program calculates *y* value of fixed point (Q).
(Frame 4)

The program draws secant lines between the fixed point (1, 0.75) and other points [referenced by (R, S) in the program]. The results of that graphics screen are seen in Figure 2.6A.4.

Figure 2.6A.4 *The secant lines are drawn in the Lbl 1 loop. Note that Y1 must be entered before running the program. (Frame 6)*

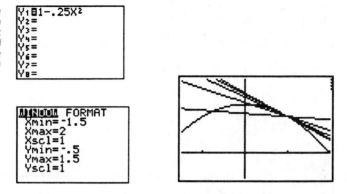

The final graphics screen that the program generates is the function with a tangent line drawn at (P, Q). This screen, along with the text screen stating the slope and *y* intercept of the tangent line, are depicted in Figure 2.6A.5.

Figure 2.6A.5 *The approximated tangent line (Frames 7 through 9)*

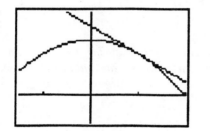

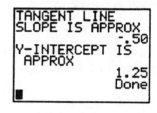

The TANGENTS program can be used to find the instantaneous rate of change of a function at a particular *x* value. For example, to approximate the rate of change of $f(x) = 2x - x^2$ at $x = 3$, we enter $2x - x^2$ for Y1 before we run the program. Upon execution, we enter a 3 for the value of P. See Figure 2.6A.6.

Figure 2.6A.6 *The function $f(x) = 2x - x^2$ and a tangent line at $x = 3$.*

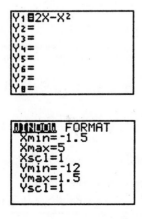

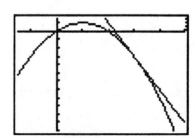

The instantaneous rate of change of f at $x = 3$ is $\lim\limits_{x \to 3} \dfrac{f(x) - f(3)}{x - 3}$. According to Figure 2.6A.7, we have $\lim\limits_{x \to 3} \dfrac{f(x) - f(3)}{x - 3} = -4$, the slope of the tangent line drawn to the curve at $x = 3$.

Figure 2.6A.7

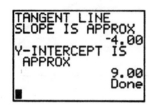

```
TANGENT LINE
SLOPE IS APPROX
            -4.00
Y-INTERCEPT IS
 APPROX
            9.00
          Done
■
```

2.6A   EXERCISES

In exercises 1 through 6, use a graphics calculator to approximate the slope of the line tangent to the given curve at the given point. You may find it useful to use the program outlined in this section or a variation of it.

1. $f(x) = x^2 - x$ at $(2, 2)$

2. $f(x) = 4 - x^2$ at $(-1, 3)$

3. $f(x) = 2x^2 - x + 3$ at $(2, 9)$

4. $f(x) = x^2 - 4$ at $(-1, -3)$

5. $f(x) = x^3 + 2$ at $(1, 3)$

6. $f(x) = x^3 + x$ at $(0, 0)$

In exercises 7 through 12, approximate the equation of the tangent line to the given curve at the given point.

7. $f(x) = x^2 - x$ at $(2, 2)$

8. $f(x) = 4 - x^2$ at $(-1, 3)$

9. $f(x) = \dfrac{4}{x - 2}$ at $(3, 4)$

10. $f(x) = \dfrac{x + 2}{x - 2}$ at $(-1, -1/3)$

11. $f(x) = x^3 + x^2 + 1$ at $(1, 3)$

12. $f(x) = x^3 + x$ at $(0, 0)$

In exercises 13 through 16, approximate the instantaneous rate of change of the function at the indicated value.

13. $f(x) = 2x^3 + 1$, $x = 0$

14. $f(x) = 2x^3 + 1$, $x = 1$

15. $f(x) = x^3 + x^2 + \dfrac{1}{x}$, $x = 1$

16. $f(x) = x^3 + x - 3$, $x = 1$

17. Why would it make no sense to try to find the tangent line drawn to $y = \dfrac{1}{x^2}$ at $x = 0$?

18. Try to use the TANGENTS program (or your own version of it) to find the slope of the tangent line drawn to $(0, 0)$ for $y = \sqrt[3]{x}$. Zoom in on the graph of $y = \sqrt[3]{x}$ around $x = 0$. What do you conjecture about the slope of the tangent line?

19. **a.** Use a graphics calculator to graph the function $y = x^{2/3}$. (*Hint:* In order to get an accurate graph, y should be entered as $(X^2)^{(1/3)}$ on many calculators.)
    b. From the graph, what do you think is the value of the slope of the tangent line drawn at $x = 0$?
    c. Why does the program TANGENTS not give the correct result for this situation?

20. Repeat exercise 19 for the function $y = |x|$ at $x = 0$. The calculator again gives erroneous results, but the situation is slightly different here. Comment on those differences.

2.7 THE DERIVATIVE

In Section 2.6, we presented two interpretations for the quantity represented by

$$\lim_{x \to a} \frac{f(x) - f(a)}{x - a}$$

as follows:

1. A formula for computing the *slope* of a line tangent to the curve $y = f(x)$ at the point $(a, f(a))$.
2. The *instantaneous rate of change* of the function $y = f(x)$ at $x = a$.

This particular limit is one of the most fundamental concepts in calculus and is called the **derivative.**

Derivative of a function

The **derivative** of $y = f(x)$ at the point $(a, f(a))$ is given by

$$f'(a) = \lim_{x \to a} \frac{f(x) - f(a)}{x - a} \tag{1}$$

provided the limit exists.

Example 1 For $f(x) = x^2$, find $f'(1)$.

Solution We are asked to calculate the derivative of the function $f(x) = x^2$ for $x = 1$. Using the definition, we have (see Figure 2.53)

$$f'(1) = \lim_{x \to 1} \frac{f(x) - f(1)}{x - 1}$$
$$= \lim_{x \to 1} \frac{x^2 - 1}{x - 1}$$
$$= \lim_{x \to 1} \frac{(x + 1)(x - 1)}{x - 1}$$
$$= \lim_{x \to 1} (x + 1)$$
$$= 2$$

Figure 2.53

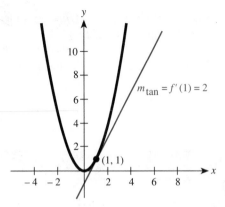

$m_{tan} = f'(1) = 2$

(1, 1)

Before using this definition to calculate any more derivatives, we will make a change in notation in order to expand the usefulness of the definition. Look again at Example 1. If we now wanted to calculate $f'(5)$ in that example, we would have to start over because the preceding definition is designed to help you calculate a derivative for a function at a particular point. Figure 2.54 illustrates this first notation. Notice that we want the slope of the tangent line at a specific point labeled $(a, f(a))$.

Figure 2.54

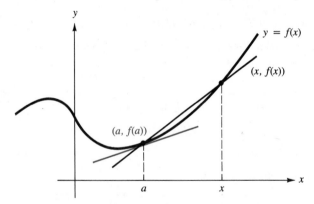

Now we want to generalize this definition. That is, we want a way of computing a derivative that is more generic—one for which we need not pick a specific point (x value). To do this, we label the fixed point on the curve as $(x, f(x))$ and consider a second point h units away labeled $(x + h, f(x + h))$. See Figure 2.55.

Figure 2.55

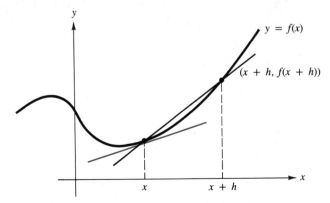

Now the slope of the secant line would be written as:

$$m_{sec} = \frac{f(x + h) - f(x)}{(x + h) - x}$$

$$= \frac{f(x + h) - f(x)}{h}$$

To find the slope of the tangent line (the derivative), we need the limit as h gets small (written $h \to 0$):

$$m_{tan} = \lim_{h \to 0} \frac{f(x + h) - f(x)}{h}$$

We now have a new form for the derivative of a function.

Derivative of a function

The **derivative** of $y = f(x)$ is given by

$$f'(x) = \lim_{h \to 0} \frac{f(x + h) - f(x)}{h} \tag{2}$$

provided the limit exists. This can be referred to as the symbolic definition.

Note: The derivative may also be symbolized using $\dfrac{dy}{dx}$, y', $D_x y$, $\dfrac{d[f(x)]}{dx}$, or $\dfrac{d}{dx}[f(x)]$ in place of $f'(x)$.

Example 2   For $f(x) = x^2$, find $f'(3)$.

Solution   First we find $f'(x)$ in general:

$$f'(x) = \lim_{h \to 0} \frac{f(x + h) - f(x)}{h}$$

$$= \lim_{h \to 0} \frac{(x + h)^2 - x^2}{h}$$

$$= \lim_{h \to 0} \frac{x^2 + 2xh + h^2 - x^2}{h}$$

$$= \lim_{h \to 0} \frac{2xh + h^2}{h}$$

$$= \lim_{h \to 0} (2x + h)$$

$$= 2x$$

Now, substituting 3 for x gives $f'(3) = 2(3) = 6$.   ∎

When we compare this with the result in Example 1, we find that they are the same. This illustrates that only the *notation* (not the concept) has changed.

Returning to our original function from Section 2.6, $f(x) = 1 - \frac{1}{4}x^2$, we compute the *formula* for the "slope of the curve," which we now know is called the *derivative*.

Example 3   For $f(x) = 1 - \frac{1}{4}x^2$, find:

   a. $f'(x)$ **b.** $f'(1)$

Solution   **a.** $f'(x) = \lim\limits_{h \to 0} \dfrac{f(x + h) - f(x)}{h}$

$= \lim\limits_{h \to 0} \dfrac{[1 - \frac{1}{4}(x + h)^2] - [1 - \frac{1}{4}x^2]}{h}$

$= \lim\limits_{h \to 0} \dfrac{1 - \frac{1}{4}x^2 - \frac{1}{2}hx - \frac{1}{4}h^2 - 1 + \frac{1}{4}x^2}{h}$

$= \lim\limits_{h \to 0} \dfrac{-\frac{1}{4}h(2x - h)}{h}$

$= \lim\limits_{h \to 0} [-\frac{1}{4}(2x - h)]$

$= \dfrac{-x}{2}$

Notice that this gives a "formula" for computing the derivative (slope of the tangent line to the curve) for *any x* value (at any point).

b.   Substituting $x = 1$ into the derivative formula gives

$$f'(1) = \frac{-1}{2} \quad \text{or} \quad -0.5$$

This is the slope of the line tangent to the curve at the point (1, 0.75) that we calculated in Section 2.6 (see Figure 2.56).

Figure 2.56

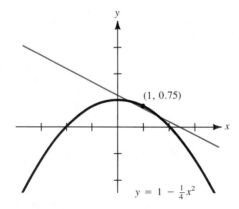

$(1, 0.75)$

$y = 1 - \frac{1}{4}x^2$

■

Example 4   If $f(x) = x^2$, find $f'(x)$.

Solution   Using definition (2), we get

$$f'(x) = \lim_{h \to 0} \frac{f(x + h) - f(x)}{h}$$

$$= \lim_{h \to 0} \frac{(x + h)^2 - x^2}{h}$$

$$= \lim_{h \to 0} \frac{x^2 + 2xh + h^2 - x^2}{h}$$

$$= \lim_{h \to 0} \frac{2xh + h^2}{h}$$

$$= \lim_{h \to 0} (2x + h) = 2x \qquad \blacksquare$$

A note on vocabulary is included here. The original function is $f(x) = x^2$ and the result $f'(x) = 2x$ is called its **derivative** (or derivative with respect to x). We **differentiate** $f(x)$ to get $f'(x)$, and the whole process in getting from $f(x)$ to $f'(x)$ is called **differentiation.** If the derivative exists, then the function is said to be **differentiable.** It is important to note that for the function $f(x) = x^2$, the derivative, $f'(x) = 2x$, is also a function.

Example 5   If $f(x) = x^2$, find each of the following:
   a. $f'(-1)$ **b.** $f'(0)$ **c.** $f'(2)$ **d.** $f'(6)$

Solution   Since, by the previous example, $f'(x) = 2x$, we have
   a. $f'(-1) = 2(-1) = -2$ **b.** $f'(0) = 2(0) = 0$
   c. $f'(2) = 2(2) = 4$ **d.** $f'(6) = 2(6) = 12$ $\qquad \blacksquare$

From Example 5, we can see that the derivative of a function at a point can be negative, positive, or zero with the respective tangent lines rising to the right with a positive slope, falling to the right with a negative slope, or horizontal with zero slope (see Figure 2.57 for $f(x) = x^2$).

Figure 2.57

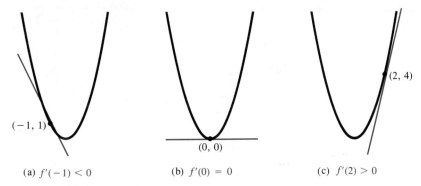

(a) $f'(-1) < 0$ (b) $f'(0) = 0$ (c) $f'(2) > 0$

It is also possible for the derivative of a function to be undefined at a point. This can happen in any of three ways:

1. A function that is undefined at a point produces a derivative that is undefined at that same point, as with the function $f(x) = \dfrac{1}{x-2}$, which is undefined at $x = 2$. (See Example 6.)

2. The limit from the left side ($h \to 0^-$) of the point is not the same as the limit from the right side ($h \to 0^+$) of the point, so we say the derivative is undefined at that point. (See exercise 28.)

3. The tangent line may be vertical at a point as in the function $y = x^{2/3}$. Recall that vertical lines do not have a slope. (See exercise 49, Section 3.1.)

Example 6   For $f(x) = \dfrac{1}{x-2}$,

    a.   Find $f'(x)$.

    b.   Find the equation of the line tangent to the curve at the point $(3, 1)$.

Solution   **a.** $f'(x) = \lim\limits_{h \to 0} \dfrac{f(x+h) - f(x)}{h}$

$$= \lim_{h \to 0} \frac{\dfrac{1}{x+h-2} - \dfrac{1}{x-2}}{h}$$

$$= \lim_{h \to 0} \frac{(x-2) - (x+h-2)}{h(x-2)(x+h-2)}$$

$$= \lim_{h \to 0} \frac{-h}{h(x-2)(x+h-2)}$$

$$= \lim_{h \to 0} \frac{-1}{(x-2)(x+h-2)}$$

$$= \frac{-1}{(x-2)(x-2)} = \frac{-1}{(x-2)^2}$$

Notice that the original function is not defined for $x = 2$ and $f'(x)$ (the derivative) is not defined for $x = 2$.

    b.   The slope of the tangent line will be $f'(3)$.

$$f'(3) = \frac{-1}{(3-2)^2} = -1$$

Using the point-slope form for the equation of the line, we have

$$y - y_1 = m(x - x_1)$$
$$y - 1 = -1(x - 3)$$
$$y - 1 = -x + 3$$
$$\text{or} \qquad y = 4 - x$$

See Figure 2.58.

Figure 2.58

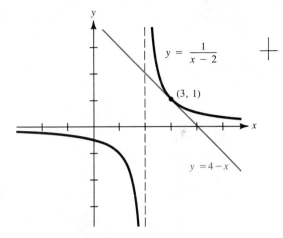

$$y = \frac{1}{x - 2}$$

(3, 1)

$$y = 4 - x$$

Example 7   If $y = x^2 - 7x + 12$, find $\dfrac{dy}{dx}$.

Solution $\dfrac{dy}{dx} = \lim_{h \to 0} \dfrac{f(x + h) - f(x)}{h}$

$= \lim_{h \to 0} \dfrac{[(x + h)^2 - 7(x + h) + 12] - [x^2 - 7x + 12]}{h}$

$= \lim_{h \to 0} \dfrac{x^2 + 2xh + h^2 - 7x - 7h + 12 - x^2 + 7x - 12}{h}$

$= \lim_{h \to 0} \dfrac{2xh + h^2 - 7h}{h}$

$= \lim_{h \to 0} \dfrac{h(2x + h - 7)}{h}$

$= \lim_{h \to 0} (2x + h - 7)$

$= 2x - 7$

Therefore, $\dfrac{dy}{dx} = 2x - 7$.

If you are evaluating a derivative for a specific value, say $x = 2$, and using the $\dfrac{dy}{dx}$ notation, then the notation is as follows:

$$\left. \frac{dy}{dx} \right|_{x = 2}$$

We use this notation in the next example.

Example 8   Suppose that a manufacturer has determined that the demand for her product can be predicted by the equation $p = 500 - 2x^2$, where p represents the price (measured in dollars) for each unit and x represents the number of units demanded (measured in hundreds). Find the instantaneous rate of change of price (with respect to x) when $x = 7$.

Solution   We must find $\left. \dfrac{dp}{dx} \right|_{x=7}$.

$$\frac{dp}{dx} = \lim_{h \to 0} \frac{500 - 2(x + h)^2 - (500 - 2x^2)}{h}$$

$$= \lim_{h \to 0} \frac{500 - 2x^2 - 4xh - 2h^2 - 500 + 2x^2}{h}$$

$$= \lim_{h \to 0} \frac{-4xh - 2h^2}{h}$$

$$= \lim_{h \to 0} (-4x - 2h)$$

$$= -4x$$

Thus, $\left. \dfrac{dp}{dx} \right|_{x=7} = -4(7) = -28.$

Notice that the derivative is negative, which indicates a negative rate of change. In this particular case, it means that when the number of units demanded (x) is 700, the price (p) is *decreasing* at a rate of $28.

Table 2.5 summarizes the notation presented in this section and in Section 2.6 for the average rate of change of a function and the instantaneous rate of change (or derivative).

Table 2.5

Average rate of change (slope of secant)	Instantaneous rate of change (derivative or slope of tangent)
from $(a, f(a))$ to $(x, f(x))$	at the point $(a, f(a))$
$\dfrac{f(x) - f(a)}{x - a}$	$\lim\limits_{x \to a} \dfrac{f(x) - f(a)}{x - a}$
from $(x, f(x))$ to $(x + h, f(x + h))$	at the point $(x, f(x))$
$\dfrac{f(x + h) - f(x)}{(x + h) - x}$ or $\dfrac{f(x + h) - f(x)}{h}$	$\lim\limits_{h \to 0} \dfrac{f(x + h) - f(x)}{h}$

2.7 EXERCISES

SET A

In exercises 1 through 12, calculate the derivative using the definition

$$f'(x) = \lim_{h \to 0} \frac{f(x + h) - f(x)}{h}$$

(See Examples 3, 4, 6, and 7.)

1. $f(x) = 3x$
2. $f(x) = 2x + 5$
3. $y = 4 - x$
4. $y = 10x - 8$
5. $f(x) = 4 - x^2$
6. $f(x) = x^2 - 6x$
7. $f(x) = x^2 + x$
8. $y = \frac{1}{2}x^2 + 3x$
9. $f(x) = 2x^2 + 3x - 1$
10. $f(x) = 3x - 2x^2$
11. $f(x) = x^3$
12. $f(x) = x^3 + x$

In exercises 13 through 16, use the symbolic definition of the derivative to find the specified values. (See Examples 2 and 3.)

13. $f(x) = 3x^2$
    a. $f'(1)$
    b. $f'(3)$
    c. $f'(-1)$

14. $f(x) = -2x^2$
    a. $f'(0)$
    b. $f'(2)$
    c. $f'(\frac{1}{2})$

15. $y = x^2 + 3x$
    a. $\left.\dfrac{dy}{dx}\right|_{x = -1}$
    b. $\left.\dfrac{dy}{dx}\right|_{x = 2}$
    c. $\left.\dfrac{dy}{dx}\right|_{x = 3}$

16. $y = 3x - x^2$
    a. $\left.\dfrac{dy}{dx}\right|_{x = 0}$
    b. $\left.\dfrac{dy}{dx}\right|_{x = -\frac{1}{2}}$
    c. $\left.\dfrac{dy}{dx}\right|_{x = 4}$

In exercises 17 through 22, find the equation of the line tangent to the given curve at the given point. Then draw the tangent line as a check. (See Example 6.)

17. $f(x) = 4x^2$ at the point $(1, 4)$

18. $f(x) = -3x^2$ at the point $(1, -3)$

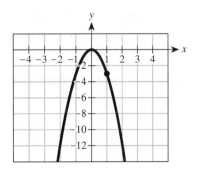

19. $f(x) = x^2 - x - 6$ at the point $(3, 0)$

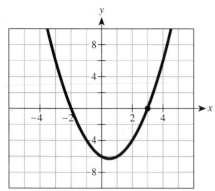

20. $f(x) = x^2 + 7x + 12$ at the point $(-1, 6)$

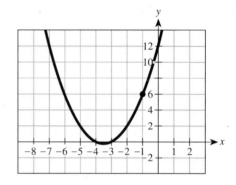

21. $f(x) = \dfrac{1}{x + 2}$ at the point $(-1, 1)$

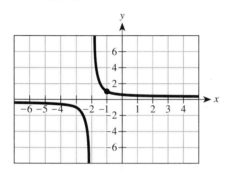

22. $f(x) = \dfrac{1}{2 - x}$ at the point $(1, 1)$

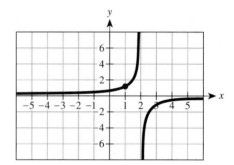

SET B

23. **a.** Use the definition of the derivative to find $f'(x)$ for $f(x) = \dfrac{1}{x}$.

    b. Why can't $f'(0)$ be evaluated?

24. If $f(x) = \sqrt{x}$, find $f'(x)$.

    $\left(\text{Hint: Rationalize the numerator in the expression} \dfrac{\sqrt{x + h} - \sqrt{x}}{h}.\right)$

25. **a.** If m and b are any constants, find y' for $y = mx + b$.

    b. Since y' represents the slope of the tangent at x, and the slope of $y = mx + b$ is m, your answer
    to part a, in effect, says that the slopes (of the original line and the tangent line at any point) will
    be equal. Verify this result by comparing it with your answers to exercises 1–4.

26. If $f(x) = 6 - 8x + x^2$, for what value(s) of x is $f'(x) = 0$?

27. **a.** Suppose that $f(x) = 24 + 2x - x^2$. On a sheet of graph paper, graph $y = f(x)$.
    Using the geometrical interpretation of $f'(x)$ as an expression that represents the slope of the tangent
    line at the point $(x, f(x))$, answer the following questions:

b. At what point(s) on the graph is $f'(x) = 0$?

c. At what point(s) on the graph is $f'(x) < 0$?

d. At what point(s) on the graph is $f'(x) > 0$?

28. a. Graph $y = |x|$.

b. What is the slope of any tangent line drawn to $y = |x|$ for $x < 0$?

c. What is the slope of any tangent line drawn to $y = |x|$ for $x > 0$?

d. $y = |x|$ is **not differentiable** at $x = 0$. Why? (*Hint:* Review the definition of derivative.)

29. a. Find the equation of the tangent line at the point $(1, 3)$ on the graph of the function $f(x) = 4 - x^2$.

b. A **normal line** to a curve at a point is the line perpendicular to the tangent line at the point (see the accompanying figure). Find the equation of the normal line for $f(x) = 4 - x^2$ at $(1, 3)$. (*Hint:* Perpendicular lines have negative reciprocal slopes.)

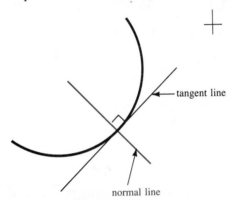

tangent line

normal line

30. Calculator Exercise

a. Consider $f(x) = 0.5x - 0.25x^2$. Use a calculator to evaluate the expressions in the following table:

x	$f(x)$	h	$f(x + h)$	$\dfrac{f(x + h) - f(x)}{h}$
3		−0.5		
3		−0.25		
3		−0.1		
3		−0.01		
3		0.001		
3		0.01		
3		0.1		
3		0.25		
3		0.5		

b. Using the table, approximate $f'(3)$.

31. From the accompanying graph of the function $y = f(x)$, use the slope of the tangent lines drawn to estimate the value of each of the following:

a. $f'(0)$ **b.** $f'(1)$ **c.** $f'(3)$ **d.** $f'(-2)$

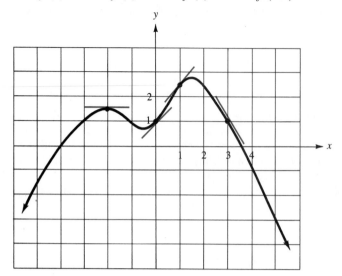

BUSINESS
PROFIT EQUATIONS

32. A company has determined that its pofit, P (in dollars), is related to the number of units produced, x, by the equation

$$P(x) = 300x - 3x^2$$

a. Find $P'(x)$.   **b.** Find $P'(40)$.

COST EQUATIONS

33. The Edgallo Corporation has determined that a function describing the cost for a particular product expressed in terms of the number of units produced is

$$C = 900 + 0.8x^2$$

where x is measured in thousands of units and C represents cost, in dollars.

a. Find $\dfrac{dC}{dx}$.

b. Find $\dfrac{dC}{dx}\bigg|_{x=5}$

c. Find C when $x = 5$.

DEMAND

34. The demand (x) for a certain product in terms of the price (p) is

$$p = 125 - \frac{x}{4}$$

Find $\dfrac{dp}{dx}$.

PROFIT

35. The profit for producing and selling x units of a product is

$$P(x) = 40x - x^2 - 300$$

Determine the number of units for which $P'(x) = 0$.

BIOLOGY
BODY FAT

36. A group of overweight males was put on an experimental diet. The group's average body fat was measured each month for eight months and is given by

$$F = 0.125t^2 - 2t + 30, \qquad 0 \le t \le 8$$

where F represents the body fat content as a percentage and t is measured in months.

  a. What was the body fat content for the group when it started?

  b. Find $\dfrac{dF}{dt}$ for $t = 3$ and interpret the result.

BOTANY

37. A certain type of tree increases in height according to the model

$$h = 5\sqrt{t}$$

where h is measured in feet and t is measured in years. Find the rate of change in the height for $t = 9$ years.

SOCIAL SCIENCE
DRUG REHABILITATION

38. A social service agency is running a drug rehabilitation program and estimates its number of clients, N, at any time, t (measured in months), to be

$$N(t) = -t^2 + 10t + 15, \qquad 0 \le t \le 12$$

  a. How many people were in the program initially?

  b. Find $N'(t)$.

  c. Find $N'(5)$.

  d. Find $N'(6)$ and interpret the result.

CHARITABLE
CONTRIBUTIONS

39. Donations to area shelters for the homeless are higher during the winter months. The following graph shows the dollar amount for the month of December for the past 10 years. Notice that the amount has been increasing. Has it been increasing at an increasing rate or at a decreasing rate? Justify your answer.

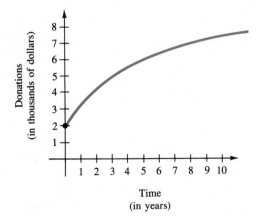

PHYSICAL
SCIENCE
VELOCITY

40. A ball is thrown upward with an intial velocity of 40 feet per second. Its position at any time t is given by

$$s = -16t^2 + 40t$$

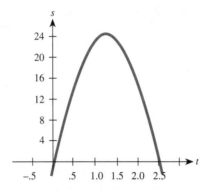

a. Find the average rate of change in its position from 0 to 2 seconds.

b. Find its velocity, $\dfrac{ds}{dt}$.

c. Find its velocity at $t = 1$.

d. For what value of t is the velocity zero?

*For exercises 41 through 44, use the following graph of $y = f(x)$ defined on $-5 \le x \le 6$. Remember: Think of the derivative, $f'(x)$, as the **rate of change** of the function at a specified point.*

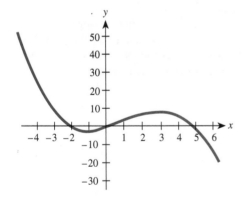

41. What is the value of $f'(x)$ at $x = -1$?

42. What can be said of $f'(x)$ for $-5 \le x < -1$?

43. What can be said of $f'(x)$ for $-1 < x < 3$?

44. Which is greater: $f'(-4)$ or $f'(-2)$?

WRITING AND CRITICAL THINKING

1. It is possible for a function to be continuous at a particular point but not be differentiable at that point. Explain why this can happen.

2. For $f(x) = \begin{cases} 1 - x, & x \leq 1 \\ x^2 - 1, & x > 1 \end{cases}$, calculate $\lim\limits_{h \to 0^+} \dfrac{f(1 + h) - f(1)}{h}$ and

   $\lim\limits_{h \to 0^-} \dfrac{f(1 + h) - f(1)}{h}$. Is the derivative of this function defined at $x = 1$? Explain why or why not.

CHAPTER 2 REVIEW

VOCABULARY/FORMULA LIST

relation
function
domain
range
composition of functions
inverse function
piecewise functions
parabola
absolute value function

greatest integer function
limit
left-hand limit
right-hand limit
indeterminate form
continuous
discontinuous
infinite limit
limit at infinity

vertical asymptote
horizontal asymptote
secant line
average rate of change
tangent line
instantaneous rate of change
first derivative
$$\lim_{h \to 0} \frac{f(x + h) - f(x)}{h}$$

CHAPTER 2 REVIEW EXERCISES

[2.1] 1. Find the domain of each of the following functions:

   a. $f(x) = \sqrt{x - 1}$

   b. $g(x) = \dfrac{1}{x + 4}$

[2.1] 2. Given $f(x) = 2x^2 - x - 1$, find:

   a. $f(0)$

   b. $f(-2)$

[2.2] 3. Sketch a graph of $y = |x|$.

In exercises 4 through 8, evaluate each limit that exists.

[2.3] 4. $\lim\limits_{x \to 6} (1 - x^2)$

[2.3] 5. $\lim\limits_{x \to 0} \dfrac{x}{x + 5}$

[2.5]   **6.** $\lim\limits_{x \to \infty} \dfrac{4x}{2x - 7}$

[2.3]   **7.** $\lim\limits_{x \to 1^+} \sqrt{x^2 - 1}$

[2.3]   **8.** $\lim\limits_{x \to 1} [\![x]\!]$

[2.4]   *In exercises 9 through 12, use the definition of continuity to determine whether or not the given function is continuous at $x = 2$.*

9. $f(x) = 2x^2 - 5x + 1$

10. $f(x) = \dfrac{10x}{x - 2}$

11. $f(x) = \begin{cases} -x - 1, & x \le 2 \\ x^2 - 7, & x > 2 \end{cases}$

12. $f(x) = \begin{cases} -x - 1, & x \le 2 \\ x^2 - 9, & x > 2 \end{cases}$

[2.5]   *In exercises 13 through 16,*

a.   *Find any vertical asymptotes.*

b.   *Find any horizontal asymptotes by calculating* $\lim\limits_{x \to -\infty} f(x)$ *and* $\lim\limits_{x \to +\infty} f(x)$.

13. $f(x) = \dfrac{x}{x - 3}$

14. $f(x) = \dfrac{1}{x - 3}$

15. $f(x) = \dfrac{x}{x^2 - 9}$

16. $f(x) = \dfrac{2x^2 - 1}{x^2 - 2}$

[2.5]   **17.**   The High-View Log Cabin Company manufactures prefab resort homes. It spends x dollars (measured in thousands) on advertising, and the income from sales $S(x)$ (in thousands of dollars) is

$$S(x) = \frac{160x^2 - 800x + 4}{x^2 - 5x + 1}$$

Determine the absolute ceiling for its sales.

[2.6]   **18.**   **a.**   Find the slope of the tangent line drawn to the curve $y = 4 - x^2$ at the point $(-1, 3)$.

b.   Find the equation of that tangent line.

[2.6]   **19.**   Find the average rate of change for $f(x) = 3x^2 + 1$ from $x = 0$ to $x = 5$.

[2.6]   **20.**   Find the instantaneous rate of change for $f(x) = 2x - x^2$ at $x = 3$.

BUSINESS   [2.6]   **21.**   The demand for a new product is given by
DEMAND

$$p = 400 - 0.3x^2$$

where x is the number of units and p is the price per unit.

a.   Find the average rate of change of the price when x is increased from 10 to 20 units.

b.   Find the instantaneous rate of change in the price for $x = 10$.

REVENUE   [2.7] **22.** The revenue, $R(x)$ (in dollars), from producing and selling x keychains is given by

$$R(x) = 3x - 0.02x^2$$

    a. Find the revenue from producing and selling 100 keychains.
    b. Find the average rate of change of $R(x)$ if the number of keychains sold is increased from 100 to 120.
    c. Find $R'(x)$.
    d. Find $R'(120)$.

COST FUNCTION   [2.1] **23.** Suppose the total cost of producing x units of a product is given as $C(x) = 12x^2 + 60x + 1200$. Graph this function for $x \geq 0$. What is the cost of producing zero units, called the **fixed cost**?

CHAPTER 2 TEST

PART ONE
Multiple choice. Put the letter of your correct response in the blank provided.

_____ **1.** Given $f(x) = 3x^2$ and $g(x) = 5 - x$, find $f[g(3)]$.

    a. 27   **b.** 12   **c.** 36   **d.** -22

_____ **2.** Evaluate: $\lim\limits_{x \to 4} (2x^2 + 1)$.

    a. 33   **b.** 65   **c.** 17   **d.** Does not exist

_____ **3.** Find the domain for the function whose equation is $y = \sqrt{2x - 1}$.

    a. $x < \frac{1}{2}$ **b.** $x \geq 1$ **c.** $x \geq 0$ **d.** $x \geq \frac{1}{2}$

_____ **4.** Evaluate: $\lim\limits_{x \to -2} \dfrac{x^2 - 4}{3x + 6}$.

    a. 0   **b.** $\dfrac{4}{3}$ **c.** $\dfrac{-4}{3}$ **d.** Does not exist

_____ **5.** Consider the function $f(x) = \begin{cases} x + 1, & x \leq 2 \\ x^2 - 1, & x > 2 \end{cases}$.

    a. $\lim\limits_{x \to 2^-} f(x) = -1$

    b. The function is discontinuous at $x = 2$.

    c. The function is continuous for all x.

    d. $\lim\limits_{x \to 2^+} f(x) = +\infty$

_____ **6.** What is the equation of the tangent line for the curve $y = \dfrac{2}{x - 1}$ at the point $(2, 2)$?

    a. $y = 4 - x$ **b.** $y = 6 - 2x$
    c. $y = 2x - 2$ **d.** $y = 2 - 2x$

_____ 7. For $f(x) = -x^2 - 2x + 1$, find $f(-1)$.

     a. 4   **b.** -2 **c.** 3   **d.** 2

_____ 8. Which of the following represents the graph of the function $f(x) = -x^2 - 1$?

    a.

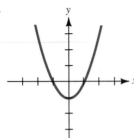

    b.

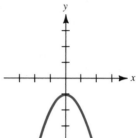

    c.

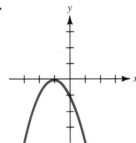

    d.

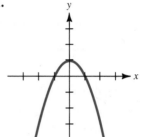

_____ 9. Find the instantaneous rate of change for $f(x) = -3x^2$ at $x = 4$.

     a. -48 **b.** -24 **c.** -36 **d.** -12

_____ 10. For $f(x) = 5x - x^2$, find $f'(x)$.

     a. $5 + 2x$ **b.** $5 - 2x - h$

     c. $5 - 2x$ **d.** $2x - 5$

PART TWO
Solve and show all work.

11. For $f(x) = 3x^2 + x - 5$, find the average rate of change of f from $x = 1$ to $x = 3$.

12. For $f(x) = x^2 + 4x$, find $\dfrac{f(x + h) - f(x)}{h}$.

13. For $f(x) = 1 - 2x^2$, find:

    a. $\dfrac{f(x + h) - f(x)}{h}$ **b.** $\lim\limits_{h \to 0} \dfrac{f(x + h) - f(x)}{h}$

14. For $f(x) = 4x^2 - x$, find the instantaneous rate of change of f for $x = 2$.

15. Find any points on the curve $y = x^2 - 3x$ where the tangent line is horizontal.

16. Sketch the graph of $y = x^2 + 4$.

BUSINESS
COST FUNCTION

17. The total cost of producing x units of an item is given by the function

$$C(x) = 500 + 3x + 0.2x^2$$

where $C(x)$ is measured in dollars. Determine the total cost of producing 150 units.

MARGINAL COST

18. The Spartan Company, which manufactures scissors, has determined that the cost of producing x pairs of scissors is given by the function

$$C(x) = 0.01x^2 + 400$$

where $C(x)$ is measured in dollars.

a. Find $C'(x)$.   **b.** Find $C'(100)$.

BIOLOGY
FISH POPULATION

19. The number of fish in a certain lake is given by the function

$$N(t) = \frac{2100t^2 + 300}{t^2 + 1}$$

where t is measured in months. Determine the limit of this function as $t \to \infty$, which represents the upper limit for the fish population.

**PHYSICAL
SCIENCE**
MOTION

20. If the position of an object at any time t is given by

$$s(t) = -16t^2 + 16t + 96$$

where t is measured in seconds and $s(t)$ is measured in feet, find the average rate of change of its position over the time interval from 10 to 20 seconds.

21. Choose five of the vocabulary words from the list given for this chapter and write a sentence using them.

3

DIFFERENTIATION AND APPLICATIONS

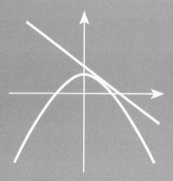

3.1 RULES OF DIFFERENTIATION

Imagine that one day you finally find a shorter route to work—one that has far less traffic and avoids three school speed zones. You are likely to start using this shorter route even though the longer route passes by some stores where you occasionally shop.

This section addresses just such a situation. We have already learned how to take the derivative of a function using the definition—the long route—and it's now time to find a shortcut. We can group our functions into certain categories, for which we develop differentiation formulas (rules) in this section and the next two sections. Keep in mind that we have not abandoned the long route (the definition) forever and will find that we may need to use it occasionally. In fact, to prove that our shortcut formulas work, we must use the long route!

The first category of functions we examine are the constant functions. From Chapter 1, we know that the graph of a constant function is a horizontal line with slope zero. Because, in the case of any straight line, the tangent line is the graph of the function itself, Rule 1 states the obvious: The slope of the tangent line of any constant function must be zero.

Rule 1
Constant Rule

> If c is any constant and $f(x) = c$, then $f'(x) = 0$. Equivalently, we can write:
>
> $y = c$ implies $\dfrac{dy}{dx} = 0$.

Proof By the definition of derivative,

$$f'(x) = \lim_{h \to 0} \frac{f(x + h) - f(x)}{h}$$

But $f(x + h) = c$ and $f(x) = c$, so

$$f'(x) = \lim_{h \to 0} \frac{c - c}{h} = \lim_{h \to 0} 0 = 0 \qquad \square$$

In Section 2.6, we found that for the parabola, $f(x) = x^2$, the derivative is $f'(x) = 2x$. For the function $f(x) = x^3$, the derivative is $f'(x) = 3x^2$. The next rule gives a shorter method for calculating the derivative for these functions or any function of the form $f(x) = x^n$, where n is any real number.

Rule 2
Power Rule

> If $f(x) = x^n$, then $f'(x) = nx^{n-1}$ for any real number n. This rule is commonly referred to as the **Power Rule**. Equivalently, we can write: $y = x^n$ implies
>
> $\dfrac{dy}{dx} = nx^{n-1}$.

To use this rule, it is necessary to have the function written in exponential form; for example,

$$f(x) = \frac{1}{x^2} \quad \text{becomes} \quad f(x) = x^{-2}$$

$$f(x) = \sqrt[3]{x} \quad \text{becomes} \quad f(x) = x^{1/3}$$

Some examples follow.

Example 1 If $f(x) = x^6$, find $f'(x)$.

Solution The Power Rule applied to $f(x)$ yields:

$$f'(x) = 6x^{6-1} = 6x^5$$ ∎

Example 2 If $g(x) = x^{-3}$, find $g'(x)$.

Solution $g'(x) = -3x^{-3-1} = -3x^{-4}$ ∎

Example 3 If $y = \frac{1}{\sqrt[3]{x^2}}$, find $\frac{dy}{dx}$.

Solution First, we rewrite the function in the form $y = x^n$:

$$y = \frac{1}{\sqrt[3]{x^2}} = x^{-2/3}$$

Then

$$\frac{dy}{dx} = -\frac{2}{3}x^{-5/3}$$

or equivalently,

$$\frac{dy}{dx} = \frac{-2}{3\sqrt[3]{x^5}}$$ ∎

Rule 3 gives us a way of differentiating a constant c times a function $g(x)$. The resulting derivative is c times $g'(x)$.

Rule 3	If $f(x) = c \cdot g(x)$, where c is any real number and $g(x)$ is differentiable, then
Constant Multiple Rule	$f'(x) = c \cdot g'(x)$.

Proof Again, by the definition of derivative, we have

$$f'(x) = \lim_{h \to 0} \frac{f(x + h) - f(x)}{h}$$

$$= \lim_{h \to 0} \frac{c \cdot g(x + h) - c \cdot g(x)}{h}$$

By a property of limits from Chapter 2, we can rewrite this as

$$f'(x) = \lim_{h \to 0} \frac{c[g(x + h) - g(x)]}{h} = c \cdot \lim_{h \to 0} \frac{g(x + h) - g(x)}{h}$$

$$= c \cdot g'(x)$$ □

Example 4   If $y = 7x^3$, find y'.

Solution   With $c = 7$ and $g(x) = x^3$ in Rule 3, we have

$$y' = 7 \cdot g'(x) = 7(3x^2) = 21x^2$$ ■

Example 5   If $f(x) = \dfrac{10}{\sqrt{x}}$, find $f'(9)$.

Solution
$$f(x) = 10x^{-1/2}$$
$$f'(x) = 10\left(-\frac{1}{2}x^{-3/2}\right)$$
$$f'(x) = -5x^{-3/2}$$
$$f'(9) = -5(9)^{-3/2} = -\frac{5}{27}$$ ■

The last rule we present in this section gives us the capability of differentiating a function that consists of more than one term.

Rule 4

Sum or

Difference Rule

If $f(x) = u(x) \pm v(x)$, where $u(x)$ and $v(x)$ are both differentiable, then $f'(x) = u'(x) \pm v'(x)$.

In other words, Rule 4 states that "the derivative of a sum (or difference) of functions is the sum (or difference) of the derivatives." The verification of Rule 4 is presented following some examples using this rule.

Example 6   If $f(x) = x^2 + 7x$, find $f'(x)$.

Solution   Think of $u(x)$ as x^2 and $v(x)$ as $7x$. Then $u'(x) = 2x$ and $v'(x) = 7$, so that applying Rule 4 yields

$$f'(x) = u'(x) + v'(x)$$
$$= 2x + 7$$

∎

Rule 4 can be extended to any finite sum of differentiable functions, as Example 7 illustrates.

Example 7   If $y = 1000 + 8x - \dfrac{150}{x^2}$, find $\dfrac{dy}{dx}$.

Solution $y = 1000 + 8x - 150x^{-2}$

$$\frac{dy}{dx} = 0 + 8 - 150(-2x^{-3})$$

$$= 8 + \frac{300}{x^3}$$

∎

Example 8   The monthly profit from the sale of x stereos per month for a certain company is given by

$$P(x) = 800 - x^2 + 0.2x^3$$

where $P(x)$ is measured in dollars. Determine the rate of change in the monthly profit for 100 stereos.

Solution   The rate of change in the monthly profit is

$$P'(x) = -2x + 0.6x^2$$

For $x = 100$ stereos, we have

$$P'(100) = -2(100) + 0.6(100)^2$$
$$= \$5800$$

∎

VERIFICATION OF THE SUM RULE

For $f(x) = u(x) + v(x)$, we have, by the definition of derivative,

$$f'(x) = \lim_{h \to 0} \frac{f(x + h) - f(x)}{h}$$

$$= \lim_{h \to 0} \frac{[u(x + h) + v(x + h)] - [u(x) + v(x)]}{h}$$

$$= \lim_{h \to 0} \frac{[u(x + h) - u(x)] + [v(x + h) - v(x)]}{h}$$

$$= \lim_{h \to 0} \left[\frac{u(x + h) - u(x)}{h} + \frac{v(x + h) - v(x)}{h} \right]$$

By a property of limits from Chapter 2, we can rewrite this as

$$f'(x) = \lim_{h \to 0} \frac{u(x + h) - u(x)}{h} + \lim_{h \to 0} \frac{v(x + h) - v(x)}{h}$$
$$= u'(x) + v'(x)$$

3.1   EXERCISES

SET A

In exercises 1 through 29, differentiate each function. (See Examples 1–4, 6, and 7.)

1. $f(x) = x^2$

2. $f(x) = x^4$

3. $f(x) = 7$

4. $f(x) = 8$

5. $y = x$

6. $g(x) = 8x^2$

7. $h(x) = -6x$

8. $f(x) = -x^{-2}$

9. $s(t) = 3t^2 + 9$

10. $R(x) = 20x - 7x^2$

11. $f(x) = 3x^2 + 9x - 6$

12. $g(x) = \sqrt{x}$

13. $y = 3\sqrt{x}$

14. $y = \sqrt{3x}$
(*Hint:* $y = \sqrt{3} \cdot \sqrt{x}$.)

15. $y = \sqrt{x} + 3$

16. $y = \sqrt{x} - 3$

17. $f(x) = \sqrt[3]{x}$

18. $f(x) = 19x^2 - 5x + \dfrac{1}{x}$

19. $y = \dfrac{x + 5x^2}{7}$

20. $g(x) = \dfrac{x + 5x^2}{x}$
(*Hint:* Write as a sum.)

21. $y = \dfrac{1}{x^2} + \dfrac{1}{x^3}$

22. $y = \dfrac{2}{x^2} + \dfrac{3}{x^3}$

23. $y = \dfrac{x^2}{2} + \dfrac{x^3}{3}$

24. $f(x) = 2x^{-3/2}$

25. $f(x) = 2x^{-3/2} - 4x^{1/2}$

26. $f(x) = 2x^{-3/2} - 4x^{1/2} + x^{-1/2}$

27. $f(x) = (2x - 7)^2$
(*Hint:* Expand.)

28. $h(x) = 3x^4 - 7x^2 + x - 11$

29. $y = \sqrt[3]{x^2} - \dfrac{5}{\sqrt{x}}$

In exercises 30 through 40, find the derivative for the indicated value. (See Example 5.)

30. $f(x) = 3x^2, f'(1)$

31. $f(x) = -3x^2, f'(1)$

32. $f(x) = x^3 - x + 4, f'(2)$

33. $f(x) = 2x^3 + 4x^2, f'(-1)$

34. $f(x) = x^2 + \dfrac{1}{x^2}, f'(1)$

35. $f(x) = x + \dfrac{2}{x}, f'(-1)$

36. $f(x) = 2\sqrt[3]{x^2}, f'(8)$

37. $f(x) = \dfrac{1}{3\sqrt{x}}, f'(4)$

38. a. $f(x) = \dfrac{x+1}{x^2}, f'(-2)$

     b. Verify your answer to part a numerically by finding $\displaystyle\lim_{x \to -2} \dfrac{f(x) - f(-2)}{x + 2}$.

39. a. $f(x) = \dfrac{x^2 + 2}{2x^4}, f'(-1)$

     b. Verify your answer to part a numerically by finding $\displaystyle\lim_{x \to -1} \dfrac{f(x) - f(-1)}{x + 1}$.

40. a. $f(x) = \dfrac{-2}{x\sqrt{x}}, f'(4)$

     b. Verify your answer to part a numerically by finding $\displaystyle\lim_{x \to 4} \dfrac{f(x) - f(4)}{x - 4}$.

In exercises 41 through 44, find the equation for the tangent line pictured for the given function.

41. $f(x) = x^2$

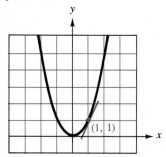

42. $f(x) = x^3$

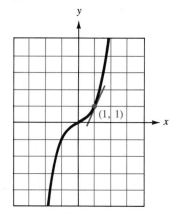

43. $f(x) = 1 - x^3$

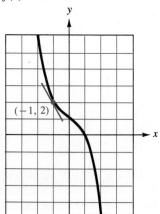

44. $f(x) = \sqrt{x}$

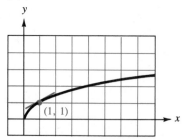

SET B

45. a.   Suppose $g(x) = 4x^2 + 4x + 1$. Find $g'(x)$.
    b.   Find $g(0)$.
    c.   Find $g'(0)$.
    d.   Find the value(s) of x such that $g'(x) = 0$.
    e.   Using the value(s) of x found in part d, find $g(x)$.

46. a.   Suppose that $f(x) = \frac{1}{3}x^3 - x^2 - 24x + 20$. Find $f'(x)$.
    b.   Find $f(0)$.
    c.   Find $f'(0)$.
    d.   Find the value(s) of x such that $f'(x) = 0$.
    e.   Using the value(s) of x found in part d, find $f(x)$.

47. a.   Refer to exercise 45. Graph $y = g(x)$.
    b.   What is the equation of the tangent line to the curve at the point $(-\frac{1}{2}, 0)$?

48. **Calculator Exercise**
    Consider the function $f(x) = \sqrt{x}$, which is defined only for $x \geq 0$. With $(0, 0)$ as a fixed point, complete the following table for the various chosen points $(x, f(x))$. You will have to approximate the radicals as decimals on your calculator.

Variable point $(x, f(x))$	Fixed point $(a, f(a))$	Slope of the line $\dfrac{f(x) - f(a)}{x - a}$
$(2, \sqrt{2})$	$(0, 0)$	
$(1, \sqrt{1})$	$(0, 0)$	
$(0.5, \sqrt{0.5})$	$(0, 0)$	
$(0.1, \sqrt{0.1})$	$(0, 0)$	
$(0.01, \sqrt{0.01})$	$(0, 0)$	
$(0.001, \sqrt{0.001})$	$(0, 0)$	
$(0.0001, \sqrt{0.0001})$	$(0, 0)$	
$(0.00001, \sqrt{0.00001})$	$(0, 0)$	

    a.   Sketch the graph of $y = \sqrt{x}$.
    b.   What can be concluded from the table about the derivative at $(0, 0)$?

49. **Calculator Exercise**
    For the function $f(x) = x^{2/3}$, find $f'(x)$. Complete the following table of values:

x	-1	-0.1	-0.01	-0.001	0.001	0.01	0.1	1
$f(x)$								
$f'(x)$								

    a.   Sketch the function.
    b.   What can be concluded about $f'(0)$ from the table?
    c.   Write the equation of the line tangent to the curve $y = x^{2/3}$ at the point $(0, 0)$.

BUSINESS
COST FUNCTIONS

50. The unit cost for a commodity is given by $C(x) = \dfrac{10,000}{x} + x + 20$, where x is the number of units. Find $C'(x)$ for this function. For what value of x is $C'(x) = 0$?

COST FUNCTIONS

51. The Aquarius Manufacturing Company has determined that the cost function for producing a particular type of cement block is given by

$$C(x) = 80 + 600x^{2/3}$$

where x is measured in number of units and $C(x)$ is in dollars.
  a. Find $C'(x)$.
  b. Determine the rate of change in the cost for $x = 1000$.

PROFIT FUNCTIONS

52. The monthly profit from the sale of x compact disc players is given by

$$P(x) = 1200 - 0.02x^2 + 0.5x^3$$

where $P(x)$ is measured in dollars. Determine the rate of change in the monthly profit for 20 stereos.

REVENUE

53. The revenue for a certain product is given by

$$R(x) = 20x - 0.03x^2$$

Find the rate of change in revenue for 100 units.

DEMAND

54. The price per item of a new product is given by $p = \dfrac{50}{x}$. Find $\dfrac{dp}{dx}$.

BIOLOGY
MEDICINE

55. The concentration of a drug in the bloodstream is given by

$$C = 400 - t^2$$

where t is measured in minutes. Find the rate of change in the concentration at 1 hour.

MIGRATION

56. The migration of a certain type of warbler has been followed over a ten-year period. Scientists are afraid that it is destined for extinction. If the population is estimated by

$$P = t^3 - 30t^2 + 200t + 4040$$

where t is measured in years, find the rate of change in the population for $t = 15$.

PHYSICAL SCIENCE
VELOCITY

57. If the position of an object, s, at any time, t, is given by

$$s(t) = -16t^2 + 128t + 100$$

where s is measured in feet and t is measured in seconds,
  a. Find the velocity, $v(t) = s'(t)$.
  b. Find $v(0)$, the initial velocity.
  c. Find the velocity at 3 seconds.

WRITING AND CRITICAL THINKING

1. For both the functions $f(x) = 5$ and $f(x) = -2$, the derivative is $f'(x) = 0$. Sketch the graph of both functions. Using your sketch, explain why the derivative is zero for both functions.

2. For both the functions $f(x) = x^2 + 1$ and $f(x) = x^2 - 4$, the derivative is $f'(x) = 2x$. In particular, $f'(1) = 2$. Sketch the graph of both functions on the same set of axes. Draw a tangent line to both curves at $x = 1$. What do you notice about the tangent lines? Explain.

3. Write out, in complete sentences, Rules 1 through 4 for finding derivatives.

3.2   PRODUCT AND QUOTIENT RULES

In this section, we continue with our "shortcut" rules for differentiation. You may find the rules presented here to be slightly more complicated, but that stems from the fact that the functions with which we will be dealing are a little more involved than those of Section 3.1.

We are interested in establishing rules for differentiating a product of two (or more) functions and the quotient of two functions. The natural inclination at this point may be to assume that the outcome of differentiating a product is similar to the outcome of differentiating a sum. The derivative of a sum is the sum of the derivatives, but the derivative of a product is *not* the product of the derivatives, as we see in Rule 5.

Rule 5
Product Rule

If $f(x) = u(x)v(x)$, where u and v are differentiable, then

$$f'(x) = u(x)v'(x) + u'(x)v(x)$$

This rule is called the **Product Rule**.

The verification of the Product Rule is given at the end of this section.

This rule states that for the product of two functions, the derivative is the first function times the derivative of the second function plus the second function times the derivative of the first function.

Example 1   If $f(x) = (x^2 + x)(2x^3 - 3)$, find $f'(x)$.

Solution   Applying the Product Rule, we let

$$u(x) = x^2 + x \qquad \text{and} \qquad v(x) = 2x^3 - 3$$

Then $\qquad u'(x) = 2x + 1 \qquad$ and $\qquad v'(x) = 6x^2$

$$\begin{aligned}
f'(x) &= u(x)v'(x) + u'(x)v(x) \\
&= (x^2 + x)(6x^2) + (2x + 1)(2x^3 - 3) \\
&= 6x^4 + 6x^3 + 4x^4 - 6x + 2x^3 - 3 \\
&= 10x^4 + 8x^3 - 6x - 3
\end{aligned}$$ ∎

For the function $f(x) = (x^2 + x)(2x^3 - 3)$ of Example 1, we could have used an alternate technique of multiplying the factors and then differentiating. When the functions get more involved, however, as in Section 3.3, the Product Rule is virtually indispensable.

Example 2   If $f(x) = \dfrac{1}{4x}\left(x^2 + \dfrac{1}{x^2}\right)$, find $f'(x)$.

Solution   Let

$$u(x) = \frac{1}{4x} \qquad \text{and} \qquad v(x) = x^2 + \frac{1}{x^2}$$

First, we rewrite $u(x)$ and $v(x)$ in the more convenient forms:

$$u(x) = \frac{1}{4}x^{-1} \qquad\qquad v(x) = x^2 + x^{-2}$$

Then $\qquad u'(x) = -\dfrac{1}{4}x^{-2} \qquad v'(x) = 2x - 2x^{-3}$

or $\qquad u'(x) = -\dfrac{1}{4x^2} \qquad v'(x) = 2x - \dfrac{2}{x^3}$

So

$$\begin{aligned}
f'(x) &= u(x)v'(x) + u'(x)v(x) \\
&= \frac{1}{4x}\left(2x - \frac{2}{x^3}\right) + \frac{-1}{4x^2}\left(x^2 + \frac{1}{x^2}\right) \\
&= \frac{1}{2} - \frac{1}{2x^4} - \frac{1}{4} - \frac{1}{4x^4} \\
&= \frac{1}{4} - \frac{3}{4x^4} \qquad \text{or} \qquad \frac{x^4 - 3}{4x^4}
\end{aligned}$$ ∎

Example 3   If $f(x) = \dfrac{1}{4x}\left(x^2 + \dfrac{1}{x^2}\right)$, find $f'(2)$.

Solution   From Example 2, we have

$$f'(x) = \frac{x^4 - 3}{4x^4}$$

$$\text{So} \quad f'(2) = \frac{2^4 - 3}{4(2)^4}$$

$$= \frac{16 - 3}{64}$$

$$= \frac{13}{64} \qquad \blacksquare$$

Now we proceed to the rule for differentiating the quotient of two functions.

Rule 6
Quotient Rule

If $f(x) = \dfrac{u(x)}{v(x)}$, where u and v are differentiable, then

$$f'(x) = \frac{v(x)u'(x) - u(x)v'(x)}{[v(x)]^2}$$

provided $v(x) \neq 0$. This is called the **Quotient Rule.**

The Quotient Rule states that for the quotient of two functions, the derivative is the denominator times the derivative of the numerator minus the numerator times the derivative of the denominator divided by the denominator squared. See Example 6 for the use of an alternate notation for the Quotient Rule. The verification of the Quotient Rule is found at the end of this section.

Example 4   If $g(t) = \dfrac{t^3}{1 - t^4}$, find $g'(t)$.

Solution   Let

$$u(t) = t^3 \qquad \text{and} \qquad v(t) = 1 - t^4$$
$$\text{Then} \quad u'(t) = 3t^2 \qquad \text{and} \qquad v'(t) = -4t^3$$

By the Quotient Rule,

$$g'(t) = \frac{v(t)u'(t) - u(t)v'(t)}{[v(t)]^2}$$

$$= \frac{(1 - t^4)(3t^2) - (t^3)(-4t^3)}{(1 - t^4)^2}$$

$$= \frac{3t^2 - 3t^6 + 4t^6}{(1 - t^4)^2}$$

$$= \frac{3t^2 + t^6}{(1 - t^4)^2} = \frac{t^2(3 + t^4)}{(1 - t^4)^2} \qquad \blacksquare$$

Example 5   For $f(x) = 3x^2(1 - x^4)$, find $f'(x)$.

Solution   Let $u(x) = 3x^2$ and $v(x) = 1 - x^4$. An alternate notation that can be used for the Product Rule is

$$f'(x) = u(x)\frac{d[v(x)]}{dx} + v(x)\frac{d[u(x)]}{dx}$$

For our function we have

$$f'(x) = 3x^2\frac{d(1 - x^4)}{dx} + (1 - x^4)\frac{d(3x^2)}{dx}$$
$$= 3x^2(-4x^3) + (1 - x^4)(6x)$$
$$= 6x - 18x^5 \qquad\blacksquare$$

Example 6   For $f(x) = \dfrac{2x + 3}{x^2 + 1}$, find $f'(x)$.

Solution   The Quotient Rule can be rewritten as follows:

$$f'(x) = \frac{v(x)\dfrac{d[u(x)]}{dx} - u(x)\dfrac{d[v(x)]}{dx}}{[v(x)]^2}$$

So, for our function, with $u(x) = 2x + 3$ and $v(x) = x^2 + 1$, we have

$$f'(x) = \frac{(x^2 + 1)\dfrac{d(2x + 3)}{dx} - (2x + 3)\dfrac{d(x^2 + 1)}{dx}}{(x^2 + 1)^2}$$
$$= \frac{(x^2 + 1)(2) - (2x + 3)(2x)}{(x^2 + 1)^2}$$
$$= \frac{2 - 6x - 2x^2}{(x^2 + 1)^2} \qquad\blacksquare$$

Example 7   The Chateau Vin Winery has found that the number of bottles sold, x, for the new wine it is producing can be expressed in terms of the price, p, in dollars, by the following function:

$$x = \frac{500(2 - p)}{p^2}$$

a.   Find $\dfrac{dx}{dp}$, the rate of change of the number of bottles sold with respect to price.

b.   Find $\dfrac{dx}{dp}\bigg|_{p = 5}$, the rate of change when the price is $5.

Solution **a.** $\dfrac{dx}{dp} = \dfrac{p^2 \dfrac{d[500(2-p)]}{dp} - 500(2-p)\dfrac{d[p^2]}{dp}}{(p^2)^2}$

$= \dfrac{p^2(-500) - 500(2-p)(2p)}{p^4}$

$= \dfrac{-500p^2 - 2000p + 1000p^2}{p^4}$

$= \dfrac{500(p-4)}{p^3}$

b. $\left.\dfrac{dx}{dp}\right|_{p=5} = \dfrac{500(5-4)}{5^3}$

$= \dfrac{500}{125} = 4$ ∎

Table 3.1 summarizes all of the differentiation rules given thus far and includes both forms of notation.

Table 3.1

SUMMARY OF DIFFERENTIATION RULES

Function	Derivative	Derivative (alternate notation)
$f(x) = c$	$f'(x) = 0$	$\dfrac{d(c)}{dx} = 0$
$f(x) = cg(x)$	$f'(x) = cg'(x)$	$\dfrac{d[cg(x)]}{dx} = c\dfrac{d[g(x)]}{dx}$
$f(x) = x^n$	$f'(x) = nx^{n-1}$	$\dfrac{d(x^n)}{dx} = nx^{n-1}$
$f(x) = u(x) \pm v(x)$	$f'(x) = u'(x) \pm v'(x)$	$\dfrac{d[u(x) \pm v(x)]}{dx} = \dfrac{d[u(x)]}{dx} \pm \dfrac{d[v(x)]}{dx}$
$f(x) = u(x)v(x)$	$f'(x) = u(x)v'(x) + v(x)u'(x)$	$\dfrac{d[u(x)v(x)]}{dx} = u(x)\dfrac{d[v(x)]}{dx} + v(x)\dfrac{d[u(x)]}{dx}$
$f(x) = \dfrac{u(x)}{v(x)}$	$f'(x) = \dfrac{v(x)u'(x) - u(x)v'(x)}{[v(x)]^2}$	$\dfrac{d\left[\dfrac{u(x)}{v(x)}\right]}{dx} = \dfrac{v(x)\dfrac{d[u(x)]}{dx} - u(x)\dfrac{d[v(x)]}{dx}}{[v(x)]^2}$

VERIFICATION OF THE PRODUCT RULE

Let $f(x) = u(x)v(x)$, where $u(x)$ and $v(x)$ are differentiable functions of x. From the definition of derivative,

$$f'(x) = \lim_{h \to 0} \frac{f(x+h) - f(x)}{h}$$

$$= \lim_{h \to 0} \frac{u(x+h)v(x+h) - u(x)v(x)}{h}$$

Add and subtract $u(x + h)v(x)$ in the numerator.

$$f'(x) = \lim_{h \to 0} \frac{u(x + h)v(x + h) - u(x + h)v(x) + u(x + h)v(x) - u(x)v(x)}{h}$$

$$= \lim_{h \to 0} \left[\frac{u(x + h)v(x + h) - u(x + h)v(x)}{h} + \frac{u(x + h)v(x) - u(x)v(x)}{h} \right]$$

Using a limit property from Chapter 2, we have

$$f'(x) = \lim_{h \to 0} \frac{u(x + h)v(x + h) - u(x + h)v(x)}{h} + \lim_{h \to 0} \frac{u(x + h)v(x) - u(x)v(x)}{h}$$

Factoring the numerators gives

$$f'(x) = \lim_{h \to 0} \frac{u(x + h)[v(x + h) - v(x)]}{h} + \lim_{h \to 0} \frac{v(x)[u(x + h) - u(x)]}{h}$$

Using another limit property gives

$$f'(x) = \left[\lim_{h \to 0} u(x + h) \right]\left[\lim_{h \to 0} \frac{v(x + h) - v(x)}{h} \right]$$

$$+ \left[\lim_{h \to 0} v(x) \right]\left[\lim_{h \to 0} \frac{u(x + h) - u(x)}{h} \right]$$

Evaluating these limits, we have

$$f'(x) = u(x)v'(x) + v(x)u'(x)$$

VERIFICATION OF THE QUOTIENT RULE

Let $f(x) = \dfrac{u(x)}{v(x)}$, where $u(x)$ and $v(x)$ are differentiable functions of x. From the definition of derivative,

$$f'(x) = \lim_{h \to 0} \frac{f(x + h) - f(x)}{h}$$

$$= \lim_{h \to 0} \frac{\dfrac{u(x + h)}{v(x + h)} - \dfrac{u(x)}{v(x)}}{h}$$

$$= \lim_{h \to 0} \frac{u(x + h)v(x) - v(x + h)u(x)}{h[v(x + h)v(x)]}$$

Add and subtract $u(x)v(x)$ in the numerator.

$$f'(x) = \lim_{h \to 0} \frac{u(x + h)v(x) - u(x)v(x) - v(x + h)u(x) + u(x)v(x)}{h[v(x + h)v(x)]}$$

$$= \lim_{h \to 0} \frac{v(x)[u(x + h) - u(x)] - u(x)[v(x + h) - v(x)]}{h[v(x + h)v(x)]}$$

Divide the numerator and denominator by h.

$$f'(x) = \lim_{h \to 0} \frac{v(x)\left[\dfrac{u(x+h) - u(x)}{h}\right] - u(x)\left[\dfrac{v(x+h) - v(x)}{h}\right]}{v(x+h)v(x)}$$

Using several limit properties from Chapter 2, we have

$$f'(x) = \frac{\left[\lim_{h \to 0} v(x)\right]\left[\lim_{h \to 0} \dfrac{u(x+h) - u(x)}{h}\right] - \left[\lim_{h \to 0} u(x)\right]\left[\lim_{h \to 0} \dfrac{v(x+h) - v(x)}{h}\right]}{\lim_{h \to 0} v(x+h)v(x)}$$

Evaluating these limits gives us

$$f'(x) = \frac{v(x)u'(x) - u(x)v'(x)}{[v(x)]^2}$$

3.2 EXERCISES

SET A

In exercises 1 through 19, find the derivative for each function using either the Product Rule or the Quotient Rule. (See Examples 1, 2, and 4.)

1. $f(x) = x^2(3x^2 - 1)$

2. $y = 6x^4(x^3 + 2x)$

3. $f(x) = (1 - x^2)(1 + x^2)$

4. $g(x) = (3x^4 - 1)(5x^2 - 6x + 11)$

5. $g(x) = \dfrac{1}{x^3}(x^2 - x)$

6. $y = \left(\dfrac{2}{x} + 1\right)\left(\dfrac{1}{x^2} - 3\right)$

7. $s = (t^2 + 1)(2t^2 - t + 3)$

8. $y = (1 - \sqrt{x})(2 + \sqrt{x})$

9. $f(x) = \dfrac{2}{x^3}$

10. $y = \dfrac{x}{x + 1}$

11. $y = \dfrac{2x}{x + 1}$

12. $f(x) = \dfrac{x - 3}{2x + 5}$

13. $g(x) = \dfrac{5x + 1}{x^2 - 1}$

14. $f(x) = \dfrac{x^2}{1 - x^2}$

15. $y = \dfrac{x^2}{x^3 + 4}$

16. $f(x) = \dfrac{x^2 - x + 1}{x^2 + 2x - 3}$

17. $g(x) = \dfrac{\sqrt{x}}{x + 1}$

18. $f(x) = \dfrac{x + 1}{\sqrt{x}}$

19. $f(x) = \left(\dfrac{x}{x + 1}\right)^2$ (*Hint:* At present, there is no rule to find $f'(x)$ for $f(x)$ in this form. Expand $\left(\dfrac{x}{x + 1}\right)^2$ first and then differentiate.)

In exercises 20 through 22, find the indicated value. (See Examples 3 and 5.)

20. If $f(x) = \dfrac{1}{2x}(x^3 - 3x)$, find $f'(1)$.

21. If $g(x) = \dfrac{x+1}{x-1}$, find $g'(2)$.

22. If $f(x) = \dfrac{1}{1 - \sqrt{x}}$, find $f'(9)$.

SET B

BUSINESS
COST

23. The cost of producing x units of a product is

$$C(x) = 0.2x^2(x + 3000)$$

   a. Find $C'(x)$.
   b. Find $C'(100)$.

PROFIT

24. The monthly profit from producing and selling x microwave ovens (measured in hundreds) is

$$P(x) = \frac{50x}{3x^2 + 1}$$

where $P(x)$ is measured in thousands of dollars. Find the rate of change of the profit for 100 microwave ovens.

CONSUMER DEMAND

25. The price, p (in dollars), for a new toy is dependent on the consumer demand, x (in hundreds), and is

$$p = \frac{20}{2x + 1}$$

Find the rate at which the price is changing with respect to the demand.

PROFIT FUNCTION

26. The Lightner Company has found that when it produces $x \geq 1$ fixtures, its profit is given by the function

$$P(x) = 0.003(x + 2)\left(x - \frac{1}{x}\right)$$

Find $P'(x)$.

REVENUE FUNCTION

27. The revenue for a certain product at the Boxrite Company is given by the equation

$$R(x) = \frac{8x}{4 + x^2}$$

   a. Find $R'(x)$.
   b. Find $R'(2)$.

BIOLOGY
MUSCLE EFFICIENCY

28. It has been determined that, under certain conditions, the efficiency, E, of a muscle performing a contraction in terms of the time of contraction, t (in seconds), is

$$E = \frac{4 - t^2}{8 + 7t}$$

Find the rate of change of efficiency with respect to time when $t = 1$.

POLLUTION CONTROL

29. An environmentalist takes samples of river water from various distances, x, measured in miles, from a river that passes a sanitary landfill. She has determined that the concentration, in parts per million, $C(x)$, of a particular pollutant is

$$C(x) = \frac{500 + x}{x^3 + x + 1}$$

Find $C'(x)$.

PHYSICAL SCIENCE
MOTION

30. An object is moving and travels a distance, $s(t)$, in t seconds according to

$$s(t) = 1 + \frac{0.5t^3 + 4t^2 - 0.1}{t + 1}$$

Assume $s(t)$ is measured in meters and $0 \le t \le 10$.
a. Find $s'(t)$.
b. Find $s'(2)$.

WRITING AND CRITICAL THINKING

1. Write out, in complete sentences, the rule for finding the derivative of a product. Do the same for the Quotient Rule.

2. For $f(x) = \dfrac{x^3}{x^2 + 1}$, explain what is wrong with the calculation for the derivative:

$$f'(x) = \frac{3x^2}{2x}$$

$$f'(x) = \frac{3}{2}x$$

USING TECHNOLOGY IN CALCULUS

We can use technology to check our work when finding complicated derivatives. One way to do this on a graphics calculator is to graph the answer as well as the nDeriv, the numerical derivative of a function. If the two graphs coincide, we can have confidence in our answer.

Notice below that we enter the original function (from Example 6) as Y_1 but do not "select it" in the $\boxed{Y=}$ list. Y_2 is the numerical calculation of the derivative at each pixel in the graphing window, and Y_3 is the first derivative we are checking.

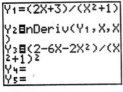

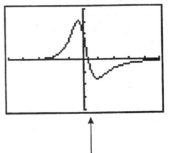

The graphs of Y_2 and Y_3 coincide.

1. Use this approach to test if $-\dfrac{1}{x+1}$ is the first derivative of $y = \dfrac{x}{x+1}$.

2. If your calculator is not equipped with an nDeriv function, you can use the difference quotient (Y_4 in the following graphic) for the test. The storage location H should hold a small value (recall, H approaches 0 in the definition of the first derivative).

```
Y1=(2X+3)/(X²+1)
Y2=nDeriv(Y1,X,X
)
Y3◻(2-6X-2X²)/(X
²+1)²
Y4◻(Y1(X+H)-Y1(X
))/H
```

Use this method to determine whether or not $\dfrac{-2}{(1+x)^2}$ is the derivative of $y = \dfrac{1-x}{1+x}$.

3.3 THE CHAIN RULE

Thus far, we have only needed to examine the derivative for one function at a time. We would now like to explore a situation where we have two (or more) functions that are related to or dependent on each other and examine the effect this relationship has on their respective derivatives.

Suppose the price, p, a company charges for a particular plastic product is a function of the material costs, M, and the material costs are a function of the price of petroleum, r. Symbolically we can write

$$p = f(M) \quad \text{and} \quad M = g(r)$$

Looking at their respective derivatives, the rate of change in price with respect to material costs is $\dfrac{dp}{dM}$, and the rate of change in material costs with respect to the price of petroleum is given by $\dfrac{dM}{dr}$.

The question we now want to answer is how we can compute the rate of change in the price of the product with respect to the price of petroleum, $\dfrac{dp}{dr}$. This derivative is related to the derivatives $\dfrac{dp}{dM}$ and $\dfrac{dM}{dr}$ and can be computed in the following way:

$$\frac{dp}{dr} = \frac{dp}{dM} \cdot \frac{dM}{dr}$$

This is an example of what is called the Chain Rule for derivatives.

Rule 7
Chain Rule

For the functions $y = f(u)$ and $u = g(x)$,

$$\frac{dy}{dx} = \frac{dy}{du} \cdot \frac{du}{dx}$$

provided f is differentiable at u, and u is differentiable at x.

Example 1   If $y = u^8$ and $u = x^3 + 1$, find $\dfrac{dy}{dx}$.

Solution   We first need to calculate $\dfrac{dy}{du}$ and $\dfrac{du}{dx}$.

$$\frac{dy}{du} = 8u^7 \quad \text{and} \quad \frac{du}{dx} = 3x^2$$

Now, applying the Chain Rule, we have

$$\frac{dy}{dx} = \frac{dy}{du} \cdot \frac{du}{dx}$$
$$= 8u^7 \cdot 3x^2$$
$$= 24u^7 x^2$$

If we want the derivative to be expressed explicitly in terms of x, we can substitute $x^3 + 1$ for u. This gives

$$\frac{dy}{dx} = 24x^2(x^3 + 1)^7$$

■

Example 2   If $y = \dfrac{1}{u}$ and $u = \sqrt{x}$, find $\dfrac{dy}{dx}$.

Solution $y = u^{-1}$ and $u = x^{1/2}$

So $\dfrac{dy}{du} = -u^{-2}$ and $\dfrac{du}{dx} = \dfrac{1}{2}x^{-1/2}$

$\qquad\qquad = \dfrac{-1}{u^2}$ $\qquad\qquad = \dfrac{1}{2\sqrt{x}}$

$\dfrac{dy}{dx} = \dfrac{dy}{du} \cdot \dfrac{du}{dx}$

$\qquad = \dfrac{-1}{u^2} \cdot \dfrac{1}{2\sqrt{x}}$

$\qquad = \dfrac{-1}{2u^2\sqrt{x}}$ Substitute $\sqrt{x}$ for u.

$\qquad = \dfrac{-1}{2x\sqrt{x}}$

■

Example 3   The demand for a particular type of fruit is dependent on the price per bushel, p, and is given by

$$x = 20 - 4p$$

where x is measured in hundreds of bushels. It has also been found that the price per bushel charged by the growers is dependent on the crop yield, n, measured in bushels. This relationship can be expressed by

$$p = 500 - 40n^2$$

Find the rate of change of the demand with respect to the yield.

Solution   We need to find $\dfrac{dx}{dn}$. We have

$$\frac{dx}{dp} = -4 \qquad \text{and} \qquad \frac{dp}{dn} = -80n$$

Using the Chain Rule, since $\dfrac{dx}{dn} = \dfrac{dx}{dp} \cdot \dfrac{dp}{dn}$, we get

$$\frac{dx}{dn} = (-4)(-80n)$$

$$= 320n$$

∎

We can combine the two separate functions $y = f(u)$ and $u = g(x)$ into a single function by substituting $g(x)$ for u in $y = f(u)$. This gives us

$$y = f(u)$$
$$\uparrow$$
$$g(x)$$

or

$$y = f[g(x)]$$

You should recognize this as the notation for the composition of the two functions f and g. This means that we can use the Chain Rule to calculate the derivative for the composition of functions. To help us see how it works, we make some changes in the notation we used earlier for the Chain Rule. First, changing to the "prime" notation, we have

$$\frac{dy}{dx} = f'(u) \cdot g'(x)$$

Substituting $g(x)$ for u then gives us the desired result.

Rule 8
Derivative for
Composition of
Functions
(Chain Rule)

Given $y = f[g(x)]$, where g is differentiable at x and f is differentiable at $g(x)$, then

$$\frac{dy}{dx} = f'[g(x)] \cdot g'(x)$$

At present, the functions we use that can be thought of as a composition of functions are limited. In Chapter 5 we will see how the Chain Rule can be used to find the derivatives for exponential and logarithmic functions such as

$$y = e^{x^2} \quad \text{and} \quad y = \log(x^4 + 1)$$

Here, the compositions of functions we deal with all look like "some quantity raised to a power." For example,

$$y = (3x - 4)^5 \quad \text{or} \quad y = \sqrt{1 - 2x} \quad \text{which can be rewritten as}$$
$$y = (1 - 2x)^{1/2}$$

Because this is true, we can write a formula called the Generalized Power Rule that computes their derivatives directly. The Power Rule we used previously was for a single variable raised to a power, and this rule is for some function raised to a power.

Rule 9 *Generalized* *Power Rule*	If $u(x)$ is a differentiable function of x and $y = [u(x)]^n$, then $$\frac{dy}{dx} = n[u(x)]^{n-1}u'(x)$$ This may also be written as $\dfrac{d(u^n)}{dx} = nu^{n-1}\dfrac{du}{dx}$.

In words, the Generalized Power Rule states that to find the derivative of a function $u(x)$ raised to a power n, apply the Power Rule to $[u(x)]^n$ and then multiply by $u'(x)$.

Example 4   Suppose that $f(x) = \sqrt{1 - 2x}$. Find $f'(x)$.

Solution   Rewrite $f(x)$ as

$$f(x) = (1 - 2x)^{1/2}$$

Think of $1 - 2x$ as $u(x)$ and $n = \frac{1}{2}$. Then $u'(x) = -2$ and $n - 1 = -\frac{1}{2}$. We have

$$f'(x) = n[u(x)]^{n-1}u'(x)$$
$$= \tfrac{1}{2}(1 - 2x)^{-1/2}(-2)$$
$$= \frac{-1}{\sqrt{1 - 2x}} \qquad \blacksquare$$

Example 5   If $f(x) = \dfrac{5}{(x^3 + 1)^4}$, find $f'(x)$.

Solution   Rewrite $f(x)$ as

$$f(x) = 5(x^3 + 1)^{-4}$$
$$f'(x) = 5[-4(\underset{\underset{\textstyle u(x)}{\uparrow}}{x^3 + 1})^{-5}(\underset{\underset{\textstyle u'(x)}{\uparrow}}{3x^2})]$$
$$= -60x^2(x^3 + 1)^{-5}$$
$$= \frac{-60x^2}{(x^3 + 1)^5} \qquad \blacksquare$$

The next example demonstrates that it may be necessary to use more than one differentiation rule in the same problem.

Example 6   If $f(x) = x^3\sqrt[3]{x^2 + 1}$, find $f'(x)$.

Solution   Using the Product Rule, we get

$$f'(x) = x^3 \frac{d[(x^2 + 1)^{1/3}]}{dx} + (x^2 + 1)^{1/3} \frac{d[x^3]}{dx}$$

Then using the Chain Rule, we have

$$= x^3[\tfrac{1}{3}(x^2 + 1)^{-2/3}(2x)] + (x^2 + 1)^{1/3}(3x^2)$$

To simplify, we use some factoring techniques from Chapter 1. In particular, we extract the common factor, $\tfrac{1}{3}x^2(x^2 + 1)^{-2/3}$.

$$= \tfrac{1}{3}x^2(x^2 + 1)^{-2/3}[2x^2 + 9(x^2 + 1)]$$

$$= \frac{x^2[11x^2 + 9]}{3(x^2 + 1)^{2/3}}$$

$$f'(x) = \frac{11x^4 + 9x^2}{3\sqrt[3]{(x^2 + 1)^2}}$$ ∎

3.3 EXERCISES

SET A

In exercises 1 through 10, find $\dfrac{dy}{du}, \dfrac{du}{dx}$, and $\dfrac{dy}{dx}$. Express $\dfrac{dy}{dx}$ in terms of x (not u). (See Examples 1 and 2.)

1. $y = u^2 + 1, u = 2x - 5$

2. $y = 1 - u^3, u = 4x + 1$

3. $y = 3u + 1, u = \dfrac{1}{x}$

4. $y = 4 - 2u, u = \sqrt{x}$

5. $y = \dfrac{1}{u + 1}, u = \sqrt{x}$

6. $y = \sqrt{u}, u = \dfrac{1}{x + 1}$

7. $y = 2u^3 - 3u + 4, u = 1 + x^2$

8. $y = \sqrt{u + 1}, u = (x - 1)^2$

9. $y = \dfrac{u}{u + 1}, u = 2x^2 - 1$

10. $y = \dfrac{1}{\sqrt{u}}, u = 4x^2 - 9$

In exercises 11 through 28, find the derivative. (See Examples 4, 5, and 6.)

11. $f(x) = (x + 2)^{10}$

12. $f(x) = (1 - x)^5$

13. $f(x) = 5(x^4 - 1)^3$

14. $f(x) = -3(x^2 - 4)^6$

15. $f(x) = \sqrt[3]{5 - 3x}$

16. $y = \sqrt[4]{5 - x^2}$

17. $y = -2x^2(3x + 1)^5$

18. $y = x^3(2x - 3)^6$

19. $f(x) = (x + 1)^3(2x - 3)^4$

20. $f(x) = (1 - 2x)^3(1 - x)^5$

21. $f(x) = \dfrac{2x}{(1 - x)^4}$

22. $f(x) = \dfrac{-3x}{(2x + 1)^3}$

23. $y = \dfrac{5}{\sqrt{x^2 + 1}}$

24. $y = (x + 2)\sqrt{x - 3}$

25. $y = \dfrac{-3x}{\sqrt[4]{x + 1}}$

26. $y = \dfrac{1}{\sqrt[3]{x^2 + x + 1}}$

27. $f(x) = \left(\dfrac{x - 1}{x + 1}\right)^4$

28. $f(x) = \dfrac{x^2}{\sqrt{x^2 + 1}}$

29. Let $f(x) = \dfrac{-2x}{\sqrt{x^2 + 9}}$. Find $f'(0)$.

30. Let $y = \dfrac{x}{\sqrt{x^2 + 9}}$. Find $\dfrac{dy}{dx}\bigg|_{x=4}$.

31. Let $y = -2(x^2 + x + 1)^6$. Find $\dfrac{dy}{dx}\bigg|_{x=0}$.

32. Let $g(t) = \dfrac{4t}{(2t^2 - 1)^4}$. Find $g'(1)$.

33. Let $f(y) = (y^3 - 1)^4$. Find $f'(1)$.

SET B

34. **a.** Find the slope of the tangent line to the graph of

$$f(x) = (3x^2 - 2x + 1)^2$$

at the point $(0, 1)$.
   b. What is the equation of the tangent line?

35. Suppose $y = y(t)$ and $x = x(t)$ (referred to as **parametric equations** with **parameter** t). Then

$$\frac{dy}{dx} = \frac{y'(t)}{x'(t)} \quad \text{or} \quad \frac{\dfrac{dy}{dt}}{\dfrac{dx}{dt}}$$

Find $\dfrac{dy}{dx}$ for each of the following and leave your answers in terms of t:

   a. $y = t^2 + 1$ and $x = 4t^3$
   b. $y = \sqrt{2t + 1}$ and $x = \sqrt{2t}$

   c. $y = \dfrac{1}{2 - t}$ and $x = \dfrac{1}{t}$.

BUSINESS
COST ANALYSIS

36. A manufacturer of car stereos has determined that an appropriate cost, $C(x)$ (in dollars), to produce car stereos is

$$C(x) = (3x - 6)^2 + 30$$

Here, $C(x)$ is the cost of producing x thousand car stereos.
   a. Use the Generalized Power Rule to determine $C'(x)$.
   b. Find $C'(2)$.
   c. Find $C'(6)$.

DEMAND

37. The demand, x, for a certain type of T-shirt is dependent on the price, p, and is given by the equation

$$x = 45 - 0.2p^2$$

where x is measured in hundreds. It has been determined that the price will be increased as the popularity of the T-shirt increases, according to the function

$$p = 8 + 0.1t^2$$

where t is measured in weeks. At what rate will the demand for the T-shirt be changing (with respect to time) six weeks from now?

PROFIT FUNCTION

38. The Lamsburg Company has determined that its profit function is

$$P(x) = 8x - \tfrac{2}{3}(0.1x + 44)^{3/2}$$

a. Find $P'(x)$.
b. Find $P'(1000)$.

PRODUCT DEMAND

39. The demand for a product is related to the price by the function

$$p = \frac{128}{\sqrt{2x + 6}}$$

where x is measured in hundreds. Determine how p is changing with respect to x when $x = 5$; that is, find $\dfrac{dp}{dx}$ and evaluate it for $x = 5$.

EMPLOYEE EARNINGS

40. For a certain company, a salesperson's earnings, E, depend on the total sales per week for that person and are given by the function

$$E = 200 + 0.15x$$

It has been determined that the total sales, x, can be related to the number of hours, t, spent "out on the road" per week and are given by

$$x = 100t + t^3, \qquad 0 \le t \le 40$$

Determine the rate at which a salesperson's earnings are changing when 20 hours are spent on the road.

PROFIT

41. The profit from the sale of x hundred lawnmowers is given by

$$P(x) = 10(x - 3)^3 + 500$$

where $P(x)$ is measured in dollars.

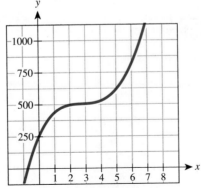

a. Find the average rate of change in profit from 100 to 300 lawnmowers. (*Hint:* $x = 1$ to $x = 3$.)
b. Find $P'(2)$.
c. The rate of change in profit would be zero for how many lawnmowers?

BIOLOGY
NUTRITION

42. A study was conducted to determine how eating or not eating breakfast affects the hunger level at lunchtime. A hunger level index (H) of 0 to 10 was established with

$$H = 10 - 0.0001(x - 200)^2$$

where x is the number of calories consumed for breakfast.
a. What is the hunger level if 400 calories are consumed for breakfast?
b. Find the rate of change in the hunger level for 400 calories.

BACTERIOLOGY

43. The number of bacteria in a certain culture is multiplying according to the function

$$N(t) = 50\sqrt{t + 1} + 2400$$

where t is measured in hours.
a. How many bacteria were present initially?
b. Find the rate of change at three hours.

SOCIAL SCIENCE
DEMOGRAPHICS

44. A city planner decided to do a study on the number of white-collar workers who worked downtown and commuted more than 15 miles one way to work each day. According to his data

$$N = 365 - 72\sqrt[3]{m - 5}, \qquad m > 15$$

where N is the number of workers and m is the number of miles. A graph of this function is given here.

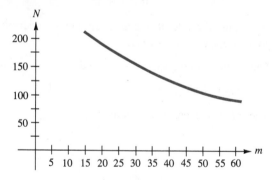

a. Approximately how many workers commute 25 miles one way?
b. Find $\dfrac{dN}{dm}$.

c. From the graph, determine whether $\dfrac{dN}{dm}$ is positive or negative for $m = 20$ miles.

ACTUARIAL SCIENCE

45. An actuary has determined that in a certain population, the number of people surviving, $P(x)$, to age x years is given by

$$P(x) = 400\sqrt{101 - x}, \qquad 0 \le x \le 101$$

a. Find $P'(x)$.
b. Find $P'(76)$.

46. Calculator Exercise

Make a table of values for the function $P(x)$ in exercise 45, using $x = 0, 10, 20, \ldots, 100$.

a. Sketch the graph. Be sure to use an appropriate scale for the vertical axis.

b. For what values of x on the given interval is the function increasing and for what values of x is it decreasing?

c. For what values of x on the given interval is the derivative $P'(x)$ positive and for what values of x is $P'(x)$ negative?

WRITING AND CRITICAL THINKING

1. Explain in words how the Generalized Power Rule introduced in this section differs from the Power Rule introduced in Section 3.1.

2. Explain how the Chain Rule relates rates of change for one variable with respect to another. Use y as a function of u and u as a function of x in your explanation. You may also want to include a discussion of the composition of these two functions.

3.4 HIGHER-ORDER DERIVATIVES

The derivatives of functions that we have calculated in this chapter thus far can be referred to as **first derivatives** or first-order derivatives. In some applications, it is often necessary to calculate the derivative of the first derivative, called the **second derivative**. If $f'(x)$ is used to denote the first derivative, then $f''(x)$ can be used for the second derivative. If $\dfrac{dy}{dx}$ is used to denote the first derivative, then

$$\frac{d\left[\dfrac{dy}{dx}\right]}{dx} \quad \text{or} \quad \frac{d^2y}{dx^2}$$

is used to denote the second derivative.

Example 1   If $f(x) = 5x^3 - x^2 + 3$, find $f''(x)$.

Solution   First, we must find $f'(x)$.

$$f'(x) = 15x^2 - 2x$$

Now $f''(x)$ is the derivative of this new function.

$$f''(x) = 30x - 2$$

∎

Example 2   If $f(x) = (2x - 3)^6$, find $f''(1)$.

Solution $f'(x) = 6(2x - 3)^5 \dfrac{d(2x - 3)}{dx}$

$= 6(2x - 3)^5(2)$
$= 12(2x - 3)^5$
$f''(x) = 12(5)(2x - 3)^4(2)$
$= 120(2x - 3)^4$
So $f''(1) = 120[2(1) - 3]^4 = 120$ ∎

For most applications, we do not need any derivative beyond the second derivative. Any derivative beyond the first derivative is called a *higher-order derivative* (for example, second-order derivative, third-order derivative). In Table 3.2, you will find notation commonly used for higher-order derivatives.

Table 3.2

HIGHER-ORDER DERIVATIVES

First derivative	$f'(x)$	y'	$\dfrac{dy}{dx}$
Second derivative	$f''(x)$	y''	$\dfrac{d^2y}{dx^2}$
Third derivative	$f'''(x)$	y'''	$\dfrac{d^3y}{dx^3}$
Fourth derivative	$f^{(4)}(x)$	$y^{(4)}$	$\dfrac{d^4y}{dx^4}$
nth derivative	$f^{(n)}(x)$	$y^{(n)}$	$\dfrac{d^ny}{dx^n}$

Example 3   If $y = \dfrac{2x}{x - 1}$, find $\dfrac{d^3y}{dx^3}$.

Solution   Using the Quotient Rule, we get

$$\frac{dy}{dx} = \frac{(x - 1)(2) - (2x)(1)}{(x - 1)^2}$$

$$= \frac{-2}{(x - 1)^2}$$

Rewrite $\dfrac{dy}{dx}$ as $-2(x - 1)^{-2}$ before differentiating again.

$$\frac{d^2y}{dx^2} = -2(-2)(x - 1)^{-3}(1)$$

$$= \frac{4}{(x - 1)^3} \quad \text{or} \quad 4(x - 1)^{-3}$$

So $\dfrac{d^3y}{dx^3} = 4(-3)(x-1)^{-4}(1)$

$\qquad = \dfrac{-12}{(x-1)^4}$ ∎

Before concluding this section, we will use alternate notation for derivatives called **differential operator notation**. The derivative of y with respect to x is symbolized by D_xy, the second derivative by D_x^2y, and so on. We can also use this notation for rewriting the differentiation rules, such as the Product Rule:

$$D_x(uv) = uD_xv + vD_xu$$

Example 4   Suppose $y = x^3 - 4\sqrt{x}$. Find D_x^2y.

Solution $D_xy = 3x^2 - 2x^{-1/2}$
$\qquad\quad D_x^2y = 6x + x^{-3/2}$ ∎

3.4 EXERCISES

SET A

In exercises 1 through 6, find $f''(x)$ for each function. (See Example 1.)

1. $f(x) = x^4 + 2x^2 + 1$

2. $f(x) = (2x + 5)^4$

3. $f(x) = \dfrac{1}{\sqrt{2x+1}}$

4. $f(x) = x^2 + \dfrac{1}{x^2}$

5. $f(x) = (3x - 2)^4$

6. $f(x) = \dfrac{x-1}{\sqrt{x}}$

In exercises 7 through 12, find the indicated value. (See Example 2.)

7. $f(x) = (2x + 1)^5, f''(0)$

8. $f(x) = (x^2 + 1)^5, f''(1)$

9. $f(x) = x^3 - 7x^2 + 5, f''(-1)$

10. $g(x) = x^2 + x^{-2}, g''(1)$

11. $y = \dfrac{1}{\sqrt{x}}; \dfrac{d^2y}{dx^2}\bigg|_{x=4}$

12. $s = \dfrac{1}{\sqrt[3]{t-1}}; \dfrac{d^2s}{dt^2}\bigg|_{t=9}$

SET B

In exercises 13 through 24, find the indicated derivative for each function. (See Examples 3 and 4.)

13. $y = (3x + 1)^4, \dfrac{d^2y}{dx^2}$

14. $y = (1 - 2x)^4, \dfrac{d^3y}{dx^3}$

15. $f(x) = \dfrac{1}{x^2}, f^{(4)}(x)$

16. $y = (x^2 + 1)^6, y''$

17. $y = \dfrac{x}{x-2}, \dfrac{d^2y}{dx^2}$

18. $f(x) = \dfrac{1}{\sqrt{x^2+1}}, f''(x)$

19. $y = x^4 - 4x^3, D_x^5y$

20. $y = \sqrt[3]{x}, \dfrac{d^4y}{dx^4}$

21. $f(x) = \dfrac{1}{\sqrt{x}}, f^{(4)}(x)$

22. $f(x) = (x - 3)^5, f'''(2)$ **23.** $y = \dfrac{2x}{\sqrt[3]{x - 1}}, y''$ **24.** $y = (x + 1)^{7/3}, D_x^4(0)$

In exercises 25 through 28, determine the values of x for which $f''(x) = 0$.

25. $f(x) = x^3 - x + 1$ **26.** $f(x) = x^4 - 2x^3 - 36x^2 + 5$ **27.** $f(x) = \sqrt{x} + \dfrac{1}{\sqrt{x}}$

28. $f(x) = x\sqrt{x - 3}$

WRITING AND CRITICAL THINKING

1. Explain in words why, if you repeatedly differentiate a polynomial function, the result will eventually be zero.
2. For the function $f(x) = (x^2 + 1)^{15}$, a student used the Generalized Power Rule to find the first derivative. Explain why the Product Rule would then be used to find the second derivative.

3.5 APPLICATIONS

This section discusses the *application* of information about the derivative we have acquired thus far. The two main areas from which we draw our examples are physics and economics. Keep in mind, however, that the derivative has applications in many other areas. Some of these will be dealt with in the exercises and in the application sections of later chapters. You may find it helpful to reread the section on Problem Solving on pages 66–75.

Recall that in Section 2.6 we discussed both the **average** rate of change and the **instantaneous** rate of change of $y = f(x)$ with respect to x. Geometrically speaking, we interpreted the average rate of change as the slope of the *secant* line over a given interval and the instantaneous rate of change as the slope of the *tangent* line at a point and called it the *derivative*.

Before presenting any applications, we would like to review the symbolism associated with both rates of change. The *average* rate of change of $f(x)$ over an interval from x to $x + h$ is given by the expressions

$$\frac{f(x + h) - f(x)}{(x + h) - x} \quad \text{or simply} \quad \frac{f(x + h) - f(x)}{h}$$

The *instantaneous* rate of change of $f(x)$ is given by

$$\lim_{h \to 0} \frac{f(x + h) - f(x)}{h} = f'(x)$$

ECONOMICS: MARGINAL ANALYSIS

Businesses, for the most part, produce items in the hope of making a profit. We will analyze manufacturers' income and production mathematically, using the following notation:

---

x: The number of units produced (or sold) in some time interval.

$C(x)$:   The cost function, where $C(x)$ is the company's cost for producing x units.

$R(x)$:   The revenue function, where $R(x)$ is the company's total revenue (or income) for producing (or selling) x units. We assume all that is produced is sold.

$P(x)$:   The profit function, where $P(x) = R(x) - C(x)$; that is, profit is revenue less cost.

Let's first examine the cost function. Suppose Ronolog, Inc., has determined that the cost of producing x electronic calculators is

$$C(x) = 0.04x^2 + 4000$$

If Ronolog produces 100 calculators, the cost is

$$C(100) = 0.04(100)^2 + 4000 = \$4400$$

If $x = 101$,

$$C(101) = 0.04(101)^2 + 4000 = \$4408.04$$

We can say that the cost of producing the 101st calculator is

$$\$4408.04 - \$4400.00 = \$8.04$$

Economists, however, use the instantaneous rate of change of C with respect to x, $C'(x)$, to *approximate* the increase in cost of producing one more unit. They refer to this instantaneous rate of change of C with respect to x as the **marginal cost function.**

In this illustration, notice that if

$$C(x) = 0.04x^2 + 4000$$
then $\quad C'(x) = 0.08x$
and $\quad C'(100) = \$8.00$

which is a close approximation to the actual $8.04 increase, $C(101) - C(100)$.

Example 1   The cost, $C(x)$ (in thousands of dollars), of producing x thousand math textbooks is given by

$$C(x) = 30 + 20x - 0.5x^2, \quad 0 \le x \le 15$$

a.   Find the marginal cost function.

b.   Find the marginal cost for producing 12,000 textbooks.

Solution   **a.**   The graph of $C(x)$ is shown in Figure 3.1.

$$C(x) = 30 + 20x - 0.5x^2$$
$$C'(x) = 20 - x$$

b.   For $x = 12$,

$$C'(12) = 20 - 12 = 8$$

Figure 3.1

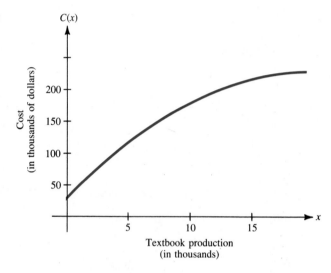

Thus, the increase in cost for producing an additional 1000 texts at the level of production of 12,000 texts is approximately $8000.   ∎

Economists similarly define **marginal revenue** and **marginal profit** as a rate of change; that is:

Marginal revenue is the instantaneous rate of change of revenue with respect to the number of units sold. Symbolically, marginal revenue is $R'(x)$ or $\dfrac{dR}{dx}$.

Marginal profit is the instantaneous rate of change of profit with respect to the number of units sold. Symbolically, marginal profit is $P'(x)$ or $\dfrac{dP}{dx}$.

Some examples follow.

Example 2   The cost in dollars of producing x electric hair dryers is given by $C(x) = 200 + 7x$, and the revenue for selling all x hair dryers is

$$R(x) = 20x - \frac{x^2}{50}$$

a. Find the profit function.
b. Find the marginal profit at $x = 200$.

Solution   a.   Recall that $P(x) = R(x) - C(x)$.

$$\text{So} \quad P(x) = \left(20x - \frac{x^2}{50}\right) - (200 + 7x)$$

$$= \frac{-x^2}{50} + 13x - 200$$

b.
$$P'(x) = \frac{-x}{25} + 13$$

$$P'(200) = \frac{-200}{25} + 13 = \$5$$

That is, the increase in profit for one additional hair dryer at a production level of 200 is \$5.   ■

To actually *construct* a revenue function or equation, economists examine a **demand function,** which relates the quantity, x, demanded (or sold) of an item with its price, p.

For example, suppose the marketing research division of Acme Tool has determined that the relationship between the asking price, p, for an air wrench and the quantity sold, x, is

$$p = 125 - \frac{x}{20}, \quad 0 < x < 2400$$

This is called the **demand function** or **demand equation.**

Now, the revenue, which is the total income from the product, can be calculated by multiplying the number of items sold times the price per item; that is, total revenue $R(x) = p \cdot x$.

Substituting $125 - \frac{x}{20}$ for p in $R(x) = p \cdot x$, we have

$$R(x) = p \cdot x = \left(125 - \frac{x}{20}\right)x$$

$$= 125x - \frac{x^2}{20}$$

$R(x)$ is the revenue function and $R'(x) = 125 - \frac{x}{10}$ is the **marginal revenue function.**

Example 3   Suppose that the demand equation is given by $p = 120 - \frac{x}{10}$ and $C(x) = 1500 + 4x$.
Find:
a. $R(x)$ b. $P(x)$ c. $P'(x)$

Solution   **a.** $p = 120 - \dfrac{x}{10}$

$$R(x) = p \cdot x = \left(120 - \frac{x}{10}\right)x$$

$$= 120x - \frac{x^2}{10}$$

b. $P(x) = R(x) - C(x)$

$$= 120x - \frac{x^2}{10} - (1500 + 4x)$$

$$= 116x - \frac{x^2}{10} - 1500$$

c. $P'(x) = 116 - \dfrac{x}{5}$ ∎

It is very useful for a company to determine at what point it is making money, losing money, or breaking even. This can be analyzed using revenue, $R(x)$, cost, $C(x)$, and profit, $P(x)$, relationships as follows:

For $P(x) = R(x) - C(x)$:

1. If $R(x) > C(x)$, then $P(x) > 0$ and the company makes money.
2. If $R(x) < C(x)$, then $P(x) < 0$ and the company loses money.
3. If $R(x) = C(x)$, then $P(x) = 0$ and the company breaks even. This point is called the **break-even point.**

A diagram of this analysis is given in Figure 3.2.

Figure 3.2

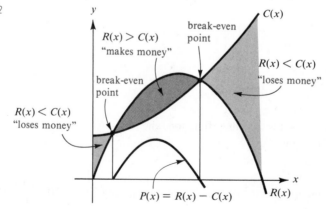

Example 4   The cost function for a particular item is given by

$$C(x) = x^2 + 5000$$

and the revenue function is

$$R(x) = 250x - x^2$$

Determine the number of units, x, that must be produced and sold to break even.

Solution   The break-even point occurs for $R(x) = C(x)$:

$$
\begin{aligned}
R(x) &= C(x) \\
250x - x^2 &= x^2 + 5000 \\
0 &= 2x^2 - 250x + 5000 \\
&= x^2 - 125x + 2500 \\
&= (x - 25)(x - 100)
\end{aligned}
$$

$$x - 25 = 0 \qquad \text{or} \qquad x - 100 = 0$$
$$x = 25 \qquad\qquad\qquad x = 100$$

The company can produce and sell either 25 or 100 units to break even. A graphical model of this problem appears in Figure 3.3.

Figure 3.3

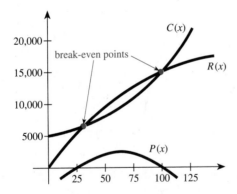

PHYSICS: STRAIGHT-LINE MOTION

In physics, the average rate of change of position with respect to time of an object moving in a straight line is called **average velocity** (denoted by v_{av} or $\bar{v}$). The position of the object s (usually measured in feet or meters) will be represented as a function of time, t (usually measured in seconds).

Example 5   A ball is thrown vertically upward from ground level with an initial velocity (velocity at time $t = 0$) of 64 ft/sec. If its distance above ground level is given by

$$s(t) = 64t - 16t^2, \qquad 0 \le t \le 4$$

find the average velocity from $t = 1$ to $t = 2$.

Solution $v_{av} = \dfrac{s(2) - s(1)}{2 - 1}$

$\qquad\quad = \dfrac{64 \text{ ft} - 48 \text{ ft}}{1 \text{ sec}}$

$\qquad\quad = 16 \text{ ft/sec}$

Figure 3.4 shows the distance of the ball from the ground at any time t. In particular, it shows the average rate of change or average velocity over the interval from $t = 1$ to $t = 2$ depicted by the secant line.

Figure 3.4

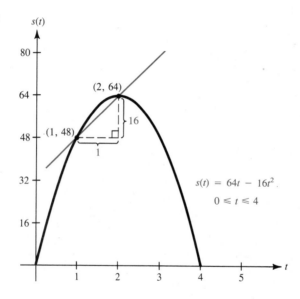

(*Note:* The parabola is the graph of the function, not the path of the ball. The ball was thrown vertically and will fall vertically.)   ■

We define the **instantaneous velocity** $v(t)$ as the rate of change of $s(t)$ with respect to t; that is,

$$v(t) = \lim_{h \to 0} \frac{s(t + h) - s(t)}{h}$$

or, more simply, $v(t) = s'(t)$.

Example 6   Using the same information from Example 5, find the instantaneous velocity of the ball for

a. $t = 1$ **b.** $t = 3.5$

Solution Since $v(t) = s'(t)$,

$$v(t) = 64 - 32t$$

a. $v(1) = 64 - 32(1) = 32$ ft/sec
b. $v(3.5) = 64 - 32(3.5) = -48$ ft/sec ∎

Notice that the answer to part a is positive, indicating an upward direction of motion; the answer to part b is negative, indicating a downward direction. Sometimes the direction is not important, in which case we refer to the **speed** of the object, defined by $|v(t)|$, which is never negative. We should also mention here that *instantaneous velocity* is often referred to as just *velocity*.

We now define the **instantaneous acceleration** $a(t)$, or simply the *acceleration*. Acceleration is the instantaneous rate of change of velocity with respect to time—that is, $v'(t)$, which is also $s''(t)$.

Example 7 Once more using the ball from Example 5, find the acceleration for
a. $t = 1$ b. $t = 3.5$

Solution From Example 6, we have

$$v(t) = 64 - 32t$$

Since $a(t) = v'(t)$,

$$a(t) = -32$$

a. $a(1) = -32$ ft/sec^2
b. $a(3.5) = -32$ ft/sec^2

In fact, the acceleration is always -32 ft/sec^2. The only acceleration the ball experiences is the acceleration due to the force of gravity, which is -32 ft/sec^2. ∎

Example 8 Suppose an object moves along a straight line and, at t seconds, its distance (in feet) from a fixed point is given by

$$s(t) = 3t + \frac{1}{t + 1}$$

a. Find the object's velocity after 3 sec (that is, at $t = 3$).
b. Find the object's acceleration at $t = 3$.

Solution a. We must find the object's velocity at $t = 3$, $v(3)$.

$$s(t) = 3t + (t + 1)^{-1}$$
$$v(t) = s'(t) = 3 - (t + 1)^{-2}$$
$$v(3) = 3 - 4^{-2}$$
$$= 3 - \tfrac{1}{16} = 2\tfrac{15}{16} \text{ ft/sec}$$

b. $a(t) = v'(t) = s''(t)$
$$= 2(t + 1)^{-3}$$
$$a(3) = 2(4)^{-3} = \tfrac{1}{32} \text{ ft/sec}^2$$

■

Example 9   An object is dropped from the top of a 320-ft building and its height, s (in feet), above the ground in time t (in seconds) is given by

$$s = 320 - 16t^2$$

See Figure 3.5.

Figure 3.5

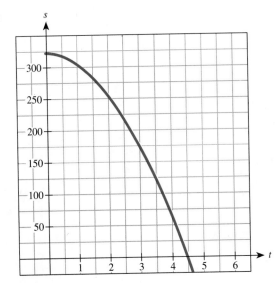

a.   Find $\left.\dfrac{ds}{dt}\right|_{t=4}$. What does this represent?

b.   For what value(s) of t is $s = 0$?

c.   For what value(s) of t is the object's velocity zero?

Solution   **a.** $s = 320 - 16t^2$

$$\frac{ds}{dt} = -32t$$

$$\left.\frac{ds}{dt}\right|_{t=4} = -32(4) = -128$$

It represents the velocity 4 sec after it is dropped. The negative sign indicates movement in a downward direction.

b. $s = 320 - 16t^2 = 0$
$$-16(t^2 - 20) = 0$$

So $t = \pm\sqrt{20} = \pm 2\sqrt{5}$

Because only positive values of t pertain, $t = 2\sqrt{5} \approx 4.472$ sec. Thus, the object hits the ground ($s = 0$ means 0 distance above the ground) after approximately 4.472 sec.

c. $v = \dfrac{ds}{dt} = -32t$

So when $t = 0$, $v = 0$. ∎

We summarize the notation for motion problems in Table 3.3.

Table 3.3

SUMMARY OF NOTATION FOR MOTION PROBLEMS		
Position:	$s(t)$	s
Velocity:	$v(t) = s'(t)$	$v = \dfrac{ds}{dt}$
Speed:	$\lvert v(t) \rvert$	$\lvert v \rvert$
Acceleration:	$a(t) = v'(t)$	$a = \dfrac{dv}{dt}$

3.5 EXERCISES

SET A

BUSINESS
MARGINAL COST

1. The cost, $C(x)$ (in dollars), of producing x electric pencil sharpeners is $C(x) = 20 + 40x - 0.50x^2$.
    a. Find the marginal cost function.
    b. Find the marginal cost for producing 30 pencil sharpeners, and write a sentence to explain your result.
    c. Find $C(31) - C(30)$. Compare with part b.

MARGINAL COST

2. An electronics manufacturer has determined that the daily cost, $C(x)$, of producing x microcomputers on an assembly line is

    $$C(x) = 0.1x^3 - 12x^2 + 300x + 150$$

    a. Find the marginal cost for $x = 100$, and write a sentence to explain your result.
    b. Find $C(101) - C(100)$. Compare with part a.

MARGINAL PROFIT

3. The profit in thousands of dollars for producing x hundred televisions daily has been estimated to be $P(x) = 0.02x^3 + 0.04x$.
    a. Determine the marginal profit function.

    b. What is the marginal profit at the 500 televisions-per-day level of production ($x = 5$)? Write a sentence to explain your result.

    c. Find $P(6) - P(5)$ and compare with part b.

MARGINAL ANALYSIS

4. The demand equation for x cassette recorders at a price of p dollars is $p = 32 - 0.04x$, and the cost function for producing x recorders is

$$C(x) = 4000 + 12x$$

    a. Find the revenue function.

    b. Find the marginal cost function.

    c. Find the marginal revenue function.

    d. Find the profit function, $P(x)$.

    e. Find the marginal profit function.

    f. Find $P'(200)$, $P'(250)$, $P'(300)$.

    g. Use part f to determine how many recorders the company should produce (and sell) so that the rate of change of profit is zero.

MARGINAL ANALYSIS

5. A company has determined that the demand equation for x hand-held electronic games at a price of p dollars is given by $p = 40 - 0.05x$. The company has determined that the cost function for producing x games is $C(x) = 3500 + 20x$.

    a. Find the revenue function.

    b. Find the marginal cost function.

    c. Find the profit function, $P(x)$.

    d. Find the marginal profit function, $P'(x)$.

FIXED COSTS

6. *Fixed costs* are the costs encountered by a company even when they produce and sell no goods. In effect, fixed cost equals $C(0)$. In exercise 5, the fixed costs for the company are $3500. Assume the company's fixed costs increase to $4500 and rework the four parts of exercise 5 on this basis.

PROFIT FUNCTION
ANALYSIS

7. Refer to Example 2.

    a. Find $P(201) - P(200)$. Compare with the value of $P'(200)$ found in the example.

    b. Complete the following table.

x	$P(x)$	$P'(x)$
200		
250		
300		
325		
350		

    c. Using your results from part b, find how many units should be produced in order to make a profit as great as possible.

REVENUE FUNCTION

8. Given that the demand function for a particular product is $p = (120 - x)^{1/2}$, find the revenue function $R(x) = px$ and the marginal revenue function $R'(x)$.

MARGINAL AVERAGE COST

9. Another notion used by economists is the **average cost function, $\bar{C}(x)$**. Basically, $\bar{C}(x) = \dfrac{C(x)}{x}$; that is, $\bar{C}(x)$ is the cost per unit produced. Also, the **marginal average cost function** is $\bar{C}'(x)$. Suppose that

$$C(x) = \frac{200}{x} + 5.5x$$

a. Find the average cost function, $\bar{C}(x)$.
b. Find the marginal cost function, $C'(x)$.
c. Find the marginal average cost function, $\bar{C}'(x)$.

MARGINAL AVERAGE COST

10. Given the cost function $C(x) = x^2 + 5x + 25$, find $\bar{C}'(x)$. Also find $\bar{C}'(5)$, $C'(5)$, and $\bar{C}(5)$. Note that $C'(x) = \bar{C}(x)$ for $x = 5$, which is the value that results in $\bar{C}'(x) = 0$.

COST FUNCTIONS

11. Assume that the total cost of producing x items of a product is given by $C(x) = 2x - 3x^2 + x^3$.
a. Find the marginal cost function, $C'(x)$.
b. Determine the average cost function, $\bar{C}(x) = \dfrac{C(x)}{x}$.
c. Find the marginal average cost function, $\bar{C}'(x)$.

NEW-PRODUCT ANALYSIS

12. A manufacturer of video recorders is in the process of setting the price on a new recorder to be introduced. The firm's marketing department, on the basis of preliminary research, estimates that the demand for the new recorder can be described by the demand equation

$$p = 400 - \frac{x}{3}$$

The firm's cost accountants estimate that the total cost for the production of x units can be described by

$$C(x) = 5000 + 40x$$

a. Determine x as a function of p.
b. Find the cost as a function of price, $C(p)$.
c. Find revenue as a function of price, $R(p)$.
d. Express profit as a function of price, $P(p)$.
e. Find $P'(p)$ for $p = 300$.

REVENUE

13. A company finds that its total sales revenue is represented by the function $R(x) = 8x - x^2$, where x is the number of units sold (in thousands).
a. Find $R'(x)$, the marginal revenue function.
b. What is the marginal revenue when $x = 2$?
c. For what value of x is the marginal revenue equal to zero?

MARGINAL COST

14. Given the total cost function $C(x) = 2x^3 - 12x^2 + 30x + 15$, determine the marginal cost function $C'(x)$. Also calculate $C'(1)$, $C'(2)$, and $C'(3)$.

BREAK-EVEN ANALYSIS

15. Find the number of units necessary to break even if $R(x) = 50x$ and $C(x) = 1000 + 10x$.

BREAK-EVEN ANALYSIS

16. Find the number of units needed for a company to break even if $R(x) = 800x - x^2$ and $C(x) = 50x$.

BREAK-EVEN ANALYSIS

17. For $R(x) = 40x - x^2$ and $C(x) = 75 + 12x$, there are two break-even points. Find the value of x for each point.

BREAK-EVEN ANALYSIS

18. Suppose $R(x) = 100x - 0.5x^2$ and $C(x) = 1.5x^2$.
  a. For what values of x is $P(x) > 0$?
  b. How many units must be produced and sold to break even?

REVENUE AND COST

19. The following graph shows a revenue function and a cost function for a certain product on the same set of axes. Use the graph to answer the questions, in each case giving your answer in terms of the number of units.

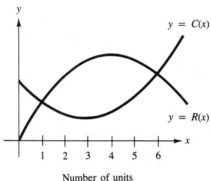

Number of units
(in hundreds)

  a. How many units should be produced and sold to break even?
  b. How many units should be sold to maximize revenue?
  c. Beyond what level of production does cost increase?
  d. What production level range represents a profit for the company?

PROFIT ANALYSIS

20. Calculator Exercise
The profit, $P(x)$ (in dollars), for producing x units per day of a product is given by

$$P(x) = -1000 - 501.2x + 24x^2 - 0.217x^3, \qquad 31 \le x \le 81$$

  a. Complete the following table:

x	$P(x)$	$P'(x)$
31		
41		
51		
61		
71		
81		

  b. How many units per day would you advise the company to produce?

BIOLOGY
EFFECTIVENESS OF A DRUG

21. In a laboratory experiment, rats have been injected with a drug and the effectiveness of the drug t hours after entering the bloodstream is given by

$$E(t) = 12t + 4t^2 - t^3, \qquad 0 \le t \le 6$$

  a. Find the average rate of change with respect to t over the time interval from $t = 1$ to $t = 3$.

    b.   Find the instantaneous rate of change for $t = 1, 2, \ldots, 6$.

    c.   At what point in time (approximately) does the effectiveness of the drug begin to diminish?

PHYSICAL CONDITIONING   **22.** A runner has been training for distance running. Her distance, D (in miles), in terms of time, t (in weeks), is given by

$$D = 2\sqrt[3]{(t - 1)^2} + 15$$

    a.   How many miles did she run during week one?

    b.   What is the average rate of change in her mileage from week one to week three?

    c.   Find $\dfrac{dD}{dt}$ for $t = 4$ and interpret the result.

PROTEIN POLYMERIZATION   **23.** The following graph shows the time course of polymerization of a protein as measured by its increasing viscosity. Polymerization does not proceed at a uniform rate but begins with a lag phase, in which there is no observable increase in viscosity. This is thought to occur because the first step requires two or three molecules to come together in a specific and particularly different geometric conformation. Once this has been achieved, however, the addition of further molecules proceeds rapidly.

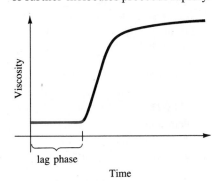

    a.   Describe how the slope of tangent lines to the curve changes during the lag phase.

    b.   From part a, describe the effect this has on the rate of change in viscosity during this period.

    c.   What happens to the rate of change in viscosity after the lag phase? Explain.

SOCIAL SCIENCE
POPULATION GROWTH   **24.** Sociologists have been studying the population growth rate of a tribe in a remote area. They have determined that the following function gives a good estimate of the population at any time t (in years):

$$P(t) = 400 + 20t^2 - 18t^{3/2}$$

    a.   Find the growth rate (rate of change of P with respect to t).

    b.   Find the population for $t = 4$.

    c.   Find the growth rate for $t = 4$.

FAMILY STRUCTURE

25. The number of single-parent households has been increasing in a certain city and is estimated by

$$N(t) = \frac{400t + 250}{t + 2}$$

where t is measured in years. Find the rate at which the number of single-parent households has been changing.

INCOME SURVEY

26. A survey was conducted to determine how family yearly income was related to the number of hours, T, of television watched by the family per week. This relationship was found to be

$$T(i) = \frac{400}{i} + 10, \qquad 10 \le i \le 100$$

where the yearly income, i, is measured in thousands of dollars. Find the rate of change in the number of hours of television watched per week for a family with a yearly income of $40,000.

SET B

PHYSICAL SCIENCE
TUNING A HARP

27. In tuning a harp, the tighter the string, the more vibrations per second are produced. Let V represent the vibrations per second and t the tension measured in pounds in the function $V(t) = 300t^{1/3}$. Find the rate of change of $V(t)$ with respect to t when $t = 8$ lb.

FIELD OF VISION

28. The distance to an object on the horizon that a person can see is a function of how far above ground level a person is located. The distance, D (in miles), can be approximated by $D = \sqrt{1.5x}$, where x represents the person's height above the ground measured in feet.
   a. Find the average rate of change of D for $x = 90$ to $x = 100$.
   b. Find the rate of change of D with respect to x for $x = 100$.

PROJECTILE

29. A projectile is fired vertically upward from a platform 320 ft above ground level with an initial velocity of 128 ft/sec. Its position at any time t is

$$s(t) = 320 + 128t - 16t^2, \qquad 0 \le t \le 10$$

   a. Find the average velocity from $t = 2$ to $t = 3$ sec.
   b. Find the velocity at time $t = 3$ sec.
   c. For what value of t is $v(t) = 0$?
   d. Over what time interval is the projectile traveling upward? (*Hint:* $v(t) > 0$.)
   e. Over what time interval is it traveling downward? (*Hint:* $v(t) < 0$.)

FREE-FALLING BODY

30. A coin is dropped from the top of a bridge and its height above the surface of the water is given by $s(t) = 96 - 16t^2$, where s is measured in feet and t is measured in seconds.
   a. How long will it take for the coin to hit the surface of the water?
   b. How fast will the coin be traveling when it hits the water?

SPEED AND ACCELERATION   **31.**   An object moves in a straight line so that its distance in feet from a fixed point after t sec is given by

$$s(t) = 5t + \frac{1}{t^2}$$

a.   Find the object's speed after 2 sec.
b.   Find the object's acceleration at $t = 1$.

STRAIGHT-LINE MOTION   **32.**   A slow-moving object moving in a straight line travels $s(t)$ meters in t hours, where $s(t) = 10t^2 + 2\sqrt{t}$.
a.   What is the object's velocity after four hours?
b.   What is the object's acceleration when $t = 1$?
c.   How far has the object traveled after nine hours?
d.   How far does the object travel in its tenth hour of travel?

FREE-FALLING BODY   **33.**   A ball is dropped from the top of a 480-ft building, and its height h (in feet) after t sec is given by $h = 480 - 16t^2$.

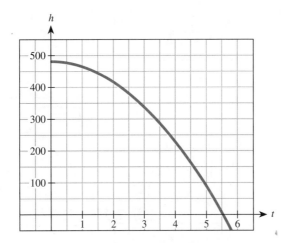

a.   Find $\dfrac{dh}{dt}\bigg|_{t=1}$. What does this represent?
b.   For what value(s) of t is $h = 0$?
c.   What is the object's velocity when $h = 0$ (that is, when it hits the ground)?

PROJECTILE   **34.**   A rocket hobbyist fires a model upward so that its height $h(t)$ (in feet) after t sec is given by $h(t) = 320t - 16t^2$.
a.   Find $h'(t)$.
b.   Find $h''(t)$.
c.   When does the model hit the ground?
d.   What is the velocity when $t = 0$?
e.   What is the velocity when the model hits the ground?

VELOCITY   **35.**   When a ball is thrown upward from a building rooftop, its height above ground level, $s(t)$ (in feet), after t sec is given by

$$s(t) = -16t^2 + 64t + 112$$

a.   When is the ball's velocity zero?
b.   At what height is the ball when its velocity is zero?
c.   When does the ball hit the ground?
d.   What is the ball's initial velocity?

USING TECHNOLOGY IN CALCULUS

We can use the zooming capability of graphics software packages and calculators to help determine the optimum number of units to produce in exercise 20. Consider these two views of graphs done on a graphics calculator; we perform a BOX zoom on the first one to obtain the second:

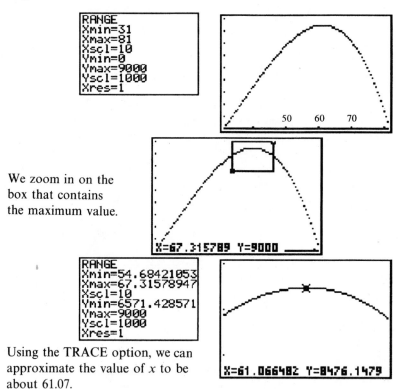

We zoom in on the box that contains the maximum value.

Using the TRACE option, we can approximate the value of x to be about 61.07.

It appears the company should produce about 61 units.

1.   What do you think the approximate value of $P'(61)$ is?
2.   What is the company's greatest possible profit?
3.   Draw the tangent line at the curve's maximum point. What is the equation of this line?

3.6 IMPLICIT DIFFERENTIATION AND RELATED RATES

All the equations for which we have calculated the derivative thus far have been functions and can be expressed in the form $y = f(x)$. This form, in which the equation is solved for one variable (usually y), is called an **explicit** form. In each of the following equations, we say that y is expressed **explicitly** in terms of x:

$$y = x^2 + 2x \qquad y = \sqrt{1 - x} \qquad y = \frac{x^2}{x^2 + 1}$$

We now establish a technique for calculating the derivative for an equation in which y is expressed **implicitly** in terms of x, as in each of the following:

$$x^2 + y^2 = 4 \qquad \sqrt{x} + \sqrt{y} = 1 \qquad x^2 y - 2y^3 = 7$$

The process we use to calculate the derivative is called **implicit differentiation.** Keep in mind that we still want $\dfrac{dy}{dx}$ (or $f'(x)$), which means that we are still differentiating y with respect to x. To differentiate when y is defined implicitly in terms of x, think of y as a function of x, but don't solve for y before differentiating. When differentiating a term such as y^3, think of it as $[f(x)]^3$ and use the Generalized Power Rule so that the derivative of y^3 will be

$$3[f(x)]^2 \cdot f'(x)$$

or $\quad 3y^2 \dfrac{dy}{dx}$

In Example 1, we demonstrate that differentiating implicitly or solving for y and then differentiating produce the same result.

Example 1   For $y^3 = x^2$, find $\dfrac{dy}{dx}$.

Solution   **Method 1**

Solve for y and then differentiate.

$$y = x^{2/3}$$
$$\frac{dy}{dx} = \frac{2}{3}x^{-1/3}$$
$$= \frac{2}{3x^{1/3}}$$

Method 2

Use implicit differentiation.

$$y^3 = x^2$$

$$3y^2 \frac{dy}{dx} = 2x$$

$$\frac{dy}{dx} = \frac{2x}{3y^2}$$

Now substitute $x^{2/3}$ for y to show that the results are the same.

$$\frac{dy}{dx} = \frac{2x}{3(x^{2/3})^2}$$

$$= \frac{2x}{3x^{4/3}}$$

$$= \frac{2}{3x^{1/3}}$$ ∎

In the remainder of the examples, we use only implicit differentiation.

Example 2   For $x^2 + y^2 = 4$, find $\frac{dy}{dx}$.

Solution   Using Rule 4, we can differentiate each term on the left side separately. Use Rule 1 for the right side. Notice that to differentiate y^2, we use the Generalized Power Rule.

$$\frac{d[x^2]}{dx} + \underbrace{\frac{d[y^2]}{dx}}_{\substack{\text{Generalized} \\ \text{Power Rule}}} = \frac{d[4]}{dx}$$

↓

$$2x + 2y\frac{dy}{dx} = 0$$

Now, to find $\frac{dy}{dx}$, we treat the preceding as an equation that we solve for $\frac{dy}{dx}$.

$$2y\frac{dy}{dx} = -2x$$

$$\frac{dy}{dx} = \frac{-2x}{2y}$$

$$= \frac{-x}{y}$$ ∎

Example 3   For $x^3 + x^2y^2 - 2y = 0$, find $\frac{dy}{dx}$.

Solution   Note that we use the Product Rule on the x^2y^2 term.

Product Rule

$$3x^2 + \left[x^2\left(2y \cdot \frac{dy}{dx}\right) + y^2(2x) \right] - 2\frac{dy}{dx} = 0$$

$$3x^2 + 2x^2y\frac{dy}{dx} + 2xy^2 - 2\frac{dy}{dx} = 0$$

Now solve for $\dfrac{dy}{dx}$.

$$2x^2y\frac{dy}{dx} - 2\frac{dy}{dx} = -3x^2 - 2xy^2$$

$$(2x^2y - 2)\frac{dy}{dx} = -3x^2 - 2xy^2$$

$$\frac{dy}{dx} = \frac{-3x^2 - 2xy^2}{2x^2y - 2}$$

$$= \frac{3x^2 + 2xy^2}{2 - 2x^2y}$$

■

Example 4   For $\sqrt{x} + \sqrt{y} = 1$, find $\dfrac{dy}{dx}\bigg|_{x=0.25}$.

Solution   First, rewrite in exponential form.

$$x^{1/2} + y^{1/2} = 1$$

$$\frac{1}{2}x^{-1/2} + \frac{1}{2}y^{-1/2}\frac{dy}{dx} = 0$$

$$x^{-1/2} + y^{-1/2}\frac{dy}{dx} = 0 \qquad \text{Multiply both sides by 2.}$$

$$y^{-1/2}\frac{dy}{dx} = -x^{-1/2}$$

$$\frac{dy}{dx} = \frac{-x^{-1/2}}{y^{-1/2}}$$

$$= \frac{-y^{1/2}}{x^{1/2}} = \frac{-\sqrt{y}}{\sqrt{x}}$$

We need to substitute $x = 0.25$, but we also need the corresponding y value. We can find the y value using the original equation.

$$\sqrt{0.25} + \sqrt{y} = 1$$

$$0.5 + \sqrt{y} = 1$$

$$\sqrt{y} = 0.5$$

$$y = 0.25$$

Now, $\dfrac{dy}{dx}\bigg|_{x\,=\,0.25} = \dfrac{-\sqrt{0.25}}{\sqrt{0.25}}$

$$= -1 \qquad\blacksquare$$

If necessary, we can calculate the second derivative for an equation that is expressed in the implicit form, as illustrated in Example 5.

Example 5   If $x^2 + 3y^2 = 12$, find $\dfrac{d^2y}{dx^2}$.

Solution   First, we find $\dfrac{dy}{dx}$ using implicit differentiation.

$$2x + 3\left(2y\,\dfrac{dy}{dx}\right) = 0$$

$$6y\,\dfrac{dy}{dx} = -2x$$

$$\dfrac{dy}{dx} = \dfrac{-x}{3y}$$

Now, to find $\dfrac{d^2y}{dx^2}$, we have to use the Quotient Rule.

$$\dfrac{d^2y}{dx^2} = \dfrac{3y(-1) - (-x)\left(3\,\dfrac{dy}{dx}\right)}{(3y)^2}$$

$$= \dfrac{-3y + 3x\,\dfrac{dy}{dx}}{9y^2}$$

To finish, we must substitute for $\dfrac{dy}{dx}$.

$$\dfrac{d^2y}{dx^2} = \dfrac{-3y + 3x\left(\dfrac{-x}{3y}\right)}{9y^2}$$

$$= \dfrac{-3y^2 - x^2}{9y^3}$$

$$= \dfrac{-1(x^2 + 3y^2)}{9y^3}$$

$$= \dfrac{-(12)}{9y^3} \qquad \text{Substitute 12 for } x^2 + 3y^2 \text{ from the original equation.}$$

$$= \dfrac{-4}{3y^3} \qquad\blacksquare$$

In Section 2.7, we saw that the derivative represents an instantaneous rate of change of a function, y, at a particular value, x. For the remainder of this section, we differentiate with respect to time, t. These problems fall into the category of problems called **related rates.** To differentiate, think of both x and y as functions of t.

Example 6 Differentiate $x^2 - 9y^2 = 16$ with respect to t using implicit differentiation and solve for $\dfrac{dy}{dt}$.

Solution
$$\frac{d[x^2]}{dt} - 9\frac{d[y^2]}{dt} = \frac{d[16]}{dt}$$

$$2x\frac{dx}{dt} - 18y\frac{dy}{dt} = 0 \qquad \text{Using the Generalized Power Rule on both } x^2 \text{ and } y^2$$

$$-18y\frac{dy}{dt} = -2x\frac{dx}{dt}$$

$$\frac{dy}{dt} = \frac{x}{9y}\frac{dx}{dt}$$ ■

Examples 7 and 8 are application problems, and you may find it helpful to reread the section on Problem Solving on pages 66–75 before looking at these examples or doing the associated exercises.

Example 7 A spherical hot-air balloon is being filled (see Figure 3.6). If the volume of the balloon increases at the rate of 50 ft^3/sec, at what rate does the radius increase when the radius is 10 ft?

Figure 3.6

Solution Recall the formula for the volume of a sphere: $V = \frac{4}{3}\pi r^3$. In this example, both V and r are functions of t because both are changing with respect to time. Now we differentiate with respect to t:

$$\frac{dV}{dt} = \frac{4}{3}\pi\left(3r^2\frac{dr}{dt}\right)$$

$$\frac{dV}{dt} = 4\pi r^2\frac{dr}{dt}$$

Using the information from the problem, we have

$$\frac{dV}{dt} = 50 \qquad\qquad \text{(“... the volume is increasing at a \textit{rate} of 50 ...”)}$$

$$r = 10 \qquad\qquad \text{(“... the radius is 10 feet”)}$$

$$\frac{dr}{dt} \text{ is the unknown} \qquad \text{(“... at what \textit{rate} is the radius increasing ...”)}$$

Solving for the unknown and substituting values give

$$\frac{dr}{dt} = \frac{dV}{dt} \cdot \frac{1}{4\pi r^2}$$

$$= 50 \cdot \frac{1}{4\pi \cdot 100}$$

$$= \frac{1}{8\pi} \approx 0.0398 \text{ ft/sec}$$

■

Notice that in Example 7 the "when" information ("... when $r = 10$") is not substituted in until after differentiation has been done.

Example 8   Two ships leave point A at the same time; one is headed due east at 21 mph and one is headed due south at 28 mph. At what rate is the distance between them changing at the end of 2 hours?

Solution   Let x and y be the distance traveled by the eastbound and southbound ships, respectively, at any given time (see Figure 3.7).

Figure 3.7

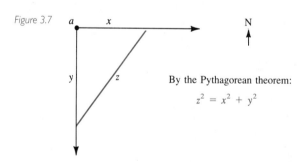

By the Pythagorean theorem:

$$z^2 = x^2 + y^2$$

We need to find $\dfrac{dz}{dt}$, the rate at which the distance between them is changing. Also, we know that $\dfrac{dx}{dt} = 21$ and $\dfrac{dy}{dt} = 28$. We now differentiate $z^2 = x^2 + y^2$ with respect to time:

$$z^2 = x^2 + y^2$$

$$2z\frac{dz}{dt} = 2x\frac{dx}{dt} + 2y\frac{dy}{dt}$$

$$\frac{dz}{dt} = \left(x\frac{dx}{dt} + y\frac{dy}{dt}\right)\frac{1}{z}$$

In two hours, $x = 21 \cdot 2 = 42$ miles; $y = 28 \cdot 2 = 56$ miles; and $z = \sqrt{42^2 + 56^2} = 70$ miles. Substituting all known values, we have

$$\frac{dz}{dt} = \left(x\frac{dx}{dt} + y\frac{dy}{dt}\right)\cdot\frac{1}{z}$$

$$= (42 \cdot 21 + 56 \cdot 28)\cdot\frac{1}{70}$$

$$= 35 \text{ mph}$$

Thus, in two hours, the ships are moving apart from each other at the relative rate of 35 mph. ∎

In Example 8, there were three variables involved in the set-up of the problem (x, y, and z). We were given the rates of change with respect to time for x and y and were asked to calculate a rate of change for z. In Example 9, we again use three variables in the set-up of the problem. This time, however, we are given the rate of change for one of them and wish to calculate the rate of change for another but have no information about the rate of change for the third variable. In this situation, enough information is given so that we can eliminate this variable before differentiation.

Example 9   The area of a triangle is increasing at the rate of 4 cm²/min. If the height is always twice the base, find the rate at which the base is increasing when the base is 6 cm.

Solution   The equation for the area of a triangle is $A = \frac{1}{2}bh$. We are given that $\frac{dA}{dt} = 4$ cm²/min and we wish to find $\frac{db}{dt}$ when $b = 6$. We aren't given $\frac{dh}{dt}$ and are not asked to calculate $\frac{dh}{dt}$, so we eliminate h.

$\quad\quad h = 2b$ Height is twice the base.

So $A = \frac{1}{2}(b)(2b)$

$\quad\quad A = b^2$

Differentiating with respect to t, we get

$$\frac{dA}{dt} = 2b\frac{db}{dt}$$

Now substituting, we have

$$4 = 2(6)\frac{db}{dt}$$

or $\dfrac{db}{dt} = \dfrac{1}{3}$ cm/min   ∎

See exercise 26 for a similar problem, but one in which we do not need to eliminate h.

3.6 EXERCISES

SET A

In exercises 1 through 16, find $\dfrac{dy}{dx}$ by using implicit differentiation. (See Examples 1, 2, and 3.)

1. $x - 3y + 1 = 0$ **2.** $x - 3y^2 + 1 = 0$ **3.** $y^2 = 2x - 1$

4. $y^2 = 2x^2 - 1$ **5.** $x^2 - 4y^2 = 4$ **6.** $9x^2 - 16y^2 = 144$

7. $x^3 + y^3 = 1$ **8.** $4x^3 - 8y^3 = 32$ **9.** $x^2y^2 - 4y^3 + 3x = 0$

10. $x^2y^2 - 4y^3 + 3x^2 = 0$ **11.** $2xy - 4xy^2 + x^2y = 0$ **12.** $5xy - y^4 = x$

13. $\sqrt{xy} = 1$ **14.** $\sqrt{x} + \sqrt{y} = 1$ **15.** $(x + y)^2 = 1$

16. $\dfrac{1}{x} + \dfrac{1}{y} = 1$

In exercises 17 and 18, find the indicated value. (See Example 4.)

17. Find $\dfrac{dy}{dx}\bigg|_{x=2}$ if $x^2y^2 - 2x = 0$.   **18.** Find $\dfrac{dy}{dx}\bigg|_{x=4}$ if $\sqrt{x} + \sqrt{y} = 11$.

19. The graph of $9x^2 + 16y^2 = 144$ is an ellipse and is pictured as shown here.

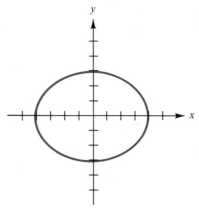

    a. Find $\dfrac{dy}{dx}$ using implicit differentiation.

    b. Find the slope of the tangent lines at $x = 2$. (*Beware:* There are two such lines!)

    c. Find the equation of the tangent line at $\left(2, \dfrac{3\sqrt{3}}{2}\right)$.

    d. Find $\dfrac{d^2y}{dx^2}$. (See Example 5.)

20. The graph of $x^2y - 4y = x$ is depicted as shown here.

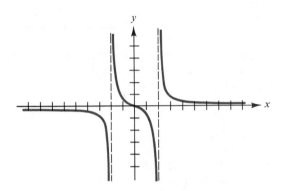

    a. Find $\dfrac{dy}{dx}$ using implicit differentiation.

    b. Find the slope of the tangent at $(3, \frac{3}{5})$.

    c. Find the equation of the tangent line at $(3, \frac{3}{5})$.

21. If $x^2 - 9y^2 = 16$, find $\dfrac{dx}{dt}$ when $\dfrac{dy}{dt} = 10$, $x = -5$, and $y = 1$.

22. If $9x^2 - 16y^2 = 144$, find $\dfrac{dy}{dt}$ when $\dfrac{dx}{dt} = -32$, $y = \dfrac{9}{4}$, and $x = 5$.

23. If $x^2 + y^2 = 25$, find $\dfrac{dy}{dt}$ when $\dfrac{dx}{dt} = 2$, $x = 3$, and $y = -4$.

SET B

Use Examples 7, 8, and 9 as a guide for exercises 24 through 26.

24. The area of a circle is increasing at the rate of 16 in.^{2}/min. Find the rate at which the radius is increasing when the radius is 2 in.

25. Water is being drained from an upright cylindrical tank at the rate of 3.5 ft^3/hr (approximately 10 liters/hr). The height of the tank is 20 ft and the diameter is 6 ft. At what rate is the water level falling?

26. The area of a triangle is increasing at a rate of 6 cm²/sec. If the height is increasing at the rate of 2 cm/sec, find the rate at which the base is decreasing when the height is 10 cm and the base is 12 cm. (*Hint:* Use the Product Rule to differentiate the area formula.)

BUSINESS
DEMAND

27. The quantity demanded, x, and price, p, of a product are related by the equation $px + 2x = 1000$. Use implicit differentiation to determine $\dfrac{dp}{dx}$.

PROFIT

28. A manufacturer has found that the daily profit from the sale of x items is $P(x) = -2x^2 + 200x$. At the same time, it has been determined that the number of items, x, that can be produced is related to the time of the day by $x = t^2 + 2t$. Find how profit is changing with respect to time when $t = 3$.

PROFIT

29. Suppose that a manufacturer finds that the daily profit from the manufacture and sale of x items is given by

$$P(x) = -2x^2 + 140x - 4$$

At the same time, it is determined that 18 items are sold daily; that is, $\dfrac{dx}{dt} = 18$. How is the profit changing when the 20th item is sold?

DEMAND

30. The demand function for a particular product is given by

$$px^2 = 1500$$

where p is measured in dollars and x is measured in hundreds. Find the rate of change of demand with respect to price, $\dfrac{dx}{dp}$, for 200 units. Interpret the results.

COST

31. The cost function for a product is given by

$$C = 400 + 2x$$

and the demand is given by

$$p^2 + 2x^2 = 100, \qquad 0 \le x \le 7$$

where x is measured in hundreds and p is measured in dollars. Find $\dfrac{dC}{dp}$.

BIOLOGY
ATHLETIC PERFORMANCE

32. A cyclist has found that the relationship between his speed (mph) and the miles covered per day is given by

$$0.1s^2M - 2s^2 = 450$$

Find the rate of change of the miles covered with respect to time, $\dfrac{dM}{dt}$.

SOCIAL SCIENCE
AGENCY MANAGEMENT

33. The number of clients, C, at a local agency is composed of walk-ins, W, and appointments, A. That is,

$$C = W + A$$

The number of clients varies according to the time of day. Find $\dfrac{dC}{dt}$ if $\dfrac{dW}{dt} = 4$ clients per hour and $\dfrac{dA}{dt} = -1$ client per hour. Interpret this result.

PHYSICAL SCIENCE
BOYLE'S LAW

34. Boyle's Law states that if the amount of gas and the temperature are held constant, then we have $PV = c$, where P represents pressure, V represents volume, and c is a constant. If the volume is decreasing at the rate of 5 in.³/min, at what rate will the pressure be changing when the volume is 60 in.³ and the pressure is 20 lb/in.²? Is the pressure increasing or decreasing? (*Hint:* Use the Product Rule to differentiate the formula.)

BOYLE'S LAW

35. Using Boyle's Law from exercise 34, find the rate at which the pressure is changing if the volume of the gas is increasing at a rate of 10 in.³/sec, when the volume is 120 in.³ and the pressure is 15 lb/in.².

VOLUME

36. A metal sphere is being exposed to extreme temperatures in a controlled environment. If the volume of the sphere is decreasing at a rate of 150 cm³/sec, at what rate is the radius decreasing when the radius is 14 cm?

AREA

37. Oil is leaking from a piece of equipment and dripping to form a circular pattern. If the oil is dripping so that the radius of the spot is increasing at a rate of 2 in./min, how fast is the area of the spot changing when the radius is 6 in.?

WRITING AND CRITICAL THINKING

In words (complete sentences), explain how to find $\dfrac{dy}{dx}$ for the function $x^3 + y^3 = 4$ using implicit differentiation.

3.7 THE DIFFERENTIAL

In Chapter 1, we defined the slope of a line m to be

$$m = \frac{\text{change in } y}{\text{change in } x}$$

In the beginning of this chapter, we returned to that idea and examined the rate of change of y with respect to x for any function, called the *derivative*. Suppose we have a particular change in x given for a function and want to calculate the corresponding change in y. As an example, consider the function $f(x) = \frac{1}{3}x^3 - 7$ and let x change from 3 to 3.1, a net change of 0.1. What will be the corresponding change for y?

Because $f(3) = 2$ and $f(3.1) = 2.930\overline{3}$, the actual change in y is 0.930\overline{3}. Our goal here, however, is not to find the exact change but to show how the derivative can be used to give a good *approximation* for the change in y.

Figure 3.8 helps us with this process. Let the coordinates of the point P represent the original values for x and y and let the coordinates of Q represent the *changed* values. Notice in Figure 3.8 that the change in x is h and that the change in y is $f(x + h) - f(x)$.

Figure 3.8

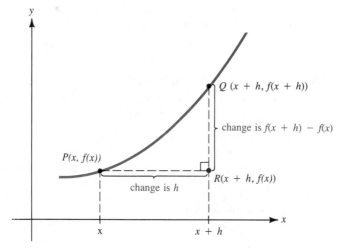

Recall that the derivative can be represented symbolically by $\dfrac{dy}{dx}$. It is now desirable to consider dy and dx as two separate quantities having dy, called the **differential of y,** associated with the change in y and dx, called the **differential of x,** associated with the change in x. Refer to Figure 3.9 to see how these quantities can be represented graphically.

The representation in Figure 3.9 is consistent with earlier definitions of slope in Chapter 1 for a line and in Chapter 2 for a curve, because the slope of the line $\overleftrightarrow{PS}$

Figure 3.9

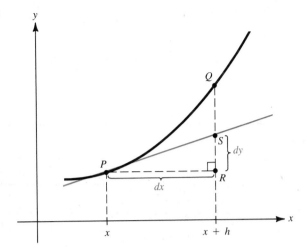

can be found by computing $\dfrac{SR}{PR}$ or by computing the derivative $\dfrac{dy}{dx}$ at the point P. In Figure 3.8, we saw that QR, or $f(x + h) - f(x)$, gives the actual change in y. In Figure 3.9, notice that if Q is very close to P, SR or dy will be a good approximation for QR. For this reason, the actual change in y, $f(x + h) - f(x)$, will be approximated using dy. Also notice from Figure 3.9 that the change in x, given by h, is the same as dx. The following definition gives us the technique for computing dy:

Differential of y

> The **differential of y** is
>
> $$dy = f'(x)\, dx$$

Before returning to our original problem, let's consider an example demonstrating the calculation of dy.

Example 1   Find dy for each of the following functions:
a. $f(x) = 2x + 5$ **b.** $f(x) = x^3 - x^2$ **c.** $f(x) = \dfrac{1}{x}$

Solution   From the definition,

$$dy = f'(x)\, dx$$

a.   Since $f'(x) = 2$
$$dy = 2\, dx$$

b. $f'(x) = 3x^2 - 2x$
$$dy = (3x^2 - 2x)\, dx$$

c. $f'(x) = \dfrac{-1}{x^2}$
$$dy = \dfrac{-1}{x^2}\, dx$$

∎

Going back to the original function $f(x) = \frac{1}{3}x^3 - 7$, $f'(x) = x^2$, so

$$dy = x^2\, dx$$

The change in x from 3 to 3.1 means h or $dx = 0.1$, which gives

$$dy = 3^2(0.1)$$
or $dy = 0.9$

Recall that the actual change in y was $0.930\overline{3}$ so that the value calculated here is a relatively good approximation.

Example 2   If x changes from 16 to 16.01 for $f(x) = \sqrt{x}$, approximate the change in y using differentials.

Solution

$$dx = 16.01 - 16 = 0.01$$

Since $dy = f'(x)\, dx$

we have $dy = \dfrac{1}{2\sqrt{x}}\, dx$

$$= \dfrac{1}{2\sqrt{16}}\,(0.01)$$

$$dy = 0.00125$$ ∎

Example 3   If x changes from 3 to 3.1 for $f(x) = x^2$, find:
a.   The actual change in y, $f(x + h) - f(x)$.
b.   The approximate change in y, dy.

Solution   a. $f(x + h) - f(x) = f(3 + 0.1) - f(3)$
$$= f(3.1) - f(3)$$
$$= (3.1)^2 - 3^2$$
$$= 9.61 - 9$$
$$= 0.61$$

b. $dy = f'(x)\, dx$
$$= 2x\, dx$$
$$= 2(3)(0.1)$$
$$= 0.6$$

Notice that this is a good approximation to the actual change.   ∎

Example 4   The Outdoor Pleasures Company makes redwood picnic tables. The cost of producing x picnic tables per week is given by

$$C(x) = 1200 - 10x + 0.1x^2$$

a.   Use the differential to approximate the change in the cost if production is increased from 20 tables per week to 21 tables per week.

b. Calculate the actual change, $C(21) - C(20)$, and compare it to your result in part a.

Solution **a.** Since $C'(x) = -10 + 0.2x$ and $dx = 21 - 20 = 1$, the change in cost is given by

$$dC = C'(20)dx$$
$$= [-10 + 0.2(20)](1)$$
$$= -6$$

This means that the cost goes down approximately $6 when the production level is increased from 20 to 21 tables per week.

b. $C(20) = 1200 - 200 + 40\ \ = 1040$
$C(21) = 1200 - 210 + 44.1 = 1034.1$

Therefore, the actual drop in cost is

$$C(21) - C(20) = 1034.1 - 1040 = -\$5.90$$

The approximate change differs from the actual change by only 10¢. ∎

Example 5 A company is redesigning the cans used for packaging their product. Height restrictions on the store shelves mean that the cans must remain 10 in. in height. The plan is to increase the radius of the can from 4 in. to 4.2 in.
  a. Approximately how much will the volume be affected?
  b. Each cubic inch of the substance going into the cans weighs 0.5 oz. If there is a restriction on weight increases of no more than 2 lb for each can, will the new cans meet this restriction?

Solution **a.** The volume of a can is given by

$$\text{volume} = \pi\ (\text{radius})^2\ (\text{height})$$
$$V = \pi r^2\ (10)$$
$$= 10\pi r^2$$

Using the differential with $dr = 4.2 - 4 = 0.2$, we have

$$dV = 20\pi r\ dr$$
$$= 20\pi\ (4)(0.2)$$
$$= 50.24$$

The volume will increase by approximately 50.24 in.^{3}.

b. An increase of 50.24 in.^{3} will mean an increase in weight of

$$(50.24)(0.5) = 25.12 \text{ oz per can}$$

Because 2 lb is equivalent to 32 oz, the 25.12 oz meets the weight restriction. ∎

Suppose it is the y coordinate of Q given by $f(x + h)$ that we wish to approximate. To do this, we need to add the change in y, dy, to the original y coordinate $f(x)$. This gives

$$f(x + h) \approx f(x) + dy$$

Example 6 illustrates this idea.

Example 6   Approximate $\sqrt[3]{29}$ using differentials.

Solution   The function we need is $f(x) = \sqrt[3]{x}$. Since $\sqrt[3]{27} = 3$, we can use 27 for our original x (coordinate of P) and $f(27) = 3$ for the corresponding y.

$$dx = 29 - 27 = 2$$
$$dy = f'(x)\, dx$$
$$= \frac{1}{3x^{2/3}}\, dx$$
$$= \frac{1}{3(27)^{2/3}}\, (2)$$
$$dy \approx 0.074074$$

The y coordinate of Q is $\sqrt[3]{29}$ and

$$\sqrt[3]{29} \approx f(27) + dy$$
$$\text{or} \quad \sqrt[3]{29} \approx 3.074074$$

■

In Chapter 6, the idea of differentials will be used once again. At that time, we will be changing derivatives to what is called the *differential form* for a derivative. For the function $f(x) = 2x^5 - 4x^3 + x - 1$, the differential form for $\dfrac{dy}{dx}$ is given by $dy = (10x^4 - 12x^2 + 1)\, dx$.

3.7   EXERCISES

SET A

In exercises 1 through 6, find dy, the differential of y. (See Example 1.)

1. $f(x) = 7 - 4x$
2. $f(x) = 4 - 3x^2$
3. $f(x) = x^3 + x$

4. $f(x) = \sqrt{x + 1}$
5. $y = \dfrac{1}{2 - x}$
6. $y = x + \dfrac{1}{x^2}$

For exercises 7 through 12, approximate the values using differentials. (See Examples 2, 3, and 6.)

7. $\sqrt{26}$
8. $\sqrt[4]{17}$
9. $\sqrt[3]{8.1}$

10. $\sqrt[3]{7.9}$
    (*Hint: dx* is negative.)
11. $\dfrac{1}{\sqrt{9.1}}$
12. $\dfrac{1}{\sqrt{15.99}}$

In exercises 13 through 20, for the given function find:

a. *The actual change in y*
b. *The approximate change in y using differentials (See Example 3.)*

13. $f(x) = x^3, x = 2, h = 0.1$ **14.** $f(x) = 3x^2, x = 2, h = 0.1$

15. $f(x) = 5x^2, x = 3, h = -0.1$ **16.** $f(x) = \sqrt{x}, x = 9, h = -0.01$

17. $f(x) = 4 - \sqrt{x}, x = 4, h = 0.01$ **18.** $f(x) = 0.5x^2 + 3, x = 2, h = -0.3$

19. $f(x) = \dfrac{1}{x^2}, x = 2, h = -0.01$ **20.** $f(x) = 1 - \dfrac{1}{x^2}, x = 3, h = 0.2$

For exercises 21 through 24 calculate $\dfrac{dy}{dx}$ and put in differential form.

21. $y = 5 + x - x^2$ **22.** $y = x^3 - x + 2$ **23.** $y = \sqrt{x - 3}$ **24.** $y = \dfrac{1}{\sqrt{1 - 2x}}$

SET B

25. Suppose the length of the edge of a cube has been increased from 5 to 5.1 cm. Compute the approximate change in the volume of the cube.

26. Suppose the length of the edge of a cube has been decreased from 3 to 2.9 in. Compute the approximate change in the volume of the cube.

27. The radius of a circle has been decreased from 6 in. to 5.8 in. Approximate the change in the area of the circle.

BUSINESS
MANUFACTURING

28. A manufacturer of ball bearings has determined that his machinery produces ball bearings of radius 3 cm with a maximum error of 0.02 cm in the radius. Because the weight of a ball bearing is sometimes important, it is necessary to control the volume of the ball bearings. What will be the maximum error for the volume? (*Hint:* The volume of the sphere equals $\frac{4}{3}\pi r^3$.)

29. Repeat exercise 28 assuming the maximum error in the radius is 0.01 cm.

MANUFACTURING

30. An engineering company is redesigning a part that is circular with a hole in the middle, as pictured. The radius of the hole, r_1, is to remain constant at 2 cm. The outer radius, r_2, is being increased from 6 cm to 6.1 cm. Approximate the change in the surface area of the part. [*Hint:* $S = \pi(r_2^2 - r_1^2)$.]

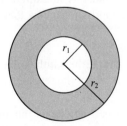

MANUFACTURING

31. Rework exercise 30 assuming the outer radius is increased from 6 cm to 6.2 cm.

COST FUNCTION

32. The cost of producing x clock radios is

$$C(x) = 0.02x^2 + 800$$

a. Find the cost of producing 1000 radios.
b. Estimate (using the differential) the cost of producing the 1001st clock radio.

PRICING

33. The price for a product is $p = 50 - \sqrt{x}$, where x is the number of units demanded at price p. Use differentials to approximate the price when 80 units are demanded.

REVENUE AND COST

34. A company manufactures and sells x items each month. The cost and revenue functions are, respectively,

$$C(x) = 75 + 5x + \frac{x^2}{500}$$

and

$$R(x) = 10x - \frac{x^2}{100}$$

Using differentials, find the approximate changes in revenue and cost if production is increased from 100 to 101 units.

BIOLOGY
MEDICINE

35. A certain type of tumor has been found to be nearly circular in shape. A new medicine causes the tumor to shrink from a diameter of 4 mm to 3.8 mm. Approximate the change in the area of the tumor.

DRUG RESEARCH

36. The time it takes for symptoms of a certain disease to reappear depends on the amount of a certain drug administered. The relationship is given by

$$t = 0.2x^2$$

where x is measured by cc's and t is measured in hours. If the dosage is increased from 20 to 21 cc's, approximate how long it will take for the symptoms to reappear.

PHYSICAL SCIENCE
AREA

37. The length of a rectangle is 12 in. If the width is changed from 4 in. to 4.25 in., approximate the change in the area of the rectangle.

VOLUME

38. The radius of a spherical-shaped object for an important research project needs to be 5 cm, with a maximum error of ± 0.1 mm. Approximate the corresponding maximum error in volume. (*Hint:* You must convert all units to cm or mm.)

WRITING AND CRITICAL THINKING

In the accompanying diagram two distances are labeled; one is labeled as d_1 and one is labeled as d_2. One of these is the actual change in y for the function and the other is the approximate change in y using the differential. Decide which is which and write out an explanation for your choice.

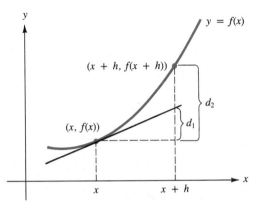

CHAPTER 3 REVIEW

VOCABULARY/FORMULA LIST

$f(x) = c$ implies $f'(x) = 0$
Power Rule
Product Rule
Quotient Rule
Generalized Power Rule
Chain Rule
second derivative
third derivative

higher-order derivative
explicit differentiation
implicit differentiation
related rates
marginal cost function
marginal revenue function
marginal profit function
break-even point

demand function
average velocity
instantaneous velocity
speed
instantaneous acceleration
differential of y
differential of x

$$\frac{d}{dx}(x^n) = nx^{n-1}$$

$$\frac{d}{dx}(cf(x)) = cf'(x)$$

$$\frac{d}{dx}[g[u(x)]] = g'[u(x)]u'(x)$$

$$\frac{d}{dx}(uv) = u\frac{dv}{dx} + v\frac{du}{dx}$$

$$\frac{d}{dx}(u/v) = \frac{v\frac{du}{dx} - u\frac{dv}{dx}}{v^2}$$

$$\frac{d}{dx}[g(x)]^n = n[g(x)]^{n-1}g'(x)$$

CHAPTER 3 REVIEW EXERCISES

In exercises 1 through 6, find $\dfrac{dy}{dx}$.

[3.1]   **1.** $y = 5x^2 - 7x + 12$

[3.1]   **2.** $y = -x^3 + 7$

[3.2]   **3.** $y = \dfrac{x+1}{x^2+2}$

[3.3]   **4.** $y = x^2\sqrt{1-x^2}$

[3.1]   **5.** $y = x^2 - \dfrac{1}{\sqrt{x}}$

[3.6]   **6.** $x^2 - 3y^2 = 5x$

[3.2]   **7.**   **a.**   Find the slope of the tangent line drawn to the curve $y = \dfrac{12x}{x-3}$ at the point $(-1, 3)$.

            b.   Find the equation of that tangent line.

[3.1]   **8.**   If $f(x) = -2x^2 + 11x - 1$, find $f'(0)$.

[3.3]   **9.**   **a.**   Suppose $y = \dfrac{1}{u+2}$ and $u = \sqrt{x}$. Find $\dfrac{dy}{du}$ and $\dfrac{du}{dx}$ and apply the Chain Rule to find $\dfrac{dy}{dx}$. Express $\dfrac{dy}{dx}$ in terms of x (not u).

            b.   Use the composition of functions to find $y(x)$ given $y(u) = \dfrac{1}{u+2}$ and $u(x) = \sqrt{x}$ and then find $y'(x)$ directly. Compare this result with part a.

[3.4]   **10.**   Find $f''(x)$ if $f(x) = \dfrac{x}{x^2-1}$.

[3.4]   **11.**   Find $\left.\dfrac{d^3y}{dx^3}\right|_{x=0}$ if $y = (2x+1)^4$.

BUSINESS   [3.5]   **12.**   If the cost, $C(x)$ (in dollars), of producing x typewriters is
COST FUNCTION

$$C(x) = 100 + 25x + 0.1x^2$$

find each of the following:
a.   The cost of producing 50 typewriters.
b.   The marginal cost function.
c.   The marginal cost of producing a typewriter at the 50-typewriter level of production.

PRODUCT ANALYSIS   [3.5]   **13.**   The demand for x cans of a new type of paint is

$$p = 16 - 0.02x$$

where p is measured in dollars. The total cost of production, $C(x)$ (in dollars), for x cans is

$$C(x) = 0.004x^2 + 1000$$

a. Find the revenue function, $R(x)$.
b. Find the marginal revenue, $R'(x)$.
c. Find the profit function, $P(x)$.
d. Find the marginal profit function, $P'(x)$.

PROFIT   [3.5]   **14.** The Greystone Company has determined that its profit from producing and selling x hundred concrete benches is given by

$$P(x) = \frac{700}{\sqrt{x^2 + 16}} - 175$$

Find the marginal profit for 300 benches.

SOCIAL   [3.5]   **15.** The number of people, N, in line at the post office in a big city at any time, t,
SCIENCE   over a three-hour period during the Christmas holidays is approximated by
QUEUING THEORY

$$N(t) = 20 - t + 0.025t^2, \qquad 0 \le t \le 180$$

where t is measured in minutes. Determine the instantaneous rate of change for $N(t)$ at $t = 10$ min.

PHYSICAL   [3.6]   **16.** If the volume of a spherical balloon is increasing at the rate of 20 ft^3/sec, at
SCIENCE   what rate is the radius increasing when the radius is 5 ft (volume of sphere is
VOLUME $\frac{4}{3}\pi r^3$)?

MOTION   [3.5]   **17.** When a ball is dropped from a height of 32 ft, its height, h (in feet), at t sec is $h = 32 - 16t^2$.

a. Find $\dfrac{dh}{dt}\bigg|_{t = 0.5}$.

b. When will the ball hit the ground?

[3.7]   **18.** Approximate $\sqrt{4.1}$ using differentials.

CHAPTER 3 TEST

PART ONE

Multiple choice. In exercises 1 through 10, put the letter of your correct response in the blank provided.

_____   **1.** If $y = 3x^2 + \dfrac{2}{x^2}$, then $\dfrac{dy}{dx} = \underline{\ ?\ }$.

a. $6x - \dfrac{4}{x^3}$ b. $6x - \dfrac{4}{x}$

c. $6x - \dfrac{1}{4x^3}$ d. $6x + \dfrac{4}{x^3}$

_____ 2. If $f(x) = \sqrt{2x + 5}$, then $f'(2) =$ _?_ .

a. $\dfrac{1}{6}$ b. -3 c. $\dfrac{-1}{6}$ d. $\dfrac{1}{3}$

_____ 3. If $f(x) = 2x^3(1 - 2x)^5$, then $f'(1) =$ _?_ .

a. 4   b. -16 c. -26 d. 14

_____ 4. If $f(x) = 4[g(x)]^3$, then $f'(x) =$ _?_ .

a. $12[g(x)]^2$ b. $12[g(x)]^2 \cdot g'(x)$

c. $12[g'(x)]^2$ d. $3[4g(x)]^2 \cdot g'(x)$

_____ 5. Find $\dfrac{dy}{dx}$ for $x^3 + y^3 = 2y$.

a. $\dfrac{-x^2}{y^2 - 2}$ b. $\dfrac{3x^2}{2 - 3y^2}$

c. $\dfrac{2 - 3x^2}{3y^2}$ d. $\dfrac{-3x^2}{2y}$

_____ 6. If $y = \dfrac{x^2}{1 - x}$, find $\dfrac{dx}{dx}$.

a. $\dfrac{2x - x^2}{(1 - x)^2}$ b. $\dfrac{2x - 3x^2}{(1 - x)^2}$

c. $\dfrac{x^2 - 2x}{(1 - x)^2}$ d. $\dfrac{-2x}{(1 - x)^2}$

_____ 7. For $f(x) = (x^3 + 2)^4$, find $f''(x)$.

a. $12(x^3 + 2)^2$

b. $72x(x^3 + 2)^2$

c. $12x(x^3 + 2)^2(2x^3 + 3x + 4)$

d. $12x(x^3 + 2)^2(11x^3 + 4)$

_____ 8. What is the equation of the tangent line for the curve $y = \dfrac{2}{x - 1}$ at the point $(2, 2)$?

a. $y = 4 - x$ b. $y = 6 - 2x$

c. $y = 2x - 2$ d. $y = 2 - 2x$

_____ 9. Given $f(x) = \dfrac{x - 1}{x^2}$, for what value of x does $f'(x) = 0$?

a. $x = 0$ b. $x = 2$ c. $x = -2$ d. $x = 1$

_____ 10. If $y = u^3 + u$ and $u = 2\sqrt{x}$, then $\dfrac{dy}{dx} =$ _?_ .

a. $\dfrac{12x + 1}{\sqrt{x}}$ b. $\dfrac{12x + 2\sqrt{x}}{\sqrt{x}}$

c. $\dfrac{6x + 1}{\sqrt{x}}$ d. $- 12x\sqrt{x} - \sqrt{x}$

PART TWO

Solve and show all work.

11. For $f(x) = \dfrac{1}{\sqrt{1 - 2x}}$, find $f'(-4)$.

12. For $f(x) = x - \dfrac{1}{x^2}$, find $f'(x)$.

13. For $f(x) = x^3 - 3x$, determine any value for x such that $f'(x) = 0$.

14. For $f(x) = (1 - x^3)^4$, find $f''(0)$.

15. For $y = \dfrac{3}{\sqrt[3]{x}}$, find:

    a. $\dfrac{dy}{dx}$ **b.** $\dfrac{d^2y}{dx^2}$ **c.** $\dfrac{d^3y}{dx^3}$

BUSINESS
MARGINAL COST

16. The cost of producing x hundred units of a product is

$$C(x) = 0.05x^3 + 500$$

where $C(x)$ is measured in thousands of dollars. Find the marginal cost of producing 500 units.

MARGINAL PROFIT

17. The Waterich Insurance Company has determined that the monthly profit from selling x policies is

$$P(x) = 3\sqrt{10x - 100}$$

where P is measured in thousands of dollars.
a. Find the marginal profit.
b. Find $P'(100)$.

BIOLOGY
MEDICINE

18. An experiment involving a new drug is being conducted to determine how long it remains in the bloodstream. The amount, $Q(t)$, measured in milligrams, at any time, t, is

$$Q(t) = 40 - \frac{t^2}{120}$$

where t is measured in minutes. Determine the rate at which $Q(t)$ is changing when $t = 30$ min.

SOCIAL SCIENCE
DEMOGRAPHY

19. A study has shown that since 1983, the number of people purchasing new homes in a certain city is decreasing and can be estimated by

$$N(t) = 12\sqrt{35 - 2t}$$

where t is measured in years and $N(t)$ is measured in hundreds. At what rate was the number of new-home buyers decreasing in 1988 ($t = 5$)?

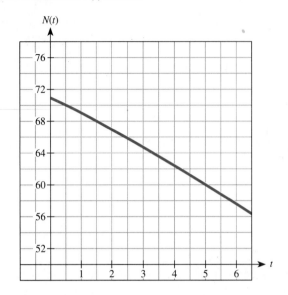

20. An object moves along a straight line according to the function

$$s(t) = t^4 + t^2 + 1$$

where $s(t)$ is measured in feet and t is measured in seconds.
a.   Find the velocity of the object for $t = 2$.
b.   Find the acceleration of the object for $t = 2$.

21. Choose five of the vocabulary words from the list given for this chapter and write a sentence using them.

4

CURVE SKETCHING AND OPTIMIZATION

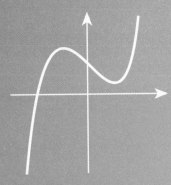

4.1 INCREASING AND DECREASING FUNCTIONS

The derivatives and their properties, introduced in Chapter 2, are further explored in this chapter. In particular, we use the first and second derivatives of a function to examine the graph of the function.

First, we want to determine where a function is increasing or decreasing. If a manufacturer has determined its profit equation to be $P(x) = 5x - 0.02x^2$, it would certainly be advantageous for the company to know the values of x for which profit is *increasing* and the values of x for which profit is *decreasing*. We have the following definition:

Increasing function

> A function is **increasing** over an interval if, whenever $x_1 < x_2$, then $f(x_1) < f(x_2)$ for any x_1 and x_2 in that interval.

Graphically this means that as we move from left to right across the graph, the y values *increase* (see Figure 4.1).

Figure 4.1

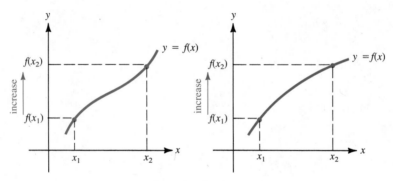

Increasing functions

We are also interested in defining what it means for a function to be decreasing over an interval.

Decreasing function

> A function is **decreasing** over an interval if, whenever $x_1 < x_2$, then $f(x_1) > f(x_2)$ for any x_1 and x_2 in that interval.

Graphically this means that as we move from left to right across the graph, the y values *decrease* (see Figure 4.2).

It is also possible that a function neither increases nor decreases on an interval. An example of such a function is $f(x) = 3$, whose graph is a horizontal line (on *any* interval). We say that f is a *constant* function for a given interval.

Figure 4.2

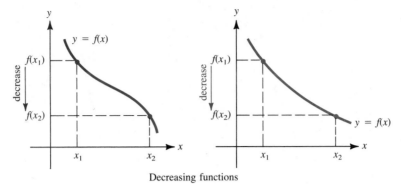

Decreasing functions

Example 1 For $f(x) = x^2$, sketch the graph and determine for what intervals the function is increasing or decreasing.

Solution First, we construct a table of values.

x	-2	-1	0	1	2
y	4	1	0	1	4

Next, we make a sketch, as in Figure 4.3.

Figure 4.3

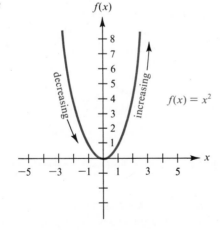

We can see from the graph that for $x < 0$, the function is decreasing and for $x > 0$, the function is increasing. ■

Example 2 Using the graph of the function given in Figure 4.4, determine visually where the function is increasing and color-code those portions of the graph.

Figure 4.4

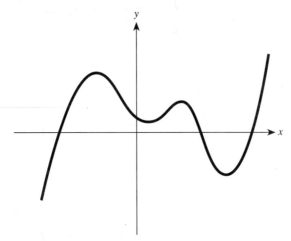

Solution   We must move our eyes from left to right across the graph, looking for the portions of the curve that are increasing or "rising." There are three such portions of the graph, and they are color-coded in Figure 4.5.

Figure 4.5

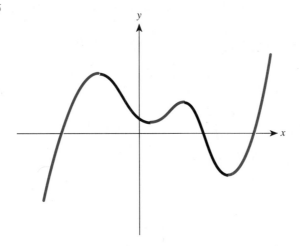

In the remainder of this section, we examine how to use the first derivative to determine where a function is increasing and where it is decreasing. Returning to the function from Example 1, we now examine its first derivative.

Example 3   For $f(x) = x^2$, calculate $f'(x)$ and determine the intervals of x for which
a. $f'(x) < 0$ **b.** $f'(x) > 0$

Solution   For $f(x) = x^2, f'(x) = 2x.$

a. If $f'(x) < 0$ **b.** If $f'(x) > 0$
    then $2x < 0$ then $2x > 0$
    or $x < 0$ or $x > 0$ ■

Notice, by comparing examples 1 and 3, that the function f is decreasing precisely where $f'(x)$ is negative, and f is increasing precisely where $f'(x)$ is positive. Recall that the derivative can be interpreted as the slope of the tangent line to a curve. You can see from the sketch of $f(x) = x^2$ (Figure 4.6) that any tangent line drawn to the curve for $x < 0$ has a negative slope ($f'(x)$ is negative) and that any tangent line drawn to the curve for $x > 0$ has a positive slope ($f'(x)$ is positive).

Figure 4.6

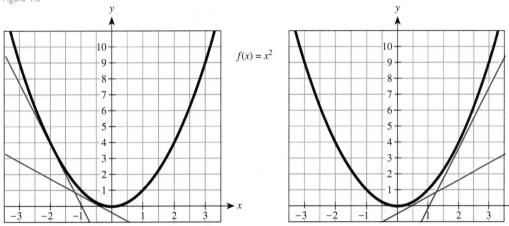

slopes of tangent lines are negative
$f'(x) < 0$

slopes of tangent lines are positive
$f'(x) > 0$

We are now ready for a more formal statement of our observations.

Property 1

For a function $f(x)$ that is differentiable for all x on an interval, we say:

1. $f(x)$ is *decreasing* over the interval if $f'(x) < 0$
2. $f(x)$ is *increasing* over the interval if $f'(x) > 0$

We should also note that if $f'(x) = 0$ for all x on an interval, then the function is constant on that interval.

Example 4   Find the intervals on which $f(x) = x^2 + 6x + 10$ is increasing or decreasing.

Solution   If $f(x) = x^2 + 6x + 10$, then

$$f'(x) = 2x + 6$$

$f'(x) > 0$	$f'(x) < 0$
$2x + 6 > 0$	$2x + 6 < 0$
$2x > -6$	$2x < -6$
$x > -3$	$x < -3$

From Property 1, this implies that the function is decreasing for $x < -3$ and increasing for $x > -3$. The graph of $y = x^2 + 6x + 10$ appears in Figure 4.7.

Figure 4.7

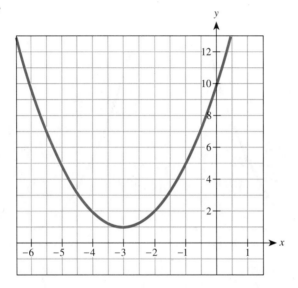

In Example 5 we use an alternate method for determining where the derivative is negative and positive. You may find this technique easier to use than solving inequalities, as in the previous example, especially as the derivatives become more involved.

The new technique uses the values of x for which the first derivative is zero. We refer to the values where $f'(x) = 0$ as **type I critical values** for the function.

Type I critical values

The values of x for which $f'(x) = 0$ are called **type I critical values.**

Example 5   For $f(x) = x^3 - 12x + 4$, find the type I critical values and the intervals on which the function is increasing or decreasing.

Solution   First, find critical values by letting $f'(x)$ equal zero.

$$f'(x) = 3x^2 - 12$$
$$\text{So } 3x^2 - 12 = 0$$
$$3x^2 = 12$$
$$x^2 = 4$$
$$x = \pm 2$$

Now write intervals using these critical values $x = 2$ and $x = -2$ as follows:

$$x < -2 \qquad\qquad -2 < x < 2 \qquad\qquad x > 2$$

Next, arbitrarily pick a test point from each interval and substitute it into $f'(x)$ to determine whether it is positive or negative. The results can be put in tabular form.

Interval	$x < -2$	$-2 < x < 2$	$x > 2$
Sign of $f'(x)$	positive (use $x = -3$)	negative (use $x = 1$)	positive (use $x = 3$)
Result: f(x) is	increasing	decreasing	increasing

Be careful in using this new technique. Do not assume that the sign of $f'(x)$ will always change at the critical values. See exercise 17 for a function with a critical value at $x = 2$, but for which the sign of $f'(x)$ does not change there.

Example 6   For $y = x^4 + \frac{4}{3}x^3 - 4x^2 + 2$, find the type I critical values and the intervals on which the function is increasing or decreasing.

Solution $\dfrac{dy}{dx} = 4x^3 + 4x^2 - 8x$

To find the critical values, we set $\dfrac{dy}{dx}$ equal to zero.

$$4x^3 + 4x^2 - 8x = 0$$
$$4x(x^2 + x - 2) = 0$$
$$4x(x + 2)(x - 1) = 0$$

Therefore, the critical values are

$$x = -2 \qquad\qquad x = 0 \qquad\qquad x = 1$$

The intervals are

$$x < -2 \qquad -2 < x < 0 \qquad 0 < x < 1 \qquad x > 1$$

Interval	$x < -2$	$-2 < x < 0$	$0 < x < 1$	$x > 1$
Sign of $\dfrac{dy}{dx}$	−	+	−	+
Result: y is	decreasing	increasing	decreasing	increasing

See Figure 4.8.

Figure 4.8

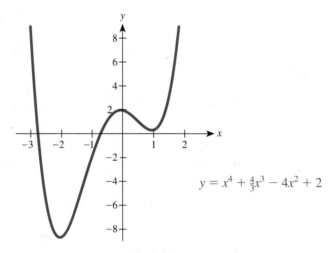

$$y = x^4 + \tfrac{4}{3}x^3 - 4x^2 + 2$$

Thus far, the examples have dealt with functions defined for all real numbers. We would also like to include some other types of functions. Before doing so, however, we will extend the definition of *critical value* to include values of x for which $f'(x)$ is not defined but $f(x)$ is defined.

Critical values

> 1. [Type I] If $f'(a) = 0$, then a is called a **critical value.**
> 2. [Type II] If $f'(b)$ is undefined but $f(b)$ is defined, then b is a **critical value.**

Example 7   For $f(x) = \sqrt[3]{x - 1}$, find the intervals on which the function is increasing or decreasing.

Solution   Rewrite as

$$f(x) = (x - 1)^{1/3}$$

So $f'(x) = \frac{1}{3}(x - 1)^{-2/3}(1)$

or $f'(x) = \dfrac{1}{3\sqrt[3]{(x - 1)^2}}$

In this case, $f'(x)$ is never zero but $f'(x)$ is undefined for $x = 1$. Since $f(1)$ is defined ($f(1) = 0$), $x = 1$ is a critical value. This yields only two intervals

$x < 1$ $x > 1$

Interval	$x < 1$	$x > 1$
Sign of $f'(x)$	+	+
Result: $f(x)$ is	increasing	increasing

We could say the function is increasing for $x < 1$ and $x > 1$, or, more simply, the function is *always increasing*.

A table of values and a sketch of this function are given below. (See Figure 4.9.)

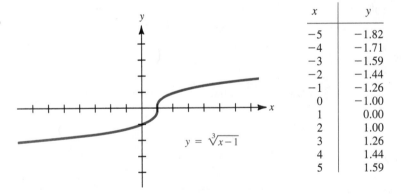

Figure 4.9

$y = \sqrt[3]{x - 1}$

x	y
−5	−1.82
−4	−1.71
−3	−1.59
−2	−1.44
−1	−1.26
0	−1.00
1	0.00
2	1.00
3	1.26
4	1.44
5	1.59

If a function has no critical values, it can still change from decreasing to increasing, or vice versa. Such may be the case for a function with vertical asymptotes.

For example, consider the function $f(x) = \dfrac{1}{x^2}$.

$$f(x) = x^{-2}$$
$$f'(x) = -2x^{-3}$$

Notice that f has no critical values. It has a vertical asymptote at $x = 0$ and the function is increasing if $x < 0$ ($f'(x) > 0$ there) and decreasing if $x > 0$ ($f'(x) < 0$ there). The graph of $y = f(x)$ appears in Figure 4.10.

Figure 4.10

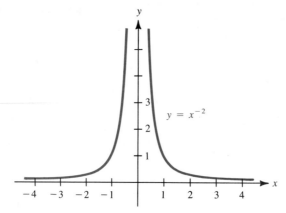

$$y = x^{-2}$$

We conclude this section with an applied problem and a summary.

Example 8   The function $R = 60 + 10t - t^2$ ($0 \leq t \leq 11$) represents the average heart rate, R, in beats per minute for a group of males in an experiment t minutes after a new drug being tested has been administered. Over what time interval did the heart rate increase and over what interval did it decrease?

Solution   For $R = 60 + 10t - t^2$,

$$\frac{dR}{dt} = 10 - 2t$$

Then, setting $\frac{dR}{dt}$ equal to zero, we have

$$10 - 2t = 0$$
$$t = 5$$

Interval	$0 < t < 5$	$5 < t < 11$
Sign of $\dfrac{dR}{dt}$	$+$	$-$
Result: R is	increasing	decreasing

This implies that the heart rate increased during the first 5 minutes and decreased during the last 6 minutes of the experiment.   ■

We summarize the work done in this section as follows:

Summary:
Finding intervals
on which a function
is increasing
or decreasing

1. Find the function's *critical values*:
    a. Find values of x for which $f'(x) = 0$.   [Type I]
    b. If appropriate, determine values of x for which $f'(x)$ is undefined but $f(x)$ is defined.   [Type II]
2. Determine the sign of $f'(x)$ in the intervals designated by the function's critical values.
3. The function is increasing where $f'(x) > 0$ and decreasing where $f'(x) < 0$.
4. The function may change from decreasing to increasing, or vice versa, at its vertical asymptotes. The asymptotes should be used when graphing the function and may determine the intervals on which the function is increasing or decreasing.

4.1 EXERCISES

SET A

In exercises 1 through 4, determine from the graph the intervals where each function is increasing, decreasing, or is constant. (See Example 2.)

1.

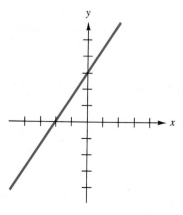

2.

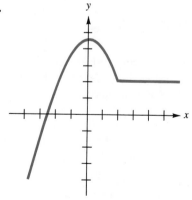

3.

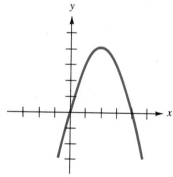

4.

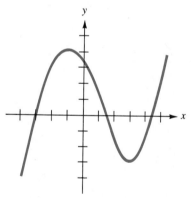

In exercises 5 through 16, find the intervals on which each function is increasing or decreasing. (See Examples 1, 3, 4, 5, 6, and 7.)

5. $y = 3x + 1$ **6.** $f(x) = 5$ **7.** $f(x) = 6 - x^2$ **8.** $y = x^2 + x - 4$

9. $f(x) = x^3$ **10.** $f(x) = 12x - x^3$ **11.** $y = \sqrt[3]{x - 4}$ **12.** $f(x) = \sqrt[3]{x + 2}$

13. $f(x) = \dfrac{2x}{x - 4}$ **14.** $y = \dfrac{1}{x^2 + 1}$ **15.** $y = \dfrac{1}{\sqrt{x + 2}}$ **16.** $y = x^2 + \dfrac{1}{x}$

17. For $f(x) = (x - 2)^3$, find any critical values and determine the intervals where f is increasing or decreasing. Make a table of values and sketch to verify your result.

SET B

In exercises 18 through 27:
a. *Determine the intervals where f is increasing or decreasing.*
b. *Sketch the graph of $y = f(x)$.*
(See Examples 4 and 6.)

18. $f(x) = x^2 - 4$ **19.** $f(x) = 4 - x^2$

20. $f(x) = 5 - 7x$ **21.** $f(x) = 5 + 7x$

22. $f(x) = x^3 - 5x^2 + 8x$ **23.** $f(x) = x^3 - 6x^2 + 11x - 6$

24. $f(x) = 6 + 15x + 6x^2 - x^3$ **25.** $f(x) = x^3 - 3x - 2$

26. $f(x) = (x - 2)^2(x - 1)$ **27.** $f(x) = (x - 2)^2(1 - x)$

28. Sketch the graph of

$$f(x) = \begin{cases} -x^2, & x \le 0 \\ x, & 0 < x < 1 \\ 2 - x, & x \ge 1 \end{cases}$$

and from the graph, find the intervals where the function is increasing or decreasing.

29. Consider $f(x) = \dfrac{2}{3 - x}$.
  a. Find $f'(x)$.
  b. Does f have any critical values?
  c. Such a function, one that is increasing for every x in its domain, is said to be *increasing where defined*. Graph the function.

In exercises 30 through 33, a function is graphed, along with its derivative, using a graphics calculator; the two are labeled "f" and "g," but not necessarily respectively. Using the relationship discussed in this section, determine which (of "f" and "g") is the function and which is the derivative.

30.

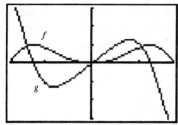

31.

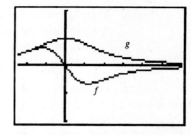

32.

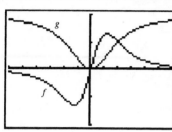

33.

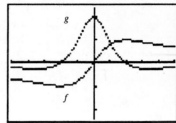

For exercises 34 through 38, use the graph given for y = f(x) defined on −8 ≤ x ≤ 6 to answer the questions about f′(x). (Hint: Think of the derivative as the slope of a tangent line to the curve at a specified point.)

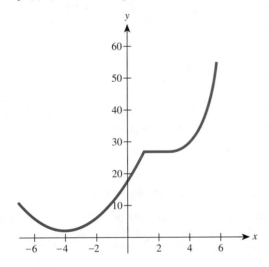

34. What is the value of $f'(x)$ at $x = -4$?

35. What can be said of $f'(x)$ for $-8 < x < -4$?

36. What can be said of $f'(x)$ for $1 < x < 3$?

37. What is $f'(1)$?

38. For what values of x is $f'(x) > 0$?

BUSINESS
PROFIT

39. The Magnifeek Company's profit, P, is depicted in the accompanying graph. Approximate the value of x where profit goes from increasing to decreasing. What is the corresponding value (in dollars) for the profit?

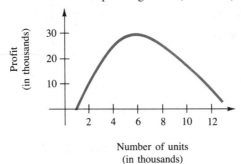

Number of units
(in thousands)

MANUFACTURING

40. A manufacturer has determined the following cost function:

$$C(x) = 18x - 15x^2 + 4x^3$$

where x is measured in thousands and $C(x)$ is measured in dollars.
a. Determine the marginal cost function.
b. When is cost decreasing? Increasing?

MARGINAL ANALYSIS

41. Suppose the cost, $C(x)$, for producing x units of a certain running shoe is

$$C(x) = 0.4x^2 - 1.6x + 800$$

where x is measured in hundreds.
a. Determine the marginal cost function.
b. When is cost increasing? Decreasing?

DEMAND AND REVENUE
EQUATIONS

42. The demand function for torque wrenches at Acme Tools is the following relationship between quantity sold, x, and asking price, p:

$$p = 80 - 0.2x, \qquad 0 \le x \le 375$$

a. Determine the revenue equation. (See Section 3.5.)
b. For what values of x is revenue increasing? Decreasing?

PROFIT

43. The owner of an electronics firm has determined that the cost and revenue functions for her new product, a memory board for a mainframe computer, are given by

$$C(x) = 0.3x^2 + 200x + 500$$
$$R(x) = 700x - 0.2x^2$$

a. Determine the profit function. (See Section 3.5.)
b. For what values of x is profit increasing? Decreasing?

BIOLOGY
PULSE RATE

44. During one phase of a physical fitness test, participants were asked to ride a stationary bicycle at varying speeds while their pulse rate was monitored. For a particular age group, the pulse rate, R, was determined to be

$$R = 60 + 10t - 0.25t^2, \qquad 0 \le t \le 15$$

where t is measured in minutes and R is measured in beats per minute. Over what time interval was the pulse rate for this particular age group found to be increasing?

SOCIAL SCIENCE
VOTER TURNOUT

45. Since 1970 in a particular town, it has been possible to approximate voter turnout by the function

$$N(t) = \frac{200t}{t^2 + 4} + 8, \qquad 0 \le t \le 15$$

where N is the number of people (in thousands) and t is the time (in years) since 1970. (For 1970, $t = 0$. For 1971, $t = 1$, and so on.) Determine the years for which the voter turnout was increasing.

WRITING AND CRITICAL THINKING

Explain in words the relationship between the sign of the first derivative and the intervals where a function is increasing or decreasing. Also discuss the behavior of the function when the derivative is zero. It might be helpful to include a discussion of the tangent lines to the curve at various points.

4.1A INCREASING AND DECREASING FUNCTIONS ON A GRAPHICS CALCULATOR

If we use a graphics calculator to examine the graph of the function $f(x) = x^3 - 3x + 5$, we notice that the graph rises (from left to right) for some intervals and falls (from left to right) for some intervals. In the interval $-4.5 \le x \le -1$, for example, the curve rises, and we say the function *increases* in that interval. See Figure 4.1A.1.

Figure 4.1A.1
$y = x^3 - 3x + 5,$
$-4.5 \le x \le -1$

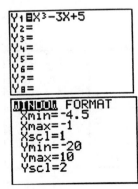

Technically, we define an increasing function over an interval as follows:

Increasing function

A function is **increasing** over an interval if, whenever $x_1 < x_2$, then $f(x_1) < f(x_2)$ for any x_1 and x_2 in that interval.

In Figure 4.1A.2, we graph the same function in a different interval and notice that the curve is *decreasing* in that interval:

Figure 4.1A.2
$y = x^3 - 3x + 5$,
$-1 \le x \le 1$

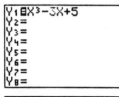

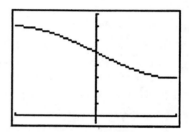

Decreasing function

A function is **decreasing** over an interval if, whenever $x_1 < x_2$, then $f(x_1) > f(x_2)$ for any x_1 and x_2 in that interval.

By altering the WINDOW values once again, we get Figure 4.1A.3, a more "global" look at the graph of $y = x^3 - 3x + 5$, and can observe the intervals over which the function is increasing and those over which it is decreasing:

Figure 4.1A.3
$y = x^3 - 3x + 5$,
$-6 \le x \le 6$

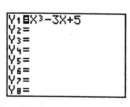

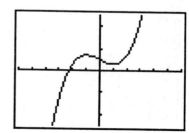

Whether a function is increasing or decreasing is directly related to the sign of its first derivative because the derivative represents the *rate of change* of the function. Thus, on intervals where the function is increasing, the derivative is positive; on intervals where the function is decreasing, the derivative is negative. This relationship is examined graphically in Figures 4.1A.4 and 4.1A.5.

Figure 4.1A.4 $f(x)$ is increasing and $f'(x)$ is above the x axis (greater than 0).

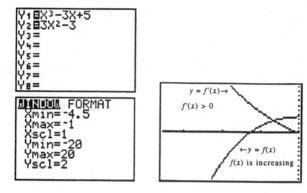

Notice that any tangent line drawn to the graph of the function in Figure 4.1A.4 (where the function is increasing) will have a positive slope. Any tangent line drawn to the graph of the function in Figure 4.1A.5 (where the function is decreasing) will have a negative slope.

Figure 4.1A.5 $f(x)$ is decreasing and $f'(x)$ is below the x axis (less than 0).

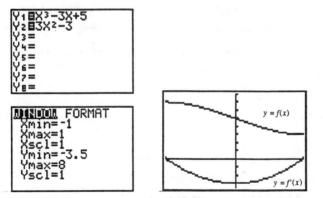

Property 1

For a function $f(x)$ that is differentiable for all x on an interval, we say:

1. $f(x)$ is *decreasing* over the interval if $f'(x) < 0$
2. $f(x)$ is *increasing* over the interval if $f'(x) > 0$

See Figure 4.1A.6.

Figure 4.1A.6 $f(x)$ is increasing and $f'(x) > 0$ if $x < -1$ or $x > 1$; $f(x)$ is decreasing and $f'(x) < 0$ if $-1 < x < 1$.

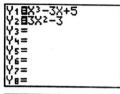

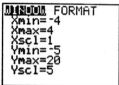

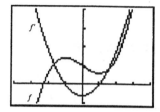

Algebraically, to determine where a function is increasing, we find the first derivative and try to solve the inequality $f'(x) > 0$. For the function $f(x) = x^3 - 3x + 5$, we have:

$$f'(x) = 3x^2 - 3$$
$$3x^2 - 3 > 0$$
$$x^2 > 1$$
$$x > 1 \quad \text{or} \quad x < -1$$

In practice, we usually find values for which the first derivative is zero, called **type I critical values.**

Type I critical values | The values of x for which $f'(x) = 0$ are called **type I critical values.**

To see how the type I critical values can be used to find the intervals on which a function is increasing and decreasing, we consider the function $y = \frac{3}{4}x^4 + x^3 - 3x^2 + \frac{3}{2}$.

Step 1.   Find $\dfrac{dy}{dx}$:

$$\frac{dy}{dx} = 3x^3 + 3x^2 - 6x$$

Step 2.   To find the critical values, set $\dfrac{dy}{dx}$ equal to zero:

$$3x^3 + 3x^2 - 6x = 0$$
$$3x(x^2 + x - 2) = 0$$
$$3x(x + 2)(x - 1) = 0$$
$$x = 0, \quad x = -2, \quad x = 1$$

Step 3.   Find the intervals created by the critical values:

$$x < -2 \qquad -2 < x < 0 \qquad 0 < x < 1 \qquad x > 1$$

Step 4. Next, arbitrarily pick a test point from each interval and substitute it into $f'(x)$ to determine whether it is positive or negative. The results, in tabular form, are as follows:

Interval	$x < -2$	$-2 < x < 0$	$0 < x < 1$	$x > 1$
Sign of $f'(x)$	−	+	−	+
Result: $f(x)$ *is*	decreasing	increasing	decreasing	increasing

A graph of $y = \frac{3}{4}x^4 + x^3 - 3x^2 + \frac{3}{2}$ appears in Figure 4.1A. 7.

Figure 4.1A.7
$y = \frac{3}{4}x^4 + x^3 - 3x^2 + \frac{3}{2}$.
The function is increasing for $-2 < x < 0$
and $x > 1$*; it is decreasing for* $x < -2$
and $0 < x < 1$*.*

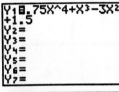

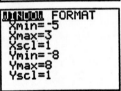

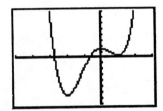

A second type of critical value occurs when there is a value of x for which the first derivative is undefined but where the function *is* defined. We call this a type II critical value.

Type II critical values | The values of x for which $f'(x)$ is undefined but $f(x)$ is defined are called **type II critical values.**

Consider $y = \sqrt[3]{2x + 5}$. The first derivative is undefined at $x = -2.5$ because $\frac{dy}{dx} = \frac{2}{3}(2x + 5)^{-2/3} = \frac{2}{3(\sqrt[3]{2x + 5})^2}$, and the intervals we examine are $x < -2.5$ and $x > -2.5$.

Interval	$x < -2.5$	$x > -2.5$
Sign of $f'(x)$	+	+
Result: $f(x)$ *is*	increasing	increasing

Notice that $\dfrac{dy}{dx}$ is positive *regardless* of x. (See Figure 4.1A.8.)

Figure 4.1A.8 Notice that as x approaches −2.5, the tangent lines become more vertical.

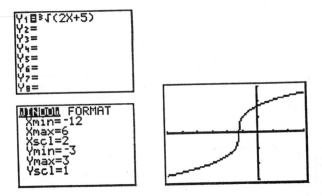

In addition to critical values, a function can change from increasing to decreasing (or vice versa) on either side of an asymptote. For example, consider $y = \dfrac{1}{(x-1)^2}$, a function whose graph has a vertical asymptote at $x = 1$. The derivative, $\dfrac{dy}{dx} = \dfrac{-2}{(x-1)^3}$, produces no critical values because the original function is undefined for $x = 1$. (See Figure 4.1A.9.)

Figure 4.1A.9 Notice that the function increases on the left of the asymptote x = 1 and decreases on the right.

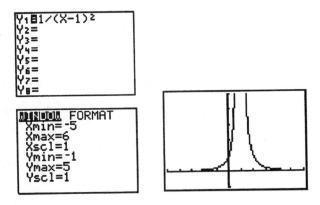

We summarize the work done in this section as follows:

Summary:
Finding intervals
on which a function
is increasing
or decreasing

1. Find the function's critical values:
   a. Find values of x for which $f'(x) = 0$.   [Type I]
   b. If appropriate, determine values of x for which $f'(x)$ is undefined but $f(x)$ is defined.   [Type II]
2. Determine the sign of $f'(x)$ in the intervals designated by the function's critical values.
3. The function is increasing where $f'(x) > 0$ and decreasing where $f'(x) < 0$.
4. The function may change from decreasing to increasing, or vice versa, at its vertical asymptotes. These asymptotes should be used when graphing the function and may determine the intervals on which the function is increasing or decreasing.

4.1A   EXERCISES

In exercises 1 through 10, use calculus and the graphics calculator to graph each function and determine the intervals on which the function is increasing and the intervals on which the function is decreasing.

1. $y = 8 - x^2$

2. $y = 8 - x^3$

3. $y = x^4 + \frac{4}{3}x^3 - 4x^2 + 2$

4. $y = 12x - x^3$

5. $f(x) = (x^2 - 2)(x^2 - 1)$

6. $f(x) = (x^2 - 2)(x^2 + 2)$

7. $y = \sqrt[3]{x - 3}$

8. $y = \dfrac{x}{x - 2}$

9. $f(x) = \dfrac{1}{\sqrt{x + 1}}$

10. $y = \dfrac{1}{x^2 + 1}$

11. Consider the function $f(x) = \dfrac{3}{x - 4}$.
    a. Find $f'(x)$.
    b. Does f have any critical values?
    c. Such a function, one that is decreasing for every x in its domain, is said to be *decreasing where defined.* Graph $y = f(x)$.

In exercises 12 through 15, a function is graphed, along with its derivative, using a graphics calculator; the two are labeled "f" and "g," but not necessarily respectively. Using the relationship discussed in this section, determine which (of "f" and "g") is the function and which is the derivative.

12.

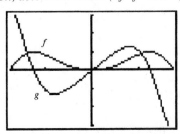

13.

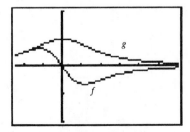

14.

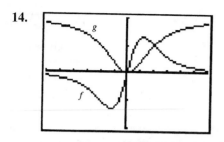

15.

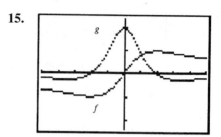

4.2   CONCAVITY AND POINTS OF INFLECTION

In Section 4.1, we examined the first derivative and how it could be used to determine where a function is increasing and where it is decreasing. Now we see how the second derivative can be used to determine *concavity* for a curve. First, let's illustrate what is meant by a curve being concave up or concave down. If the graph of a function is **concave up,** we say that it "curves up" or "holds water"; if it is **concave down,** we say it "curves down" or "spills water" (see Figure 4.11).

Figure 4.11

Concave up
"holds water"

Concave down
"spills water"

Just as with increasing and decreasing, we describe concavity for a function using intervals. From Figure 4.12, we see that a curve may have intervals where it is concave up and intervals where it is concave down. The point on the curve where $x = b$ is a point where the concavity changes and is called a **point of inflection.** Some curves are always concave up or always concave down. Such is the case with the functions $f(x) = x^2$ and $y = x^{2/3}$ (see exercises 1 and 25 at the end of this section). The graph of $f(x) = x^2$ is always concave up, and the graph of $y = x^{2/3}$ is always concave down (except at the point where $x = 0$, the origin, which is called a **cusp**) (see Figure 4.13).

Figure 4.12

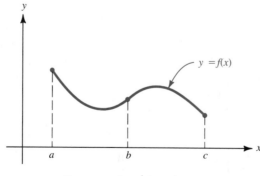

$y = f(x)$

Concave up for $a < x < b$
Concave down for $b < x < c$

Figure 4.13

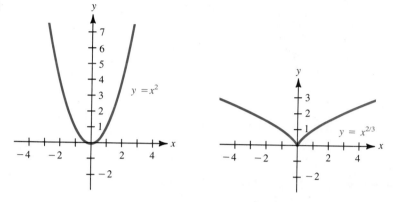

Concavity can be determined using the second derivative of a function. For $f(x) = x^2$, the second derivative $f''(x) = 2$, which is a positive number. Because the second derivative for this function is always positive, the curve is always concave up. A formal statement concerning use of the second derivative to determine concavity follows:

Property 2

> If the second derivative of a function f exists over an interval, then:
>
> 1. The function is *concave up* on that interval whenever $f''(x) > 0$.
> 2. The function is *concave down* on that interval whenever $f''(x) < 0$.

When we use the first derivative to determine whether a function is increasing or decreasing, we find that the values of x for which $f'(x) = 0$ are very useful. Similarly, the values of x for which $f''(x) = 0$ may be useful when determining concavity.

Example 1 For $f(x) = 2x^2 + x - 6$, determine where the function is concave up and where it is concave down.

Solution If $f(x) = 2x^2 + x - 6$
then $f'(x) = 4x + 1$
and $f''(x) = 4$

Since $f''(x) > 0$ for all x, the graph of the function is always concave up. ∎

Example 2 For $f(x) = x^3 + 2x^2 - x + 1$, determine the concavity.

Solution $f'(x) = 3x^2 + 4x - 1$
$f''(x) = 6x + 4$

$f''(x) > 0$	$f''(x) < 0$
$6x + 4 > 0$	$6x + 4 < 0$
$6x > -4$	$6x < -4$
$x > \dfrac{-2}{3}$	$x < \dfrac{-2}{3}$

This means the graph is concave up for $x > \dfrac{-2}{3}$ and concave down for $x < \dfrac{-2}{3}$.

Because the concavity changes at $x = \dfrac{-2}{3}$, the point $\left(\dfrac{-2}{3}, \dfrac{61}{27} \right)$ is called a *point of inflection*. The graph of $y = x^3 + 2x^2 - x + 1$ is shown in Figure 4.14.

Figure 4.14

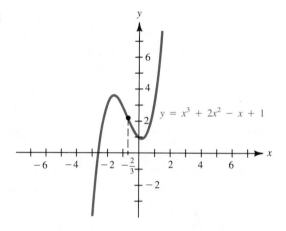

$y = x^3 + 2x^2 - x + 1$

To determine concavity, we can use a technique similar to the one we used in Section 4.1. Looking back at the previous example, Example 2, we see that $f''(x) = 0$ for $x = \dfrac{-2}{3}$, and this is precisely where the concavity changes.

Example 3   For $f(x) = 6x^2 - x^4$, determine the concavity and find any points of inflection.

Solution $f'(x) = 12x - 4x^3$
$f''(x) = 12 - 12x^2$

Now find the values of x for which $f''(x) = 0$.

$$12 - 12x^2 = 0$$
$$12x^2 = 12$$
$$x^2 = 1$$
$$x = \pm 1$$

We can use these values to write intervals and make a table.

Interval	$x < -1$	$-1 < x < 1$	$x > 1$
Sign of f"(x)	$-$	$+$	$-$
Result: f(x) is	concave down	concave up	concave down

Because the concavity changes at $x = 1$ and at $x = -1$, the points of inflection are $(1, 5)$ and $(-1, 5)$. ■

Example 4   For $y = \dfrac{1}{x + 2}$, determine the concavity and find any points of inflection.

Solution $\dfrac{dy}{dx} = \dfrac{-1}{(x + 2)^2}$

$\dfrac{d^2y}{dx^2} = \dfrac{2}{(x + 2)^3}$

Because $\dfrac{d^2y}{dx^2}$ for this function can never be zero, we determine where $\dfrac{d^2y}{dx^2}$ is positive and negative by inspection. Notice that the numerator is always positive and that the denominator is positive when $x > -2$ and negative when $x < -2$. This means that $\dfrac{d^2y}{dx^2} > 0$ and the curve is concave up for $x > -2$. For $x < -2$, $\dfrac{d^2y}{dx^2} < 0$ and the curve is concave down. The function is not defined for $x = -2$, so even though the concavity changes there, there is no point of inflection. ■

We see from Example 4 that it is possible for the concavity to change but for there to be no point of inflection there; the function has to be defined there in order to have a point of inflection. It is also possible that the second derivative will be zero for some value of x and the concavity will not change there, which of course means there is no point of inflection. See exercise 9 for an example of such a function.

We summarize the information of this section and the previous one in the following table:

Summary	$f'(x) > 0$	curve is *increasing*
	$f'(x) < 0$	curve is *decreasing*
	$f''(x) > 0$	curve is *concave up*
	$f''(x) < 0$	curve is *concave down*
	$f'(c) = 0$	$x = c$ is a *critical value* for $y = f(x)$
	$f'(c)$ undefined	$x = c$ is a *critical value* for $y = f(x)$ provided $f(c)$ is defined
	$f''(c) = 0$	$(c, f(c))$ is a *point of inflection* provided $f(c)$ is defined and concavity changes at $x = c$

4.2 EXERCISES

SET A

In exercises 1 through 12, use the second derivative to find the intervals for which each of the following is concave up and concave down. (See Examples 1 and 2.)

1. $y = x^2$

2. $y = -2x^2$

3. $y = 2x - x^2$

4. $f(x) = x + \dfrac{1}{x}$

5. $y = x^2 - 4x$

6. $y = x - \dfrac{1}{x}$

7. $y = x^3 + 3x^2 - 4$

8. $y = (x - 2)^3$

9. $y = x^4 + 4x$

10. $y = \dfrac{2x}{x - 1}$

11. $f(x) = x^2 + \dfrac{1}{x}$

12. $y = \dfrac{1}{x^2}$

In exercises 13 through 20, find any points of inflection. (See Examples 3 and 4.)

13. $y = 5 - 2x^2 - 2x^3$

14. $y = 2 - \sqrt[3]{x}$

15. $y = \sqrt[3]{x}$

16. $y = \dfrac{1}{x^2 + 1}$

17. $y = 4x - x^3$

18. $y = 10 + 4x - x^3$

19. $y = x^3 + 7x^2 + 11x + 5$

20. $y = -x^3 - 7x^2 - 11x - 5$

SET B

In exercises 21 through 26, find:

a. *the intervals where the function is increasing and decreasing using the first derivative*

b. *the intervals where the curve is concave up and concave down using the second derivative*

c. *any points of inflection*

21. $f(x) = (x + 1)^2 - 3$

22. $y = 4x - x^2$

23. $f(x) = \frac{1}{3}x^3 + x^2 - 8x + 1$

24. $y = (2 - x)^3$

25. $y = x^{2/3}$

26. $f(x) = \dfrac{1}{x}$

27. Consider the (straight-line) graph of the linear function

$$f(x) = mx + b$$

What can be said of the curve's concavity?

28. Graph $y = \dfrac{2x}{x^2 + 1}$. Be sure to use the regions where the function increases and decreases, and to use the function's concavity. Label any points of inflection.

29. Repeat exercise 28 for $y = \dfrac{2x}{x^2 - 1}$.

30. The function $f(x) = x^3$ is always increasing. However, the rate at which it increases changes. See the graph.

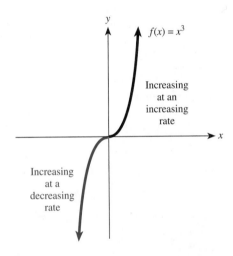

The second derivative, $f''(x) = 6x$, gives the **rate** at which it increases. For $x < 0$, $f''(x) < 0$, so it increases at a **decreasing** rate. For $x > 0$, $f''(x) > 0$, meaning it increases at an **increasing** rate. At the point of inflection, (0, 0), the rate of increase is zero. Sketch the graph of a function that increases at an increasing rate for $x < 0$ and increases at a decreasing rate for $x > 0$.

BUSINESS
PROFIT

31. The profit function for a certain company is given by

$$P(x) = (x - 2)^3 + 8$$

where x is measured in hundreds and the profit is measured in thousands of dollars.

a. For what value of x is the rate of increase in the profit equal to zero?

b. What is the profit at that point?

COST FUNCTION

32. A firm has determined that its total cost of producing x items is

$$C(x) = 2x^3 - 12x^2 + 26x + 20$$

a. Determine the marginal cost function, $C'(x)$.
b. Show that the point of inflection of $C(x)$ corresponds to the minimum point on $C'(x)$.

COST FUNCTION

33. A firm has determined that the total cost of producing x items is

$$C(x) = 120 + \sqrt{x}$$

Show that $C(x)$ is concave down for all values of $x > 0$.

COST FUNCTION

34. Assume a total cost function of

$$C(x) = 3x^3 - 18x^2 + 45x$$

a. Find the inflection point for $C(x)$.
b. How does the concavity of $C(x)$ change around the inflection point?
c. By using the first derivative as well as the concavity, determine how the rate of increase of $C(x)$ changes at the inflection point.

PROFIT

35. The monthly profit from the sales of a new type of chair is given by

$$P(x) = 6x^2 - 3x^3$$

where x is measured in hundreds and $P(x)$ is measured in thousands of dollars.
a. For what values of x is the rate of change of the profit, $P'(x)$, increasing?
b. For what values of x is the rate of change of the profit decreasing?

BIOLOGY
CELL REPRODUCTION

36. In an experiment involving tissue samples taken at certain time intervals, the number of cells, N, at any time, t, can be closely approximated by the function

$$N(t) = 5 + 6t^2 + 4t^3 - t^4$$

where t is measured in hours and $N(t)$ is measured in thousands.
a. For what values of t is the rate of change in the number of cells increasing? Decreasing?
b. Sketch the graph of this function for $t \geq 0$.
c. From the graph, approximate the time at which the number of cells appears to be at a maximum.

WRITING AND CRITICAL THINKING

1. Write (in complete sentences) an explanation of how to find a point of inflection for a curve (if one exists).
2. Do you agree with the following statement:

If a curve is concave up on a particular interval, then the function represented by its first derivative is increasing on that interval.

Justify your answer in words.

4.3 LOCAL EXTREME VALUES: FIRST AND SECOND DERIVATIVE TESTS

In the first two sections of this chapter we laid the groundwork for finding maximum and minimum values for a function. In this section, we limit ourselves to what we call *local* maximum and minimum values, collectively referred to as **local extreme values** (or **local extrema**).

Local extrema

> 1. A point $(c, f(c))$ is called a **local minimum** for a function if there exists an interval around c such that $f(x) \geq f(c)$ for all x in that interval.
> 2. A point $(c, f(c))$ is called a **local maximum** for a function if there exists an interval around c such that $f(x) \leq f(c)$ for all x in that interval.

Figure 4.15 illustrates graphically what we mean by local maximums and minimums.

Figure 4.15

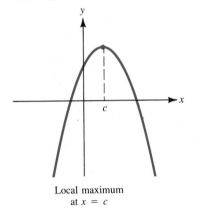

Local maximum
at $x = c$

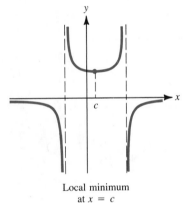

Local minimum
at $x = c$

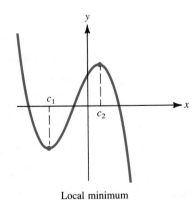

Local minimum
at $x = c_1$
Local maximum
at $x = c_2$

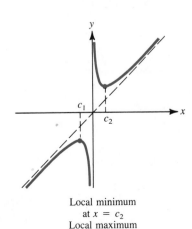

Local minimum
at $x = c_2$
Local maximum
at $x = c_1$

In Section 4.1, we used the first derivative to determine where a function is increasing and where it is decreasing. We also introduced something called a *critical value*, which we will see *may* represent an extreme value for a function. To see how this all fits together, we use a procedure called the **First Derivative Test,** which can be used to decide whether a critical value is indeed a local extreme value.

First Derivative Test

For a critical value c, the point $(c, f(c))$ can be classified as a

1. **local minimum** if $f'(x) < 0$ to the left of c and $f'(x) > 0$ to the right of c, or
2. **local maximum** if $f'(x) > 0$ to the left of c and $f'(x) < 0$ to the right of c.

Informally, the First Derivative Test states that a function decreases to the left of a local minimum and increases to the right of it. The opposite is true for a local maximum.

Example 1   For $f(x) = 27x - x^3$, use the First Derivative Test to find any local extreme values.

Solution

$$f(x) = 27x - x^3$$
$$f'(x) = 27 - 3x^2 \qquad \text{Determine critical values by finding } f'(x).$$
$$27 - 3x^2 = 0 \qquad \text{Set } f'(x) \text{ equal to zero.}$$
$$3x^2 = 27$$
$$x^2 = 9$$
$$x = \pm 3 \qquad \text{There are two critical values, } x = 3 \text{ and } x = -3.$$

We use the First Derivative Test and test the values $x = 3$ and $x = -3$ in the following table:

Interval	$x < -3$	$-3 < x < 3$	$x > 3$
Sign of $f'(x)$	negative	positive	negative
Location	left of -3	right of -3 left of 3	right of 3

The First Derivative Test applied to the information in the table tells us that we have a local minimum at $x = -3$ and a local maximum at $x = 3$. The graph appears in Figure 4.16.

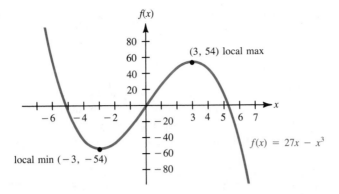

Figure 4.16

(3, 54) local max

local min $(-3, -54)$

$f(x) = 27x - x^3$

A note on vocabulary is in order here. When we defined "local maximum" and "local minimum," we defined them to be *points* because we have been dealing with characteristics of the graph of a function. If, graphically, the *point* $(c, f(c))$ is a local maximum, then analytically the *value* $f(c)$ is a local maximum value for the function. As in Example 1, we may also say that the function has a local maximum *at* $x = c$.

Example 2   For $f(x) = x^4 - 4x^3 - 8x^2 + 7$, find any local extreme values.

Solution
$$f'(x) = 4x^3 - 12x^2 - 16x \qquad \text{Set } f'(x) \text{ equal to zero.}$$
$$4x^3 - 12x^2 - 16x = 0$$
$$4x(x^2 - 3x - 4) = 0$$
$$4x(x - 4)(x + 1) = 0$$

We test the critical values, $x = -1$, $x = 0$, and $x = 4$ in the following table:

Interval	$x < -1$	$-1 < x < 0$	$0 < x < 4$	$x > 4$
Sign of $f'(x)$	negative	positive	negative	positive
Location	left of -1	right of -1 left of 0	right of 0 left of 4	right of 4

The following conclusions can be drawn:

local minimum at $x = -1$
local maximum at $x = 0$
local minimum at $x = 4$

They are represented graphically by the points $(-1, 4)$, $(0, 7)$, and $(4, -121)$.   ■

Example 3   For $f(x) = x + \dfrac{1}{x}$, find any local extreme values.

Solution   First calculate $f'(x)$.

$$f'(x) = 1 - \frac{1}{x^2}$$

Now determine any critical values. Keep in mind that even though $f'(x)$ is undefined for $x = 0$, so is the original function, which means that $x = 0$ is not a critical value. We will, however, use $x = 0$ when setting up our intervals because the function has a vertical asymptote there. Setting $f'(x)$ equal to zero, we have

$$1 - \frac{1}{x^2} = 0$$
$$\frac{1}{x^2} = 1$$
$$x^2 = 1$$
$$x = \pm 1$$

Interval	$x < -1$	$-1 < x < 0$	$0 < x < 1$	$x > 1$
Sign of $f'(x)$	positive	negative	negative	positive
Location	left of -1	right of -1 left of 0	right of 0 left of 1	right of 1

From the table, we have a local maximum at the point $(-1, -2)$ and a local minimum at the point $(1, 2)$. The graph appears in Figure 4.17.

Figure 4.17

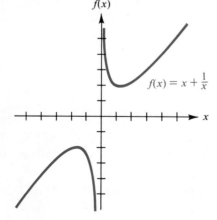

$$f(x) = x + \frac{1}{x}$$

We now explore another method, called the **Second Derivative Test,** for determining the location of local extrema.

Second Derivative Test

For a critical value c, the point $(c, f(c))$ can be classified as a

1. **local minimum** if $f''(c) > 0$, or
2. **local maximum** if $f''(c) < 0$.

If $f''(c) = 0$ or $f''(c)$ is undefined, then the test fails.

Example 4   Use the Second Derivative Test to find any local extreme values for $f(x) = 2x^3 - x^2$.

Solution   We find $f'(x)$ and set it equal to zero to determine critical values.

$$f'(x) = 6x^2 - 2x$$
$$6x^2 - 2x = 0$$
$$2x(3x - 1) = 0$$

$$x = 0 \quad \text{and} \quad x = \frac{1}{3} \quad \text{We need to find } f''(0) \text{ and } f''\left(\frac{1}{3}\right).$$

Calculate $f''(x)$ for the Second Derivative Test.

$$f''(x) = 12x - 2$$

Now,

$$f''(0) = 12(0) - 2 = -2$$

The fact that $f''(0)$ is *negative* implies a local *maximum* at $x = 0$. Substituting $x = \frac{1}{3}$, we have

$$f''\left(\frac{1}{3}\right) = 12\left(\frac{1}{3}\right) - 2 = 2$$

The fact that $f''(\frac{1}{3})$ is *positive* implies a local *minimum* at $x = \frac{1}{3}$. The graph of $f(x) = 2x^3 - x^2$ appears in Figure 4.18.

Figure 4.18

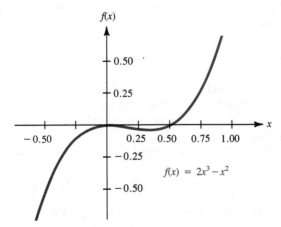

$f(x) = 2x^3 - x^2$

Example 5   For $y = (x + 1)^3$, find:
   a.   The intervals where the function is increasing and decreasing
   b.   The concavity and any points of inflection
   c.   Any local extreme values

Solution   **a.** $f'(x) = 3(x + 1)^2$

Setting $f'(x) = 0$, we have

$$3(x + 1)^2 = 0$$
$$x + 1 = 0$$
$$x = -1$$

Interval	$x < -1$	$x > -1$
Sign of f'(x)	positive	positive
Result: f(x) is	increasing	increasing

The function is always increasing.
   b. $f''(x) = 2 \cdot 3(x + 1)$
   $= 6(x + 1)$

Setting $f''(x) = 0$, we have

$$6(x + 1) = 0$$
$$x = -1$$

Interval	$x < -1$	$x > -1$
Sign of f"(x)	negative	positive
Result: f(x) is	concave down	concave up

Because the concavity changes at $x = -1$, the point $(-1, 0)$ is a point of inflection.
   c.   To use the First Derivative Test for possible local extreme values, we examine the table from part a. We can see that the function is increasing to the right and to the left of $x = -1$, so that even though -1 is a critical value, it is not an extreme value.

Notice that if we try to use the Second Derivative Test and evaluate $f''(-1)$, we have

$$f''(x) = 6(x + 1)$$
$$f''(-1) = 0$$

The Second Derivative Test *fails* (because $f''(-1)$ is neither positive nor negative). The conclusion from the First Derivative Test is sufficient (see Figure 4.19).

Figure 4.19

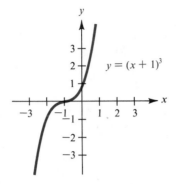

A new fertilizer was tested for a select group of farms as an alternative to crop rotation over a five-year period. The function

$$P(t) = \frac{80t}{t^2 + 4}, \qquad 0 \le t \le 5$$

gives the production level, $P(t)$ (in thousands of bushels), and t represents time (in years). When was the production level at a maximum, and what was the maximum production level?

Example 6

Solution   First, we find the critical values for the function by calculating $P'(t)$ using the Quotient Rule.

$$P'(t) = \frac{(t^2 + 4)(80) - (80t)(2t)}{(t^2 + 4)^2}$$

$$= \frac{80t^2 + 320 - 160t^2}{(t^2 + 4)^2}$$

$$= \frac{320 - 80t^2}{(t^2 + 4)^2}$$

Now setting $P'(t) = 0$, we have

$$\frac{320 - 80t^2}{(t^2 + 4)^2} = 0$$

$$320 - 80t^2 = 0$$

$$80t^2 = 320$$

$$t^2 = 4$$

$$t = \pm 2$$

We consider only $t = 2$, because t represents time (which is nonnegative). Using a First Derivative Test, we find that the maximum production level occurs at

$t = 2$ years. To find the maximum production level, we need only substitute $t = 2$ into the original function, $P(t)$:

$$P(2) = \frac{80(2)}{2^2 + 4}$$
$$= 20$$

The maximum production level is 20,000 bushels (see Figure 4.20).

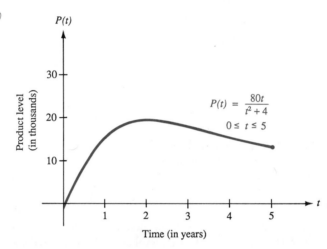

Figure 4.20

4.3 EXERCISES

SET A

In exercises 1 through 14, use the First Derivative Test to find the local maximum and minimum points for each function. (See Examples 1, 2, 3, and 5.)

1. $f(x) = x^2 - 2x$

2. $f(x) = 2x^2 + 2x - 1$

3. $f(x) = x^3 - 2x^2$

4. $f(x) = x^3 - 3x^2 - 9x + 18$

5. $f(x) = 3x^4 - 8x^3 + 2$

6. $f(x) = x^2 - x^4$

7. $f(x) = x^4 - 6x^2 + 2$

8. $f(x) = x^5 - 5x^4 - 1$

9. $f(x) = (x + 3)^{2/3}$

10. $f(x) = 5 - (x - 1)^{2/3}$

11. $y = \dfrac{-x}{x^2 + 1}$

12. $y = \dfrac{1}{x^2 - 9}$

13. $f(x) = \dfrac{x^2}{x - 1}$

14. $f(x) = x^2 + \dfrac{16}{x}$

In exercises 15 through 20, use the Second Derivative Test (where possible) to find the local maximum and minimum points for each function. (See Example 4.)

15. $f(x) = \frac{2}{3}x^3 - x^2 - 12x + 2$

16. $f(x) = x^5 + x^4$

17. $f(x) = 2x^3 + \frac{27}{2}x^2 - 15x + 2$

18. $f(x) = x^5 - x^4$

19. $f(x) = 3x^5 - 20x^3$

20. $f(x) = x - 2x^{2/3}$

SET B

BUSINESS
REVENUE FUNCTIONS

21. The demand equation for a dot-matrix printer ribbon manufacturer is $p = 20 - 0.2x$, where p is measured in dollars.
   a. Find the revenue function, $R(x)$.
   b. Graph $y = R(x)$.
   c. For what value(s) of x is $R(x)$ maximum?
   d. What is the associated price, p?
   e. What is the maximum revenue?

COST FUNCTION

22. The cost of producing x direct-drive turntables is given by

$$C(x) = 0.2x^2 - 1.6x + 9$$

where x is measured in hundreds and $C(x)$ is measured in thousands of dollars. Determine the number of turntables that minimizes the cost.

PROFIT FUNCTION

23. The Design One Furniture Company has determined that the monthly profit from producing and selling x dining tables is

$$P(x) = -2x^2 + 12x - 7$$

where x is measured in hundreds and $P(x)$ is measured in thousands of dollars. Determine the maximum profit.

ADVERTISING

24. A small company has determined that its profit, P, is dependent on the amount of money spent on advertising, x. This relationship is given by

$$P(x) = \frac{20x}{x^2 + 16} + 7$$

where P and x are measured in thousands of dollars. What amount should the company spend on advertising to ensure maximum profit?

COST

25. Assume a total cost function of

$$C(x) = 4x^3 - 6x^2 - 24x + 100$$

where x is measured in thousands of units. Find the value of x that minimizes the marginal cost, $C'(x)$.

VARIABLE COST

26. The total cost of producing x items is given by

$$C(x) = 2x^3 - 12x^2 + 26x + 20$$

the **total variable cost** is given by

$$\text{TVC}(x) = 2x^3 - 12x^2 + 26x$$

and the **average variable cost** is given by

$$\text{AVC}(x) = \frac{\text{TVC}(x)}{x} = 2x^2 - 12x + 26$$

Find the value of x for which AVC(x) is a minimum. Show that for this value of x the marginal cost, $C'(x)$, equals AVC(x).

REVENUE

27. If the demand for a product is given by $p = 48 - 4x^2$, find the value of x that maximizes revenue, $R(x)$, and find the maximum revenue. What is the price for the maximum revenue?

PROFIT

28. The cost function for a particular product is given by

$$C(x) = 400 + 3x$$

and the revenue by

$$R(x) = 6x - x^2$$

where x is measured in thousands. Determine the number of units that should be produced and sold to maximize the profit.

ADVERTISING

29. A company has noticed that after a certain period, spending more money on advertising did not bring in any new customers. In fact, the number of new customers began to decrease. Previously, the number of customers, N, was given by

$$N = 120x - x^2$$

where x was the amount spent on advertising (in thousands of dollars). How much was the company spending on advertising when its number of new customers was at a maximum?

BIOLOGY
MICROBIOLOGY

30. The number of microorganisms in a sample of seawater is given by

$$N(t) = 1.5 + 5t - t^2, \qquad 0 \le t \le 5$$

where N is measured in tens of thousands of microorganisms and t is measured in hours after the sample was drawn. What is the maximum number of microorganisms present?

SOCIAL SCIENCE
PAROLE

31. A sociologist has determined that, in a certain midwestern city, the number of parolees will increase according to the formula:

$$P = 1000 + 270t^2 - 27t^3, \qquad 0 \le t \le 10$$

where t is time (measured in years) and P represents the number of parolees. When will the growth rate of parolees in the city be at a maximum?

32. What will be the maximum number of people on parole, according to the information in exercise 31?

PHYSICAL SCIENCE
MOTION

33. A ball is thrown upward and its position at any time t is given by

$$s(t) = -16t^2 + 64t$$

where s is measured in feet and t in seconds. What is the maximum height the ball reaches?

MOTION

34. An object moves according to the function

$$s(t) = 48t - t^3, \qquad 0 \le t \le 6$$

where t is measured in minutes and s represents its position in inches relative to some starting point. At what time is the object the farthest from the starting point? What is the corresponding distance?

VELOCITY

35. The velocity of a particle is given by

$$v(t) = \frac{10t}{t^2 + 1}$$

where v is measured in cm/sec and t is measured in sec. At what time does the particle reach its maximum velocity?

For exercises 36 through 41, consider the graphs of these six distinct functions as they have been graphed on a graphics calculator:

Function f_1:

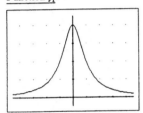

Function f_2:

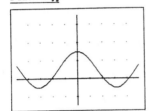

Function f_3:

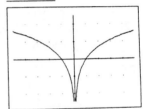

Function f_4:

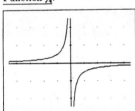

Function f_5:

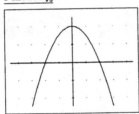

Function f_6:

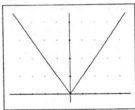

The First Derivative Test tells us that when a function is increasing, its derivative is positive and when a function is decreasing, its derivative is negative. Each of the figures in exercises 36–41 represents the graph of the first derivative of the functions above, but in a different order! Match the function above with its derivative below. (Hint: Look at the slopes of lines tangent to the curves at various points.)

36. Derivative of function _____

37. Derivative of function _____

38. Derivative of function _____

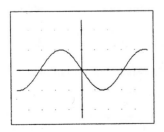

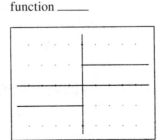

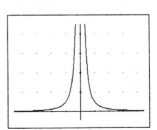

39. Derivative of
function _____

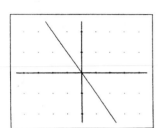

40. Derivative of
function _____

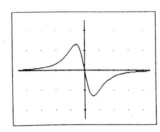

41. Derivative of
function _____

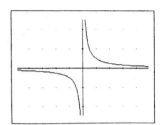

WRITING AND CRITICAL THINKING

Explain the difference between the First and Second Derivative Tests for
finding local maximum and minimum values. Does either test ever fail for any
particular situation? Explain why or why not.

4.4   ABSOLUTE EXTREMA

Consider the function $y = 5 + 8x^2 + 4x^3 - x^4$ and the question: What is the
absolute largest value y can be? The analytical tools we have developed thus far do
not address the question of absolute extreme values; we have examined ways of
finding only *local* (that is, values within certain intervals) maximum and minimum
values of y. In most applications, it is usually the **absolute maximum value** or
absolute minimum value that interests us rather than just the local extrema. By
looking at the entire graph and not just sections of it, we are able to examine absolute
extremes of a function.

We have the following definition:

Absolute
maximum value

> A number $f(c)$ is called an **absolute maximum value** for a function if
> $f(x) \leq f(c)$ for all x in the domain of the function.

Study Figure 4.21 for a graphical display of absolute maximum points on the
graphs of various functions.

Figure 4.21

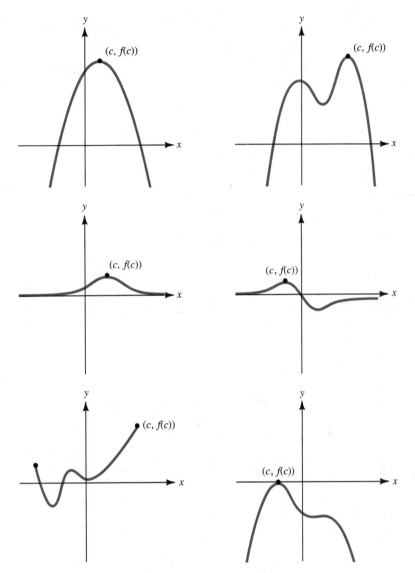

In trying to calculate the absolute maximum value, if it exists, we consider as candidates the *local* maximum values. This is explored in Example 1.

Example 1   Find the absolute maximum value for the function

$$f(x) = 3 + 2x - x^2$$

Solution   First, we determine any local extreme values using the First Derivative Test.

$$f'(x) = 2 - 2x \qquad \text{Set } f'(x) \text{ equal to zero.}$$
$$2 - 2x = 0$$
$$-2x = -2$$
$$x = 1$$

Interval	$x < 1$	$x > 1$
Sign of f'(x)	$+$	$-$
Result: f(x) is	increasing	decreasing

From the table, we can conclude that we do indeed have a local maximum at $x = 1$. Because the curve was increasing to the left of the point $(1, 4)$ and decreasing to the right and the curve never *increases* again, it can never go back up to a point higher than the point $(1, 4)$, which means that this point represents the *absolute* maximum point. The absolute maximum *value* is 4, the *y* coordinate. To verify our result for this particular function, we need only examine its graph in Figure 4.22.

Figure 4.22

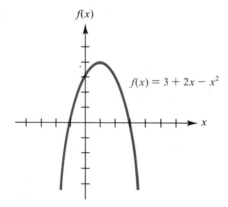

$$f(x) = 3 + 2x - x^2$$

Let's now return to our original question in this section.

Example 2   Find the absolute maximum value for $y = 5 + 8x^2 + 4x^3 - x^4$.

Solution   Determine any critical values by setting $\dfrac{dy}{dx}$ equal to zero.

$$\frac{dy}{dx} = 16x + 12x^2 - 4x^3$$
$$0 = 16x + 12x^2 - 4x^3$$
$$0 = -4x(x^2 - 3x - 4)$$
$$0 = -4x(x - 4)(x + 1)$$

We now have

$$-4x = 0 \qquad \text{or} \qquad x - 4 = 0 \qquad \text{or} \qquad x + 1 = 0$$
$$x = 0 \qquad\qquad\qquad x = 4 \qquad\qquad\qquad x = -1$$

The table is as follows:

Interval	$x < -1$	$-1 < x < 0$	$0 < x < 4$	$x > 4$
Sign of $\dfrac{dy}{dx}$	$+$	$-$	$+$	$-$
Result: y is	increasing	decreasing	increasing	decreasing

We see that we have a local maximum for $x = -1$ and for $x = 4$. There is a local minimum at $x = 0$ that does not interest us. To determine which of the local maximums gives the absolute maximum value, we must calculate the corresponding y values.

For

$$x = -1, \qquad y = 5 + 8(-1)^2 + 4(-1)^3 - (-1)^4 = 8$$

and for

$$x = 4, \qquad y = 5 + 8(4)^2 + 4(4)^3 - (4)^4 = 133$$

From this, we conclude that 133 is the absolute maximum value, which is "the largest y can be," the answer to our original question (see Figure 4.23).

Figure 4.23

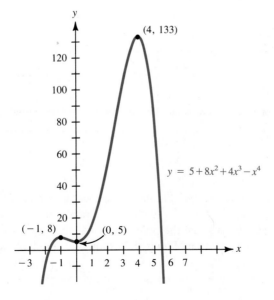

■

In both of the previous examples, we used polynomial functions that graphically turned down on the extreme left and on the extreme right. For this reason, both functions had an absolute maximum value. Example 3 shows that a function that graphically turns up on one side and down on the other need not have an absolute maximum.

Example 3   For $f(x) = 2x^3 - 2x^2 + 5$, find the absolute maximum value.

Solution   Calculate $f'(x)$ and set it equal to zero.

$$f'(x) = 6x^2 - 4x$$
$$0 = 6x^2 - 4x$$
$$= 2x(3x - 2)$$

Setting both factors equal to zero, we have

$$2x = 0 \qquad 3x - 2 = 0$$
$$x = 0 \qquad x = \frac{2}{3}$$

Interval	$x < 0$	$0 < x < \frac{2}{3}$	$x > \frac{2}{3}$
Sign of $f'(x)$	$+$	$-$	$+$
Result: $f(x)$ *is*	increasing	decreasing	increasing

From the table, we see that there is a local maximum at $x = 0$ and $f(0) = 5$. We see that to the right of $x = \frac{2}{3}$ the curve is increasing and continues to increase as $x \to +\infty$. This means that the point $(0, 5)$ remains just a local maximum and we have no absolute maximum. Figure 4.24 shows the graph of this function.

Figure 4.24

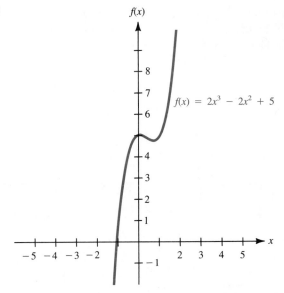

$f(x)$

$f(x) = 2x^3 - 2x^2 + 5$

We now define *absolute minimum value* in a way similar to that in which we defined absolute maximum value.

Absolute minimum value

A number $f(c)$ is called an **absolute minimum value** for a function if $f(x) \geq f(c)$ for all x in the domain of the function.

Example 4   Find the absolute minimum value for the function $f(x) = (x - 3)^4 + 2$.

Solution   First, find any local minimum values. Using the First Derivative Test, we get

$$f'(x) = 4(x - 3)^3$$

and setting $f'(x) = 0$, we have

$$4(x - 3)^3 = 0$$
$$x - 3 = 0$$
$$x = 3$$

Interval	$x < 3$	$x > 3$
Sign of $f'(x)$	$-$	$+$
Result: $f(x)$ is	decreasing	increasing

The results imply that there is a local minimum at $x = 3$. From the graph of this function in Figure 4.25, we see that this is also the absolute minimum value. The curve turns up on both sides of this point. The absolute minimum *value* is $f(3) = 2$.

Figure 4.25

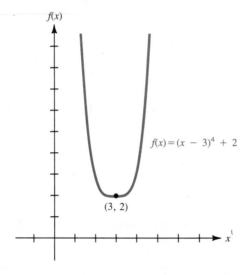

$f(x) = (x - 3)^4 + 2$

$(3, 2)$

■

Example 5   Find any absolute extreme values for $f(x) = \dfrac{4x}{x^2 + 1}$.

Solution   Determine the critical numbers. Using the Quotient Rule to find $f'(x)$, we get

$$f'(x) = \frac{(x^2 + 1)(4) - (4x)(2x)}{(x^2 + 1)^2}$$
$$= \frac{4x^2 + 4 - 8x^2}{(x^2 + 1)^2}$$
$$= \frac{4 - 4x^2}{(x^2 + 1)^2}$$

Setting $f'(x)$ equal to zero, we have

$$\frac{4 - 4x^2}{(x^2 + 1)^2} = 0$$
$$4 - 4x^2 = 0$$
$$4(1 - x)(1 + x) = 0$$
$$1 - x = 0 \qquad 1 + x = 0$$
$$x = 1 \qquad\qquad x = -1$$

Use the First Derivative Test for these critical values.

Interval	$x < -1$	$-1 < x < 1$	$x > 1$
Sign of f'(x)	$-$	$+$	$-$
Result: f(x) is	decreasing	increasing	decreasing

There is a local minimum at $x = -1$, which is also the absolute minimum. Even though the function decreases again, it decreases so that as $x \to \infty$, $y \to 0^+$, and the absolute minimum value is $f(-1) = -2$, which is less than zero.

There is a local maximum at $x = 1$, which becomes the absolute maximum value $f(1) = 2$ (see Figure 4.26).

Figure 4.26

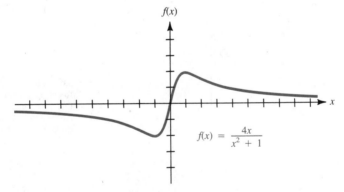

$$f(x) = \frac{4x}{x^2 + 1}$$

Because many applications involve functions for which there is some restriction on a variable (either implied or stated), we now turn our attention to examining functions defined on some *closed interval*. Examples of such functions include profit equations, where the number of items sold, x, must be restricted, or physical experiments where the variable time, t, must begin and end at specified times.

When we have such restricted domains, we have what are graphically referred to as endpoints. These endpoints *may* represent extrema for the function, and when they do, they are referred to as **endpoint extrema.** Notice that if f is a function defined on the closed interval, $a \le x \le b$, then it is always true that $f'(a)$ and $f'(b)$ are undefined (the two-sided limit definition of derivative cannot be satisfied); technically, a and b are type II critical values.

> The steps for finding **absolute extrema** for a continuous function f on a closed interval $[a, b]$ are:
>
> 1. Find the critical values of f in the interval $[a, b]$. These will always include a and b.
> 2. Evaluate f at each value found in Step 1.
> 3. The greatest value found in Step 2 is the absolute maximum; the smallest value found in Step 2 is the absolute minimum value for the function on $[a, b]$.

Figure 4.27 should help visualize the possibilities for absolute extrema on a closed interval.

Figure 4.27

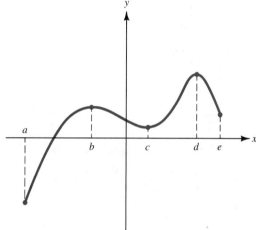

Absolute (or endpoint) minimum at $x = a$
Local maximum at $x = b$
Local minimum at $x = c$
Absolute maximum at $x = d$
Endpoint minimum at $x = e$

Example 6   Consider $f(x) = x^3 - 9x$ on the interval $-3 \leq x \leq 4$. Find the absolute maximum and absolute minimum of f.

Solution   First, we find any critical values. Setting $f'(x) = 0$ gives:

$$f'(x) = 3x^2 - 9$$
$$0 = 3x^2 - 9$$
$$9 = 3x^2$$
$$x = \pm\sqrt{3}$$

We use the Second Derivative Test to test these values:

$$f''(x) = 6x$$
$$f''(\sqrt{3}) = 6\sqrt{3} > 0 \qquad \text{Implies a local minimum at } x = \sqrt{3}$$
$$f''(-\sqrt{3}) = -6\sqrt{3} < 0 \qquad \text{Implies a local maximum at } x = -\sqrt{3}$$

Since the derivative is undefined at the endpoints, $x = -3$ and $x = 4$, we have four critical values to check: $\sqrt{3}$, $-\sqrt{3}$, -3, and 4.

$$f(\sqrt{3}) = -6\sqrt{3} \approx -10.4 \qquad \text{Thus } -6\sqrt{3} \text{ is a relative minimum value.}$$
$$f(-\sqrt{3}) = 6\sqrt{3} \approx 10.4 \qquad \text{Thus, } 6\sqrt{3} \text{ is a relative maximum value.}$$
$$f(-3) = 0 \qquad\qquad\quad \text{The function's value at the left endpoint is 0.}$$
$$f(4) = 28 \qquad\qquad\quad\; \text{The function's value at the right endpoint is 28.}$$

We see that the greatest of the function's values is 28, and thus f achieves an absolute maximum value of 28 at $x = 4$. The smallest of the function's values is $-6\sqrt{3}$, and thus f achieves an absolute minimum value of $-6\sqrt{3}$ at $x = \sqrt{3}$. We conclude with a graph of f in $[-3, 4]$ (see Figure 4.28).

Figure 4.28

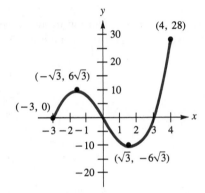

Example 7

The level of radiation, R, in a particular area of a nuclear plant has been found to fluctuate over a 12-hr interval according to the equation

$$R(t) = t^3 - 13.5t^2 + 42t + 60, \qquad 0 \le t \le 12$$

where t represents the number of hours elapsed. Find the time during the 12-hr interval when the radiation level is the lowest (absolute minimum) and when it is the highest (absolute maximum).

Solution

First, determine the critical values.

$$R'(t) = 3t^2 - 27t + 42$$
$$0 = 3t^2 - 27t + 42$$
$$= 3(t^2 - 9t + 14)$$
$$= 3(t - 7)(t - 2)$$
$$t - 2 = 0 \qquad t - 7 = 0$$
$$t = 2 \qquad\quad t = 7$$

The values $t = 2$ and $t = 7$ are critical values; we should also include $t = 0$ and $t = 12$, the endpoints, among our candidates for extreme values. Now we must test these values. We test $t = 2$ and $t = 7$ using the Second Derivative Test (see Section 4.3).

$$R''(t) = 6t - 27$$
$$R''(2) = 6(2) - 27 = -15$$
$$R''(7) = 6(7) - 27 = 15$$

The fact that $R''(2)$ is negative implies a local maximum; $R''(7)$ is positive, which implies a local minimum.

To determine the absolute maximum and absolute minimum, we must calculate the R values for all four t values and compare.

$$R(0) = 0^3 - 13.5(0)^2 + 42(0) + 60 = 60$$
$$R(2) = 2^3 - 13.5(2)^2 + 42(2) + 60 = 98$$
$$R(7) = 7^3 - 13.5(7)^2 + 42(7) + 60 = 35.5$$
$$R(12) = 12^3 - 13.5(12)^2 + 42(12) + 60 = 348$$

We can conclude that the absolute minimum occurs when $t = 7$ hr and the absolute maximum occurs when $t = 12$ hr (see Figure 4.29 for a graph of the function).

Figure 4.29

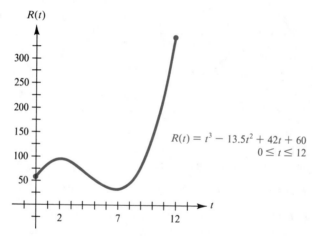

$$R(t) = t^3 - 13.5t^2 + 42t + 60$$
$$0 \le t \le 12$$

Does a function *have* to have an absolute maximum or absolute minimum on a closed interval? We conclude this section with a theorem, called the **Extreme Value Theorem,** that answers the question in the affirmative provided that the function is continuous on the closed interval.

*Extreme Value
Theorem*

| If a function f is continuous on a closed interval, then it has an absolute maximum value and an absolute minimum value on that interval. |

The two key words in this theorem are *continuous* and *closed* interval. Exercise 29 at the end of this section illustrates the importance of continuity, and exercise 30 illustrates the importance of the interval being closed.

4.4   EXERCISES

SET A

In exercises 1 through 8, find the absolute maximum value for each function. (See Examples 1, 2, and 6.)

1. $y = 4 - x^2$

2. $y = 4x - x^2$

3. $f(x) = 1 - (x + 3)^4$

4. $f(x) = 2 + 8x^2 - x^4$

5. $f(x) = x^3 - x^2, 0 \le x \le 3$

6. $f(x) = x - \dfrac{2}{x^2}, x < 0$

7. $f(x) = 2x^3 - 2x^2 + 5, x \le 5$

8. $f(x) = \dfrac{-1}{\sqrt{4 - x^2}}$

In exercises 9 through 15, find the absolute mimimum values for each function. (See Examples 4 and 5.)

9. $y = x^2 - 2x - 8$

10. $y = 2x^4 + 6x^3 + 2x^2 + 1$

11. $y = \dfrac{-1}{x^2 + 4}$

12. $f(x) = \dfrac{2}{\sqrt{1 - x^2}}$

13. $y = \dfrac{x^2}{x^2 + 1}$

14. $y = (x - 3)^4$

15. $y = x^4 - 5x^2 + 4$

In exercises 16 through 28, find any absolute extrema on the specified closed intervals for the functions. (See Examples 6 and 7.)

16. $f(x) = 3x^2 - x^3 + 1$ on $[-2, 5]$

17. $f(x) = 3x^2 - x^3 + 1$ on $[-0.5, 5]$

18. $f(x) = 3x^2 - x^3 + 1$ on $[-0.5, 2.5]$

19. $f(x) = \dfrac{x^3}{3} - \dfrac{x^2}{2} - 6x + 4$ on $[-3, 3]$

20. $f(x) = \dfrac{x^3}{3} - \dfrac{x^2}{2} - 6x + 4$ on $[-4, 4]$

21. $f(x) = 12x^2 - 3x^4$ on $[-1, 1]$

22. $f(x) = 12x^2 - 3x^4$ on $[-2, 2]$

23. $f(x) = 12x^2 - 3x^4$ on $[-2, 3]$

24. $S(t) = (t - 1)^3 - 4$ on $[-2, 2]$

25. $g(w) = w^3 - 2w^2 + w + 2$ on $[-2, 3]$

26. $g(w) = w^3 - 2w^2 + w + 2$ on $[-1, 1.5]$

27. $g(w) = w^3 - 2w^2 + w + 2$ on $[-3, 4]$

28. $g(w) = w^3 - 2w^2 + w + 2$ on $[-0.5, 2.5]$

SET B

29. For the function $f(x) = \dfrac{x^2 + 1}{x - 2}$, $0 \le x \le 4$, sketch the graph and explain why there is no absolute maximum or absolute minimum.

30. For the function $f(x) = (x - 2)^3$, $0 < x < 4$, make a sketch and explain why there is no absolute maximum or absolute minimum.

BUSINESS
PROFIT

31. A manufacturer has found that weekly profit for the company depends on x, the weekly promotion budget, according to the equation

$$P(x) = 200 \sqrt{x} - x - 1000$$

where x and $P(x)$ are measured in dollars.

    a.   Find the value of x that maximizes profit and find the maximum profit.

    b.   If the weekly promotional budget is later cut to \$8000, what will the profit be?

PROFIT

32. A company has determined that its profit from the manufacture and sale of x items is

$$P(x) = -2x^3 + 24x^2 - 42x - 20$$

where x is measured in hundreds and $0 \le x \le 12$.

    a.   Find the number of items that results in maximum profit. Is this a relative or absolute maximum?

    b.   Find the number of items that results in minimum profit. Is this an absolute minimum?

AVERAGE COST

33. The total cost function for a commodity is

$$C(x) = 2x^3 - 4x^2 + 4x$$

where market conditions limit the quantity produced, x, to $3 \le x \le 10$. Find the value of x that minimizes average cost, $\overline{C}(x)$.

BIOLOGY
POLLUTION CONTROL

34. An environmentalist, testing levels of dioxin in a town water supply, has found that the levels, D, change over a 30-hr period according to the relationship

$$D = -t^3 + 32t^2 - 80t + 600, \qquad 0 \le t \le 30$$

where D is in parts per million and t is in hours.

    a.   Find the time (in the 30-hr period) when the dioxin level is the lowest.

    b.   Find the time (in the 30-hr period) when the dioxin level is the highest.

    c.   What is the lowest level of dioxin?

    d.   What is the highest level?

WRITING AND CRITICAL THINKING

Consider the piecewise function $f(x) = \begin{cases} -x, & x < 0 \\ x^2, & x > 0 \end{cases}$.

Draw the graph and explain in words why the function has no absolute maximum. Also, explain why it has no absolute minimum.

4.4A   ABSOLUTE EXTREMA ON A GRAPHICS CALCULATOR

We can use a graphics calculator to help us understand the concepts of calculus and help us with such questions as "What is the *absolute largest* value the function $f(x) = 5 + 8x^2 + 4x^3 - x^4$ can be?" In fact, a graphics calculator can help us better appreciate the fact that a question about a function's absolute maximum or minimum values is best asked within a specific domain.

Consider the graph of $f(x) = 5 + 8x^2 + 4x^3 - x^4$ in Figure 4.4A.1. We are used to considering the domain of a function *before* we graph it. For this function, the domain is all real numbers, however, and we wish to choose a restricted domain on which to "view" it. If we use $-2 \leq x \leq 5$ for our domain, what is the largest value of $f(x)$ in that interval?

Figure 4.4A.1

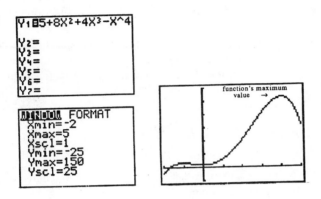

It appears that that maximum value is approximately 130 and appears around $x = 4$; to get a better approximation, we ZOOM in on the point in Figure 4.4A.2.

Figure 4.4A.2   After ZOOMing in, we TRACE to the closest pixel representing the maximum point.

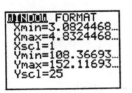

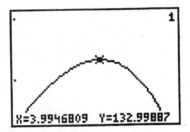

When we ZOOM in using TRACE, we find that the maximum y value is given as 132.99887. Try it on your calculator, then move the trace point to the right and to the left of the apparent maximum. Do you get a larger y value? We are certainly close to the maximum y value, but how can we be sure we have the *exact* value?

Before we begin to refine this procedure, we define what is meant by the **absolute maximum value** for a function on an interval:

Absolute maximum value

> A number $f(c)$ is called an **absolute maximum value** for a function f if $f(x) \leq f(c)$ for all x in the domain of the function.

We define the **absolute minimum value** for a function in a similar way:

> A number $f(d)$ is called an **absolute minimum value** for a function f if $f(x) \geq f(d)$ for all x in the domain of the function.

The power of calculus, along with the graphics calculator tool, can help us get the *exact* value of the function's maximum on $-2 \leq x \leq 5$, as well as answer a more global question and determine whether this is the function's absolute maximum on $-\infty < x < \infty$.

If an absolute maximum (or minimum) value occurs for a continuous function on a specified domain, it occurs at a value of x that is either an endpoint or a local maximum (or minimum). Thus, we use the first derivative to find critical values:

$$\frac{dy}{dx} = 16x + 12x^2 - 4x^3$$

$$0 = 16x + 12x^2 - 4x^3$$

$$4x(x - 1)(x + 1) = 0 \qquad \text{implies } x = 0 \quad \text{or} \quad x = 4 \quad \text{or} \quad x = -1$$

The following table helps us determine that a local maximum occurs at $x = -1$ and at $x = 4$; a local minimum occurs at $x = 0$:

Interval	$x < -1$	$-1 < x < 0$	$0 < x < 4$	$x > 4$
Sign of $\dfrac{dy}{dx}$	$+$	$-$	$+$	$-$
Function is	increasing	decreasing	increasing	decreasing

	↑	↑	↑	
	local max (increasing to decreasing)	local min (decreasing to increasing)	local max (increasing to decreasing)	

$$f(-1) = 5 + 8(-1)^2 + 4(-1)^3 - (-1)^4 = 8$$
$$f(0) = 5 + 8(0)^2 + 4(0)^3 - (0)^4 = 5$$
$$f(4) = 5 + 8(4)^2 + 4(4)^3 - (4)^4 = 133$$

Thus, the combination of graphs from the graphics calculator, along with the power of calculus, helps us determine that the function $f(x) = 5 + 8x^2 + 4x^3 - x^4$ *never* gets any greater than 133, which occurs when $x = 4$. We say $f(x) = 5 + 8x^2 + 4x^3 - x^4$ achieves an absolute maximum of 133 on $-\infty < x < \infty$ at $x = 4$. That is, $f(x) \leq 133$ for all x.

To approximate the *minimum* value of the function $f(x) = 5 + 8x^2 + 4x^3 - x^4$ on another interval, say $-4 \leq x \leq 6$, we use a graphics calculator to get the graph in Figure 4.4A.3.

Figure 4.4A.3

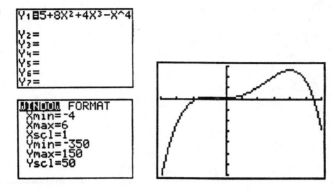

Now, using calculus, we need only determine the value of the local minimum, $f(0)$, and the values of the endpoints, $f(-4)$ and $f(6)$:

$$f(0) = 5 + 8(0)^2 + 4(0)^3 - (0)^4 = 5$$
$$f(-4) = 5 + 8(-4)^2 + 4(-4)^3 - (-4)^4 = -379$$
$$f(6) = 5 + 8(6)^2 + 4(6)^3 - (6)^4 = -139$$

We say $f(x) = 5 + 8x^2 + 4x^3 - x^4$ achieves an absolute minimum of -379 on $-4 \leq x \leq 6$ at $x = -4$. That is, $f(x) \geq -379$ for all x in $[-4, 6]$. The graphics calculator graph in Figure 4.4A.3 helps substantiate that fact, but a TRACE of the graph, shown in Figure 4.4A.4, helps even more.

Figure 4.4A.4

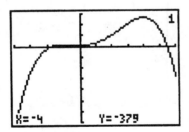

Let's examine the absolute extrema of another function, $g(x) = \dfrac{2x}{x^2 + 4}$, over the interval $-10 \leq x \leq 10$. This time we perform the calculus first in order to get a good idea of WINDOW values for our calculator to use.

$$g'(x) = \frac{(x^2 + 4)(2) - (2x)(2x)}{(x^2 + 4)^2} = \frac{8 - 2x^2}{(x^2 + 4)^2}$$

$$g'(x) = 0 \qquad \text{implies } 8 - 2x^2 = 0 \qquad \text{implies } x = \pm 2$$

Interval	$-10 < x < -2$	$-2 < x < 2$	$2 < x < 10$
Sign of $\dfrac{dy}{dx}$	$-$	$+$	$-$
Function is	decreasing	increasing	decreasing

$$g(-10) = \frac{2(-10)}{(-10)^2 + 4} = \frac{-20}{104} \approx -0.19$$

$$g(-2) = \frac{2(-2)}{(-2)^2 + 4} = \frac{-4}{8} = -0.5$$

$$g(2) = \frac{2(2)}{(2)^2 + 4} = \frac{4}{8} = 0.5$$

$$g(10) = \frac{2(10)}{(10)^2 + 4} = \frac{20}{104} \approx 0.19$$

We see that on the interval $-10 \le x \le 10$ this function achieves an absolute minimum of -0.5 at $x = -2$ and an absolute maximum of 0.5 at $x = 2$. We set our WINDOW values for y to be between -1 and 1 (a bit larger than the absolute extrema) and graph it in Figure 4.4A.5.

Figure 4.4A.5
$g(x) = \dfrac{2x}{x^2 + 4}$
TRACEd to its absolute maximum point: (2, 0.5)

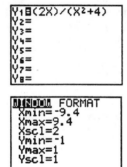

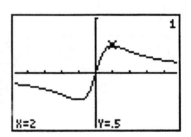

The **Extreme Value Theorem** tells us that we are assured of a function having an absolute maximum value and an absolute minimum value provided the function is continuous on a closed interval:

Extreme Value Theorem | If a function f is continuous on a closed interval, then it has an absolute maximum value and an absolute minimum value on that interval.

To see the necessity for "continuous" in the theorem, let's examine the function $f(x) = x + \dfrac{1}{x}$ on the interval $-3 \le x \le 3$. Clearly, there is an asymptote at $x = 0$, and therefore the function is discontinuous at $x = 0$ (see Figure 4.4A.6).

Figure 4.4A.6

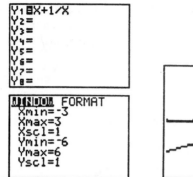

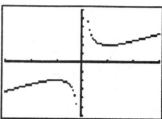

Notice that there is a local maximum at $x = -1$ and a local minimum at $x = 1$, but because the values of $f(x)$ get extremely large (and extremely small) near $x = 0$, there is no absolute maximum (or minimum).

4.4A   EXERCISES

In exercises 1 through 13, find any absolute extrema on the specified closed interval. Use a graphics calculator to substantiate your answer.

1. $f(x) = 3x^2 - x^3 + 1$ on $[-2, 5]$

2. $f(x) = 3x^2 - x^3 + 1$ on $[-0.5, 5]$

3. $f(x) = 3x^2 - x^3 + 1$ on $[-0.5, 2.5]$

4. $f(x) = \dfrac{x^3}{3} - \dfrac{x^2}{2} - 6x + 4$ on $[-3, 3]$

5. $f(x) = \dfrac{x^3}{3} - \dfrac{x^2}{2} - 6x + 4$ on $[-4, 4]$

6. $f(x) = 12x^2 - 3x^4$ on $[-1, 1]$

7. $f(x) = 12x^2 - 3x^4$ on $[-2, 2]$

8. $f(x) = 12x^2 - 3x^4$ on $[-2, 3]$

9. $S(t) = (t - 1)^3 - 4$ on $[-2, 2]$

10. $g(w) = w^3 - 2w^2 + w + 2$ on $[-2, 3]$

11. $g(w) = w^3 - 2w^2 + w + 2$ on $[-1, 1.5]$

12. $g(w) = w^3 - 2w^2 + w + 2$ on $[-3, 4]$

13. $g(w) = w^3 - 2w^2 + w + 2$ on $[-0.5, 2.5]$

14. Consider the function $f(x) = (x - 2)^3$ on the open interval $0 < x < 4$. Can it be said that f has an absolute maximum on this interval?

15. Use a graphics calculator to graph the function $f(x) = \dfrac{x^2 + 1}{x - 2}$ on $0 \le x \le 4$. Explain why there is no absolute maximum or absolute minimum value for the function on this interval.

4.5   CURVE SKETCHING

We are all aware that the understanding of a concept or idea can be greatly enhanced with a picture. As one instance, think of how many statistical data are displayed using graphs and charts so that we can visualize and make comparisons of a lot of material or information at a glance.

For an example of how much information can be derived from a graph, we examine Figure 4.30 and make some conclusions about the effects that money spent on advertising, x, can have on profits, y, for a company.

Figure 4.30

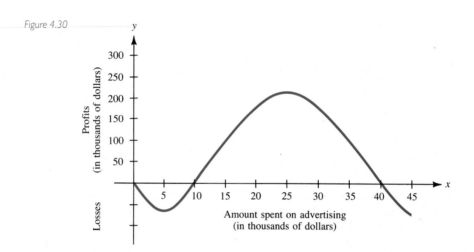

As we move from left to right across the graph, we see that the graph of the function starts to decrease, indicating the beginning of the company's promotional period, when it spends money on advertising but does not realize a profit from it. In fact, at $x = 5$, there is a local minimum representing the point at which "profit," represented by a negative value, is actually a loss. The company spends $5000 on advertising and still loses money. Looking at the interval from $x = 5$ to $x = 10$, we see that the function increases. At $x = 10$, which is an x intercept, we see that the company now spends $10,000 on advertising and breaks even (no loss, no gain, as the "profit" is zero). The curve continues to increase from $x = 10$ to $x = 25$.

The point on the curve where $x = 25$ is probably the most significant point for the company. It is at this point, where the company spends $25,000 on advertising, that it achieves its maximum profit of $210,000. As the company continues to spend more on advertising, it begins to feel the effect of a phenomenon that is very common in marketing a product. There is usually a point where the added costs of advertising that would be required to increase sales outweigh the contribution to profits. Beyond this point, notice that the curve decreases, indicating that the profits are decreasing. In fact, if a company becomes extremely extravagant on advertising without increasing sales, the profit may eventually reach the zero level again, as it does on our graph at $x = 40$.

In our discussions of curve sketching in Chapter 2, you learned about intercepts, symmetry, and asymptotes. We now combine that information with our new material using the first and second derivatives to help us in sketching curves. In Figure 4.31, we have compiled a summary of information to use where appropriate for sketching graphs of functions.

Figure 4.31   Summary of aids to curve sketching

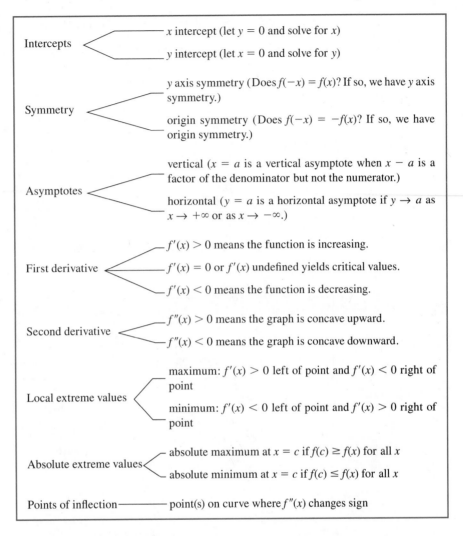

Intercepts
- x intercept (let y = 0 and solve for x)
- y intercept (let x = 0 and solve for y)

Symmetry
- y axis symmetry (Does $f(-x) = f(x)$? If so, we have y axis symmetry.)
- origin symmetry (Does $f(-x) = -f(x)$? If so, we have origin symmetry.)

Asymptotes
- vertical ($x = a$ is a vertical asymptote when $x - a$ is a factor of the denominator but not the numerator.)
- horizontal ($y = a$ is a horizontal asymptote if $y \to a$ as $x \to +\infty$ or as $x \to -\infty$.)

First derivative
- $f'(x) > 0$ means the function is increasing.
- $f'(x) = 0$ or $f'(x)$ undefined yields critical values.
- $f'(x) < 0$ means the function is decreasing.

Second derivative
- $f''(x) > 0$ means the graph is concave upward.
- $f''(x) < 0$ means the graph is concave downward.

Local extreme values
- maximum: $f'(x) > 0$ left of point and $f'(x) < 0$ right of point
- minimum: $f'(x) < 0$ left of point and $f'(x) > 0$ right of point

Absolute extreme values
- absolute maximum at $x = c$ if $f(c) \geq f(x)$ for all x
- absolute minimum at $x = c$ if $f(c) \leq f(x)$ for all x

Points of inflection — point(s) on curve where $f''(x)$ changes sign

---

Example 1   Sketch the graph of $f(x) = x^4 + 4x^3$ using the appropriate information from the summary table.

Solution   Before calculating any derivatives, we check intercepts, symmetry, and asymptotes.

Intercepts: For the x intercept, setting $f(x)$ equal to zero, we have

$$x^4 + 4x^3 = 0$$
$$x^3(x + 4) = 0$$
$$x^3 = 0 \qquad x + 4 = 0$$
$$x = 0 \qquad\quad x = -4$$

The x intercepts are $(0, 0)$ and $(-4, 0)$.
For the y intercept, setting $x = 0$, we have $y = 0$. The y intercept is $(0, 0)$.

Symmetry: All of our symmetry tests fail for this function.

Asymptotes: There are no asymptotes for this, or any, polynomial function.

Critical Values:

$$f'(x) = 4x^3 + 12x^2$$

Setting $f'(x) = 0$, we have

$$4x^3 + 12x^2 = 0$$
$$4x^2(x + 3) = 0$$
$$4x^2 = 0 \qquad x + 3 = 0$$
$$x = 0 \qquad\quad x = -3$$

Critical values are $x = 0$ and $x = -3$.

Increasing and Decreasing: The table for the intervals created by the critical values is as follows:

Interval	$x < -3$	$-3 < x < 0$	$x > 0$
Sign of $f'(x)$	$-$	$+$	$+$
Result: $f(x)$ is	decreasing	increasing	increasing

Extreme Values: From the table, we conclude that there is a local minimum at the point $(-3, -27)$, which also turns out to be the absolute minimum.

Concavity: Computing the second derivative, we have

$$f''(x) = 12x^2 + 24x$$

Setting $f''(x) = 0$ yields

$$12x^2 + 24x = 0$$
$$12x(x + 2) = 0$$
$$12x = 0 \qquad x + 2 = 0$$
$$x = 0 \qquad\quad x = -2$$

The table for $f''(x)$ is

Interval	$x < -2$	$-2 < x < 0$	$x > 0$
Sign of $f''(x)$	$+$	$-$	$+$
Result: $f(x)$ is	concave up	concave down	concave up

Points of Inflection: From the table, we see that there are two points of inflection. The coordinates of the points are $(-2, -16)$ and $(0, 0)$.

Extra Points: We will use the additional points $(-1, -3)$ and $(1, 5)$. Now see Figure 4.32 for the sketch of $f(x) = x^4 + 4x^3$.

Figure 4.32

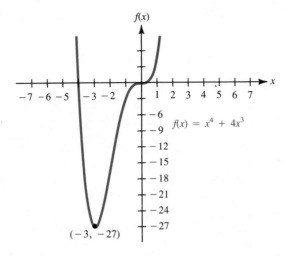

$$f(x) = x^4 + 4x^3$$

$(-3, -27)$

Example 2   Sketch the graph of $y = \dfrac{x^2 - 3}{x^3}$.

Solution   *Intercepts:* For the x intercept, set $y = 0$.

$$\frac{x^2 - 3}{x^3} = 0$$
$$x^2 - 3 = 0$$
$$x^2 = 3$$
$$x = \pm\sqrt{3}$$

So, the x intercepts are $(\sqrt{3}, 0)$ and $(-\sqrt{3}, 0)$.
For the y intercept, let $x = 0$. This is not possible because the function is not defined for $x = 0$, so there is no y intercept.

Symmetry: Replacing x with $-x$ and y with $-y$, we have

$$-y = \frac{(-x)^2 - 3}{(-x)^3}$$
$$-y = \frac{x^2 - 3}{-x^3}$$
$$y = \frac{x^2 - 3}{x^3}$$

Because this is the same as the original equation, we have symmetry with respect to the origin.

Asymptotes: (See Section 2.5.)
Vertical asymptote at $x = 0$, since, as $x \to 0$ from the left, $y \to \infty$; and as $x \to 0$ from the right, $y \to -\infty$.

Horizontal asymptote at $y = 0$, since dividing the numerator and denominator by x^3, we have

$$y = \frac{(x^2/x^3) - (3/x^3)}{x^3/x^3}$$

$$= \frac{1}{x} - \frac{3}{x^3}$$

Then, as $x \to \infty$, we have $y \to 0$.

Now we use the first and second derivatives to determine increasing, decreasing, extreme values, concavity, and points of inflection.

Increasing and Decreasing: Using the Quotient Rule, we get

$$\frac{dy}{dx} = \frac{x^3(2x) - (x^2 - 3)(3x^2)}{(x^3)^2}$$

$$= \frac{2x^4 - 3x^4 + 9x^2}{x^6}$$

$$= \frac{x^2(9 - x^2)}{x^6}$$

$$= \frac{9 - x^2}{x^4}$$

Setting $\dfrac{dy}{dx} = 0$ for critical values, we have

$$\frac{9 - x^2}{x^4} = 0$$

$$9 - x^2 = 0$$

$$x^2 = 9$$

$$x = \pm 3$$

Using these critical values 3 and -3, we set up a table. In setting up the intervals, we use $x = 0$, because the function is not defined there.

Interval	$x < -3$	$-3 < x < 0$	$0 < x < 3$	$x > 3$
Sign of $\dfrac{dy}{dx}$	$-$	$+$	$+$	$-$
Result: y is	decreasing	increasing	increasing	decreasing

Extreme Values: Because the function is decreasing to the left of -3 and increasing to the right, there is a local minimum at $x = -3$. A local maximum occurs at $x = 3$ because the function is increasing to the left of 3 and decreasing to the right.

Local minimum at $\left(-3, \dfrac{-2}{9} \right)$

Local maximum at $\left(3, \dfrac{2}{9} \right)$

Concavity: Using the Quotient Rule again, we get

$$\frac{d^2y}{dx^2} = \frac{x^4(-2x) - (9 - x^2)(4x^3)}{(x^4)^2}$$

$$= \frac{-2x^5 - 36x^3 + 4x^5}{x^8}$$

$$= \frac{2x^3(x^2 - 18)}{x^8}$$

$$= \frac{2(x^2 - 18)}{x^5}$$

Setting $\dfrac{d^2y}{dx^2} = 0$, we have

$$\frac{2(x^2 - 18)}{x^5} = 0$$

$$2(x^2 - 18) = 0$$

$$x^2 = 18$$

$$x = \pm 3\sqrt{2}$$

The table is as follows:

Interval	$x < -3\sqrt{2}$	$-3\sqrt{2} < x < 0$	$0 < x < 3\sqrt{2}$	$x > 3\sqrt{2}$
Sign of $\dfrac{d^2y}{dx^2}$	–	+	–	+
Result: y is	concave down	concave up	concave down	concave up

Points of Inflection: From the table, we see that there are points of inflection at $x = -3\sqrt{2}$ and at $x = 3\sqrt{2}$. The coordinates of these points are $\left(-3\sqrt{2}, \dfrac{-5\sqrt{2}}{36} \right)$ and $\left(3\sqrt{2}, \dfrac{5\sqrt{2}}{36} \right)$, which, we find by using a calculator, are approximately $(-4.24, -0.196)$ and $(4.24, 0.196)$.

Using all of the preceding information and plotting a few extra points, we can make a relatively good sketch, as shown in Figure 4.33.

Figure 4.33

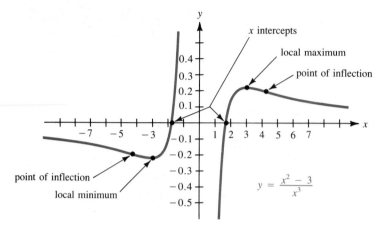

The table in Figure 4.34 summarizes the relationship between the concavity ($f''(x)$) for a function and its increasing or decreasing ($f'(x)$) behavior.

Figure 4.34

	Concave upward $f''(x) > 0$	Concave down $f''(x) < 0$
Increasing function $f'(x) > 0$	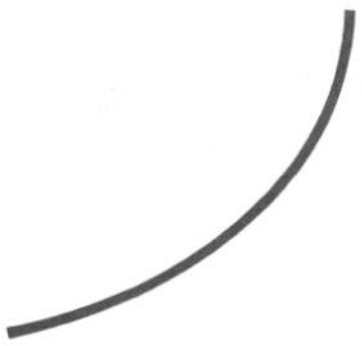	
Decreasing function $f'(x) < 0$		

Example 3   A cereal company has found that during the process of filling the cereal boxes, the average number of boxes, y, measured in tens per day, that are rejected by the company's quality control unit is related to the speed, x, measured in feet per second, at which the machinery moves, by the equation

$$y = x^{4/3} - 4x^{1/3} + 5, \qquad 0 < x \le 8$$

Sketch this equation and, from the sketch, determine the speed at which the machinery should move to minimize the number of boxes rejected.

Solution   This curve has no asymptotes and the symmetry tests fail. We calculate the y intercept; there are no x intercepts. Setting $x = 0$, we have $y = 5$ so that $(0, 5)$ is the y intercept.

Critical Values:

$$\frac{dy}{dx} = \frac{4}{3}x^{1/3} - \frac{4}{3}x^{-2/3} \qquad \text{Set } \frac{dy}{dx} = 0.$$

$$\frac{4}{3}x^{1/3} - \frac{4}{3}x^{-2/3} = 0 \qquad \text{Multiplying both sides by } \frac{3}{4}$$

$$x^{1/3} - x^{-2/3} = 0$$

$$x^{-2/3}(x - 1) = 0 \qquad \text{Factor } x^{-2/3} \text{ on the left.}$$

$$\frac{x - 1}{x^{2/3}} = 0$$

So, $$x - 1 = 0$$
$$x = 1$$

Increasing and Decreasing:

Interval	$0 < x < 1$	$1 < x < 8$
Sign of $f'(x)$	$-$	$+$
Result: $f(x)$ is	decreasing	increasing

Concavity: Calculating $\dfrac{d^2y}{dx^2}$, we get

$$\frac{d^2y}{dx^2} = \frac{4}{3}\left(\frac{1}{3}\right)x^{-2/3} - \frac{4}{3}\left(\frac{-2}{3}\right)x^{-5/3}$$

$$= \frac{4}{9}x^{-2/3} + \frac{8}{9}x^{-5/3}$$

$$= \frac{4}{9}x^{-5/3}(x + 2)$$

$$= \frac{4(x + 2)}{9x^{5/3}}$$

$\dfrac{d^2y}{dx^2} = 0$ for $x = -2$ and $\dfrac{d^2y}{dx^2}$ is undefined for $x = 0$.

Because we are interested only in the graph to the right of $x = 0$, the value $x = -2$ is not important to us. For $x > 0$, $\dfrac{d^2y}{dx^2}$ is always positive and therefore the curve is always concave up.

Points of Inflection: There are none for $x > 0$. See Figure 4.35 for a sketch of the function. We see that the minimum occurs at the point $(1, 2)$, representing a speed of 1 ft/sec and a rejection rate of 20 boxes per day for the machinery.

Figure 4.35

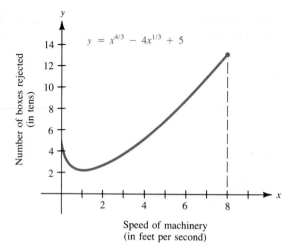

4.5   EXERCISES

SET A

In exercises 1 through 6, sketch the graph of each of the parabolas using the techniques of this section. (See Example 1.)

1. $y = x^2 + 1$

2. $y = (x - 1)^2 + 3$

3. $y = 2x^2 - 5x - 3$

4. $f(x) = 8x - 2x^2$

5. $y = 6 - x - x^2$

6. $S = 64 - 16t^2$

In exercises 7 through 20, sketch each of the polynomial functions. (See Example 1.)

7. $y = (x + 2)^3$

8. $y = 1 - (x - 2)^3$

9. $f(x) = 15 - 24x + 15x^2 - 2x^3$

10. $f(x) = 4x^3 - x^4$

11. $f(x) = x^3 - 3x^2 - 9x$

12. $y = \dfrac{x^3}{3} + \dfrac{x^2}{2} - 2x + 5$

13. $f(x) = \dfrac{x^3}{3} + x^2 - 8x - 1$

14. $f(x) = 3x^5 - 5x^3$

15. $y = 4x - x^4$

16. $y = (x + 3)^4$

17. $f(x) = 12x - x^3$

18. $f(x) = 4x^4 + x^2$

19. $y = x^2 - 2x^4$

20. $f(x) = (x + 1)^5$

In exercises 21 through 28, refer to the graph sketched here.

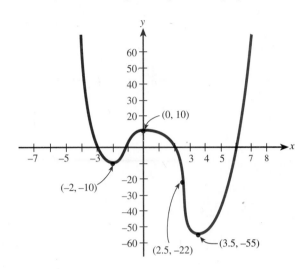

21. Is there any symmetry? If so, what is it?
22. What are the intercepts?
23. What are the asymptotes, if any?
24. In what regions is $f(x)$ positive? Negative?
25. In what regions is $f'(x)$ positive? Negative?
26. What are the local extrema, if any?
27. What are the absolute extrema, if any?
28. In what regions is the curve concave up? Concave down? (Approximate answers are acceptable.)

In exercises 29 through 31, refer to the graph of the function $y = \dfrac{1}{1 - x^2}$, *shown here.*

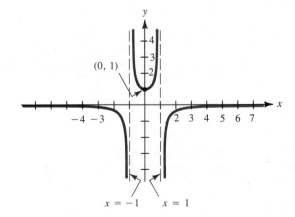

29. Discuss the graph's symmetry, intercepts, and asymptotes.

30. What are the local extrema, if any?

31. What are the absolute extrema, if any?

SET B

In exercises 32 through 38, each function has one or more asymptotes. Sketch the graph of each. (See Example 2.)

32. $f(x) = \dfrac{x^2 - x + 4}{x - 1}$

33. $f(x) = \dfrac{1}{x} + 4$

34. $f(x) = \dfrac{1}{x} - 4x$

35. $f(x) = \dfrac{1}{x^2 - 4}$

36. $f(x) = \dfrac{x - 2}{x^2}$

37. $y = \dfrac{1}{(x + 2)^2}$

38. $y = x^2 + \dfrac{1}{x}$

In exercises 39 through 42, the functions have been restricted so that there will be endpoints. Sketch the graph of each. (See Example 3.)

39. $y = 1 + \sqrt[3]{x - 1}, \quad x \geq 0$

40. $y = 16 - (x - 3)^4, \quad 0 \leq x \leq 5$

41. $y = x^{4/3} + x^{1/3}, \quad x \geq 0$

42. $y = x^3 - 6x^2 + 9x + 1, \quad 0 \leq x \leq 6$

BUSINESS
PROFIT

43. A company has found that its profit is related to the amount of money, x, it spends on advertising by the function

$$P(x) = -2x^3 + 150x + 60, \qquad 0 \leq x \leq 10$$

Graph this function and use the graph to determine the maximum profit.

AVERAGE COST

44. Assume a total cost function

$$C(x) = 0.2x^2 + 40x + 20$$

Graph the average cost function

$$\overline{C}(x) = \frac{C(x)}{x} = 0.2x + 40 + \frac{20}{x}, \qquad x > 0$$

MARGINAL ANALYSIS

45. Assume that for a company the revenue and cost functions are, respectively,

$$R(x) = -8x^3 - 4x^2 + 52x$$
$$C(x) = 2x^2 + 16x$$

   a. Determine the profit function, $P(x)$, and graph it.

   b. Graph the marginal cost and marginal revenue functions.

   c. Verify that for the value of x that yields maximum profit, marginal revenue equals marginal cost.

BIOLOGY
BACTERIA

46. The growth of a certain bacteria is given by

$$N(t) = t^3 - 6t^2 + 12t + 2$$

where N is measured in thousands and t is measured in hours. Sketch a graph of the function over the first 8-hr period.

WRITING AND CRITICAL THINKING

Suppose a polynomial function is defined on the interval $-2 \le x \le 5$. Assume that this function has x intercepts $(-1, 0)$, $(1, 0)$, and $(4, 0)$. Draw a sketch of such a function (there are an infinite number of possibilities). Now, using the terminology from this chapter (see the "Vocabulary/Formula List" on page 341), describe what your polynomial function looks like. Imagine that you are describing it to someone who cannot see your picture. Try the description out on a classmate. How does his or her picture compare to yours? Does your description need any revisions for clarity? If so, try to refine it.

USING TECHNOLOGY IN CALCULUS

Each of the following four graphs represents a different function as graphed on a TI-81 graphics calculator:

Function f_1:

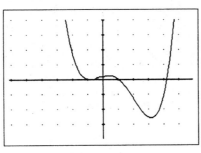

Function f_2:

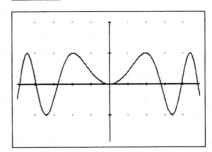

Function f_3:

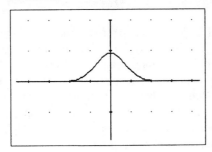

Function f_4:

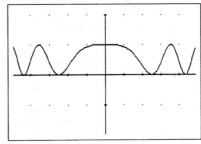

(continued)

(*continued*)

Use the following fact: On intervals where the graph of a function is concave up, its second derivative is positive, and on intervals where the graph of a function is concave down, its second derivative is negative.

Now, match a function $y = f(x)$ above with the graph of its second derivative, $y = f''(x)$, below.

1. Second derivative
of function _____

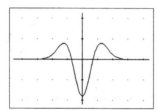

2. Second derivative
of function _____

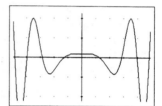

3. Second derivative
of function _____

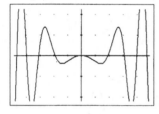

4. Second derivative
of function _____

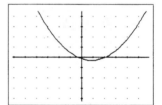

4.6 OPTIMIZATION PROBLEMS

We have arrived at our major objective for this chapter. In the previous section, we laid the groundwork for solving a type of application problem called an **optimization problem.** Basically, an optimization problem is one in which you are asked to either *maximize* or *minimize* some quantity. We dealt with some application problems previously in this chapter, but each time we were provided with the equation for the quantity to be maximized or minimized. We are now ready to continue solving that type of problem and also to solve some for which it will be our responsibility to create the function to be optimized. You may find it helpful to reread the section on Problem Solving on pages 66–75.

Suppose your neighbor has just put up a new fence around his yard and has 24 ft of fencing left over that he will sell to you. You plan to use the fencing to enclose a rectangular area next to the back of your house for storage purposes. You would like to enclose the maximum area possible with the 24 ft of fencing, using the back of your house as the fourth side. Should the area be long and narrow or more square? Figure 4.36 shows some possibilities.

Figure 4.36

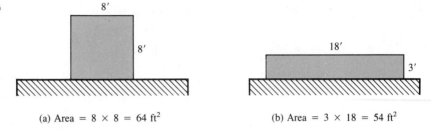

(a) Area = 8 × 8 = 64 ft²

(b) Area = 3 × 18 = 54 ft²

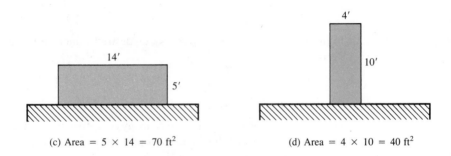

(c) Area = 5 × 14 = 70 ft²

(d) Area = 4 × 10 = 40 ft²

From the diagrams we have, it seems that choice (c) may represent something close to the maximum area. But how can we be sure? Instead of experimentation, we use the techniques developed in this chapter to find the actual maximum area.

Example 1   Find the maximum rectangular area that can be enclosed using 24 ft of fencing if fencing is required on only three of the four sides.

Solution   Since we want to *maximize* the area, we need to determine an equation for the area. Using a diagram (see Figure 4.37) with x as the width and y as the length, the area, A, will be

$$A = xy$$

Figure 4.37

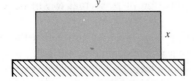

Next, we need to eliminate a variable on the right side of our equation before differentiating. Because we have only 24 ft of fencing, we can conclude that

$$2x + y = 24$$
or $\qquad y = 24 - 2x$

Substituting this into the area equation, we get

$$A = x(24 - 2x)$$
$$= 24x - 2x^2, \qquad 0 < x < 12$$

(We can assume that $0 < x < 12$ because no area would be formed otherwise.) Taking the first derivative and setting it equal to zero, we have

$$\frac{dA}{dx} = 24 - 4x$$
$$0 = 24 - 4x$$
$$4x = 24$$
$$x = 6$$

To make sure that $x = 6$ represents our desired value, we test it using the Second Derivative Test.

$$\frac{d^2A}{dx^2} = -4$$

The fact that the second derivative is always negative implies there is a *maximum* at $x = 6$. This means that the dimensions of the enclosed area will be

width $\qquad x = 6$ ft
length $\qquad y = 24 - 2x = 12$ ft

and the maximum area will be

$$A = (6)(12) = 72 \text{ ft}^2$$

■

Example 2   The Kwik-Wiff Cologne Company has been selling (on the average) 3500 bottles of its cologne per month at a price of $5 each but has found that for each $0.15 reduction in price, sales go up by 150 bottles each month. What price should the company charge to maximize its monthly revenue?

Solution   We would like to maximize the revenue, R. Recall from the previous chapter that revenue is the product of the number of items sold, x, and the price, p, per item:

$$R = xp \tag{1}$$

We need to express x as a function of p, so that R will be expressed solely as a function of price.

To express x as a function of p, we first notice that the relationship between x and p is *linear*. Thinking of x as the dependent variable and p as the independent variable and using the point-slope form for a line, we can express that relationship as

$$x - x_1 = \frac{x_2 - x_1}{p_2 - p_1}(p - p_1) \qquad (2)$$

Now, we need two points (x_1, p_1) and (x_2, p_2). Using the statement "for each $0.15 reduction in price, sales go up by 150 bottles," we have

x	p	
3500	$5.00	← (x_1, p_1)
3650	4.85	← (x_2, p_2)

Substituting 3500 for x_1, $5.00 for p_1, 3650 for x_2, and $4.85 for p_2 in equation (2) yields

$$x - 3500 = \frac{150}{-0.15}(p - 5)$$

or $x = 8500 - 1000p$

Finally, we can rewrite equation (1) as

$$\begin{aligned} R &= xp \\ &= (8500 - 1000p)p \\ &= 8500p - 1000p^2 \end{aligned}$$

To maximize this function, we need to take the derivative and set it equal to zero.

$$\begin{aligned} \frac{dR}{dp} &= 8500 - 2000p \\ 0 &= 8500 - 2000p \\ 2000p &= 8500 \\ p &= 4.25 \end{aligned}$$

To verify that this produces a maximum, we use the Second Derivative Test.

$$\frac{d^2R}{dp^2} = -2000$$

The fact that the second derivative is negative implies a maximum. Therefore, we can conclude that the company should charge $4.25 per bottle to maximize its monthly revenue. That maximum monthly revenue will be

$$\begin{aligned} R &= 8500p - 1000p^2 \\ &= 8500(4.25) - 1000(4.25)^2 \\ &= \$18{,}062.50 \end{aligned}$$

We include a Converge® graph of $y = R(p)$ in Figure 4.38. Notice that the maximum revenue occurs at a selling price of $4.25.

Figure 4.38

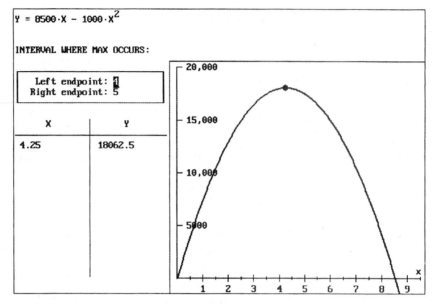

Reading and interpreting application problems seems to give students a difficult time. The following steps can help in solving optimization problems.

Steps for solving
optimization problems

1. Read the problem carefully.
2. Identify the quantity that is to be maximized or minimized.
3. Draw a picture, if appropriate.
4. Choose variables to represent the quantity to be maximized or minimized and the other unknowns.
5. Write an equation for the quantity to be maximized or minimized.
6. Eliminate extra variables (if necessary) from the equation in Step 5 so that the quantity to be maximized or minimized is expressed explicitly in terms of one variable.
7. Find the first derivative to determine critical values.
8. Use the First or Second Derivative Test to determine whether the critical values in Step 7 produce the desired maximum or minimum.
9. Reread the problem to be sure that you have answered the question.

Step 2 is important in helping you understand how to create the desired equation. It is so important, in fact, that Example 3 will be practice for Steps 1 and 2 only.

Example 3   Read each of the following problems and determine the quantity to be maximized or minimized.

a.   Find two positive numbers whose product is 121 and whose sum is as small as possible.

b. A company has determined that its profit, P, can be determined by the equation $P = 38x - 0.1x^2$, where x is the number of units sold. How many units should be sold so that the profit is greatest?

c. Find the dimensions of the largest rectangle that can be inscribed in a circle of radius 6 in.

d. A manufacturer of cylindrical containers that hold 90π in.^{3} plans to make the top and bottom of a material that costs $0.05 per square inch and the sides of a material that costs $0.03 per square inch. What dimensions should the containers have to keep the cost of material as low as possible?

Solution **a.** Minimize the sum.
b. Maximize the profit.
c. Maximize the area of the rectangle.
d. Minimize the cost (surface area times price per square inch). ∎

Now we give the complete solutions for three problems.

Example 4 A company is changing the shape of its containers for a certain product. Previously, the product was packaged in cans that held 64 in.^{3}. The company would now like to use cardboard boxes with a square base that will hold the same amount as the cans. What dimensions should the box have to minimize the amount of cardboard used for the new containers?

Solution We wish to minimize the surface area, S, of the boxes (see Figure 4.39).

Figure 4.39

$$S = \underbrace{2x^2}_{\substack{\text{top and} \\ \text{bottom}}} + \underbrace{4xy}_{\text{four sides}}$$

We need to eliminate a variable, so we use the fact that the volume, V, should be 64 in.^{3}.

$$V = x^2y$$
$$64 = x^2y$$

So $y = \dfrac{64}{x^2}$

Substituting, we have

$$S = 2x^2 + 4x\left(\frac{64}{x^2}\right)$$

$$= 2x^2 + \frac{256}{x}$$

We now need to differentiate and set equal to zero.

$$\frac{dS}{dx} = 4x - \frac{256}{x^2}$$

$$0 = 4x - \frac{256}{x^2}$$

$$4x = \frac{256}{x^2}$$

$$4x^3 = 256$$

$$x^3 = 64$$

$$x = 4$$

Now $\dfrac{d^2S}{dx^2} = 4 + \dfrac{512}{x^3}$, which is positive for $x = 4$, which implies a minimum.

Since $y = \dfrac{64}{x^2}$, we have

$$y = \frac{64}{16} = 4$$

The dimensions that minimize the amount of cardboard used are length 4 in., width 4 in., and height 4 in. ∎

Example 5   An electronics firm has found that an assembly line worker can produce an average of 600 units per month when there are no more than 12 workers on the line. Because of crowded conditions and other factors, for each additional worker added to the assembly line, the productivity per worker decreases by 30 units per month. How many workers should be on the line to maximize the number of units produced per month?

Solution   Let x represent the number of workers added to the assembly line. Then **$x + 12$** represents the **total number of workers** on the assembly line.

The new productivity per worker can now be expressed as follows:

original productivity + (decrease in productivity) (number of additional workers)
         600   + (-30) (x)

That is, the **productivity per worker** is now given by **$600 - 30x$.**

To get a representation for the number of units, N, produced, we need to multiply the total number of workers by the productivity per worker:

$$N = (x + 12)(600 - 30x)$$
$$= 7200 + 240x - 30x^2$$
$$\frac{dN}{dx} = 240 - 60x$$
$$0 = 240 - 60x$$
$$60x = 240$$
$$x = 4$$

To check this, we use the Second Derivative Test.

$$\frac{d^2N}{dx^2} = -60$$

The fact that the second derivative is negative implies that we have a maximum at $x = 4$. Because the total number of workers was represented by $x + 12$, the maximum output is achieved with $4 + 12$, or 16, workers. ∎

Example 6   The postal service in a certain city has decided to impose a restriction on the dimensions of packages to be mailed. The length and girth (perimeter of a cross section) combined can be no more than 90 in. A local manufacturer needs to determine dimensions for its shipping cartons that will be of maximum volume subject to the postal restrictions and that will also have cross sections that are twice as long as they are wide.

Solution   We need to maximize the volume of the cartons. Figure 4.40 shows a diagram and labeling of dimensions for the carton.

Figure 4.40

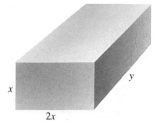

The equation for the volume is

$$V = (2x)(x)(y)$$
$$= 2x^2y$$

We need to eliminate a variable and can do so using the postal restriction that says, "the perimeter of the cross section plus the length can be 90 in."

$$90 = \underbrace{2(2x) + 2(x)}_{\substack{\text{perimeter of} \\ \text{cross section}}} + \underbrace{y}_{\text{length}}$$

or $y = 90 - 6x$

Substituting this into our volume equation, we have

$$V = 2x^2y$$
$$= 2x^2(90 - 6x)$$
$$= 180x^2 - 12x^3$$

To find the maximum value, calculate $\dfrac{dV}{dx}$ and set it equal to zero.

$$\frac{dV}{dx} = 360x - 36x^2$$
$$0 = 360x - 36x^2$$
$$= 36x(10 - x)$$
$$36x = 0 \qquad 10 - x = 0$$
$$x = 0 \qquad\quad x = 10$$

Since $x = 0$ means that we would have no box, we choose $x = 10$ for the location of the maximum for the function. Figure 4.41 displays the graph of the volume function.

Figure 4.41

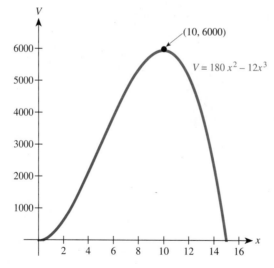

(10, 6000)

$V = 180\,x^2 - 12x^3$

The dimensions for the packing carton that will hold the maximum volume are 10 in. × 20 in. × 30 in.   ∎

PLATE I   The shape of a curve is certainly related to its first derivative. In this *DERIVE*®
2-D graphing window, 401 tangent lines have been drawn. Each line is tangent to the curve
$y = 4 - x^2$ at a different *x*-value. Although the parabola is not drawn, its shape is easily
seen as a result of graphing the tangent lines.

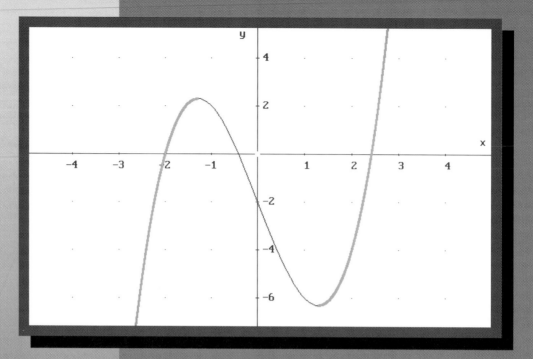

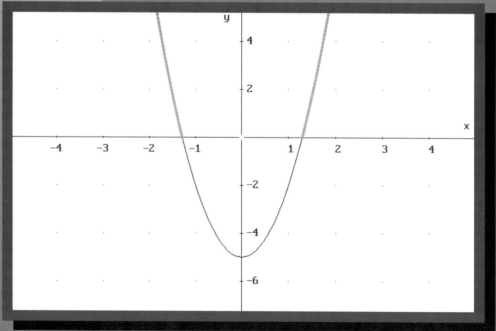

PLATE II   A curve is increasing precisely where its first derivative is positive. It is decreasing where its first derivative is negative. The top *DERIVE*® screen shows the graph of $y = x^3 - 5x - 2$. The red portion of the curve is where the function is increasing. The bottom screen is a graph of $y = 3x^2 - 5$, the function's derivative. The red portion is where the first derivative is positive. So, the red portions occur for precisely the same x-values.

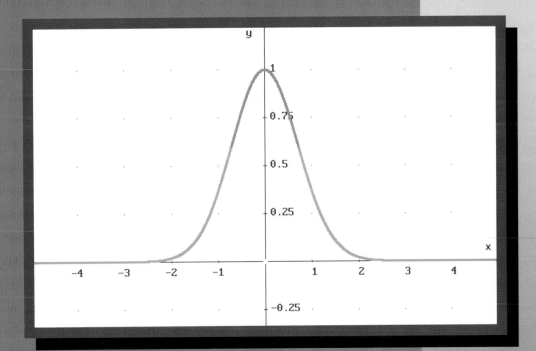

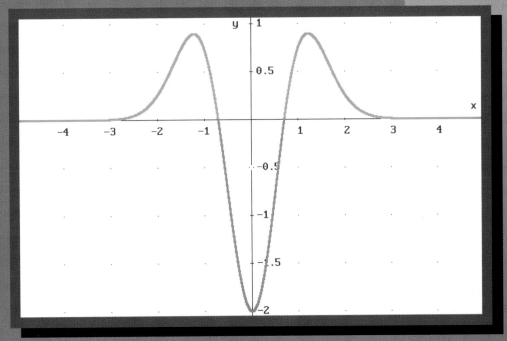

PLATE III   A curve is concave upward precisely where its second derivative is positive and concave downward where its second derivative is negative. The top *DERIVE*® screen shows the graph of $y = e^{-x^2}$. The red portion of the curve is where the function is concave upward; the blue portion is where the curve is concave downward. The bottom screen is a graph of the function's second derivative. The red portion is where the second derivative is positive, and the blue portion is where the second derivative is negative.

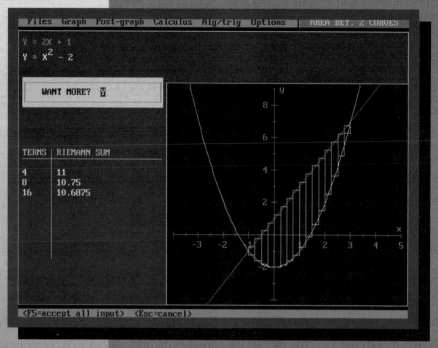

Y = 2X + 1
Y = X^2 − 2

WANT MORE? Y

TERMS	RIEMANN SUM
4	11
8	10.75
16	10.6875

<F5=accept all input>   <Esc=cancel>

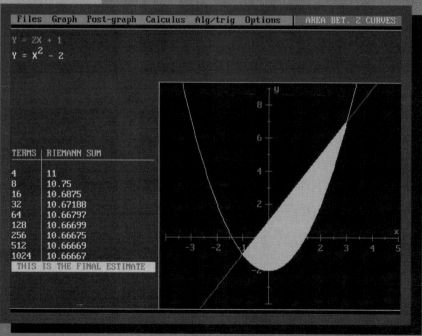

Y = 2X + 1
Y = X^2 − 2

TERMS	RIEMANN SUM
4	11
8	10.75
16	10.6875
32	10.67188
64	10.66797
128	10.66699
256	10.66675
512	10.66669
1024	10.66667

THIS IS THE FINAL ESTIMATE

PLATE IV   These two *CONVERGE*® screens show how rectangles can be used to find the area between two curves. Notice that the two approximations get closer to the exact area found in the text in Section 6.4.

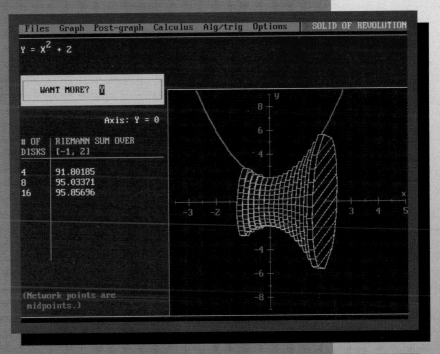

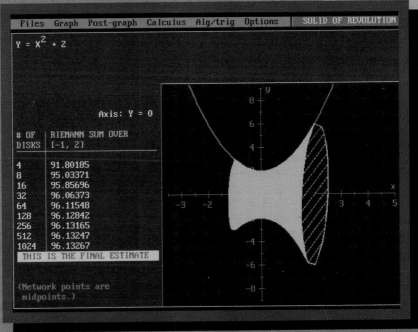

PLATE V The volume that results when a curve is revolved about a line can be approximated by revolving a large number of rectangles about the line. Each revolved rectangle is a disk. These two *CONVERGE*® screens show how a finite number of disks (top screen 16, bottom screen 1024) can be used to approximate volume. The exact value can be found in the text in Section 6.5.

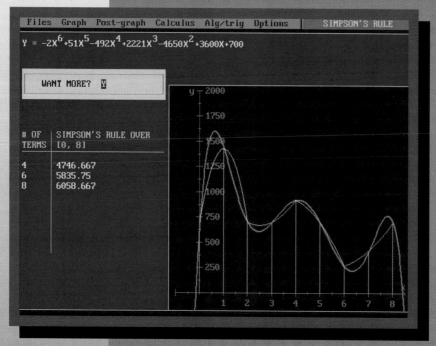

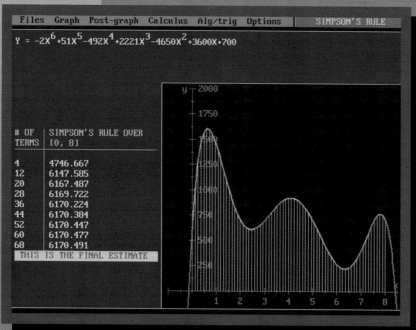

PLATE VI Simpson's rule. In the top *CONVERGE*® screen, the approximation using 8 parabolic arcs follows very closely to the development in Section 7.3A. The bottom screen shows the calculation and the accompanying graph for 68 subdivisions.

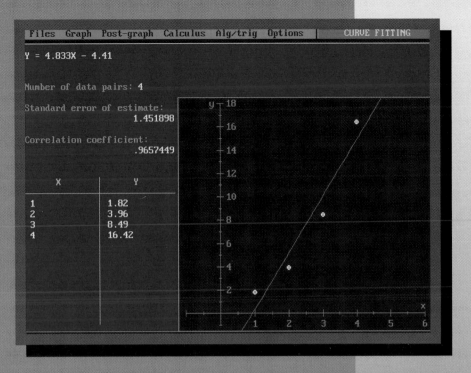

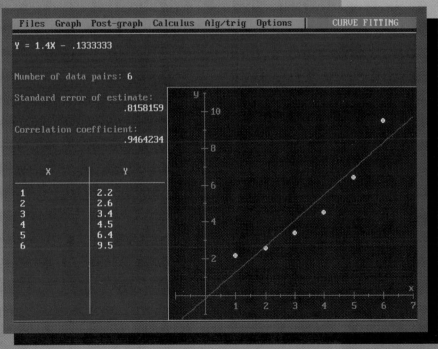

PLATE VII These two *CONVERGE*® screens calculate and draw the "line of best fit" for the data in examples in Section 9.6.

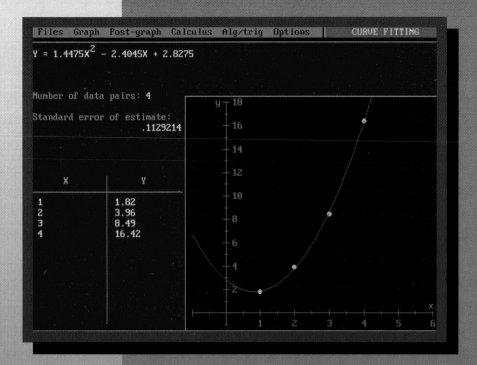

Files Graph Post-graph Calculus Alg/trig Options | CURVE FITTING

$Y = 1.4475X^2 - 2.4045X + 2.8275$

Number of data pairs: **4**

Standard error of estimate:
.1129214

X	Y
1	1.82
2	3.96
3	8.49
4	16.42

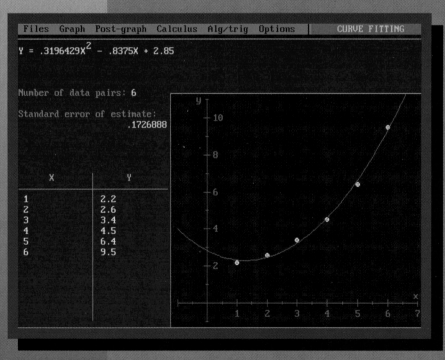

Files Graph Post-graph Calculus Alg/trig Options | CURVE FITTING

$Y = .3196429X^2 - .8375X + 2.85$

Number of data pairs: **6**

Standard error of estimate:
.1726888

X	Y
1	2.2
2	2.6
3	3.4
4	4.5
5	6.4
6	9.5

PLATE VIII   Compare these two *CONVERGE*® screens with the screens in Plate VII.
Here, we have *CONVERGE*® calculate a second-degree polynomial for the data and draw
the "parabola of best fit."

The exercises in this section consist of two main types: those for which the function is provided and those for which it will be your responsibility to determine the equation for the function. In Example 7, we provide the function to be minimized.

Example 7   A study was conducted to determine how the thermal pollution by a factory located near a lake has affected the fish population of the lake. The equation

$$N = \frac{-40t}{t^2 - 112t + 2560}, \qquad 32° < t < 80°$$

gives the number of fish dying per month, N (in hundreds), based on the temperature, t (in degrees Fahrenheit), of the water in the lake. Determine the temperature that minimizes the number of fish dying.

Solution   Because the equation is given, we calculate the derivative of the function and set it equal to zero. Using the Quotient Rule, we get

$$\frac{dN}{dt} = \frac{(t^2 - 112t + 2560)(-40) - (-40t)(2t - 112)}{(t^2 - 112t + 2560)^2}$$

$$= \frac{-40(t^2 - 112t + 2560 - 2t^2 + 112t)}{(t^2 - 112t + 2560)^2}$$

$$= \frac{-40(-t^2 + 2560)}{(t^2 - 112t + 2560)^2}$$

Setting $\dfrac{dN}{dt} = 0$, we have

$$\frac{-40(-t^2 + 2560)}{(t^2 - 112t + 2560)^2} = 0$$

$$-40(-t^2 + 2560) = 0$$

$$-t^2 + 2560 = 0$$

$$t^2 = 2560$$

$$t = \pm\sqrt{2560}$$

$$\approx 50.6°F \qquad (32° < t < 80°)$$

No negative values for t

To check to see if this produces a minimum, use a First Derivative Test. The curve is found to be decreasing to the left of $t = 50.6°F$ and increasing to the right. We can conclude that the minimum number of fish die when the temperature is 50.6°F. Figure 4.42 displays the graph of the function.

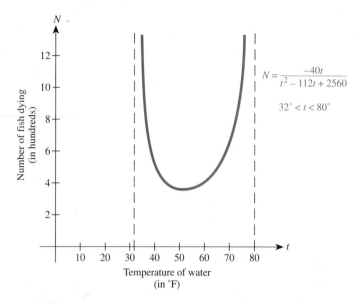

4.6   EXERCISES

SET A

Solve the following optimization problems.

1. Find two numbers whose sum is 160 and whose product is as large as possible.

2. Find two positive numbers whose product is 36 such that the sum of the first and four times the second is a minimum.

3. The product of two numbers is 16. Find the two numbers so that the sum of their squares is a minimum.

4. Solve part a of Example 3.

5. Find two positive numbers whose product is 48 such that the sum of the first and three times the second is as small as possible.

6. A rectangular garden is to be fenced in next to a garage so that fencing is needed on three sides only. If 16 feet of fencing is used, what is the maximum area that can be enclosed?

7. A rectangle is to be constructed so that the area will be 64 cm². What should the dimensions of the rectangle be so that the perimeter is a minimum?

8. A rectangle is to be constructed so that the perimeter will be 80 cm. What is the maximum possible area for the rectangle?

9. Solve part c of Example 3.

BUSINESS
REVENUE

10. The demand for a certain product is given by

$$p = 50 - 2x$$

Find the number of items that should be produced and sold to maximize the revenue.

PACKAGING

11. Boxes are being designed for a new product. Because of limited shelf space in the stores, the height must be 3 in. and the combined width and length of the box cannot be more than 10 in. What is the maximum volume for the boxes?

MANUFACTURING

12. A fast-food service supplier is constructing open boxes for takeout orders. The boxes are to be constructed from rectangular pieces of cardboard by cutting a square from each corner and bending up the sides. If pieces of cardboard are 12 in. by 18 in., what size squares should be cut from each corner to maximize the volume of boxes? See the accompanying figure.

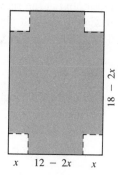

MANUFACTURING

13. Suppose that the demand equation for a certain type of tennis racquet sold by a company is

$$p = 160 - 0.2x$$

where p is the price per item. Find the number of racquets, x, that will maximize the company's revenue.

DEMAND EQUATIONS

14. The demand equation for a certain item is

$$x = 400 - \frac{p}{3}$$

where p is the price per item in dollars and x is the number of units sold. Find the price per item that will maximize the revenue.

PUBLISHING

15. A publishing company can sell its newest cookbook to bookstores for $5 each. The cost of producing x thousand cookbooks is given by the equation

$$C(x) = 500 - 118x - 36.5x^2 + 2x^3$$

Assuming the company can sell all of the cookbooks it produces, how many should it produce to maximize its profit?

PUBLISHING

16. A book is going into its second edition, and the plans are to have wider margins on each page than it had in the first edition. The margins are to be 1.5 in. at the top and bottom and 1 in. on the sides. If the total area of the page is to be 60 in.^{2}, what dimensions should each page have so that the area of the printed area is a maximum?

MANUFACTURING

17. A company that manufactures toasters estimates that the cost of producing x toasters per month is given by the equation

$$C(x) = 0.005x^2 + 0.01x + 200$$

Determine the number of toasters the company must produce per month to minimize the average cost per toaster. $\left(Hint: \text{Average cost} = \dfrac{C(x)}{x}. \right)$

SURVEYS AND RESEARCH

18. A television network is going to conduct a national survey. The network would like to keep the cost to a minimum, but at the same time they would like the results to be meaningful by including enough people. If x (in hundreds) is the number of people in the survey, then the cost, $C(x)$, is

$$C(x) = 0.005x^2 + \frac{5120}{x}, \qquad x \geq 1$$

Determine the number of people that should be included in the survey in order to minimize the cost.

MANUFACTURING

19. A toothbrush manufacturer has found that the cost for producing x toothbrushes per month is

$$C(x) = 200 - 4x + 0.008x^2$$

where $C(x)$ is measured in cents. Determine the number of toothbrushes that should be produced to minimize the cost.

PROFIT

20. The toothbrush manufacturer in exercise 19 has found that his demand per month, if he charges p cents per toothbrush, is $p = 6.8 - 0.01x$. Determine the price that should be charged per toothbrush to maximize the manufacturer's profits.

BIOLOGY
PHARMACOLOGY

21. Suppose that the concentration, C, of a certain drug in the bloodstream t hours after it has been administered is given by

$$C = \frac{5t}{t^2 + 3}, \qquad t \geq 0$$

At what time is the concentration of the drug at a maximum?

SET B

BUSINESS
TRAVEL INDUSTRY

22. A travel agency is running a special on weekend ski packages. The price it will charge per person is $275. The minimum number of people who must sign up to get this special rate is 20. For each person over 20 who goes, there will be a discount of $5 per person. If the maximum number of people the agency can accommodate is 50, find the number of people that would maximize the profits for the travel agency if their costs are $75 per person. (*Hint:* See Example 5.)

PROFIT

23. Solve part b of Example 3.

MANUFACTURING

24. Solve part d of Example 3.

MANUFACTURING

25. A company that manufactures aluminum cans has been requested to submit a bid for the production of cans that will hold 64π cm^3. The company would like to minimize the cost of the aluminum for the cylindrical cans. What dimensions should they use for the cans?

CONSTRUCTION

26. A company needs to build a small warehouse, half of which will be a refrigeration area. The floor plan of the warehouse is shown in the accompanying figure. The walls will all be 8 ft high. The four interior walls for the refrigerated area are to be covered with material that costs $6/ft^2, and the four interior walls for the remaining part are to be covered with material that costs $2/ft^2. If the total floor space needed is 3200 ft^2, what dimensions should the refrigerated area have to minimize the total cost?

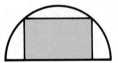

MANUFACTURING

27. Large rectangular containers with square bases are designed to hold 160 ft^3. The sides and top are constructed of a material that costs $0.08 per square foot. The material for the bottom of the container is stronger and costs $0.12 per square foot. Determine the dimensions of the most economical container.

ARCHITECTURE

28. A rectangular stained glass window is being designed to fit in a semicircular opening above a door. The remaining area will be filled with clear glass. See the accompanying diagram. If the diameter of the semicircle is 36 in., determine the dimensions of the largest stained glass window that can be used.

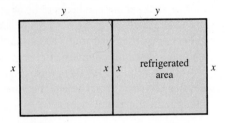

CONSTRUCTION

29. A new auditorium being built for a town will have a floor plan as illustrated in the figure. The rectangular part is the seating and stage area and the semicircular part will be the lobby area. If the distance around the outside of the building is to be 500 ft, what dimensions should the auditorium have to produce the maximum area for the rectangular part?

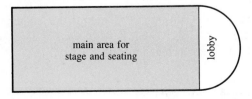

FUEL EFFICIENCY

30. A new car is being tested for fuel efficiency at varying speeds. Experimentation has shown that the fuel efficiency, F, is related to the speed, S, by the equation

$$F = \frac{S - 25}{S^2} + 40$$

Determine the speed that produces maximum fuel efficiency.

ARCHITECTURE

31. The strength of a rectangular beam is proportional to the product of its width and the square of its depth (the length of the beam is not used in computing strength). What would be the width and depth of a beam cut from a circular log 4 ft in diameter in order for the beam to be as strong as possible?

REAL ESTATE

32. A new apartment complex is charging $400 a month for each of its 200 two-bedroom apartments, and the complex is currently filled. Because of rising maintenance costs, the manager needs to raise the rent, but she knows that for each $20 she raises the rent, 5 apartments become vacant. What rent should she charge to maximize her rental income?

BIOLOGY
ECOLOGY

33. Two factories (A and B) on the outskirts of a town are 20 miles apart. Both factories are emitting pollutants into the air. Suppose that the concentration is inversely proportional to the square of the distance from the source. Suppose further that the pollution from factory A is twice as concentrated as that from factory B. Determine the approximate point between the two factories at which the total concentration of pollution from the two factories is at a minimum.

PHYSICAL SCIENCE
SOUND WAVES

34. Calculator Exercise

A scientist is doing experiments with sound. A sound must have a minimum intensity or the participant will not be able to hear it. This intensity is referred to as the *threshold* of hearing. Suppose that the following equation gives a description of threshold of hearing for a range of frequencies from 500 to 4000 Hz (the most sensitive range) for a certain participant:

$$I = (3.7143 \cdot 10^{-6})f^2 - (2.1 \cdot 10^{-2})f + 34.5714, \qquad 500 \leq f \leq 4000$$

where I is the intensity measured in decibels and f is the frequency measured in hertz (Hz). Determine the frequency for which the intensity is a minimum.

DENSITY

35. The density, s, of water is given by the equation

$$s = s_0(1 + at + bt^2 + ct^3)$$

where s_0 is the density at 0°C and t is the temperature in degrees Celsius. If $a = 5.3 \times 10^{-5}$, $b = -6.5 \times 10^{-6}$, and $c = 1.4 \times 10^{-8}$, find the temperature at which water has maximum density.

CHAPTER 4 REVIEW

VOCABULARY/FORMULA LIST

decreasing function
increasing function
critical values (type I, type II)
concave up
concave down
points of inflection

cusp
local extrema
local minimum
local maximum
First Derivative Test
Second Derivative Test

absolute extrema
absolute maximum
absolute minimum
endpoint extrema
optimization

CHAPTER 4 REVIEW EXERCISES

Exercises 1 through 6 refer to the figure shown here, the graph of $y = f(x)$ in the interval $a \le x \le b$. Your answers should be based on the graph only for this interval.

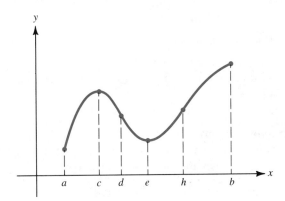

[4.3] **1.** At what point is there a relative maximum?

[4.2] **2.** What point(s) is a point of inflection?

[4.2] **3.** For what values of x is the curve concave down?

[4.2] **4.** For what values of x is the curve concave up?

[4.4] **5.** What is the (absolute) minimum value of the function?

[4.3] **6.** At what point is there a relative minimum?

[4.5] *In exercises 7 through 16, find (a) the intervals on which the function is increasing or decreasing; (b) regions where the curve is concave up, regions where it is concave down, and points of inflection; (c) local extrema; (d) absolute extrema; (e) intercepts, symmetry, and asymptotes; and (f) sketch the graph.*

   7. $y = 4x - x^3$ **8.** $f(x) = 4x^2 - x^3$

9. $f(x) = x^4 - 10x^2 + 9$

10. $y = x^4 - 5x^2 + 4, \quad -2 \le x \le 3$

11. $y = \sqrt[3]{x - 1}$

12. $f(x) = x^{2/3}$

13. $y = x^5 - 4x^3$

14. $y = \dfrac{4}{x^2} - x, \quad x > 0$

15. $y = \dfrac{x}{16 - x^2}$

16. $y = \dfrac{x}{\sqrt{1 - x^2}}$

[4.6] **17.** A rectangular yard is to be fenced in on three sides; the fourth side is a straight stream. If there is 80 m of fencing available, what dimensions will maximize the area of the yard?

[4.6] **18.** The sum of a number and three times a second number is 72. What two such numbers will yield a maximum product?

BUSINESS
DEMAND, PROFIT, COST,
REVENUE EQUATIONS

[4.6] **19.** Suppose a business has determined that the demand for its product is given by

$$p = \dfrac{40}{\sqrt{x}}$$

where x is the number of items sold and p is the price per item. Furthermore, the cost of producing x items is

$$C(x) = 0.5x + 400$$

    a. Find its revenue function, $R(x) = xp$.
    b. What value of x maximizes profit? (Recall that profit P is $R(x) - C(x)$.)
    c. What is that maximum profit?
    d. What price should be charged in order to maximize profit?

BIOLOGY
ECOLOGY

[4.6] **20. Calculator Exercise**
The level of radiation, R, in a stream outside New London, Connecticut (where nuclear submarines are made), was found to be

$$R = \tfrac{1}{16}(21x^3 - 127x^2 + 154x + 80), \quad 0 \le x \le 2$$

where x is the distance in miles from the factory.

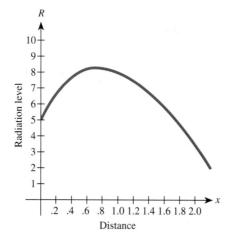

a. Use the graph to **estimate** the distance from the factory where the radiation is the lowest and where it is the highest.

b. Find the **exact** distance from the factory where the radiation is the lowest and where it is the highest.

CHAPTER 4 TEST

PART ONE

Multiple choice. In exercises 1 through 8, put the letter of your correct response in the blank provided.

_____ **1.** If $f'(c) = 0$, then f has either a local maximum or local minimum at $x = c$.

   a.   true   **b.**   false

_____ **2.** If $f''(c) = 0$, then $(c, f(c))$ is a point of inflection.

   a.   true   **b.**   false

_____ **3.** If c is a critical value, then $f'(c) = 0$.

   a.   true   **b.**   false

_____ **4.** If f has an absolute minimum at $x = c$, then $f'(c) = 0$.

   a.   true   **b.**   false

_____ **5.** Consider $f(x) = x^4 - 6x^2$. The point(s) of inflection are

   a. $(0, 0)$ **b.** $(1, -5)$

   c. $(1, -5)$ and $(-1, -5)$ **d.** $(12, 0)$

_____ **6.** Consider $y = (x + 1)^{2/3}$. The point $(-1, 0)$ is

   a.   a local minimum   **b.**   an absolute minimum

   c.   both   **d.**   neither

_____ **7.** How many critical values does $y = \sqrt{x} + (x - 1)^3$ have?

   a.   1   **b.**   2

   c.   3   **d.**   none

_____ **8.** If $f(1) = 2, f'(1) = 0$, and $f''(1) = -6$, then $(1, 2)$ must be

   a.   a local minimum   **b.**   a local maximum

   c.   a point of inflection   **d.**   an absolute extreme

PART TWO

Solve and show all work.

9. Sketch the graph of $y = -2x^3 + 5x^2 + 4x - 1$.

10. Consider $f(x) = x^3 - 5x^2 - 8x$. Find:

   a.   The intervals where the function is increasing

   b.   The intervals where the function is decreasing

   c.   The intervals where the function is concave down

   d.   The intervals where the function is concave up

   e.   The points of inflection

BUSINESS
ADVERTISING

11. A small company has determined that its profit, P, depends on the amount of money spent on advertising, x. This relationship is given by the equation

$$P(x) = \frac{8x}{x^2 + 16} + 10$$

where P and x are measured in thousands of dollars. What amount should the company spend on advertising to ensure maximum profit?

REVENUE

12. The Griprite Manufacturing Company has determined that the relationship between the quantity of flywheels sold, x, and the asking price, p, is

$$p = 75 - \frac{x}{4}, \qquad 0 \le x \le 260$$

   a. Determine the revenue equation.
   b. For what value(s) of x is revenue increasing?
   c. For what value(s) of x is revenue decreasing?
   d. What asking price will produce the maximum revenue?

PROFIT

13. Given that profit, $P(s)$, is related to the selling price, s, of an item by

$$P(s) = 12{,}000s - 100s^2$$

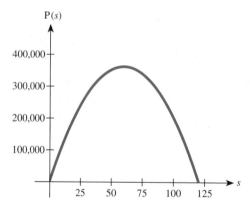

   a. Is the profit increasing or decreasing when the selling price is 40?
   b. For what range of selling prices is profit increasing?
   c. What selling price results in maximum profit?

REVENUE

14. A firm has determined that its revenue, $R(x)$, from the sale of x units of its product is

$$R(x) = 425 - \frac{2500}{x + 6} - 4x$$

   Find the value of x that maximizes revenue and find the maximum revenue.

15. A rectangle is to be constructed so that the perimeter will be 160 in. What is the maximum possible area for the rectangle?

16. Choose five of the vocabulary words from the list given for this chapter and write a sentence using them.

WRITING AND CRITICAL THINKING

Write out in words (use complete sentences) the procedure for solving an optimization problem.

5
EXPONENTIAL AND LOGARITHMIC FUNCTIONS

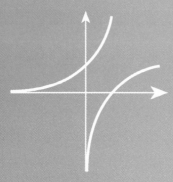

5.1 EXPONENTIAL FUNCTIONS WITH APPLICATIONS

The functions we have differentiated thus far have been confined to polynomial and rational functions. Many applications, however, cannot be modeled using rational or polynomial functions. Compound interest in banking, population models in social science, bacteria counts in biology, even things like the measure of the concentration of drugs in the bloodstream (medicine), the measure of the loudness of sound (physics), the study of the acidity of shampoo (chemistry), and the carbon-14 dating of fossils (archaeology) all have mathematical models that resemble functions studied in this chapter.

We begin by studying **compound interest.** Money earns *compound* interest when an initial investment, called the **principal** (P), earns interest for a period of time, and thereafter the principal *plus* accrued interest earns additional interest. For example, suppose an initial investment of $1000 earns 6% *annual* interest and interest is compounded 12 times per year, or monthly. We let

P = the principal or initial investment $\qquad\qquad$ = $1000

r = interest rate per year (6%) $\qquad\qquad\qquad$ = 0.06

n = the number of times per year interest is compounded = 12

Then

$$\frac{r}{n} = \text{the rate of interest } per\ period = \frac{0.06}{12} = 0.005$$

The balance at the end of the first month is

$$P\left(1 + \frac{r}{n}\right) = \$1000\left(1 + \frac{0.06}{12}\right) = \$1005$$

After the second month, the balance is the previous balance $P\left(1 + \dfrac{r}{n}\right)$ times $\left(1 + \dfrac{r}{n}\right)$:

$$\left[P\left(1 + \frac{r}{n}\right)\right]\left(1 + \frac{r}{n}\right) = P\left(1 + \frac{r}{n}\right)^2$$

$$= 1000\left(1 + \frac{0.06}{12}\right)^2$$

$$= \$1010.025$$

After three months, we have

$$\left[P\left(1 + \frac{r}{n}\right)\left(1 + \frac{r}{n}\right)\right]\left(1 + \frac{r}{n}\right) = P\left(1 + \frac{r}{n}\right)^3$$

$$= \$1015.075125$$

After one year:

$$P\left(1 + \frac{r}{n}\right)^{12} \approx \$1061.677812$$

Of course, banks round to the nearest cent, so each of the amounts would be rounded.

In general, after t years, the interest has been compounded nt times, and the formula for the balance A is

$$A = P\left(1 + \frac{r}{n}\right)^{nt}$$

Because P, r, and n are constants, we see that A is a function of t and, because t is a variable appearing as an exponent, we have our first example of an **exponential function.** More formally, we have:

Exponential function

If $b > 0$ and $b \neq 1$, then $f(x) = b^x$ is called an **exponential function** and b is called the **base** of the exponential function.

Note in the definition that we restrict b to the positive numbers to avoid situations such as $(-3)^{1/2}$, which is not a real number. We also specify $b \neq 1$ because otherwise the function would be the same as $f(x) = 1$, which is just a horizontal line.

In working with the exponential functions in this section, keep in mind that our basic laws of exponents will hold true. You may find that a calculator with a special key (usually labeled "x^y," "y^x," or "$\wedge$") will help in working some of the functions of this section.

The graphs and properties of exponential functions are examined in the next three examples.

Example 1   Graph the exponential function $y = 4^x$.

Solution   A table of values, generated with the aid of a calculator having an x^y key, is given here. The graph appears in Figure 5.1.

Figure 5.1

x	$y = 4^x$
-2	$\frac{1}{16} = 0.0625$
-1.5	0.125
-1	0.25
-0.5	0.5
0	1.0
0.5	2.0
1	4.0
1.5	8.0
2	16.0

Notice that the domain of the function is R and the range is $\{y \mid y > 0\}$. Also, as $x \to -\infty$, $y \to 0$, and as $x \to +\infty$, $y \to +\infty$, or, using the limit notation from Chapter 2, we have

$$\lim_{x \to -\infty} 4^x = 0 \qquad \text{and} \qquad \lim_{x \to +\infty} 4^x = +\infty$$

From this information and the graph in Figure 5.1, we see that the function is *increasing*. We say that the graph exhibits **exponential growth.** ∎

Example 2   Graph $y = 4^{-x}$.

Solution   Again, a table and calculator assist in graphing the function (see Figure 5.2).

Figure 5.2

x	$y = 4^{-x}$
-2	16.0
-1.5	8.0
-1	4.0
-0.5	2.0
0	1.0
0.5	0.5
1	0.25
2	0.0625

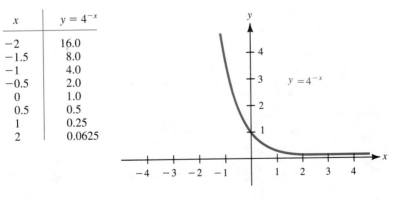

The domain of $y = 4^{-x}$ is R; the range is $\{y \mid y > 0\}$. Notice also that as $x \to -\infty$, $y \to \infty$, and as $x \to \infty$, $y \to 0$. From the graph, we see that this function is *decreasing*. We say that the graph exhibits **exponential decay.** ∎

Finally, notice that 4^{-x} is equivalent to $(4^{-1})^x = (\frac{1}{4})^x$. We say that the function $f(x) = b^x$ exhibits exponential growth if $b > 1$ and exhibits exponential decay if $0 < b < 1$.

The concepts of exponential growth and decay are summarized in Figure 5.3.

Figure 5.3

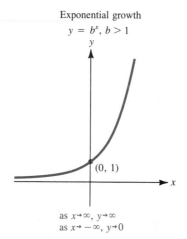

Exponential growth
$y = b^x, b > 1$

as $x \to \infty$, $y \to \infty$
as $x \to -\infty$, $y \to 0$

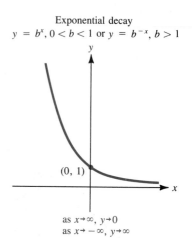

Exponential decay
$y = b^x, 0 < b < 1$ or $y = b^{-x}, b > 1$

as $x \to \infty$, $y \to 0$
as $x \to -\infty$, $y \to \infty$

Let's return to the compound interest formula with fixed values $P = \$1$, $t = 1$ year, and $r = 100\%$ and let n get *very* large. That is, instead of compounding monthly, we examine values of $n = 4$ (quarterly), 12 (monthly), 52 (weekly), 365 (daily), 8760 (hourly), and 31,536,000 (every second) and find the balance after 1 year. Note that with $r = 100\%$, if this were simple interest, the \$1 would double after 1 year and become \$2. We use a calculator to develop the following table:

n	Value of $\left(1 + \dfrac{1}{n}\right)^n$
1	2.0
4	2.44140625
12	2.61303529
52	2.69259695
365	2.71456748
8,760	2.71812669
31,536,000	2.71821394

The value of $\displaystyle\lim_{n \to \infty} \left(1 + \frac{1}{n}\right)^n$ is a very important number indeed! Its actual value is an irrational number and it is denoted by the letter e.

$$e = \lim_{n \to \infty} \left(1 + \frac{1}{n}\right)^n = 2.718281828459045\ldots$$

Later in this chapter, the significance of e will be more apparent. For now, we alert you to the table of values of e^x and e^{-x} in Appendix B and the graph of $y = e^x$ in Example 3.

Example 3 Graph $y = e^x$, the **natural exponential function.**

Solution Many calculators have an e^x key. The "Powers of e" table in Appendix B can also be used to generate the following table of values (see Figure 5.4):

Figure 5.4

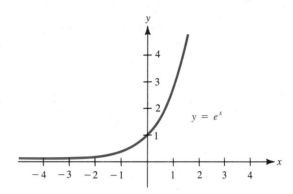

x	$y = e^x$
-3	0.0498
-2	0.1353
-1	0.3679
-0.5	0.6065
0	1.0000
0.5	1.6487
1	2.7183
2	7.3891
3	20.0855

In Examples 4 and 5, we graph more general exponential functions, those of the form $g(x) = ab^{f(x)}$, where the exponent is a function of x.

Example 4 Sketch the graph of $g(x) = -4^{x+1}$.

Solution Be careful in interpreting this function. A simple rewrite will clarify its meaning as follows:

$$g(x) = -1 \cdot 4^{x+1}$$

The table of values and graph are shown here (see Figure 5.5).

Figure 5.5

x	$g(x) = -4^{x+1}$
-3	-0.0625
-2	-0.25
-1	-1.0
0	-4
0.5	-8
1	-16
1.5	-32
2	-64

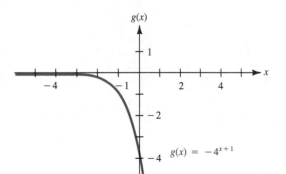

Example 5   **a.**   Graph the function $g(x) = 5(2)^{-x^2}$.
   b.   State the domain and range of the function.
   c.   Discuss the behavior of the graph as $x \to \infty$ and as $x \to -\infty$.

Solution   **a.**   We use the following table to get a "feel" for the shape of the curve (see Figure 5.6):

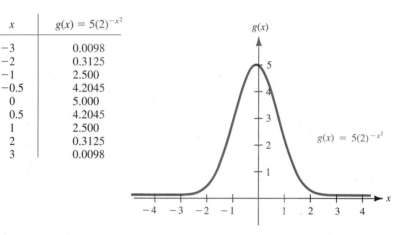

Figure 5.6

x	$g(x) = 5(2)^{-x^2}$
-3	0.0098
-2	0.3125
-1	2.500
-0.5	4.2045
0	5.000
0.5	4.2045
1	2.500
2	0.3125
3	0.0098

   b.   The domain is R; the range is $\{y | 0 < y \le 5\}$.
   c.   The x axis is an asymptote: as $x \to -\infty$, $y \to 0$, and as $x \to \infty$, $y \to 0$.   ∎

Example 6   In a study with human patients, a painkilling drug was administered. It was determined that the concentration of the medicine present in the bloodstream was a function of the time t (in hours) after the drug's administration and was given by

$$C(t) = 50e^{-0.1t}, \qquad t \ge 2$$

Here, $C(t)$ is measured in ng/ml (nanograms per milliliter). Graph the function for $2 \le t \le 36$.

Solution   We use a calculator to construct the following table of values (see Figure 5.7):

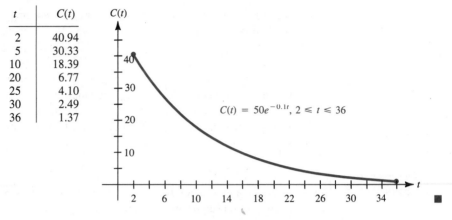

Figure 5.7

t	$C(t)$
2	40.94
5	30.33
10	18.39
20	6.77
25	4.10
30	2.49
36	1.37

$C(t) = 50e^{-0.1t}, 2 \leq t \leq 36$

In addition to being able to graph exponential functions, we need to solve exponential equations. One technique uses the following rule:

Theorem 5.1

> If $b^x = b^y$, then $x = y$ provided $b > 0$ and $b \neq 1$.

That is, if the bases are identical in an equation of two exponential expressions, then the exponents are equal. We use this principle to solve the equations in the next example.

Example 7 Solve for x: $27^{2x-1} = 9$.

Solution Both 27 and 9 are powers of 3.

$$27^{2x-1} = 9 \qquad 27 = 3^3, 9 = 3^2$$
$$(3^3)^{2x-1} = 3^2$$
$$3^{6x-3} = 3^2 \qquad \text{Since bases are the same, exponents are equal.}$$
$$6x - 3 = 2$$
$$6x = 5$$
$$x = \frac{5}{6}$$

We conclude this section with an application to business and psychology of a function called the **learning curve.** This function is defined by

$$f(x) = a(1 - e^{-bx})$$

where a and b are appropriate constants. The value a is an upper bound for the function.

In psychology, this function can be used in working with rate of learning and also in the measure of the level of performance in doing a task. Psychologists have determined that under certain conditions the rate of learning may often be initially rapid and then taper off to some maximum level. In performing a task, it has been determined that a person's performance level will at some point peak, and continued improved performance will approach unnoticeable amounts.

In business, this same class of functions can be used to represent changes in production level. Production levels may rise quickly for a new product as the bugs in the production are worked out but then eventually will peak at some optimum level due to physical limitations on the output levels. We examine such a situation in Example 8.

Example 8   Suppose that on a production line, the number of units, y, assembled per day x days after the production run began is given by

$$y = 100(1 - e^{-0.3x})$$

a.   What is the level of production after 5 days? 10 days? 20 days?
b.   Graph $y = 100(1 - e^{-0.3x})$ for $x \geq 0$.

Solution   a.   If $x = 5$, $y = 100(1 - e^{-1.5}) \approx 77.7$ units.
If $x = 10$, $y = 100(1 - e^{-3}) \approx 95$ units.
If $x = 20$, $y = 100(1 - e^{-6}) \approx 99.8$ units.
b.   A graph of the function is shown in Figure 5.8.

Figure 5.8

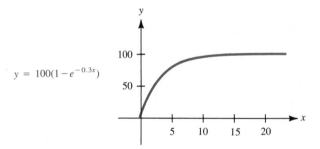

$$y = 100(1 - e^{-0.3x})$$

5.1 EXERCISES

SET A

In exercises 1 through 14, graph the given exponential functions. Does the function exhibit exponential growth or exponential decay? (See Examples 1 through 3.)

1. $y = 2^x$
2. $y = 10^x$
3. $y = 2^{-x}$
4. $y = 10^{-x}$
5. $y = 5^{-x}$
6. $y = (\frac{1}{8})^x$
7. $y = 2^{x^2}$
8. $y = 3(2)^{x^2}$
9. $y = e^{2x}$
10. $y = 3e^{-x}$
11. $y = e^{x+1}$
12. $y = e^x + 1$
13. $y = |e^x|$
14. $y = e^{|x|}$

In exercises 15 through 23, solve each equation. (See Example 7.)

15. $2^x = 32$ **16.** $2^{3x} = 64$ **17.** $10^x = \frac{1}{100}$ **18.** $2^{3x} = 4^{x-1}$

19. $81^x = 9^{x+1}$ **20.** $30^x = 1$ **21.** $3^{x^2 + 2x} = \frac{1}{3}$ **22.** $25^{x-1} = 5^{4x-3}$

23. $3 \cdot 2^x = 48$

SET B

Exercises 24 through 39 involve exponential functions in applied problems.

BUSINESS
DATA PROCESSING

24. A computer manufacturer claims that the yearly maintenance cost of its new minicomputer is related to the average monthly use, x (in hundreds of hours), by the equation $C = 14{,}000 - 12{,}000e^{-0.01x}$. Assuming that the manufacturer's claim is true, what is the yearly maintenance cost for 400 hours' average monthly use?

ECONOMICS

25. With a 6% annual inflation rate, the amount of time, t (in years), that it takes for prices to double can be found by solving the equation $(1.06)^t = 2$.
  a. Make a table of values and approximate the value of t.
  b. Approximately how long will it take prices to double at a 12% annual inflation rate?
  c. Approximately how long will it take for prices to increase by 50% at a 10% annual inflation rate?

BANKING

26. The monthly payment, P, of an amortized mortgage loan can be thought of as an exponential function of t, the total number of payments to be made, and is given by the formula

$$P = V \cdot \frac{i}{[1 - (1 + i)^{-t}]}$$

where V represents the amount of the loan (in dollars) and i represents the interest per payment period (monthly, for our purposes here).
  a. One of the authors borrowed $48,000 at 11.25%. Find his monthly payment if the loan is for 30 years. (*Hint:* $V = 48{,}000$, $i = \frac{0.1125}{12} = 0.009375$, $t = 360$).
  b. The other author, the smarter one, borrowed $34,000 at 8.5% for 30 years. Find her monthly payment.

BANKING

27. Using the formula presented in this section,

$$A = P\left(1 + \frac{r}{n}\right)^{nt}$$

find the amount of money on hand after 2 years ($t = 2$) if $2000 was placed in an account yielding *daily* (use $n = 365$) interest at an annual rate of 10%.

BANKING

28. Banks sometimes use a variation of the formula called the **Banker's Rule** for *daily* interest:

$$A = P\left(1 + \frac{r}{360}\right)^{365t}$$

and of all alternatives (including continuous interest, see exercise 29), this gives the maximum interest to the investor. Use the Banker's Rule to find the amount on deposit after $2000 was invested at a 10% annual rate for 2 years.

BANKING

29. The formula used for continuous compounding of interest is given by

$$A = Pe^{rt}$$

    a. Find A if $r = 10\%$, $P = \$2000$, and $t = 2$.
    b. Compare your result in part a with that of exercises 27 and 28.

BANKING

30. In this section we saw that the value of $1000.00 invested at a 6% annual rate compounded monthly is $1061.68 after 1 year. This is the same value that would occur at a simple interest rate of 6.168%; we call 6.168% the **effective annual rate** (or the **effective annual yield**).
    a. What is the value of an investment of $1000.00 at 10% compounded quarterly after 1 year?
    b. What is the effective annual rate of this investment?

BANKING

31. **a.** Use the Banker's Rule (see exercise 28) to determine the amount on deposit after $1.00 was invested at a 10% annual rate for 1 year.
    b. What is the effective rate of that investment?
    c. Determine the effective rate for a 10% annual rate compounded daily (use $n = 365$) and compare this with your result in part b.

PRODUCTION AND
LEARNING CURVES

32. The daily output, y, of a product x days after the start of production is given by

$$y = 50 - 50e^{-0.2x}$$

    a. After 6 days, how many units are produced daily?
    b. After 10 days, how many units are produced daily?
    c. What is the upper bound for units produced in this model?
    d. After 20 days, what percent of the upper bound is achieved?

FINANCE

33. On January 1, 1990, Ronolog, Inc. had investments with a market value of $100,000. This value has been experiencing a rate of increase of 8% per year. The value V of these investments is given by

$$V = 100{,}000(1.08)^t$$

where t is the number of years since January 1, 1990. Find the value of Ronolog's investments as of January 1, 1993.

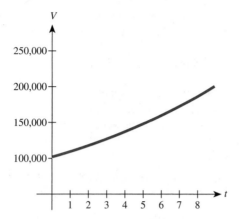

ACCOUNTING

34. A company has $6000 worth of equipment that it plans to depreciate at a rate of 20% each year; that is, the value of the equipment in any given year is 80% of the previous year's value. The formula for the value, V, of the equipment after t years is

$$V = 6000(0.8)^t$$

What is the value of the equipment after 5 years?

SALES

35. Sales of an item usually decline in the absence of advertising. Assume that a company decides to phase out a product and hence does not advertise it any longer. The decline in sales is described by

$$S(t) = 2000\left(\frac{3}{4}\right)^t$$

where t is in months.
a. Graph $S(t)$.
b. What are the initial monthly sales?
c. What are the sales when $t = 3$?

BIOLOGY
BACTERIA GROWTH

36. When bacteria are grown in a particular medium, the hourly growth rate is 5%. The number of bacteria present after t hours, $N(t)$, can be expressed by

$$N(t) = N_0(1.05)^t$$

where N_0 is the number present initially.
If $N_0 = 5000$, find $N(3)$.

ECOLOGY

37. The radioactivity, R (in roentgens per hour), that is measured after a nuclear explosion is given by

$$R = rt^{-1.2}$$

where r is the measured radiation just prior to explosion. If $R = 3$ roentgens per hour and $t = 4$ hr, find r.

SOCIAL SCIENCE
EDUCATION

38. A function that describes the number of people who scored better than the average on a standardized examination is

$$N(x) = 10{,}000e^{-0.0001x^2}$$

where x is the difference between the number of points scored and the average score and $N(x)$ is the number of people who obtained that score. If the average was 500, how many people received the following scores?

a. 420   **b.** 520   **c.** 800

MUSIC

39. The spacing of the frets on the neck of a classical guitar (see the accompanying figure) is determined by

$$d = 21.9[2^{(20-x)/12}]$$

where d = the distance in centimeters between the xth fret and the bridge
        x = the fret number

a. Determine how far the 19th fret should be from the bridge.

b. If $x = 0$, the value of d represents the distance between the nut and the bridge. Find that distance.

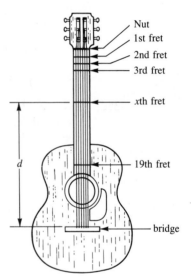

WRITING AND CRITICAL THINKING

For the exponential function $f(x) = b^x$, if $0 < b < 1$, then

$$\lim_{x \to -\infty} b^x = \infty \qquad \text{and} \qquad \lim_{x \to \infty} b^x = 0$$

Explain this behavior for the function in words. Also explain what happens if $b > 1$.

USING TECHNOLOGY IN CALCULUS

An important concept to master in calculus is the relationship between an equation in one variable, say $x^2 = 0$, and its two-variable counterpart, $y = x^2$. Remember this:

> Solution(s) to $f(x) = 0$ are the x intercept(s) of the graph of $y = f(x)$.

This is especially important for solving equations in one variable that have no closed-form, *exact* solution. One way to approximate the solutions to $6x - 1 - 2^x = 0$, for example, is to graph $y = 6x - 1 - 2^x$ with a graphics calculator or a graphing software package and then to ZOOM in on the x intercept until some reasonable precision can be found.

We use *DERIVE*® for this exercise. Notice in the following figure that the graph of $y = 6x - 1 - 2^x$ has two x intercepts. Therefore, $6x - 1 - 2^x = 0$ has two solutions.

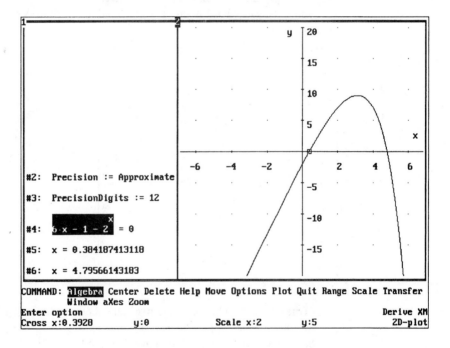

Next, we ZOOM in on those points several times. Notice the scale in the following figure (each tick mark is 0.1 unit) after we have ZOOMed in on the left-hand intercept to approximate it to be about $x \approx 0.3839$.

(*continued*)

(continued)

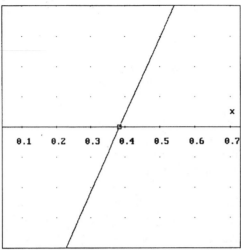

Similarly, we ZOOM in on the right-hand intercept to approximate it to be about 4.7955:

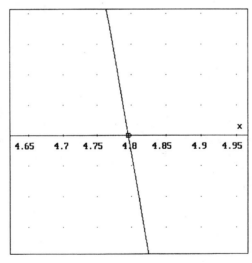

1. (True or False): The following five statements are equivalent:
   a. 4.7955 is an approximate solution to $6x - 1 - 2^x = 0$.
   b. If $f(x) = 6x - 1 - 2^x$, then $f(4.7955) \approx 0$.
   c. 4.7955 is an approximate zero of $6x - 1 - 2^x$.
   d. An x intercept of the graph of $y = 6x - 1 - 2^x$ is about $(4.7955, 0)$.
   e. The graphs of $y = 6x - 1$ and $y = 2^x$ intersect at a point whose abscissa is about 4.7955.

2. Use the technique outlined in the discussion to approximate the solution(s) to $x^2 + 4^x - 25 = 0$.

3. Approximate a zero of $\dfrac{1}{2x + 1} - x^2$.

5.1A   EXPONENTIAL FUNCTIONS ON A GRAPHICS CALCULATOR

We begin our study of exponential functions by examining the phenomenon of **compound interest.** Money earns compound interest when an initial investment, called the **principal,** P, earns interest for a period of time and thereafter the principal *plus* accrued interest earns additional interest. For example, suppose an initial investment of $1000 earns 6% annual interest compounded monthly. We let

$P =$ the principal or initial investment $\qquad\qquad = \$1000$
$r =$ interest rate per year (6%) $\qquad\qquad\qquad\;\; = 0.06$
$n =$ the number of times per year interest is compounded $= 12$

Then

$$\frac{r}{n} = \text{the rate of interest } per\ period = \frac{0.06}{12} = 0.005$$

The balance at the end of the first month is

$$P\left(1 + \frac{r}{n}\right) = \$1000\left(1 + \frac{0.06}{12}\right) = \$1005.00$$

After the second month, the balance is the previous balance, $P\left(1 + \frac{r}{n}\right)$, times $\left(1 + \frac{r}{n}\right)$:

$$\left[P\left(1 + \frac{r}{n}\right)\right]\left(1 + \frac{r}{n}\right) = P\left(1 + \frac{r}{n}\right)^2 = \$1000\left(1 + \frac{0.06}{12}\right)^2 = \$1010.025$$

After one year,

$$P\left(1 + \frac{r}{n}\right)^{12} \approx \$1061.677812$$

In general, after t years, the interest has been compounded nt times, and the formula for the balance, A, is

$$A = P\left(1 + \frac{r}{n}\right)^{nt}$$

Because P, r, and n are constants, we see that A is a function of t and, because t is a variable appearing as an exponent, we have our first example of an **exponential function.** More generally, we say:

Exponential function

If $b > 0$ and $b \neq 1$, then $f(x) = b^x$ is called an **exponential function** and b is called the **base** of the exponential function.

We use the graphics calculator to display an example of the exponential function $y = 3^x$. Before we can graph the function, we must get an appreciation for its domain and range. It should be clear that there are no restrictions on the values of x (that is, the domain is all real numbers) and that as x gets large, y gets large very quickly. Also, as x approaches negative infinity, y approaches zero. Because y can never be negative, the range appears to be the set of positive real numbers. The graph appears in Figure 5.1A.1.

Figure 5.1A.1 $y = 3^x$

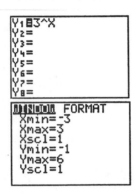
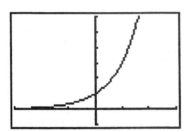

We see that the function in Figure 5.1A.1 is increasing. In fact, we can write $\lim\limits_{x \to -\infty} 3^x = 0$ and $\lim\limits_{x \to +\infty} 3^x = +\infty$; the function is said to exhibit **exponential growth.**

For a function exhibiting **exponential decay,** consider $y = 4^{-x}$. Notice from the graph in Figure 5.1A.2, that $\lim\limits_{x \to -\infty} 4^{-x} = +\infty$ and $\lim\limits_{x \to +\infty} 4^{-x} = 0$.

Figure 5.1A.2 $y = 4^{-x}$

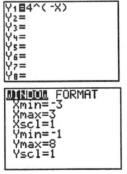
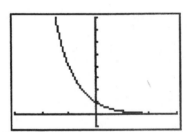

The compound interest formula we examined earlier can lead us to a special exponential function. Consider the formula $A = P(1 + \frac{r}{n})^{nt}$, with the constants $P = \$1$, $t = 1$ year, and $r = 100\%$: $A = (1 + \frac{1}{n})^n$. What happens to A as n (the number of compounding periods) gets very large? In the following table, we list the values of A for various values of n, such as $n = 4$ (quarterly compounding), $n = 12$ (monthly compounding), $n = 52$ (weekly compounding), $n = 365$ (daily compounding), $n = 8760$ (compounding every hour of every day), and $n = 31{,}536{,}000$ (compounding every second of every minute of every day of the year).

n	$A = \left(1 + \dfrac{1}{n}\right)^n$
1	2.0
2	2.44140625
12	2.61303529
52	2.69259695
365	2.71456748
8760	2.71812669
31,536,000	2.71821394

The value of $\displaystyle\lim_{n \to \infty} \left(1 + \frac{1}{n}\right)^n$ is a very important number in mathematics. It is an irrational number and is denoted by the letter e.

$$e = \lim_{n \to \infty} \left(1 + \frac{1}{n}\right)^n \approx 2.718281828459045 \ldots$$

Later in this chapter, the significance of e will be more apparent. (Exercises 22 through 25 at the end of this section walk the reader through a graphical explanation of this limit.)

The function $y = e^x$ is called the **natural exponential function** and is graphed in Figure 5.1A.3. Other exponential functions, such as $y = -3^{x+1}$, $y = 5(2)^{-x^2}$, and $y = e^{|x|}$ are graphed in Figures 5.1A.4, 5.1A.5, and 5.1A.6.

Figure 5.1A.3 The natural exponential function, $y = e^x$

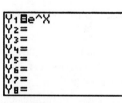

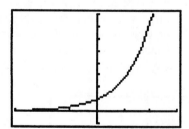

Figure 5.1A.4 $y = -1 \cdot 3^{x+1}$

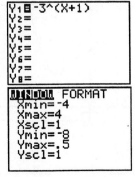

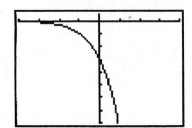

Figure 5.1A.5
$y = 5(2)^{-x^2}$

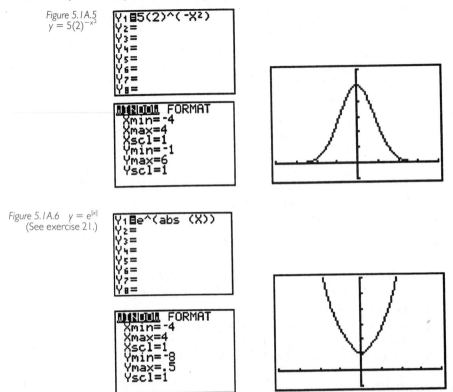

Figure 5.1A.6 $y = e^{|x|}$
(See exercise 21.)

In algebra, we learned that simple exponential equations, such as $3^x = 9$, can be solved easily. In this case, since $3^2 = 9$, the solution is $x = 2$. Generally, equations that involve exponential expressions, such as $3^x = x^2 - 1$, have solutions that can only be *approximated*. This is one place where a graphics calculator is particularly handy. The solution to $3^x = x^2 - 1$ will be the x coordinate of the point of intersection of the two graphs $y = 3^x$ and $y = x^2 - 1$. (*Why?*)

Figure 5.1A.7 $y = 3^x$
and $y = x^2 - 1$

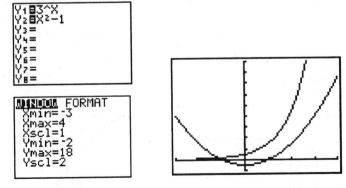

It appears from the graph in Figure 5.1A.7 that the x coordinate of the point of intersection of the two graphs is between -2 and -1. We use the ZOOM feature to obtain a better look at the graph's point of intersection. In Figure 5.1A.8, we ZOOM in, on the left screen, on $y = 3^x$ and $y = x^2 - 1$. On the right screen, we use the INTERSECTION option to see $x \approx -1.13$. That is, the solution of the equation $3^x = x^2 - 1$ is $x \approx -1.13$. As a check, we can calculate the value of 3^x and the value of $x^2 - 1$ when $x = -1.13$. That check shows $3^x \approx x^2 - 1 \approx 0.288$.

Figure 5.1A.8

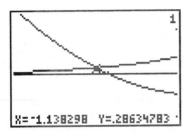
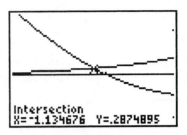

We conclude this section with a special type of exponential function called the **learning curve.** It is a popular model in psychology and business and is defined by

$$f(x) = a(1 - e^{-bx})$$

where a and b are constants. Letting $a = 100$ and $b = 0.3$, we have the function $f(x) = 100(1 - e^{-0.3x})$. This function is graphed in Figure 5.1A.9. Notice that $a = 100$ represents an upper bound for the function. This value represents the learning "peak."

Figure 5.1A.9
$y = 100(1 - e^{-0.3x})$

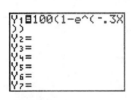

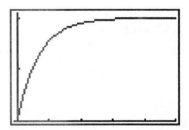

5.1A EXERCISES

In exercises 1 through 10, use a suitable set of WINDOW values to graph each exponential function.

1. $y = 2^x$

2. $y = 2^{-x}$

3. $y = 3e^{-x}$

4. $y = \left(\dfrac{1}{6}\right)^x$

5. $y = e^{x+1}$

6. $y = e^x + 1$

7. $y = 4(2)^{x^2}$

8. $y = 4(2)^{-x^2}$

9. $y = 1 - e^x$

10. $y = |1 - e^x|$

In exercises 11 through 20, use a graphics calculator to approximate the solution(s) to each exponential equation.

11. $2^x = 1 - x^2$

12. $3^x = x^2 - x$

13. $4^x = 1 - 2x^2$

14. $3^x = x^2 - x - 2$

15. $1 - x = 5^x$

16. $6^x = -x - 2$

17. $x + 2 = 2^x$

18. $e^{-x} = 3 - x$

19. $3 \cdot 2^x = 48$

20. $1 - e^{x+1} = x^3 + x$

21. Graph these two functions on the same graphics calculator graphics screen: $y = e^x$ and $y = e^{-x}$. Can you determine the "parts" of these curves that comprise the graph in Figure 5.1A.6?

In exercises 22 through 25, use the given WINDOW values to graph the function $y = \left(1 + \dfrac{1}{x}\right)^x$. In each case, TRACE to the Xmax value and record the corresponding y value or, if appropriate, use the TABLE feature of your calculator. How does that value compare with the value of e?

22.

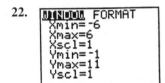
```
WINDOW FORMAT
Xmin=-6
Xmax=6
Xscl=1
Ymin=-1
Ymax=11
Yscl=1
```
Find the value of y when x = 6.

23.

```
WINDOW FORMAT
Xmin=6
Xmax=600
Xscl=100
Ymin=2.5
Ymax=2.8
Yscl=.05
```
Find the value of y when x = 600.

24.

```
WINDOW FORMAT
Xmin=600
Xmax=12000
Xscl=2000
Ymin=2.7
Ymax=2.74
Yscl=.005
```
Find the value of y when x = 12,000.

25.

```
WINDOW FORMAT
Xmin=12000
Xmax=1000000
Xscl=100000
Ymin=2.718
Ymax=2.7185
Yscl=1E-4
```
Find the value of y when x = 1,000,000.

Consider the following graph of $y = e^x$ and its inverse, using the TI-82 command DrawInv Y1:

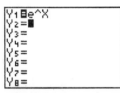
```
Y1=e^X
Y2=
Y3=
Y4=
Y5=
Y6=
Y7=
Y8=
```

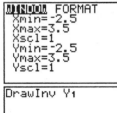
```
WINDOW FORMAT
Xmin=-2.5
Xmax=3.5
Xscl=1
Ymin=-2.5
Ymax=3.5
Yscl=1
```

```
DrawInv Y1
```

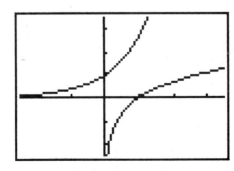

In exercises 26 through 29, use a similar approach to graph each exponential function and its inverse.

26. $y = 3^x$

27. $y = \left(\dfrac{1}{2}\right)^x$

28. $y = 10^x$

29. $y = 2 \cdot 3^{2x+1}$

30. **a.**   On a suitable collection of WINDOW settings, graph $y = 2^x$ and $y = 5^x$ on the same axes.
   b.   Use this graph to state the solution to the inequality $5^x > 2^x$.

31. Using exercise 30 as a guide, solve $4^x \le e^x$.

32. Using exercises 30 and 31 as guides, find an *approximate* solution to $3^{x+1} > x^5$.

5.2 LOGARITHMIC FUNCTIONS WITH APPLICATIONS

We further explore the study of exponents by defining and investigating **logarithms**.

Logarithm

> Let $b > 0$ and $b \ne 1$. A number u is called the **logarithm** of a positive real number v if and only if $v = b^u$. In symbols,
>
> $$u = \log_b v \qquad \text{if and only if} \qquad b^u = v$$

The most significant observation in the definition is that a *logarithm is an exponent.* In $u = \log_b v$ (read "u is the logarithm, base b, of v"), u is expressed as a logarithm; in $v = b^u$, u is an exponent.

Examples 1 and 2 further illustrate the exponent–logarithm relationship. As you read through them, remember that the number

$$\log_b v$$

represents the power b has to be raised to in order to get v.

Example 1   Rewrite each of the following in its equivalent logarithmic form:

a. $3^2 = 9$ **b.** $4^{-2} = \dfrac{1}{16}$ **c.** $10^4 = 10{,}000$

Solution   Using $u = \log_b v$ as the equivalent to $v = b^u$, we have

a. $2 = \log_3 9$ **b.** $-2 = \log_4 \dfrac{1}{16}$ **c.** $4 = \log_{10} 10{,}000$ ∎

Example 2   Rewrite each of the following in its equivalent exponential form:

a. $\log_{10} 100 = 2$ **b.** $\log_3 \dfrac{1}{9} = -2$ **c.** $\log_w R = T$

Solution   a. $10^2 = 100$ b. $3^{-2} = \dfrac{1}{9}$ c. $w^T = R$ ∎

Example 3   Solve for x: $\log_5 (2x - 3) = 2$.

Solution   We can use the definition to rewrite the equation in exponential form as

$$5^2 = 2x - 3$$
$$28 = 2x$$

or $x = 14$ ∎

As another illustration of how useful this transition from one form to another can be, we graph the function

$$y = \log_{10} x$$

by writing it equivalently in exponential form as

$$x = 10^y$$

in Example 4.

Example 4   Graph $y = \log_{10} x$.

Solution   Since $y = \log_{10} x$ is equivalent to $x = 10^y$, we choose y values and find corresponding x values as follows:

y	$x = 10^y$		x	$y = \log_{10} x$
-2	$10^{-2} = 0.01$	equivalently	0.01	-2
-1	$10^{-1} = 0.1$	$\rightarrow$	0.1	-1
-0.5	0.3162		0.3162	-0.5
-0.1	0.7943		0.7943	-0.1
0	1		1	0
0.1	1.2589		1.2589	0.1
0.5	3.162		3.162	0.5
1	10		10	1
1.5	31.623		31.623	1.5
2	100		100	2

The curve $y = \log_{10} x$ is graphed in Figure 5.9.

Figure 5.9

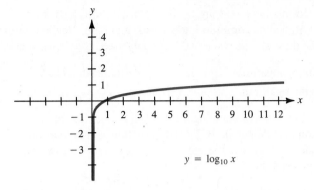

$y = \log_{10} x$

Notice that the y axis acts as a vertical asymptote for this function. We see that as $x \to \infty$, $y \to \infty$, and as $x \to 0^+$, $y \to -\infty$. The domain for this function is $\{x \mid x > 0\}$ and the range is R. ∎

If we compare the graph of $f(x) = \log_{10} x$ from Example 4 with the graph of $g(x) = 10^x$ in Figure 5.10, we see that each is a reflection of the other about the line $y = x$.

Figure 5.10

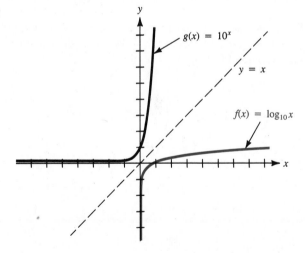

$g(x) = 10^x$

$y = x$

$f(x) = \log_{10} x$

Recall from Section 2.2 of Chapter 2 that this graphical relationship is one exhibited by inverse functions. In general, $f(x) = \log_b x$ and $g(x) = b^x$ are inverse functions and also satisfy the property from Chapter 2 that

$$f[g(x)] = g[f(x)] = x$$

This property is examined more fully in exercises 44 and 45 at the end of this section.

In this book, we are most concerned with logarithms to two different bases: base 10 (called **common logarithms**) and base e (called **natural logarithms**). Because these are the two most commonly used bases, special notation is used for each.

common logarithm	$\log_{10} x \to \log x$ (understood base 10)
natural logarithm	$\log_e x \to \ln x$

Tables of common logarithms and natural logarithms for selected numbers can be found in Appendix B. To find logarithms of numbers not listed, we can either use a calculator or use the tables in conjunction with some properties of logarithms given as follows:

Theorem 5.2
Properties of
logarithms

If $b > 0$, $b \neq 1$, and $x, y > 0$, then:

a. $\log_b (xy) = \log_b x + \log_b y$

b. $\log_b \left(\dfrac{x}{y}\right) = \log_b x - \log_b y$

c. $\log_b (x^n) = n \log_b x$

d. $\log_b N = \dfrac{\log_a N}{\log_a b}$

Note: Part d of this theorem is called a **change of base formula.**

We can use the properties listed in Theorem 5.2 to simplify expressions involving logarithms that can be useful in solving equations and later (in Section 5.3) in taking derivatives of logarithmic functions.

Example 5   Use the properties of logarithms to simplify the following:

a. $\log_b x\sqrt{y}$ **b.** $\log_b \dfrac{z}{(x + y)^2}$

Solution   **a.** $\log_b x\sqrt{y} = \log_b x + \log_b y^{1/2}$ Using part a of Theorem 5.2

$\qquad\qquad\qquad = \log_b x + \frac{1}{2} \log_b y$ Using part c

b. $\log_b \dfrac{z}{(x + y)^2} = \log_b z - \log_b (x + y)^2$

$\qquad\qquad\qquad\quad = \log_b z - 2 \log_b (x + y)$

Note: $\log_b (x + y) \neq \log_b x + \log_b y$.   ■

In Section 5.1 we solved exponential equations of a certain type that could be written so that both sides had the same base. We now use logarithms to solve exponential equations that need not have the same base on both sides. In particular, we use the following theorem:

Theorem 5.3	Suppose that $p > 0$, $q > 0$, $b > 0$, and $b \neq 1$. Then
	$p = q$ \quad if and only if \quad $\log_b p = \log_b q$

Example 6   Solve for x: $12^{3x-1} = 7$.

Solution   We use Theorem 5.3 to rewrite $12^{3x-1} = 7$ as

$$\log(12^{3x-1}) = \log 7 \qquad \text{Taking common log of both sides}$$

$$(3x - 1)\log 12 = \log 7$$

$$3x - 1 = \frac{\log 7}{\log 12}$$

$$3x = \frac{\log 7}{\log 12} + 1$$

$$x = \frac{\log 7}{3 \cdot \log 12} + \frac{1}{3}$$

$$\approx \frac{0.845098}{3.2375437} + \frac{1}{3} \qquad \text{Here, we use a calculator to find log 7 and log 12.}$$

$$\approx 0.59436395$$

Check: $12^{3x-1} = 12^{3(0.59436395)-1} \approx 6.999999$, using a calculator.   ∎

Example 7   Solve for x: $5e^{1-x} = 9$.

Solution   Taking the natural logarithm of both sides, we have

$$\ln(5e^{1-x}) = \ln 9$$

$$\ln 5 + \ln e^{1-x} = \ln 9$$

$$\ln e^{1-x} = \ln 9 - \ln 5$$

$$(1 - x)\ln e = 0.5878$$

$$1 - x = 0.5878 \qquad (\ln e = 1)$$

$$-x = -0.4122$$

$$x = 0.4122 \qquad\qquad ∎$$

In both of the previous examples, we could have used logarithms of any base. Most of the time it is best to use common or natural logarithms because of easy access to base 10 and base e tables and base 10 and base e calculator functions. Notice the use of natural logarithms in Example 7, which makes simplification easier.

Table 5.1 summarizes some important properties of the logarithmic function that are useful in solving problems:

Table 5.1

In general	In particular
$\log_a a = 1$	$\ln e = 1$
$\log_a 1 = 0$	$\ln 1 = 0$
$\log_a a^x = x$	$\ln e^x = x$
$a^{\log_a x} = x$	$e^{\ln x} = x$

Recall from Section 5.1 that the function $f(x) = a \cdot b^x$ exhibits exponential *growth* if $b > 1$ and exponential *decay* if $0 < b < 1$. The next example illustrates the application of exponential growth to population studies.

Example 8   The annual growth rate of the population of the United States is about 1%; that is, in any given year the population is 1.01 times that of the previous year. The exponential equation describing this unbounded growth is

$$N(t) = N_0(1.01)^t$$

where $N(t)$ = the number of people alive at any given time t (in years)
N_0 = the number of people alive initially (at $t = 0$)

a.   In 1984, the U.S. population was about 230,000,000 people. Using this data, what was the predicted population for 1990?
b.   How long, at this rate, will it take for the U.S. population to double?

Solution   **a.**   We must find $N(6)$ with $N_0 = 2.3 \times 10^8$.

$$N(t) = N_0(1.01)^t$$
$$N(6) = (2.3 \times 10^8)(1.01)^6$$
$$N(6) \approx 2.4415 \times 10^8 = 244{,}150{,}000$$

b.   To double means that $N(t) = 2N_0$. We must find t.

$$N(t) = N_0(1.01)^t$$

$$2N_0 = N_0(1.01)^t \qquad \text{Divide both sides by } N_0.$$

$$2 = (1.01)^t$$

$$\log 2 = t \log(1.01) \qquad \text{Here, we take the common log of both sides.}$$
Common log *or* natural log could be used!

$$t = \frac{\log 2}{\log (1.01)}$$

$$= \frac{0.301029996}{0.004321374}$$

$$\approx 69.6 \text{ yr}$$

Thus, the population will double in about 70 yr (the year 2054). ∎

Example 9   A city discovered that its water supply was polluted. To remedy this, they decided to pump polluted water out at a rate of 10,000 gal/hr and replace it with clear water until an acceptably low level of pollution was reached. An equation for the time, t (in hours), that this process takes is

$$t = \frac{M}{10,000} \ln \frac{p}{q}$$

where $M =$ the reservoir's capacity (in gallons)
$p =$ the pollution level at the beginning of the process (in milligrams per gallon)
$q =$ the acceptable pollution level that will end the process (in milligrams per gallon)

If $M = 12$ million gallons, how long will it take for the pollution to be one-third its initial value?

Solution $q = \frac{1}{3}p$

So we have

$$t = \frac{M}{10,000} \ln \frac{p}{\frac{1}{3}p}$$

$$= \frac{12,000,000}{10,000} \ln \frac{p}{\frac{1}{3}p}$$

$$= 1200 \ln 3$$

$$\approx 1318 \text{ hr} \qquad\qquad ∎$$

Scientists studying radioactive isotopes and the rate at which they change or decay into other elements use the term *radioactive decay*. This process can be described by referring to the **half-life** of an element. For example, the half-life of carbon-14, a commonly used isotope for determining the age of remains, is 5570 yr; that is, if we started with 10 g of carbon-14, after 5570 yr we would have 5 g. After another 5570 yr we would have 2.5 g, and so on. The use of carbon-14 and its half-life in determining the age of remains is examined in Example 10.

Example 10   When a plant or animal dies, the amount of carbon-14 present begins to decrease. At death, the equation

$$N = N_0 \left(\frac{1}{2}\right)^{x/5570}$$

is a model for the decay process, where

N = the amount (in grams) of carbon-14 remaining per kilogram in a plant or animal x years after its death

N_0 = the amount (in grams) of carbon-14 present per kilogram in the organism when it was living

Suppose a bone was uncovered at an archaeological dig and was found to have lost 65% of its carbon-14. How old is the bone?

Solution   Since 65% has been lost, we have 35% remaining, which means

$$N = 0.35N_0$$

Substituting into our equation gives

$$0.35N_0 = N_0 \left(\frac{1}{2}\right)^{x/5570}$$

$$0.35 = (0.5)^{x/5570} \qquad \text{Divided both sides by } N_0$$

$$\log 0.35 = \frac{x}{5570} \log (0.5)$$

$$x = (5570) \frac{\log (0.35)}{\log (0.5)}$$

$$= (5570) \frac{-0.45593}{-0.30103}$$

$$\approx 8436 \text{ yr}$$

5.2   EXERCISES

SET A

In exercises 1 through 6, rewrite each exponential equation in its equivalent logarithmic form. (See Example 1.)

1. $4^2 = 16$ **2.** $4^3 = 64$ **3.** $5^{-3} = \dfrac{1}{125}$ **4.** $\left(\dfrac{1}{2}\right)^3 = \dfrac{1}{8}$

5. $10^{-2} = 0.01$ **6.** $b^0 = 1$

In exercises 7 through 12, rewrite each logarithmic equation in its equivalent exponential form. (See Example 2.)

7. $\log_3 9 = 2$ **8.** $\log_3 \dfrac{1}{27} = -3$ **9.** $\log 100 = 2$

10. $\log 0.001 = -3$ **11.** $\ln \dfrac{1}{e} = -1$ **12.** $\log_b 1 = 0$

In exercises 13 through 18, solve for x. (See Example 3.)

13. $\log_2 (1 - x) = 4$ **14.** $\log_{1/3} 2x = -2$ **15.** $\log_3 \left(\dfrac{1}{x}\right) = -1$

16. $\log_2 \left(\dfrac{1}{8}\right) = x$ **17.** $\log_x 9 = 2$ **18.** $-\log_x 9 = 2$

In exercises 19 through 23, graph each function. (See Example 4.)

19. $y = \log_2 x$ **20.** $y = \log_3 x$ **21.** $y = \ln x$ **22.** $y = \log_{0.1} x$

23. $y = \log (x + 1)$

In exercises 24 through 27, use a calculator to approximate each expression to four decimal places.

24. $\dfrac{\log 193}{\log 14.6}$ **25.** $\dfrac{\log 892.5}{\log e}$

26. $\dfrac{\log 82.5 + \log 401}{\log 0.00312}$ **27.** $\sqrt{\ln 3007} + \ln (\sqrt{192.5})$

In exercises 28 through 33, use the properties of logarithms to simplify each of the following. (See Example 5.)

28. $\log_b \sqrt{xy}$ **29.** $\log_b xy^2$ **30.** $\log_b \dfrac{x}{y^2}$

31. $\log_b \dfrac{x}{y + z}$ **32.** $\log_b (x + y)^2$ **33.** $\log_b \sqrt[3]{x + y}$

SET B

In exercises 34 and 35, use part d of Theorem 5.2 to determine the value of each expression.

34. $\log_7 36$ **35.** $\log_5 0.012$

In exercises 36 through 43, solve each equation for x. (See Examples 6 and 7.)

36. $8^{2x} = 5$ **37.** $15^{2x+1} = 85$ **38.** $16^{4x-1} = 16$

39. $10^{2x-3} = 5$ **40.** $5 \cdot 3^{2x} = 9.7$ **41.** $e^{x^2} = 10$

42. $e^{x^2+1} = 10$ **43.** $5 \cdot (\tfrac{1}{2})^{3x-1} = 85$

44. a. Let $f(x) = \log x$ and $g(x) = 10^x$. Find $f[g(x)]$ and $g[f(x)]$ for the following values of x: 1, 10, 100, $\frac{1}{10}$.

    b. Show that $f[g(x)] = x$.

    c. Show that $g[f(x)] = x$. (*Hint:* Let $Q = 10^{\log x}$ and apply Theorem 5.3. Then, $\log Q = \log x$ and $Q = 10^{\log x} = x$.)

45. a. Repeat exercise 44a for $f(x) = \ln x$ and $g(x) = e^x$.

    b. Graph $y = f(x)$, $y = g(x)$, and $y = x$ all on the same set of axes.

46. Graph $y = \log_3 x$ and $y = 3^x$ on the same set of axes.

BUSINESS
ADVERTISING

47. From past experience, a firm knows that the relationship between the number of units sold, N, and the amount spent on advertising, x (in thousands of dollars), is given by the function

$$N(x) = 2500 + 250 \ln(x + 1)$$

    a. How many units can the firm expect to sell without any expenditure for advertising?

    b. How many units can the firm expect to sell if $8000 is spent on advertising?

    c. If the firm has budgeted $20,000 for advertising for the next fiscal year, how many units can the firm expect to sell?

FIXED COSTS

48. A company has determined that the total cost function for the manufacture of x items is given by $C(x) = 3.5 \ln (x + 2)$, where $C(x)$ is measured in thousands.

    a. Graph this function.

    b. Determine the company's *fixed cost*, $C(0)$.

INTEREST

49. A person deposits $2000 into a savings account at 6% interest compounded quarterly. The amount, A, in the account at any time, t (in years), is given by

$$A = 2000(1.015)^{4t}$$

Approximately how long will it take for the amount in the account to reach a balance of $3000?

INTEREST

50. Using the information from exercise 49, how long will it take for the money in the account to double?

SALES

51. The Furman Company has decided to phase out its new line of home-care products and hence does not advertise them any longer. The decline in sales is described by

$$S(t) = 3000\left(\frac{1}{2}\right)^t$$

where t is measured in months.

    a. Determine the initial monthly sales figure.

    b. How long will it take for sales to reach half of the initial sales?

DEMAND FUNCTION

52. The demand function for a product is described by

$$p = \frac{300}{\ln(x + 4)}$$

where x is the number of units demanded at a price p.
  a. Determine the price when no units are demanded.
  b. Determine the price when 10 units are demanded.
  c. Determine the price when 15 units are demanded.

REVENUE FUNCTION

53. Using the demand function from exercise 52, find
  a. The revenue function, $R(x)$
  b. The revenue for $x = 20$ units

BIOLOGY
UNBOUNDED GROWTH
MODEL

54. The equation $N = N_0(2.1)^{3t}$ is an unbounded growth model for the number of bacteria, N, that exist after a time, t (in hours), where N_0 is the number of bacteria present originally.
  a. If there are 200 bacteria present originally, how many are there after 4 hours?
  b. If there are N_0 bacteria present originally, how long will it take that number to quadruple?

HEART RATE

55. The heart rate, R, of a certain patient was monitored during a stress test and was found to be approximated by

$$R = 70 \ln (t + e)$$

where t is measured in minutes on the monitor.
  a. What was the heart rate initially?
  b. What was the heart rate at the end of two minutes?

HEART RATE

56. For the patient in exercise 55, approximately how long did it take for the heart rate to double?

SOCIAL SCIENCE
POPULATION

57. Use the model

$$N = \frac{N_0 L}{N_0 + (L - N_0)e^{-LKt}}$$

as a representation of the bounded population growth of a rat population in a certain city with $L = 50{,}000$.
  a. Determine K if $N = 10{,}000$ when $t = 0$ and $N = 20{,}000$ when $t = 4$.
  b. Write N as a function of t and graph the function.

LEARNING CURVES

58. A list of foreign symbols to memorize is given to a subject in an experiment. This experiment is repeated with different lists and varying times t to memorize the lists. The subject performs at a rate of

$$N = 20(1 - e^{-0.2t})$$

where N is the number of correctly memorized symbols and t is the time in minutes given for viewing the list.
  a. How many correctly memorized symbols would this person master in 10 minutes?

b.   How long would the person need in order to get five symbols correct?

PHYSICAL
SCIENCE
GEOLOGY

59.   The equation $M = \log (A/A_0)$ is a model that defines the Richter scale for measuring the magnitude, M, of an earthquake. A_0 is the amount of the earth's vibration for the faintest perceptible earthquake detectable by a seismograph. A is the amount of the earth's vibration for the earthquake in question as measured by a seismograph.

    a.   If the amount of the earth's vibration is 1 million times that of the faintest perceptible earthquake, what is M?

    b.   Suppose that $M = 7.2$. What is the relationship between A and A_0?

RADIOLOGY

60.   Workers must be properly shielded from X rays. The equation

$$x = \frac{1}{350} \ln \frac{I_0}{I}$$

is an algebraic model for the thickness of lead, x (in centimeters), required to cause an X-ray beam having initial intensity I_0 to be reduced to intensity I.

    a.   What thickness of lead is necessary to reduce X rays to one-fifth of their initial intensity?

    b.   What thickness of lead is necessary to reduce X rays to one-twentieth of their initial intensity?

ARCHAEOLOGY

61.   **a.**   Using the carbon-14 dating equation of Example 10, determine the age since death of a skull that has lost 75% of its carbon-14.

    b.   If the skull has lost 60% of its carbon-14, how long has it been since its death?

CHEMISTRY

62.   The equation $p = -\log H$ is used to define the pH of a solution in chemistry, where

    p = the pH of the solution
    H = the hydrogen ion concentration in the solution (measured in moles per liter)

    a.   If $H = 10^{-7}$ moles/liter for water, determine p.

    b.   If the pH of a shampoo is 7.6, determine H.

WRITING AND CRITICAL THINKING

The functions $y = 2^x$ and $y = \log_2 x$ are inverses of each other. Explain what is meant by inverse functions. Describe in words what is observed if we make a table of values for both functions and compare them.

USING TECHNOLOGY IN CALCULUS

Graphics calculators or graphing software packages can be very useful in approximating solutions to logarithmic equations that have no exact (or closed-form) solution. For example, consider the equation $\log (x + 3) = x^2$.

Method I   We can graph $y = \log (x + 3)$ and $y = x^2$ and ZOOM in on a point of intersection.

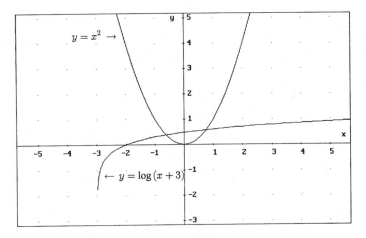

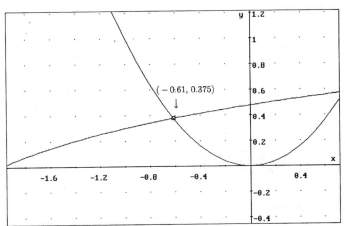

Method II   Alternatively, we can graph $y = \log (x + 3) - x^2$ and ZOOM in on the x intercept(s). At the x value of the intercept, y is zero and hence $\log (x + 3) = x^2$.

(continued)

(*continued*)

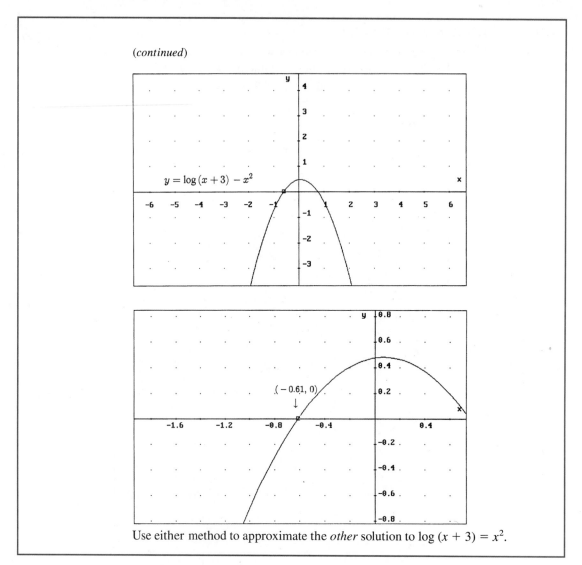

$y = \log(x + 3) - x^2$

$(-0.61, 0)$

Use either method to approximate the *other* solution to $\log(x + 3) = x^2$.

5.2A LOGARITHMIC FUNCTIONS ON A GRAPHICS CALCULATOR

A logarithm is an exponent and we write the definition formally as follows:

Logarithm

Let $b > 0$ and $b \neq 1$. A number u is called the **logarithm** of a positive real number v if and only if $v = b^u$. In symbols,

$$u = \log_b v \qquad \text{if and only if} \qquad b^u = v$$

Most calculators have two logarithm keys: one is for base 10 (or *common*) logarithms ($\log_{10} x$ is usually denoted as LOG or LOG X on a calculator) and one is for base e (or *natural*) logarithms ($\log_e x$ is usually denoted as LN or LN X on a calculator). See Figure 5.2A.1.

Figure 5.2A.1 Using the LOG and LN keys

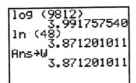

← Keep in mind that log (9812) ≈ 3.991757540 is equivalent to writing $10^{3.991757540} \approx 9812$ and ln (48) ≈ 3.871201011 is the same as $e^{3.871201011} \approx 48$.

← Here, we simply store the value of ln (48) in the storage location W.

The graph of $y = \log_{10} x$ is displayed in Figure 5.2A.2. Notice that the graph passes through the point (1, 0) since $\log_{10} 1 = 0$ because $10^0 = 1$. The graph passes through the point (10, 1) since $\log_{10} 10 = 1$ because $10^1 = 10$. Because the logarithm function is defined only for positive numbers (the domain of $\log_{10} x$ is $\{x \mid x > 0\}$), nothing is graphed in either the second or third quadrants.

Figure 5.2A.2
$y = \log_{10} x$

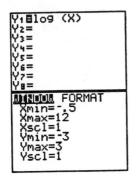
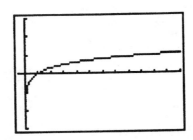

Because, by the definition of logarithm, $y = e^x$ is equivalent to $x = \ln y$, the natural exponential and natural logarithmic functions are inverses of each other. The graph of $y = \ln x$ is the reflection of the graph of $y = e^x$ about the line $y = x$, as illustrated in Figure 5.2A.3.

Figure 5.2A.3 The exponential and logarithmic functions are inverses.

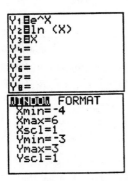
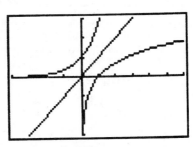

Next, we consider the graph of $y = \ln (x - 1)$. Notice that the domain of this function is $\{x \mid x > 1\}$ and $\lim\limits_{x \to 1^+} \ln (x - 1) = -\infty$. The graph appears in Figure 5.2A.4.

Figure 5.2A.4

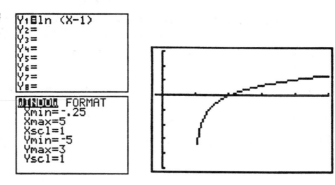

Logarithms find applications in many areas, and we use several rules to facilitate working with logarithms. Examine Theorem 5.2 on page 370 for four such rules. In particular, the Change of Base Rule (part d of Theorem 5.2) is used to find quantities such as $\log_7 589$. (See Figure 5.2A.5.)

Figure 5.2A.5 The Change of Base Rule

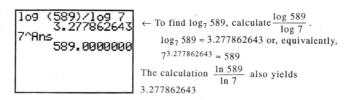

← To find $\log_7 589$, calculate $\dfrac{\log 589}{\log 7}$.

$\log_7 589 \approx 3.277862643$ or, equivalently,

$7^{3.277862643} \approx 589$

The calculation $\dfrac{\ln 589}{\ln 7}$ also yields

3.277862643

We adapt that rule here for use in calculators:

The Change of Base Rule for calculators

> If $b > 0$, $b \neq 1$, and $N > 0$, then
>
> $$\log_b N = \frac{\log N}{\log b} = \frac{\ln N}{\ln b}$$

Theorem 5.2 needs to be respected. For example, consider the exponent part of the theorem, which is often misquoted without the proper hypothesis:

Correct Form of Rule

If $b > 0$, $b \neq 1$ and $Q > 0$, then
$\log_b (Q^n) = n \log_b Q$

Incorrect Form

$\log_b (Q^n) = n \log_b Q$

Figure 5.2A.6 demonstrates the graphical impact of the rule. Consider the two functions $y = \ln (x + 3)^2$ and $y = 2 \ln (x + 3)$. Are they equivalent? According to the (correct form of the) rule, they are the same only if $x + 3 > 0$ (or equivalently if $x > -3$). Compare the graphs carefully.

Figure 5.2A.6

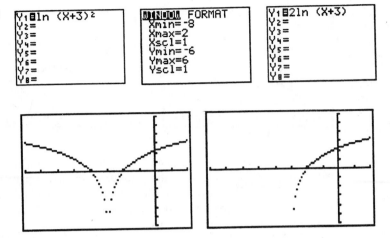

We can approximate the solutions to equations involving logarithms by graphing as we did for exponential functions in Section 5.1A. For example, consider the equation

$$\ln (2x - 1) = 3 - \frac{x}{2}$$

In Figure 5.2A.7, we graph $y = \ln (2x - 1)$ and $y = 3 - \frac{x}{2}$.

Figure 5.2A.7
$y = \ln(2x - 1)$ and
$y = 3 - \dfrac{x}{2}$

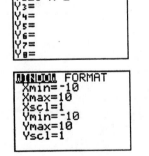

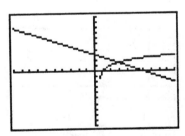

The x value of the point of intersection of the graphs of $y = \ln (2x - 1)$ and $y = 3 - \frac{x}{2}$ seems to be between 2 and 3. Further refinement by ZOOMing in yields the series of graphs in Figure 5.2A.8.

As a check, if we substitute the value of $x = 2.88$, we find $\ln (2x - 1) \approx 1.56$ and $3 - \frac{x}{2} = 1.56$. Of course, further refinement is also possible.

Figure 5.2A.8   A solution
to $\ln(2x - 1) = 3 - \dfrac{x}{2}$
is about $x \approx 2.88$.

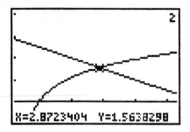

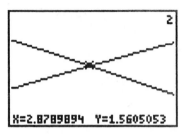

As a final example of an equation involving a logarithm, consider $4 - x^2 = \ln x$. We could proceed as we did before and graph the two equations $y = 4 - x^2$ and $y = \ln x$, ZOOM in on the point of intersection, and take the abscissa of that point as the approximate solution to our equation, $4 - x^2 = \ln x$. Instead, however, we choose an alternative approach: We graph $y = 4 - x^2 - \ln x$ and examine where y is zero (that is, the graph's x intercept). Of course, the reasoning behind this approach is that finding a solution to $4 - x^2 = \ln x$ is equivalent to finding a solution to $4 - x^2 - \ln x = 0$. (See Figures 5.2A.9 and 5.2A.10.)

Figure 5.2A.9   The graph
of $y = 4 - x^2 - \ln x$ has
an x intercept at about
$x = 1.8297872$.

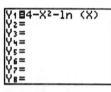

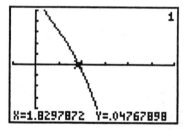

Figure 5.2A.10   After
using the "root" option,
the x intercept is shown
to be $x \approx 1.8410971$.

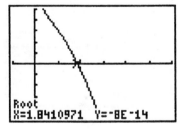

5.2A   EXERCISES

In exercises 1 through 8, use a suitable set of WINDOW values to graph each logarithmic function.

1. $y = \ln 2x$ **2.** $y = -\log x$ **3.** $y = 3 - \log x$

4. $y = \log x^3$ **5.** $y = \log_7 x$ **6.** $y = \log (x + 2)$

7. $y = \ln (x + 100)$ **8.** $y = (\ln x) + 100$

In exercises 9 through 12, use a graphics calculator to approximate the solution(s) to each equation.

9. $\ln (1 + x) = 3$ **10.** $4 - x^2 = \ln x$

11. $\ln x = \log x$ **12.** $\ln (1 - x) = \ln 6 - \ln (x + 4)$

13. a. Graph $y = \ln |x|$. How does the domain of this function differ from the domain of the natural logarithmic function?

b. Graph $y = |\log x|$. What are the domain and range of this function?

14. Recall the compound interest formula from Section 5.1: $A = P\left(1 + \dfrac{r}{n}\right)^{nt}$.

a. Use it to find the length of time, t (in years), it will take an investment to double if the annual percentage rate is 8% and the interest is compounded quarterly ($n = 4$). (*Hint: A = 2P.* To solve for t means using logarithms.)

b. Graph $y = (1.02)^{4x}$. For what value of x does this graph cross the graph of $y = 3$? Explain why this is precisely the number of years it will take an investment to triple when it is compounded quarterly at an annual rate of 8%.

15. Graph the following four functions on the same set of axes: $y = \ln x$, $y = x - 1$, $y = x$, $y = x - 2$.

16. Use your answer to exercise 15 to determine the *number of solutions* to each of the following equations:

a. $\ln x = x - 1$ **b.** $\ln x = x$ **c.** $\ln x = x - 2$

17. *The calculator can't do all the work!* Remember, the calculator is only a tool; it is not perfect. For example, consider the graph of the equation $y = x^{1/\ln x}$ on your graphics calculator.

a. What is the domain of this function?

b. Find a much simpler way of writing this function. (*Hint:* Use logarithms.)

c. Explain why there should not be a point on the graph of $y = x^{1/\ln x}$ when $x = 1$. Did your calculator's graph obey the domain of the function?

5.3 DERIVATIVES OF LOGARITHMIC FUNCTIONS

The discussion of logarithms and exponents thus far in this chapter has not included calculus; in particular, we have not examined the derivatives of these classes of functions.

We begin by differentiating $f(x) = \ln (x)$. According to the definition of derivative in Chapter 2,

$$f'(x) = \lim_{h \to 0} \frac{f(x + h) - f(x)}{h}$$

$$= \lim_{h \to 0} \frac{\ln (x + h) - \ln (x)}{h}$$

$$= \lim_{h \to 0} \frac{\ln\left(\dfrac{x+h}{x}\right)}{h}$$

$$= \lim_{h \to 0} \frac{\ln\left(1 + \dfrac{h}{x}\right)}{h} \qquad (1)$$

We are now at a point where some substitutions are required in equation (1).

Let $n = \dfrac{x}{h}$; then as $h \to 0$, $n \to \infty$. Notice also that if $n = \dfrac{x}{h}$, solving for h, we have $h = \dfrac{x}{n}$.

Equation (1) now becomes

$$f'(x) = \lim_{n \to \infty} \frac{\ln\left(1 + \dfrac{1}{n}\right)}{\dfrac{x}{n}}$$

$$= \lim_{n \to \infty} \frac{1}{x}\left[n \ln\left(1 + \frac{1}{n}\right)\right]$$

$$= \lim_{n \to \infty} \frac{1}{x} \ln\left(1 + \frac{1}{n}\right)^n$$

From Section 5.1, however, $\displaystyle\lim_{n \to \infty} \left(1 + \frac{1}{n}\right)^n = e$.

Finally, we have

$$f'(x) = \frac{1}{x} \ln e = \frac{1}{x}$$

Thus, $\boxed{\dfrac{d(\ln x)}{dx} = \dfrac{1}{x}}$

The generalized result, using the Chain Rule, is given in Theorem 5.4.

Theorem 5.4   If $f(x) = \ln u(x)$, then $f'(x) = \dfrac{1}{u(x)} \cdot u'(x)$, where $u(x) > 0$.

Alternatively, if $y = \ln u$, then $\dfrac{dy}{dx} = \dfrac{1}{u}\dfrac{du}{dx}$.

Example 1   Let $y = \ln(4x^2 + 1)$. Find $\dfrac{dy}{dx}$.

Solution   From Theorem 5.4, using $u(x) = 4x^2 + 1$, we have $u'(x) = 8x$ and

$$\frac{dy}{dx} = \frac{1}{4x^2 + 1} \cdot 8x$$

$$= \frac{8x}{4x^2 + 1}$$

■

Example 2 $f(x) = \ln (1 - x^3)$. Find $f'(-2)$.

Solution $f'(x) = \dfrac{1}{1 - x^3} \cdot (-3x^2)$

$f'(x) = \dfrac{-3x^2}{1 - x^3}$

$f'(-2) = \dfrac{-3(-2)^2}{1 - (-2)^3} = \dfrac{-4}{3}$

■

Example 3   If $f(x) = x \ln (x + 1)^3$, find $f'(x)$.

Solution   First, using one of the properties of logarithms, we can rewrite the original function as

$$f(x) = 3x \ln (x + 1)$$

Now, using the Product Rule, we have

$$f'(x) = 3x \left[\frac{1}{x + 1} \cdot 1 \right] + [\ln (x + 1)][3]$$

$$= \frac{3x}{x + 1} + 3 \ln (x + 1)$$

■

So far, we have found derivatives for natural logarithmic (base e) functions only. We can use the change of base formula from Section 5.2 (see Theorem 5.2, part d) to rewrite a logarithmic function in another base so that we may find its derivative.

Rewriting, we have

$$y = \log_b x = \frac{\ln x}{\ln b}$$

$$= \frac{1}{\ln b} \ln x \qquad \frac{1}{\ln b} \text{ is a coefficient for } \ln x.$$

Thus, $\dfrac{dy}{dx} = \dfrac{1}{\ln b} \cdot \dfrac{1}{x}$

More generally, we have

The derivative of
$\log_b u(x)$

> If $y = \log_b u(x)$, then $\dfrac{dy}{dx} = \dfrac{1}{\ln b} \cdot \dfrac{1}{u(x)} \cdot u'(x)$, where $u(x) > 0$.

Example 4 For $f(x) = \log_3 (x^2 + 1)$, find $f'(x)$.

Solution Setting $u(x) = x^2 + 1$, we have

$$u'(x) = 2x$$

so $$f'(x) = \frac{1}{\ln 3} \cdot \frac{1}{x^2 + 1} \cdot 2x$$

$$f'(x) = \frac{2x}{(x^2 + 1)\ln 3}$$ ∎

We now develop a technique of differentiation called **logarithmic differentiation.** This technique uses Theorem 5.3 together with implicit differentiation, as illustrated in Example 5.

Example 5 Use logarithmic differentiation to find $\dfrac{dy}{dx}$ for $y = (x + 2)^4(2x - 1)^3$.

Solution Taking the natural logarithm of both sides (Theorem 5.3), we have

$$\ln y = \ln (x + 2)^4(2x - 1)^3$$

Before differentiating, we use the properties of logarithms to simplify the right side:

$$\ln y = \ln (x + 2)^4 + \ln (2x - 1)^3$$
$$= 4 \ln (x + 2) + 3 \ln (2x - 1)$$

Now, differentiating implicitly, we get

$$\frac{1}{y}\frac{dy}{dx} = 4 \cdot \frac{1}{x + 2} + 3 \cdot \frac{1}{2x - 1} \cdot 2$$
$$\frac{dy}{dx} = y\left(\frac{4}{x + 2} + \frac{6}{2x - 1}\right)$$

Note: The Product Rule, together with regular differentiation, could have been used. The answers will look different but can be shown to be equivalent. ∎

Example 6 The Slowburn Corporation has determined that the revenue and cost functions for its new product are projected to be

$$R(x) = 400 \ln (5x + 1) \quad \text{and} \quad C(x) = \frac{x}{3}$$

a. Find the number of items that should be sold to maximize profit.
b. Find the maximum profit.

Solution **a.** $P(x) = R(x) - C(x) = 400 \ln (5x + 1) - \dfrac{x}{3}$

$$P'(x) = 400 \cdot \frac{1}{5x + 1} \cdot 5 - \frac{1}{3}$$

$$= \frac{2000}{5x + 1} - \frac{1}{3}$$

Setting $P'(x)$ equal to 0 gives

$$\frac{2000}{5x + 1} = \frac{1}{3}$$
$$5x + 1 = 6000$$
$$5x = 5999$$
$$x \approx 1200$$

A Second Derivative Test confirms that this value produces a maximum. The company should manufacture and sell 1200 units of its product.

b. The maximum profit is $P(1200) \approx \$3080$. ∎

Example 7 Using the graphing techniques discussed in Chapter 4, sketch the graph of $f(x) = x^3 \ln x$.

Solution The domain for the function is $x > 0$. Computing the x intercepts, by setting $f(x) = 0$, gives

$$x^3 \ln x = 0$$
$$x^3 = 0 \quad \text{or} \quad \ln x = 0$$
$$x = 0 \qquad\qquad e^0 = x$$
$$1 = x$$

Since x cannot be equal to 0, the only x intercept is $(1, 0)$.

We can find the intervals where the function is increasing and decreasing and any local maximum or minimum points by using the first derivative. The Product Rule gives

$$f'(x) = x^3 \cdot \frac{1}{x} + (\ln x)(3x^2)$$
$$= x^2 + 3x^2 \ln x$$

Setting $f'(x) = 0$ gives

$$x^2(1 + 3\ln x) = 0$$
$$x^2 = 0 \quad \text{or} \quad 1 + 3\ln x = 0$$
$$\ln x = -\tfrac{1}{3}$$
$$e^{-1/3} = x$$
$$x = \frac{1}{\sqrt[3]{e}} \approx 0.72$$

Interval	$0 < x < \dfrac{1}{\sqrt[3]{e}}$	$x > \dfrac{1}{\sqrt[3]{e}}$
Sign of $f'(x)$	$-$	$+$
Result	decreasing	increasing

The table tells us that there is a minimum for the function at $x = \dfrac{1}{\sqrt[3]{e}}$. Checking the second derivative, $f''(x) = 5x + 6x\ln x$, we find that the curve is concave up for $x > e^{-5/6}$ and concave down for $0 < x < e^{-5/6}$. The graph appears in Figure 5.11.

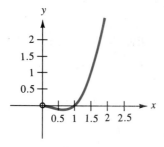

Figure 5.11

We conclude this section with an example that graphically analyzes a logarithmic function and its derivative.

Example 8   Graph the function $y = \ln x^2$, find its derivative, and graph its derivative.

Solution $\dfrac{dy}{dx} = \dfrac{1}{x^2} \cdot 2x = \dfrac{2}{x}$. We use *DERIVE*® to graph the function on the left and its derivative on the right (see Figure 5.12).

Figure 5.12

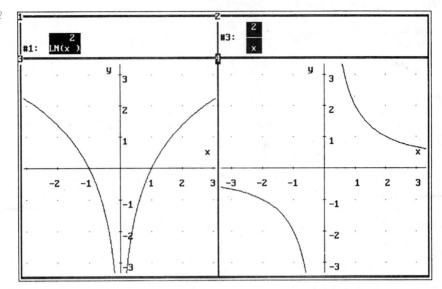

Notice that on the interval where the function is decreasing $(x < 0)$, its derivative is negative (graphically, below the x axis); on the interval where the function is increasing $(x > 0)$, the derivative is positive (above the x axis). ∎

5.3   EXERCISES

SET A

In exercises 1 through 18, find $\dfrac{dy}{dx}$. (See Examples 1, 2, and 3.)

1. $y = \ln (3x - 1)$

2. $y = \ln (x^2 + 1)$

3. $y = \ln \sqrt{x}$

4. $y = \sqrt{\ln x}$

5. $y = (\ln x)^3$

6. $y = \ln x^2$

7. $y = \ln \left(\dfrac{2x + 1}{x - 1} \right)$

8. $y = x \ln x$

9. $y = x^2 \ln x$

10. $y = \dfrac{\ln x}{x}$

11. $y = \dfrac{x}{\ln x}$

12. $y = x(\ln x)^2$

13. $y = \dfrac{\ln 2x}{\ln 3x}$

14. $y = x^2 + \ln x$

15. $y = \dfrac{1}{(\ln x)^2}$

16. $y = x \ln \sqrt{x}$

17. $y = \dfrac{\ln x}{x + 1}$

18. $y = \sqrt{1 + \ln x}$

In exercises 19 through 22, find $f'(x)$. (See Example 4.)

19. $y = \log_{10} x$

20. $y = \log_7 (1 - x^3)$

21. $y = x \log_{10} \sqrt{x}$

22. $y = x^2 \log_5 x$

In exercises 23 and 24, find $\dfrac{d^2y}{dx^2}$.

23. $y = \ln (2x + 1)$ **24.** $y = \ln \left(\dfrac{1}{x + 1} \right)$

SET B

In exercises 25 through 32, use the technique of curve sketching in Chapter 4 to graph each function. (See Example 7.)

25. $y = \ln \sqrt{x}$ **26.** $y = x \ln x$ **27.** $y = \ln x^2$

28. $y = \ln (x + 1)$ **29.** $f(x) = (\ln x)^2$ **30.** $f(x) = x^2 \ln x$

31. $y = \ln (1 - x)$ **32.** $f(x) = 1 - \ln x$

BUSINESS
DEMAND

33. The demand for a certain item is given by

$$p = 50 - 10 \ln x, \qquad x \geq 1$$

Find the rate of change of price with respect to demand.

REVENUE

34. The revenue for the sale of x hundred items is given by

$$R(x) = 60\sqrt{\ln x}, \qquad x \geq 1$$

where R is measured in hundreds of dollars. Find the marginal revenue for 400 items.

PROFIT

35. A company has found that its total revenue and cost functions are, respectively, $R(x) = 200 \ln (x + 1)$ and $C(x) = 4x + 6$.
   a. Find the value of x that maximizes profit.
   b. What is the maximum profit?

TRAVEL INDUSTRY

36. A travel agency determines that if x is the total number of hours worked per week by its employees, its revenue is given by the equation

$$R(x) = 350 \ln (x + 1)$$

and the corresponding cost function is

$$C(x) = 1800 + 2x$$

Find the total hours worked that will maximize its profit.

MARKETING

37. A company has introduced a new product, and sales of the product in thousands of dollars is represented by

$$S(t) = 60 \ln (t + 1)$$

where t is measured in months, with $t = 0$ corresponding to the introduction of the new product on the market.
   a. At what rate are sales changing after 2 months? After 4 months?
   b. Notice that sales will always be increasing using this model. Will sales be increasing at a decreasing or an increasing rate? (*Hint:* Calculate the second derivative.)

MARGINAL REVENUE   **38.** The demand for a particular item is given by

$$p = 6 - \ln (x + 1)$$

Find the marginal revenue.

SOCIAL SCIENCE   **39.** A social service agency is conducting a survey to decide whether to expand its
SURVEYS   services. Response to the survey is given by

$$N(t) = t \ln (t + 1)$$

where N is the number of respondents measured in hundreds and t is measured
in days. How is the rate of return changing after 30 days?

PHYSICAL   **40.** The following graph shows a set of data points of film density as a function of
SCIENCE   the ASA value for which the film was exposed. The curve that best estimates
PHOTOGRAPHY   this relationship is also plotted and is given by

$$y = 1.3657 - 0.46038 \log (x)$$

The rate of change of density with respect to ASA value is called the **local
contrast.** Find the local contrast for an ASA of 200.

ASA Value	Density
25.000	0.780
50.000	0.540
100.000	0.410
100.010	0.480
200.000	0.290
200.010	0.320
300.000	0.190
300.010	0.240
400.000	0.120
400.010	0.130
600.000	0.100
800.000	0.070
800.010	0.070

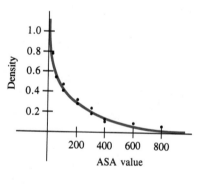

*In exercises 41 through 44, each graphics screen contains two graphs: $y = f(x) = \ln (g(x))$ and $y = f'(x)$.
For each exercise, label the graphs correctly. Keep in mind that $f'(x)$ is positive in intervals where $f(x)$ is
increasing and that $f'(x)$ is negative in intervals where $f(x)$ is decreasing. The following TI-82 parameters
were used:*

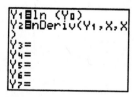

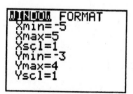

41.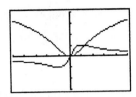

42.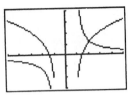

43.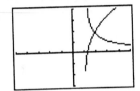

44.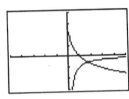

WRITING AND CRITICAL THINKING

1. For $f(x) = \ln x, f'(x) = \dfrac{1}{x}$. Does it make sense to evaluate $f'(-2)$ for this function? Explain why or why not.

2. Explain in words why $\ln e^x = x$.

USING TECHNOLOGY IN CALCULUS

Using a graphics calculator or graphing software, we examine the graphs of $y = f(x)$ for each of the following functions:

$$f(x) = \ln|x| \qquad f(x) = \frac{1}{x} \qquad f(x) = \frac{1}{|x|}$$

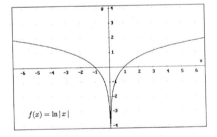

$f(x) = \ln|x|$

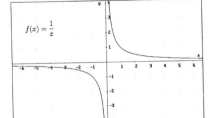

$f(x) = \dfrac{1}{x}$

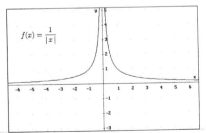

$f(x) = \dfrac{1}{|x|}$

(continued)

(*continued*)
1. Examine the graph of $f(x) = \ln |x|$ for $x < 0$. Is the function increasing or decreasing in that interval?
2. Use the fact that if a function is decreasing over a given interval then its derivative is negative over that interval to decide which of the other functions represents the derivative of $f(x) = \ln |x|$. Is the derivative of $\ln |x|$ equal to $\dfrac{1}{x}$ or $\dfrac{1}{|x|}$?

5.4 DERIVATIVES OF EXPONENTIAL FUNCTIONS

In the previous section, we demonstrated a technique of differentiation called *logarithmic differentiation*. We now use this new technique to calculate the derivative for $y = e^x$. Taking the natural logarithm of both sides, we have

$$\ln y = \ln e^x$$
$$\ln y = x$$

Using implicit differentiation, we have

$$\frac{1}{y}\frac{dy}{dx} = 1$$
$$\frac{dy}{dx} = y = e^x$$

or $$\frac{d(e^x)}{dx} = e^x$$

The reader is no longer kept in suspense as to why e is such a useful value for the base of an exponential function. For the function $f(x) = b^x$, if we let the base be e, then f is its own derivative! The Chain Rule generalization follows in Theorem 5.5.

Theorem 5.5

If $f(x) = e^{u(x)}$, then $f'(x) = u'(x)e^{u(x)}$.

Alternatively, if $y = e^u$, then $\dfrac{dy}{dx} = e^u \dfrac{du}{dx}$.

Example 1 Let $y = e^{-x^2}$. Find $\dfrac{dy}{dx}$.

Solution Using Theorem 5.5 with $u(x) = -x^2$, we have

$$u'(x) = -2x$$

So $$\frac{dy}{dx} = u'(x)e^{u(x)}$$

$$= -2xe^{-x^2}$$

∎

Example 2   Find $f'(x)$ if $f(x) = 4x^3e^{-x}$.

Solution $f(x) = 4x^3e^{-x}$ Applying the Product Rule.

$f'(x) = 4[x^3(-e^{-x}) + 3x^2e^{-x}]$ Note: $\dfrac{d(e^{-x})}{dx} = -e^{-x}$

$= -4x^3e^{-x} + 12x^2e^{-x}$

Factoring gives

$$f'(x) = 4x^2e^{-x}(-x + 3)$$

∎

Example 3   For $f(x) = \dfrac{e^x}{1 - e^x}$, find $f'(x)$.

Solution   Using the Quotient Rule gives

$$f'(x) = \frac{(1 - e^x)(e^x) - (e^x)(-e^x)}{(1 - e^x)^2}$$

$$= \frac{e^x - e^{2x} + e^{2x}}{(1 - e^x)^2}$$

$$= \frac{e^x}{(1 - e^x)^2}$$

∎

The techniques of curve sketching developed in Chapter 4 are used in the next example.

Example 4   Consider the function $f(x) = e^{-x^2}$.
   a.   What are the local extrema for this function?
   b.   Where is the function concave up? Concave down?
   c.   Graph $y = f(x)$.
   d.   Determine from the graph the value of $\lim\limits_{x \to \infty} e^{-x^2}$.

Solution   a.   We find $f'(x)$ first.

$$f(x) = e^{-x^2}$$
$$f'(x) = -2xe^{-x^2}$$

To find critical values, set $f'(x) = 0$.

$$-2xe^{-x^2} = 0$$
$$-xe^{-x^2} = 0$$
$$x = 0 \qquad \text{Since } e^{-x^2} \text{ can never be zero}$$

Calculate $f''(x)$ using the Product Rule.

$$f''(x) = -2[x(-2xe^{-x^2}) + e^{-x^2}]$$
$$= -2e^{-x^2}(-2x^2 + 1)$$

From the Second Derivative Test, when $x = 0$, $f''(x) < 0$, so the function achieves a relative maximum value of $f(0) = 1$ at $x = 0$; that is, $(0, 1)$ is a relative maximum point.

b. By setting $f''(x) = 0$, we can locate any points of inflection.

$$-2e^{-x^2}(-2x^2 + 1) = 0$$
$$e^{-x^2}(-2x^2 + 1) = 0$$
$$-2x^2 + 1 = 0$$
$$x^2 = \frac{1}{2}$$
$$x = \pm\frac{1}{\sqrt{2}} \approx \pm 0.707$$

There are two points of inflection at $x = \pm\frac{1}{\sqrt{2}}$. The curve is concave up for $x < \frac{-1}{\sqrt{2}}$ and $x > \frac{1}{\sqrt{2}}$ and concave down for $\frac{-1}{\sqrt{2}} < x < \frac{1}{\sqrt{2}}$.

c. The graph of $y = f(x)$ appears in Figure 5.13. Notice that this curve is a special form of the normal (or bell-shaped) curve from statistics.

Figure 5.13

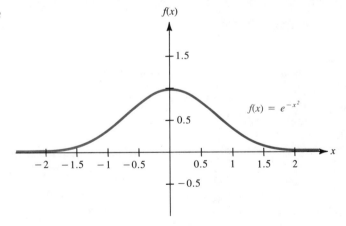

d. We see from the graph that as x approaches ∞, the y values approach zero, so we may conclude that

$$\lim_{x \to \infty} e^{-x^2} = 0$$

We now derive a formula for differentiating an exponential function whose base is not e.

If $y = b^x$, using logarithmic differentiation, we take the ln of both sides to obtain

$$\ln y = \ln b^x$$

or

$$\ln y = x \ln b \qquad \text{ln } b \text{ is a coefficient of } x.$$

Differentiating implicitly, we have

$$\frac{1}{y}\frac{dy}{dx} = \ln b$$

$$\frac{dy}{dx} = y \ln b$$

$$= b^x \cdot \ln b$$

More generally,

The derivative of $b^{u(x)}$

If $y = b^{u(x)}$, then $\dfrac{dy}{dx} = u'(x)b^{u(x)} \ln b$.

Notice that if $b = e$, then since $\ln e = 1$, the preceding formula becomes the same as Theorem 5.5.

Example 5   For $y = 7^{1-2x}$, find $\dfrac{dy}{dx}$.

Solution   Letting $u(x) = 1 - 2x$, we have $u'(x) = -2$; therefore,

$$f'(x) = -2(7^{1-2x}) \ln 7$$ ∎

Example 6   Find the absolute maximum value for $y = 5(2^{-x^2})$.

Solution   Find $\dfrac{dy}{dx}$ and set equal to zero.

$$\frac{dy}{dx} = 5[2^{-x^2}(\ln 2)(-2x)]$$

$$= (-10x \ln 2)2^{-x^2}$$

$$0 = (-10x \ln 2)2^{-x^2}$$

$$0 = -10x \ln 2$$

$$x = 0$$

Using a First Derivative Test, we see that for $x < 0$, $f'(x) > 0$; for $x > 0$, $f'(x) < 0$, which implies there is a maximum at $x = 0$. The absolute maximum value for this function is $f(0) = 5$. Refer to Example 5 of Section 5.1 for a sketch of this function.   ∎

We conclude this section with an applied problem involving differentiation of an exponential function.

Example 7 A company has determined that the demand for x units of its product based on price, p, is

$$p = 50e^{-0.04x}$$

Find the marginal revenue.

Solution Recall that revenue $R = xp$. Thus,

$$R(x) = 50xe^{-0.04x}$$

To find the marginal revenue, we differentiate using the Product Rule:

$$R'(x) = 50x(-0.04e^{-0.04x}) + e^{-0.04x}(50)$$
$$= 50e^{-0.04x}(1 - 0.04x)$$

∎

5.4 EXERCISES

SET A

In exercises 1 through 18, find $\dfrac{dy}{dx}$. (See Examples 1, 2, and 3.)

1. $y = e^{3x}$	**2.** $y = e^{-2x}$	**3.** $y = 12e^{6x}$	**4.** $y = -10e^{-4x}$
5. $y = e^{1-x^2}$	**6.** $y = e^{x^2-1}$	**7.** $y = xe^x$	**8.** $y = xe^{-x}$
9. $y = \dfrac{e^x}{e^x + 1}$	**10.** $y = \dfrac{e^x + e^{-x}}{e^x - e^{-x}}$	**11.** $y = x^2e^{-x}$	**12.** $y = -xe^{x^2}$
13. $y = \ln e^x$	**14.** $y = \ln(e^x + 1)$	**15.** $y = \ln(e^{x^2})$	**16.** $y = e^x \ln x$
17. $y = e^{\ln x}$	**18.** $y = \dfrac{e^x}{\ln x}$		

In exercises 19 through 22, find $\dfrac{d^2y}{dx^2}$.

19. $y = e^{4x}$	**20.** $y = xe^{2x}$	**21.** $y = (e^x + 1)^2$	**22.** $y = \ln(e^x + 1)$

SET B

In exercises 23 through 28, use the technique of curve sketching in Chapter 4 to graph each function. (See Example 4.)

23. $y = e^{5x-1}$	**24.** $y = e^{x^2}$	**25.** $y = xe^x$
26. $y = xe^{-x}$	**27.** $y = 4 - e^x$	**28.** $f(x) = x^2e^x$

In exercises 29 through 32, find $\dfrac{dy}{dx}$.

29. $y = \dfrac{3^x}{x + 1}$ **30.** $y = 10^x$ **31.** $y = 2^{-x^2}$ **32.** $y = x10^{-x}$

BUSINESS
BANKING

33. The formula $A = Pe^{rt}$ is the formula for continuous compounding of interest, where P and r are constants. Find $\dfrac{dA}{dt}$.

ECONOMICS

34. Suppose the demand equation for x units of a product is

$$p = 1000e^{-0.02x}$$

The revenue equation is $R(x) = x \cdot p = 1000xe^{-0.02x}$. Find the marginal revenue.

PRODUCTION RATES

35. The daily output, y, of a product x days after the start of production is given by the equation

$$y = 100 - 100e^{-0.3x}$$

At what rate is the daily output increasing when
a. $x = 5$ days?   **b.** $x = 10$ days?

AVERAGE COST

36. A company has found that the total cost of manufacturing x items is $C(x) = 20e^{x/6}$. Given that average cost is defined as $\overline{C}(x) = \dfrac{C(x)}{x}$, determine the value of x that minimizes $\overline{C}(x)$.

COST

37. The total cost function for a company is found to be

$$C(x) = 9 + \frac{e^x}{2} + 3 \ln (x + 1)$$

where $x \geq 0$.
a. How is the cost changing when $x = 3$?
b. Is $C(x)$ increasing or decreasing when $x = 10$?
c. Is $C(x)$ increasing for all values of x?

INTEREST

38. A deposit of $5000 is made into an account at 8% compounded quarterly. The amount in the account at any time, t, is given by the function

$$A = 5000(1.02)^{4t}$$

Find the rate of change of the amount with respect to time.

EMPLOYEE EFFICIENCY

39. A new employee can assemble n parts per day after x days. This relationship is given by

$$n = 50 - 50e^{-0.4x}$$

At what rate is the number of parts per day changing with respect to time at 10 days?

BIOLOGY
BACTERIAL GROWTH

40. The number of bacteria in a certain culture at any time, t, is given by

$$N(t) = 500e^{0.7t}$$

where t is measured in hours. Find the rate of change in the number of bacteria at the end of 4 hours.

VIROLOGY

41. In an experiment, a colony of bacteria was exposed to a new virus. The number of bacteria at any time, t, is given by

$$N(t) = 5000e^{(4 + 4t - t^2)}$$

where t is measured in hours. At what time was the number of bacteria at a maximum?

DRUG THERAPY

42. A number of new healthy cells is growing in response to an experimental drug at a rate given by

$$N = 500e^{0.2t}$$

where t is measured in days from the time the drug was administered. Find the rate of change in the number of new cells after 5 days.

SOCIAL SCIENCE
WELFARE

43. The number of families on welfare is decreasing in a small rural region. The number, N, of families on welfare is given by

$$N(t) = 100e^{-t} + 30$$

where t is measured in months.
a. How many families were on welfare initially?
b. How many families were on welfare after 1 month?
c. At what rate is the number of families changing after 6 months?

POPULATION

44. The population of a certain region of South America is given by

$$P(t) = \frac{192}{8 + 16e^{-0.48t}}$$

where t is measured in years and $P(t)$ is measured in thousands.
a. How many people were present at time $t = 0$?
b. At what rate is the population changing at the end of 10 years?

WRITING AND CRITICAL THINKING

It has been several chapters since we defined what is meant by a differentiable function. The objective of this exercise is to look again at the concept of differentiability, this time with respect to exponential functions.
Consider $f(x) = e^{|x|} + 1$. This function is continuous for all real numbers x.

1. Using the (limit) definition of derivative, explain in words why this function is not differentiable at $x = 0$.
2. Write f as a piecewise function and then write a piecewise expression that represents f'.
3. If you have access to computer graphing software or graphics calculators, sketch $y = f(x)$ and $y = f'(x)$.

5.5   APPLICATION TO ECONOMICS: ELASTICITY OF DEMAND

If the cost of an item you buy at the grocery store every week suddenly increased in price from 99¢ to $1.09, you might decide you like the product so well that you will continue to buy it. If most consumers of the product feel the same way you do, then the company's revenue will increase. This happens because the decrease in sales caused by the relatively few consumers who no longer buy the product is more than offset by the price increase to the rest of you.

In business and economics it is often necessary to determine what relative effect price fluctuations will have on demand for the product. A measure of this effect is called **elasticity of demand.** In general, elasticity of demand is defined as follows:

$$\text{elasticity of demand} = \frac{\text{relative change in demand}}{\text{relative change in price}}$$

To see what is meant by a *relative* change, let's return to the product in the first paragraph, which changed from 99¢ to $1.09. Here the relative change in price is the ratio of the change in price to the original price. Because the price increased or changed by 10¢, we have

$$\text{relative change in price} = \frac{0.10}{0.99} \approx 0.101$$

or we can say that the price increased by approximately 10.1%.

Before we formally define elasticity of demand using symbols, we need to return to the notion of demand equations.

Recall that a demand equation relates the quantity of an item demanded, x, to its price, p. In earlier sections, we expressed p as a function of x. For example, the price of an item is given by

$$p = 125 - \frac{x}{20}$$

In this section, we want to express x as a function of p. Solving the preceding equation for x, we have

$$p = 125 - \frac{x}{20}$$
$$20p = 2500 - x$$
$$x = 2500 - 20p$$

Using function notation, we can write

$$f(p) = 2500 - 20p$$

where $f(p)$ represents the demand, x. Notice that as price increases, demand decreases. This will be true for all illustrations in this section, and it is a fundamental assumption in elementary economic theory.

In Section 3.7 we learned that the differential can be used to measure change for a variable. Using the differential, we can write the relative change in demand as the ratio $\dfrac{dx}{x}$ and the relative change in price as the ratio $\dfrac{dp}{p}$. Now we can write elasticity of demand as follows:

$$\text{elasticity of demand} = \frac{\text{relative change in demand}}{\text{relative change in price}}$$

$$= \frac{\dfrac{dx}{x}}{\dfrac{dp}{p}}$$

We wish to rewrite it as a function of p only. If $x = f(p)$, then $\dfrac{dx}{dp} = f'(p)$ and $dx = f'(p)dp$. This gives

$$E(p) = \frac{\dfrac{f'(p)dp}{f(p)}}{\dfrac{dp}{p}}$$

$$= \frac{p}{dp} \cdot \frac{f'(p)dp}{f(p)}$$

$$= \frac{p \cdot f'(p)}{f(p)}$$

Elasticity of demand

For the demand equation $x = f(p)$, the elasticity of demand is defined by

$$E(p) = \frac{p \cdot f'(p)}{f(p)}$$

Note that $E(p)$ is a unitless quantity (see exercise 24). An example of calculating the elasticity of demand follows.

Example 1 An electronics firm's marketing division estimates that the price, p, and demand, x, for a new product can be described by the demand equation

$$p = 900 - \frac{x}{3}, \qquad 0 < x < 2700$$

a. Express demand as a function of price.
b. Find the elasticity of demand when $p = \$150$.
c. Interpret the result in part b.

Solution   **a.** $p = 900 - \dfrac{x}{3}$

$$3p = 2700 - x$$
$$x = 2700 - 3p$$
$$f(p) = 2700 - 3p$$

b. $f'(p) = -3$

$$E(p) = \frac{pf'(p)}{f(p)}$$

$$= \frac{-3p}{2700 - 3p}$$

$$E(150) = \frac{-450}{2700 - 450} = -0.2$$

c. $E(\$150) = -0.2$ means that when the price is set at $150, a small increase in price will result in a decrease in quantity demanded of about 0.2 times the relative rate of price increase. That is, if the price is increased from $150 by, say 3%, then the quantity demanded will decrease by about 0.6% since $(-0.2)(3\%) = -0.6\%$. ∎

It should be noted that $E(p)$ is always negative or zero. This is because of our underlying assumption that as price *increases*, the quantity demanded *decreases*. (Some economics texts actually define the elasticity of demand to be $\dfrac{-pf'(p)}{f(p)}$ to avoid the minus sign!)

In Example 1, $E(p) > -1$. When $E(p) > -1$, economists say that demand is **inelastic;** if $E(p) < -1$, the demand is said to be **elastic.** If $E(p) = -1$, as in the next example, we say demand has **unit elasticity.**

In an inelastic state, an increase in price produces a relatively small decrease in demand. However, in an elastic state, an increase in price produces a significant decrease in demand. Unit elasticity means that the percent decrease in demand will equal the percent increase in price.

Example 2   Using the demand equation from Example 1, find $E(450)$ and interpret the result.

Solution $E(p) = \dfrac{-3p}{2700 - 3p}$

$$E(450) = \frac{450(-3)}{2700 - 3(450)}$$

$$= -1$$

We interpret this unit elasticity to mean that if the price is increased a small amount from $450, say 2%, then the quantity demanded will decrease by about 2%. ∎

Because price fluctuations have an effect on demand that, in turn, has an effect on revenue, we will examine the relationship between elasticity of demand and revenue. We have defined revenue as $R = px$. Writing revenue as a function of p, we have

$$R(p) = p \cdot f(p) \qquad \text{Where } x = f(p)$$

Computing $R'(p)$ by taking the derivative using the Product Rule gives

$$R'(p) = p \cdot f'(p) + f(p) \cdot 1$$

For maximum revenue, $R'(p) = 0$ or

$$pf'(p) + f(p) = 0$$
$$pf'(p) = -f(p)$$
$$\frac{pf'(p)}{f(p)} = -1$$
$$E(p) = -1$$

This illustrates that unit elasticity occurs at maximum revenue. Also, if $R'(p) < 0$, then $E(p) < -1$ and demand is elastic. If $R'(p) > 0$, then $E(p) > -1$ and demand is inelastic. See Figure 5.14.

Figure 5.14

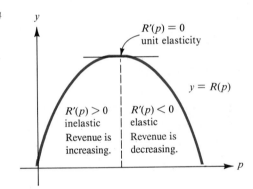

Example 3   **a.** Find $E(p)$ if $f(p) = 3600 - 4p^2$, where $0 < p < 30$.
b. Find $E(20)$.

Solution   **a.** $f(p) = 3600 - 4p^2$
$f'(p) = -8p$
$$E(p) = \frac{pf'(p)}{f(p)}$$
$$= \frac{p(-8p)}{3600 - 4p^2}$$
$$= \frac{-2p^2}{900 - p^2}$$

b. $$E(20) = \frac{-2(400)}{900 - 400}$$
$$= -1.6$$

The demand is elastic at $p = \$20$. ∎

Example 4   Suppose that the demand for a product is given by

$$f(p) = 1500 - 30p \qquad \text{Where } 0 < p < 50$$

What will be the effect on demand if a 10% price increase occurs when $p = \$30$?

Solution $E(p) = \dfrac{p \cdot f'(p)}{f(p)}$

$$= \dfrac{p(-30)}{1500 - 30p}$$

$$= \dfrac{p}{p - 50}$$

If $p = 30$, then

$$E(30) = \dfrac{30}{30 - 50} = -1.5$$

Because the elasticity of demand is -1.5 and the price increases 10%, then we have

$$(-1.5)(10\%) = -15\%$$

indicating that the demand decreases by 15%.   ∎

The demand functions used thus far have all been either linear or quadratic functions of p. In the linear case, when $f(p) = a - bp$, certain generalities can be made regarding elasticity of demand. (See exercise 21.) The next two examples explore two special types of demand functions, *hyperbolic demand* and *exponential demand*.

Example 5   Suppose the quantity demanded, $f(p)$, is a function of price, p, according to the **generalized hyperbolic demand**:

$$f(p) = kp^{-n}, \qquad k, n \text{ are constants}, \quad n > 0$$

Find $E(p)$.

Solution $f'(p) = -knp^{-n-1}$

$$E(p) = \dfrac{p(-knp^{-n-1})}{kp^{-n}}$$

$$= -n$$

We see that for hyperbolic demand, elasticity of demand is a constant.   ∎

Example 6   Suppose the quantity demanded is a function of price according to the **generalized exponential demand function**:

$$f(p) = ae^{-bp}, \qquad a, b \text{ are constants}$$

Find $E(p)$.

Solution $f'(p) = -abe^{-bp}$

$$E(p) = \frac{pf'(p)}{f(p)}$$

$$E(p) = \frac{-abpe^{-bp}}{ae^{-bp}}$$

$$= -bp$$

Thus, the elasticity of exponential demand is a constant multiple of the price.   ■

5.5   EXERCISES

SET A

In exercises 1 through 6, find the specified value for the elasticity of demand and tell whether it is elastic, inelastic, or of unit elasticity. (See Examples 1, 2, and 3.)

1. $f(p) = 400 - 10p$, $0 < p < 40$
   a.   Find $E(10)$.
   b.   Find $E(20)$.
   c.   Find $E(30)$.

2. $f(p) = 1200 - 4p$, $0 < p < 300$
   a.   Find $E(100)$.
   b.   Find $E(150)$.
   c.   Find $E(200)$.

3. $f(p) = 4500 - 5p^2$, $0 < p < 30$
   a.   Find $E(5)$.
   b.   Find $E(15)$.
   c.   Find $E(20)$.

4. $f(p) = 120 - 0.5p$, $0 < p < 240$
   a.   Find $E(100)$.
   b.   Find $E(200)$.
   c.   Find $E(120)$.

5. $f(p) = 10p^{-2}$, $0 < p < 2$
   a.   Find $E(1)$.
   b.   Find $E(1.5)$.

6. $f(p) = 4800 - 4p^2$, $0 < p < 20\sqrt{3}$
   a.   Find $E(10)$.
   b.   Find $E(15)$.

In exercises 7 through 12, calculate the specified value for the elasticity of demand and interpret the result in each case for a price increase of 5%. (See Examples 1 through 4.)

7. $f(p) = 800 - 40p$, $0 < p < 20$
   a.   Find $E(10)$.
   b.   Find $E(5)$.
   c.   Find $E(15)$.

8. $f(p) = 1800 - 90p$, $0 < p < 20$
   a.   Find $E(5)$.
   b.   Find $E(10)$.
   c.   Find $E(18)$.

9. $f(p) = 5000 - 2p^2$, $0 < p < 50$
   a.   Find $E(20)$.
   b.   Find $E(40)$.

10. $f(p) = 100 - p^2$, $0 < p < 10$
   a.   Find $E(5)$.
   b.   Find $E(6)$.

11. $f(p) = 100e^{-p}$, $0 < p < 10$

   a.   Find $E(2)$.
   b.   Find $E(0.5)$.

12. $f(p) = \dfrac{6}{p}$, $1 < p < 10$
   a.   Find $E(5)$.
   b.   Find $E(8)$.

13. Let $f(p) = 1200 - 20p$, where $0 < p < 60$. What will be the effect on demand if a 2% increase in price occurs when $p = \$40$?

14. Let $f(p) = 1800 - 20p$, where $0 < p < 90$. What will be the effect on demand if a 10% increase in price occurs when $p = \$50$?

15. Let $f(p) = 5400 - 6p^2$, where $0 < p < 30$. What will be the effect on demand if a 20% increase in price occurs when $p = \$15$?

16. Let $f(p) = 3600 - 4p^2$, where $0 < p < 30$. What will be the effect on demand if a 10% increase in price occurs when $p = \$20$?

SET B

17. Suppose $f(p) = 4200 - 6p$, where $0 < p < 700$.
    a. For what values of p is the demand elastic?
    b. For what values of p is the demand inelastic?

18. Suppose $f(p) = 6400 - 4p$, where $0 < p < 1600$.
    a. For what values of p is the demand elastic?
    b. For what values of p is the demand inelastic?

19. Recall that revenue, R, is xp. We can write $R = pf(p)$. Assume $f(p) = 2700 - 3p$, where $0 < p < 900$.
    a. Find $\dfrac{dR}{dp}$.
    b. For what value(s) of p is revenue increasing?
    c. For what value(s) of p is revenue decreasing?
    d. For what value(s) of p is demand elastic?
    e. For what value(s) of p is demand inelastic?

20. a. Generalize exercise 19 by finding $\dfrac{dR}{dp}$ for $R = pf(p)$.

    b. Show that $\dfrac{dR}{dp} = f(p)[E(p) + 1]$.

    c. Because $f(p)$ is always positive, when will $\dfrac{dR}{dp}$ be greater than zero? (Equivalently, when will revenue be increasing?)
    d. When will revenue be decreasing in terms of $E(p)$?

21. **Generalized Linear Demand**
    Let $f(p) = a - bp$, where $a > 0$, $b > 0$, and $0 < p < \dfrac{a}{b}$.
    a. Find $E(p)$.
    b. Complete the table.

p	$E(p)$
$\dfrac{a}{4b}$	
$\dfrac{a}{3b}$	
$\dfrac{a}{2b}$	
$\dfrac{2a}{3b}$	
$\dfrac{10a}{11b}$	

22. Using exercise 21, let $f(p) = 1200 - 4p$, where $0 < p < 300$.

    a. For what value(s) of p is demand elastic? $\left(\text{Hint: } p > \dfrac{a}{2b}. \right)$

    b. For what value(s) of p is demand inelastic?

    c. For what value(s) of p is demand of unit elasticity?

23. Repeat exercise 22 for $f(p) = 4800 - 2p$.

24. It was mentioned in the text that the elasticity of demand, $E(p)$, is a unitless quantity. Show that $E(p)$ has no unit attached to it by determining the units for each of the following:

    a. p **b.** $f(p)$

    c. $f'(p)$ **d.** $\dfrac{pf'(p)}{f(p)}$

CHAPTER 5 REVIEW

VOCABULARY/FORMULA LIST

exponential function	natural exponential function	learning curve
$A = P\left(1 + \dfrac{r}{n}\right)^{nt}$	logarithmic function elasticity of demand common logarithm	$\dfrac{d(e^x)}{dx} = e^x$
exponential growth exponential decay	natural logarithm bounded growth model unbounded growth model	$\dfrac{d(\ln x)}{dx} = \dfrac{1}{x}$
$e = \lim_{n \to \infty} \left(1 + \dfrac{1}{n}\right)^n$		$E(p) = \dfrac{pf'(p)}{f(p)}$

CHAPTER 5 REVIEW EXERCISES

[5.1]   **1.** Graph $y = e^x$.

[5.2]   **2.** Graph $y = \ln x$.

[5.2]   *In exercises 3 through 6, rewrite each exponential equation in its equivalent logarithmic form.*

    3. $5^2 = 25$ **4.** $10^{-2} = 0.01$

    5. $6^{-2} = \frac{1}{36}$ **6.** $e^0 = 1$

[5.2]   *In exercises 7 through 10, rewrite each logarithmic equation in its equivalent exponential form.*

    7. $\log 100 = 2$ **8.** $\ln \dfrac{1}{e} = -1$

    9. $\log_b 1 = 0$ **10.** $\log 0.001 = -3$

BUSINESS   [5.4]   **11.**   Assume that the revenue to be received from the sale of x items is given by the
REVENUE   function $R = 30xe^{-x/4}$.
         a.   Find the number of items that should be sold to maximize R.
         b.   What is the maximum value of R?

REVENUE   [5.4]   **12.**   A company has determined that the demand for one of its products is
$p = 80e^{-x/2}$. Find the revenue function, $R(x)$; the value of x that maximizes
revenue; and the maximum revenue.

DEMAND   [5.4]   **13.**   The demand for a product after t years on the market is given by the equation
$D(t) = 400 - 150e^{-t}$.
         a.   Graph this function.
         b.   How is demand changing when $t = 2$ years? 4 years? 10 years?

SOCIAL   [5.2]   **14.**   Suppose the unbounded growth rate of a country's population is
SCIENCE
DEMOGRAPHY

$$N(t) = N_0(1.06)^t$$

where $N(t)$ is the number of people alive at any given time, t (in years), and
N_0 is the number of people alive at $t = 0$. How long will it take this population
to double?

In exercises 15 through 22, find $\dfrac{dy}{dx}$.

    [5.4]   **15.** $y = e^{-x}$

    [5.3]   **16.** $y = \ln(2x + 1)$

    [5.3]   **17.** $y = \ln(4 - x^2)$

    [5.4]   **18.** $y = e^{-x^2}$

    [5.4]   **19.** $y = x^3e^x$

    [5.3]   **20.** $y = \dfrac{\ln x}{x + 1}$

    [5.4]   **21.** $y = \dfrac{e^x + e^{-x}}{2}$

    [5.3]   **22.** $y = \ln \sqrt{x^2 + 1}$

    [5.5]   **23.**   Suppose the demand–price relation for a particular item is given by
$f(p) = 2700 - 6p$, where $0 < p < 450$.
         a.   Find revenue, R.
         b.   Find $E(p)$, elasticity of demand.
         c.   For what value(s) of p is demand elastic?
         d.   For what value(s) of p is demand inelastic?

    [5.5]   **24.**   Suppose $f(p) = 4800 - 16p^2$, where $0 < p < 10\sqrt{3}$.
         a.   Find $E(p)$.
         b.   Find $E(5)$.
         c.   Find $E(10)$.

CHAPTER 5 TEST

PART ONE

Multiple choice. In exercises 1 through 10, put the letter of your correct response in the blank provided.

_____ **1.** The logarithmic equivalent of $P^Q = R$ is

    a. $P^R = Q$ **b.** $\log_P Q = R$

    c. $\log_P R = Q$ **d.** $PR = Q$

_____ **2.** The graph of which of the following exhibits exponential decay?

    a. $y = e^x$ **b.** $y = \left(\dfrac{1}{2}\right)^{-x}$

    c. $y = (0.1)^{-x}$ **d.** $y = e^{-x}$

_____ **3.** If $y = (\ln x)^2$, find $\dfrac{dy}{dx}$.

    a. $2 \ln x$ **b.** $\dfrac{2}{x}$

    c. $\dfrac{2}{x} \ln x$ **d.** $2x \ln x$

_____ **4.** Solve for x: $4^{x-1} = \dfrac{1}{8}$.

    a. $\dfrac{-1}{2}$ **b.** 16

    c. $\dfrac{-5}{2}$ **d.** $\dfrac{1}{2}$

_____ **5.** Solve for x: $\log_4 \sqrt{x} = 1$.

    a. 2 **b.** 16

    c. 4 **d.** 1

_____ **6.** For $y = \dfrac{e^{2x}}{e^{2x} + 1}$, find $\dfrac{dy}{dx}$.

    a. $\dfrac{e^{2x}}{e^{4x} + 1}$ **b.** $\dfrac{2e^{2x}}{(e^{2x} + 1)^2}$

    c. 1 **d.** $\dfrac{2e^{4x}}{e^{4x} + 1}$

_____ **7.** Solve for x: $-\log_x 9 = 2$.

    a. -3 **b.** $-\dfrac{1}{3}$

    c. 3 **d.** $\dfrac{1}{3}$

_____ **8.** Which of the following is a sketch of $y = \log_2 (x - 1)$?

a.

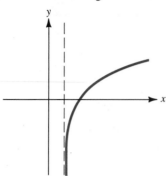

b.

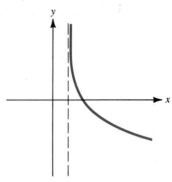

c.

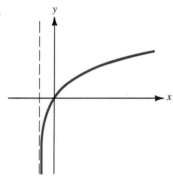

d.

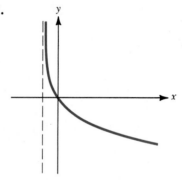

_____ **9.** Solve for x: $e^{2x+1} = 7$.

    a. $\dfrac{1}{2} \ln 6$

    b. $\dfrac{1}{2} + \dfrac{1}{2} \ln 7$

    c. $\dfrac{-1}{2} + \dfrac{1}{2} \ln 7$

    d. $\dfrac{-1}{2} + \ln 7$

_____ **10.** A relative minimum point on the curve $y = x^2 \ln x$ is

    a. $\left(\dfrac{-1}{\sqrt{e}}, \dfrac{-1}{2e} \right)$

    b. $\left(\dfrac{1}{\sqrt{e}}, \dfrac{-1}{2e} \right)$

    c. $\left(\dfrac{1}{e}, \dfrac{-1}{e^2} \right)$

    d. $\left(\dfrac{1}{\sqrt{e}}, \dfrac{-1}{2} \right)$

PART TWO

Solve and show all work.

11. Solve for x: $5e^{1-2x} = 12$.

12. Graph $y = \ln x$ and $y = e^x$ on the same set of axes.

13. Find $\dfrac{dy}{dx}$ if $y = e^{x^2 - 1} + 10x$.

14. Find $f'(x)$ if $f(x) = \ln (x^2 + 1)$.

15. If $y = \dfrac{\ln x}{x}$, find $\dfrac{dy}{dx}$.

16. What are the critical values of $f(x) = xe^x$?

17. What is the absolute maximum value of the function $f(x) = 2e^{-x^2}$?

BUSINESS
FINANCE

18. The Daytone Manufacturing Company has determined that the value, V, of its investments (which have been growing at an annual rate of increase of 8%) is

$$V = 45{,}000(1.08)^t$$

where t is the number of years since the investments were made. Find the value of Daytone's investments after 5 years.

PRODUCTION RATES

19. A small company has determined that the daily output, y, of a product x days after the start of production is

$$y = 150 - 150e^{-0.2x}$$

At what rate is the daily output increasing when
a. $x = 5$ days?   **b.** $x = 10$ days?

ECONOMICS

20. Suppose that the demand equation for x units of a product is

$$p = 1000e^{-0.01x}$$

a. Find the revenue function.
b. Find the marginal revenue.

6

INTEGRATION

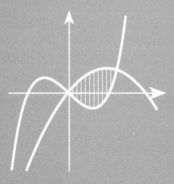

6.1   ANTIDIFFERENTIATION AND INDEFINITE INTEGRALS

We know from algebra that an operation is often associated with an "opposite" or *inverse* operation. For instance, subtraction can be thought of as the inverse of addition, division as the inverse of multiplication, and the logarithmic function as the inverse of the exponential function.

Operation	Associated inverse operation
Addition: $\boxed{5} + 9 = 14$	Subtraction: $14 - 9 = \boxed{5}$
Multiplication: $\boxed{4} \cdot 3 = 12$	Division: $12 \div 3 = \boxed{4}$
Exponent: $10^3 = 1000$	Logarithm: $\log_{10} 1000 = \boxed{3}$

Notice in each case that the inverse operation "returns" us to the original value (boxed).

Calculus involves the study of two basic "operations":

1. Finding the derivative of a function
2. Using the "inverse operation" (that is, given the derivative of a function, finding the function)

This "inverse" process is called **antidifferentiation.** We have the following definition:

Antiderivative

> Let f be a function. A function F, whose derivative is f, is called an **antiderivative** of f. In symbols,
>
> $$F'(x) = f(x)$$
>
> for all x in the domain of f.

Example 1   Show that $F(x) = x^5$ is an antiderivative of $f(x) = 5x^4$.

Solution   We must show that $F'(x) = f(x)$. Using the Power Rule, we get

$$F'(x) = 5x^4$$

which is, of course, $f(x)$.   ■

Example 2   Show that $G(x) = x^5 + 6$ is an antiderivative of $f(x) = 5x^4$.

Solution   We must show that $G'(x) = f(x)$.

$$G'(x) = 5x^4 + 0 \qquad \text{The derivative of a constant is zero.}$$
$$= 5x^4$$

∎

Notice from these examples that an antiderivative for a function is not unique; that is, x^5, $x^5 + 6$, $x^5 - 7$, $x^5 + 10$, and $x^5 - \sqrt{2}$ are all antiderivatives of $5x^4$. In fact, if $F(x)$ is an antiderivative of $f(x)$, then so is $F(x) + C$, where C represents any constant. Comparing Examples 1 and 2, we get the antiderivative $G(x) = F(x) + 6$.

Before presenting more examples, additional vocabulary and symbolism are necessary. Just as we referred to the *process* of differentiation as "finding the derivative" in Chapter 2, here we think of the *process* of antidifferentiation as "finding antiderivatives." Another name for the process of antidifferentiation is **integration.** If $F(x)$ is an antiderivative of $f(x)$, we call $F(x) + C$ the **indefinite integral** of $f(x)$. The word *indefinite* stems from the fact that the constant C is arbitrary or indefinite. More formally, we have

Indefinite integral

Suppose F is an antiderivative of f, then

$$\int f(x)\, dx = F(x) + C$$

which is read "the (indefinite) integral of $f(x)$ with respect to x is $F(x) + C$."

The symbol $\int$ is called an **integral sign,** $f(x)$ is referred to as the **integrand,** C is called the **constant of integration,** and the symbol dx indicates the variable of integration. Using this new notation for Example 1, since $F(x) = x^5$ is an antiderivative of $f(x) = 5x^4$, we can write

$$\int 5x^4\, dx = x^5 + C$$

Many of the properties (or rules) that applied to differentiation have analogous properties for integration. We present four of them in Theorem 6.1.

Theorem 6.1

1. $\displaystyle\int x^n\, dx = \frac{x^{n+1}}{n+1} + C, \qquad n \neq -1$

2. $\displaystyle\int dx = x + C$

3. $\displaystyle\int kf(x)\, dx = k\int f(x)\, dx$

4. $\displaystyle\int [f(x) + g(x)]\, dx = \int f(x)\, dx + \int g(x)\, dx$

Part 1 of Theorem 6.1 can be referred to as a **Power Rule** for integrals. It indicates that in order to integrate x^n, we must add 1 to the exponent and divide by this new exponent. (Recall that in differentiating x^n, we *multiplied* and *subtracted,* the inverse operations.) Also note that $n \neq -1$, because division by zero is undefined. We can, however, integrate x^{-1} and will do so later in this section.

Example 3 Evaluate $\int x^3 \, dx$.

Solution Using the Power Rule, we get

$$\int x^3 \, dx = \frac{x^{3+1}}{3+1} + C$$

$$= \frac{x^4}{4} + C$$

As a check, we should be able to differentiate our result, $\frac{x^4}{4} + C$, and obtain the integrand:

$$\frac{d}{dx}\left(\frac{x^4}{4} + C\right) = \frac{1}{4}(4x^3) + 0 = x^3 \qquad \blacksquare$$

Example 4 Evaluate $\int z^{-3} \, dz$.

Solution $\int z^{-3} \, dz = \frac{z^{-3+1}}{-3+1} + C$

$$= \frac{z^{-2}}{-2} + C = \frac{-1}{2z^2} + C \qquad \blacksquare$$

Example 5 Evaluate $\int \sqrt[3]{x^2} \, dx$.

Solution First rewrite the integrand $\sqrt[3]{x^2}$ in exponential form as $x^{2/3}$. Now

$$\int x^{2/3} \, dx = \frac{x^{2/3+1}}{\frac{2}{3}+1} + C$$

$$= \frac{x^{5/3}}{\frac{5}{3}} + C = \frac{3}{5}x^{5/3} + C \qquad \blacksquare$$

Part 2 of Theorem 6.1 is just a special case of the Power Rule for $n = 0$.

$$\int dx = \int x^0 \, dx = \frac{x^{0+1}}{0+1} + C = x + C$$

Part 3 of the theorem indicates that constants (multiplied times a function) may be moved across the integral sign. For a function such as $f(x) = 3x^4$, this property indicates that the coefficient, 3, can be moved across the integral sign, as indicated in the next example.

Example 6   Evaluate $\int 3x^4\, dx$.

Solution $\int 3x^4\, dx = 3\int x^4\, dx$ By part 3 of Theorem 6.1

$= 3\left[\dfrac{x^5}{5} + C_1\right]$ By part 1 of Theorem 6.1

$= \tfrac{3}{5}x^5 + 3C_1$

Since $3C_1$ is arbitrary, we replace it with just C. Now we have

$$\int 3x^4\, dx = \tfrac{3}{5}x^5 + C$$

In general, we wait and add the constant of integration separately from any computations involved in the indefinite integration problems.   ∎

Example 7   Evaluate $\int \dfrac{2}{\sqrt{x}}\, dx$.

Solution   Moving 2 across the integral sign and rewriting $\dfrac{1}{\sqrt{x}}$ as $x^{-1/2}$, we have

$$\int \dfrac{2}{\sqrt{x}}\, dx = 2\int x^{-1/2}\, dx$$

$$= 2\cdot\dfrac{x^{1/2}}{\frac{1}{2}} + C$$

$$= 4x^{1/2} + C = 4\sqrt{x} + C$$ ∎

Before examining part 4 of Theorem 6.1, it is worthwhile mentioning that you can check your answers to the integration problems by differentiating the answer, just as subtraction can be checked by doing an addition. If we let $F(x) = 4x^{1/2} + C$ be the solution in Example 7, then

$$F'(x) = 4\cdot\dfrac{1}{2}x^{-1/2} + 0$$

$$= \dfrac{2}{\sqrt{x}}$$

which is the original integrand.

Part 4 of Theorem 6.1 states that the "integral of a sum is the sum of the integrals." It also works, of course, for differences. This part allows us to integrate a function term by term, as shown in Examples 8 and 9.

Example 8   Evaluate $\int (5x^2 + 4x - 3)\, dx.$

Solution $\int (5x^2 + 4x - 3)\, dx = \int 5x^2\, dx + \int 4x\, dx - \int 3\, dx$

$$= 5 \int x^2\, dx + 4 \int x\, dx - 3 \int dx$$

$$= 5 \cdot \frac{x^3}{3} + 4 \cdot \frac{x^2}{2} - 3 \cdot x + C$$

$$= \tfrac{5}{3}x^3 + 2x^2 - 3x + C$$

∎

Example 9   Evaluate $\int \dfrac{3t^5 - t^3}{t^2}\, dt.$

Solution $\int \dfrac{3t^5 - t^3}{t^2}\, dt = \int \dfrac{3t^5}{t^2}\, dt - \int \dfrac{t^3}{t^2}\, dt$

$$= 3 \int t^3\, dt - \int t\, dt$$

$$= \frac{3t^4}{4} - \frac{t^2}{2} + C$$

∎

In Chapter 5, we found the *derivative* of exponential and logarithmic functions. Theorem 6.2 presents the corresponding integral forms for those derivatives.

Theorem 6.2

1. $\displaystyle \int e^x\, dx = e^x + C$

2. $\displaystyle \int \frac{1}{x}\, dx = \ln |x| + C$

Notice that part 2 gives a way of integrating x^{-1}. The absolute value symbols are necessary because the logarithm of a negative number is not defined.

Example 10   Evaluate $\int \left(5e^x - \dfrac{3}{x} \right) dx.$

Solution $\displaystyle\int \left(5e^x - \frac{3}{x} \right) dx = 5 \int e^x \, dx - 3 \int \frac{1}{x} \, dx$

$$= 5e^x - 3 \ln |x| + C \qquad \blacksquare$$

Recall in the definitions of antiderivative and indefinite integral, if F is an antiderivative of f, then $F'(x) = f(x)$ and

$$\int f(x) \, dx = F(x) + C$$

Another way to express these two ideas together is

$$\int F'(x) \, dx = F(x) + C$$

This particular notation really emphasizes the idea that differentiation and integration are "inverse processes."

We are now ready to move to some examples on integration where the constant of integration can be evaluated. In other words, we will be finding a *specific* antiderivative for a function.

Example 11   Find $g(x)$ if $g'(x) = 2x$ and $g(1) = 5$.

Solution   To find $g(x)$, we use the fact that

$$g(x) = \int g'(x) \, dx$$
$$= \int 2x \, dx$$
$$= x^2 + C$$

Because we are also given that $g(1) = 5$, we have

$$g(1) = 1^2 + C$$
$$5 = 1 + C \qquad \text{or} \qquad C = 4$$

So

$$g(x) = x^2 + 4 \qquad \blacksquare$$

In Example 11, the function $g(x) = x^2 + C$ represents a *family* of parabolas, a few of which are shown in Figure 6.1. We can determine specifically that $g(x) = x^2 + 4$ is the desired function (parabola) because of the additional information that $g(1) = 5$, which tells us that its graph must pass through the point $(1, 5)$. In some instances, this additional information is referred to as an **initial condition** and allows us to evaluate C.

Figure 6.1

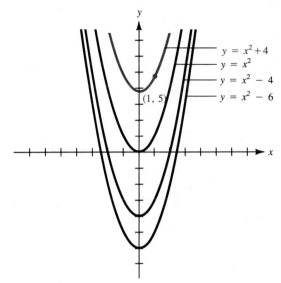

$y = x^2 + 4$
$y = x^2$
$y = x^2 - 4$
$y = x^2 - 6$

(1, 5)

Example 12 Suppose that the velocity of an object at any time, t, measured in feet per second, is given by

$$v(t) = -32t + 32$$

Find its position, $s(t)$, at any time, t, given that the initial position ($t = 0$) was 200 ft above the ground.

Solution Recall from Chapter 3 that $v(t) = s'(t)$ so that

$$s(t) = \int v(t)\, dt$$
$$= \int (-32t + 32)\, dt$$
$$= -32 \int t\, dt + 32 \int dt$$
$$= -16t^2 + 32t + C$$

Since $s(0) = 200$, we have

$$200 = -16(0)^2 + 32(0) + C$$
$$200 = C$$

So the position at any time, t, is

$$s(t) = -16t^2 + 32t + 200$$

∎

Example 13 At the Zabot Manufacturing Company, the marginal cost for producing x gears, measured in hundreds, is

$$C'(x) = 10 + 0.1x$$

The **fixed cost** (that cost attributed to overhead—or the cost of producing no items) is $3000. Find the cost for manufacturing 5000 gears ($x = 50$).

Solution   We must find $C(x)$.

$$C(x) = \int C'(x)\, dx$$
$$= \int (10 + 0.1x)\, dx$$
$$= 10x + 0.05x^2 + k \qquad \text{Using } k, \text{ because } C \text{ represents cost}$$

To find k, use the fact that $C(0) = 3000$.

$$C(x) = 10x + 0.05x^2 + k$$
$$C(0) = 10(0) + 0.05(0)^2 + k$$
$$3000 = 0 + 0 + k$$
$$3000 = k$$

Thus,

$$C(x) = 10x + 0.05x^2 + 3000$$

Finally,

$$C(50) = 10(50) + 0.05(50)^2 + 3000$$
$$= 3625$$

Therefore, the cost of producing 5000 gears is $3625.

6.1 EXERCISES

SET A

In exercises 1 through 30, evaluate each indefinite integral. (See Examples 3–10.)

1. $\int 5\, dx$

2. $\int 2\, dx$

3. $\int 4x^3\, dx$

4. $\int 3x^2\, dx$

5. $\int 4x^5\, dx$

6. $\int 6y^4\, dy$

7. $\int (3x^2 - 2x + 5)\, dx$

8. $\int (4t^3 - 3t^2 + 1)\, dt$

9. $\int (6y^5 - 30y^2 + 7y - 4)\, dy$

10. $\int (5x^5 - 20x^4 + 6x^2 - 2x + 5)\, dx$ **11.** $\int (3x^6 - 4x + 7)\, dx$

12. $\int (x^3 - x^2 + x - 1)\, dx$ **13.** $\int \dfrac{1}{x}\, dx$ **14.** $\int \dfrac{2}{x}\, dx$

15. $\int \left(5 - \dfrac{3}{x}\right) dx$ **16.** $\int \left(\dfrac{1}{x} - \dfrac{1}{\sqrt{x}}\right) dx$ **17.** $\int \dfrac{20x^3 - 6x + 1}{4}\, dx$

18. $\int \dfrac{3x^4 + x^2 - 5}{x^2}\, dx$ **19.** $\int \left(\dfrac{30}{x^3} - \dfrac{2}{x^2} - \dfrac{1}{x}\right) dx$ **20.** $\int -6e^x\, dx$

21. $\int (x - e^x)\, dx$ **22.** $\int \sqrt[3]{x}\, dx$ **23.** $\int (\sqrt[3]{x} - 4\sqrt{x})\, dx$

24. $\int \dfrac{1}{\sqrt{x^3}}\, dx$ **25.** $\int \dfrac{20}{\sqrt[4]{x^3}}\, dx$ **26.** $\int \dfrac{5 - x}{\sqrt{x}}\, dx$

27. $\int \dfrac{x}{\sqrt[3]{x^2}}\, dx$ **28.** $\int \dfrac{1 - \sqrt{x}}{x}\, dx$ **29.** $\int (2x + 1)^2\, dx$

(*Hint*: Use the law of exponents to rewrite.)   (*Hint*: Expand first.)

30. $\int 3(2 - x)^2\, dx$

In exercises 31 through 36, find the specific antiderivative, f(x), of the function with given initial condition.
(See Example 11.)

31. $f'(x) = 3x^2 - 6x + 1,\ f(0) = -10$ **32.** $f'(x) = 3x^2 - 6x + 1,\ f(2) = 15$

33. $f'(x) = 8x^3 - x^2 + x,\ f(0) = 10$ **34.** $f'(x) = 8x^3 - x^2 + x,\ f(-1) = 0$

35. $f'(x) = \dfrac{3}{x} - 5,\ f(1) = 0$ **36.** $f'(x) = 5\sqrt[3]{x^2} + 6x - 4,\ f(1) = 4$

37. The **family** of functions $\frac{1}{2}x^2 + C$ represents the collection of antiderivatives of $f'(x) = x$. In particular, three of these functions (for $C = 2, 0, -4$) are

$f(x) = \frac{1}{2}x^2 + 2$

$f(x) = \frac{1}{2}x^2$

$f(x) = \frac{1}{2}x^2 - 4$

and are graphed in the accompanying figure.

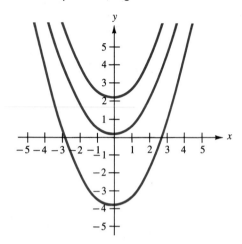

a. Show algebraically that each curve has the same slope of the tangent line drawn to it for $x = 2$. What is the slope of the tangent line?

b. Draw the tangent lines to each of the curves at $x = 2$.

38. From Chapter 5, if $f(x) = b^x$, then $f'(x) = b^x \ln b$. The corresponding integral form is

$$\int b^x \ln b \, dx = \ln b \int b^x \, dx = b^x + C$$

or, dividing by $\ln b$,

$$\int b^x \, dx = \frac{b^x}{\ln b} + C$$

Evaluate the following integrals:

a. $\displaystyle\int 2^x \, dx$

b. $\displaystyle\int (x^3 + 3^x) \, dx$

SET B

BUSINESS
COST

39. The marginal cost for a new product is given by

$$C'(x) = 2x - 6$$

a. Find $C(x)$ if the fixed cost $C(0) = 400$.

b. Find the cost of producing 50 items.

COST FUNCTION

40. The Rainbow Paint Company has determined that the marginal cost function for producing a barrel of a particular type of acrylic paint is given by

$$C'(x) = \sqrt[3]{x^2} + 3$$

If the fixed cost is $2000, find the cost of producing 1000 barrels of acrylic paint.

COST FUNCTION

41. The Edison Electronics Company's market research department has determined that the marginal cost for producing x new 64-bit processors is

$$C'(x) = \frac{600}{\sqrt[3]{x}} + 0.1x$$

If the fixed cost is $20,000, find the cost of producing 1000 processors.

PROFIT

42. The marginal profit of a company is

$$P'(x) = 80x - 4$$

where x is the number of units sold. If $P(10) = 0$, find the profit function $P(x)$.

SALES

43. The rate of change in sales for a product is given by

$$S'(t) = 0.8t$$

where S is measured in millions and t is measured in months. Find the sales function, $S(t)$, if, after 5 months, the sales figure is 10 million.

DEMAND

44. Demand for a particular product is changing as follows:

$$\frac{dx}{dp} = \frac{-500}{p^2}$$

Find the demand as a function of the price if, whenever the price is $5, the demand is 300.

PROFIT

45. If the marginal profit for a particular item is given by

$$P'(x) = 3x^2 - 360x + 10,800$$

find the profit function given that the company is losing $200 when only 50 items are sold. (*Hint:* $P(50) = -200$.)

REVENUE

46. The marginal revenue for a product is given by

$$R'(x) = 10 - 2x$$

where x is measured in hundreds and the revenue is measured in thousands of dollars. Find the revenue function if the revenue from selling 400 items is $20,000.

BIOLOGY
MARINE LIFE

47. The population of a certain species of fish is declining due to the introduction of a large predator fish into its habitat. The rate of change of the population is given by

$$N'(t) = -1000t$$

where t is measured in months.
a. Find $N(t)$ if the initial population was 80,000.
b. How long will it take at this rate for the species to become extinct?

BACTERIA

48. Bacteria are growing at a rate approximated by

$$N'(t) = 400e^t$$

where t is measured in hours. Find $N(t)$ if there were 600 bacteria present initially.

SOCIAL SCIENCE
POPULATION

49. Calculator Exercise

Suppose that the population of Bethlehem, Connecticut, is known to grow at the rate $2 + 5t^{1/3}$, *where t is measured in months.*

a. Let $P'(t) = 2 + 5t^{1/3}$; find $P(t)$, the formula for finding the population at any time, t.

b. If the population now ($t = 0$) is 2600, find the population in 5 years ($t = 60$).

PHYSICAL
SCIENCE
VELOCITY

50. Suppose an object travels in such a way that its acceleration at any time, t, is given by

$$a(t) = 2t^2 + 2t$$

and its initial velocity, $v(0)$, is 5. Find $v(t)$. (*Hint:* $v(t) = \int a(t)\, dt$.)

POSITION OR DISTANCE

51. Suppose further that $s(0) = 0$ for the object in exercise 50. Find $s(t)$.

VELOCITY AND HEIGHT

52. Using the fact that the acceleration of a free-falling body is a constant 32 ft/sec/sec and is written as

$$a(t) = -32$$

derive the formulas for the velocity and height at any time, t, of an object dropped with an initial velocity of -36 ft/sec from the top of a 600-ft-tall building.

VELOCITY AND DISTANCE

53. a. Derive the formula for velocity and distance of a free-falling object dropped from a height of 160 ft.

   b. When will the object hit the ground?

   c. What is its velocity when it hits the ground?

VELOCITY AND DISTANCE

54. a. Derive the formulas for velocity and distance of an object thrown *upward* with an initial velocity of 80 ft/sec from a height of 96 ft.

   b. When will the object hit the ground?

   c. What is the object's maximum obtained height?

   d. What is the object's velocity when it hits the ground?

WRITING AND CRITICAL THINKING

Explain the role of C, the constant of integration, when performing the process of indefinite integration.

USING TECHNOLOGY IN CALCULUS

In this section, you have learned that each of the following functions is an antiderivative of $f(x) = x^2$:

$$y = \frac{x^3}{3} + 1 \qquad y = \frac{x^3}{3} - 2 \qquad y = \frac{x^3}{3} + 4$$

How is this fact interpreted graphically? Consider the family of functions as drawn by *DERIVE*®:

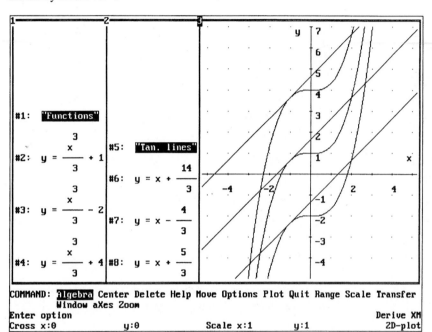

Notice for a given x value, say -1, slopes of lines drawn tangent to these curves are all equal; the tangent lines are parallel.

1. Verify the equations of the three tangent lines—$y = x + \frac{14}{3}$, $y = x - \frac{4}{3}$, and $y = x + \frac{5}{3}$—in the given graphic.
2. Use a graphics software package or a graphics calculator to graph three antiderivatives of the function $f(x) = x^2 - 2x^3$.
3. Find the equation of the three lines drawn tangent to your answers in exercise 2 at $x = 1$.

6.2   INTEGRATION BY SUBSTITUTION

In this section, we expand the types of functions we can integrate by reexamining the Chain Rule, first seen in Chapter 3. The idea of integration as the "inverse" of differentiation is fundamental in this discussion.

First, recall the Chain Rule.

If $f(x) = g[u(x)]$, where u is differentiable at x and g is differentiable at $u(x)$, then

$$f'(x) = g'[u(x)] \cdot u'(x)$$

The Chain Rule was used to generate the Generalized Power Rule (from Chapter 3) and the generalized rules for differentiating logarithmic and exponential functions (from Chapter 5), which are as follows:

1. $f(x) = [u(x)]^n \rightarrow f'(x) = n[u(x)]^{n-1} \cdot u'(x)$

   Example: $f(x) = (x^2 - 1)^5 \rightarrow f'(x) = 5(x^2 - 1)^4 \cdot 2x$

2. $f(x) = e^{u(x)} \rightarrow f'(x) = e^{u(x)} \cdot u'(x)$

   Example: $f(x) = e^{x^2} \rightarrow f'(x) = e^{x^2} \cdot 2x$

3. $f(x) = \ln(u(x)) \rightarrow f'(x) = \dfrac{1}{u(x)} \cdot u'(x)$

   Example: $f(x) = \ln(7x + 1) \rightarrow f'(x) = \dfrac{1}{7x + 1} \cdot 7$

Corresponding to each of these differentiation rules is an indefinite integral rule, as found in the following theorem:

Theorem 6.3

1. $\displaystyle\int [u(x)]^n \cdot u'(x)\, dx = \dfrac{[u(x)]^{n+1}}{n+1} + C, \qquad n \ne -1$

2. $\displaystyle\int e^{u(x)} u'(x)\, dx = e^{u(x)} + C$

3. $\displaystyle\int \dfrac{1}{u(x)} \cdot u'(x)\, dx = \ln |u(x)| + C$

Notice that part 1 of this theorem is just a generalized version of part 1 of Theorem 6.1 in the last section. For this reason, it can be referred to as a **Generalized Power Rule for Integrals.** To use it or any of the three parts, it is important to emphasize the *form* of each integrand: The *derivative* of the function $u(x)$, namely $u'(x)$, appears as a factor in each.

Example 1   Evaluate $\displaystyle\int 4x(2x^2 - 1)^7\, dx$.

Solution One method for evaluating this integral is to expand the integrand and integrate each term. However, we can do it much faster by using part 1 of Theorem 6.3. If we let $u(x) = 2x^2 - 1$, then $u'(x) = 4x$, so that by rearranging we get

$$\int 4x(2x^2 - 1)^7\, dx = \int \underbrace{(2x^2 - 1)^7} \cdot \underbrace{4x}\, dx$$

$$= \int [u(x)]^7 u'(x)\, dx \qquad (n = 7)$$

$$= \frac{[u(x)]^8}{8} + C$$

Now, by substituting $u(x) = 2x^2 - 1$, we have

$$\int 4x(2x^2 - 1)^7\, dx = \frac{(2x^2 - 1)^8}{8} + C$$ ∎

Example 2 Evaluate $\int (5x + 2)^3\, dx$.

Solution Let $u(x) = 5x + 2$; then $u'(x) = 5$. Notice that there is no 5 as a factor of the integrand to serve as $u'(x)$. We can alleviate this problem as follows:

$$\int (5x + 2)^3\, dx = \int (5x + 2)^3 \left(\frac{1}{5} \cdot 5\right) dx$$

These integrals are equal, since $\frac{1}{5} \cdot 5 = 1$. Now using part 3 of Theorem 6.1, which states that constants may be "moved across" the integral sign, we have

$$\int (5x + 2)^3\, dx = \frac{1}{5} \int \underbrace{(5x + 2)^3} \cdot \underbrace{5}\, dx$$

$$= \frac{1}{5} \int [u(x)]^3 u'(x)\, dx$$

$$= \frac{1}{5} \frac{[u(x)]^4}{4} + C$$

$$= \frac{1}{20} (5x + 2)^4 + C$$ ∎

It should be clear now why we choose to call this technique of integration a substitution technique. Some portion of the integrand is substituted with $u(x)$, a formula is then applied, and the result is obtained by a resubstitution for $u(x)$. To make this substitution process a little easier, we can simplify the notation somewhat by using a concept from Section 3.7. There, we introduced something called the *differential form* for a derivative.

$$\frac{dy}{dx} = f'(x) \quad \text{becomes} \quad dy = f'(x)\,dx$$

To use this idea, we replace $u(x)$ with u and

$$\frac{du}{dx} = u'(x) \quad \text{becomes} \quad du = u'(x)\,dx$$

Now, in each part of Theorem 6.3, we can replace $u(x)$ with u and $u'(x)\,dx$ with du. Using this to rewrite Theorem 6.3, and assuming, in each case, that u is a differentiable function of x, we have

Theorem 6.3
(Rewritten)

> **1.** $\displaystyle \int u^n\,du = \frac{u^{n+1}}{n+1} + C, \qquad n \neq -1$
>
> **2.** $\displaystyle \int e^u\,du = e^u + C$
>
> **3.** $\displaystyle \int \frac{1}{u}\,du = \ln|u| + C$

We use these new forms for the remaining examples.

Example 3   Evaluate $\displaystyle \int x^2\sqrt{x^3 + 1}\,dx$.

Solution   Let $u = x^3 + 1$; then $du = 3x^2\,dx$. We are missing a factor of 3 in the integrand, but we can multiply the integrand by 3 as long as we "compensate" with a $\frac{1}{3}$, which we move across the integral sign. We caution the reader that only constants (not variables) may be moved across the integral sign.

$$\int x^2\sqrt{x^3+1}\,dx = \frac{1}{3}\int 3x^2\,\sqrt{x^3+1}\,dx$$

$$= \frac{1}{3}\int \underbrace{(x^3+1)^{1/2}}\,\underbrace{3x^2\,dx}$$

$$= \frac{1}{3}\int u^{1/2} \qquad\quad du$$

$$= \frac{1}{3}\cdot\frac{u^{3/2}}{\frac{3}{2}} + C$$

$$= \frac{2}{9}(x^3+1)^{3/2} + C \qquad\qquad \blacksquare$$

In each of the previous three examples, $u(x)$ or u was chosen based on the fact that it was a function being raised to some constant power. In using part 2 of Theorem 6.3, it is important to note that we will now be choosing u as the exponent for a constant base e.

$$f(x) = u^n \leftarrow \text{constant} \qquad f(x) = e^u \leftarrow \text{variable}$$

$$\uparrow \qquad\qquad\qquad \uparrow$$

variable $\qquad\qquad$ constant

Example 4   Evaluate $\displaystyle\int e^{3x}\,dx$.

Solution   Let $u = 3x$, then $du = 3\,dx$.

$$\int e^{3x}\,dx = \frac{1}{3}\int e^{3x} \cdot 3\,dx$$

$$= \frac{1}{3}\int e^u\,du$$

$$= \frac{1}{3}e^u + C \qquad \text{Using part 2 of Theorem 6.3}$$

$$= \frac{1}{3}e^{3x} + C \qquad\qquad\qquad \blacksquare$$

Example 5   Evaluate $\displaystyle\int x^3 e^{x^4}\,dx$.

Solution $u = x^4$ and $du = 4x^3\,dx$

$$\int x^3 e^{x^4}\,dx = \frac{1}{4}\int e^{x^4} \cdot 4x^3\,dx$$

$$= \frac{1}{4}\int e^u\,du$$

$$= \frac{1}{4}e^{x^4} + C \qquad\qquad\qquad \blacksquare$$

Example 6   Evaluate $\displaystyle\int \frac{x}{1 - x^2}\,dx$.

Solution   Using part 3 of Theorem 6.3 with $u = 1 - x^2$ and $du = -2x\,dx$, we can multiply by -2 on the "inside" and by $-\frac{1}{2}$ on the "outside" to obtain

$$\int \frac{x}{1-x^2}\,dx = -\frac{1}{2}\int \frac{1}{1-x^2}(-2x)\,dx$$

$$= -\frac{1}{2}\int \frac{1}{u}\,du$$

$$= -\frac{1}{2}\ln|u| + C$$

$$= -\frac{1}{2}\ln|1-x^2| + C \qquad\blacksquare$$

Example 7

Evaluate $\displaystyle\int \frac{3x^3}{\sqrt{x^4-1}}\,dx$.

Solution Let $u = x^4 - 1$ and $du = 4x^3\,dx$. We have an extra 3 we don't need, and we are missing a 4 we do need. Move the 3 to the outside and multiply by 4 on the inside and $\frac{1}{4}$ on the outside.

$$\int \frac{3x^3}{\sqrt{x^4-1}}\,dx = 3\cdot\frac{1}{4}\int (x^4-1)^{-1/2}\cdot 4x^3\,dx$$

$$= \frac{3}{4}\int u^{-1/2}\,du$$

$$= \frac{3}{4}\cdot\frac{u^{1/2}}{\frac{1}{2}} + C$$

$$= \frac{3}{2}(x^4-1)^{1/2} + C \quad\text{or}\quad \frac{3}{2}\sqrt{x^4-1} + C \qquad\blacksquare$$

As we mentioned in the last section, all of the answers to these integration examples can be checked using differentiation. (See exercises 45 through 49.) Example 8 reemphasizes the fact that for this method or any method of integration, the constant of integration can be determined if enough information is given.

Example 8

A certain piece of jewelry originally sold for $900 and has been increasing in value at the rate of $V'(t)$ dollars per year, where

$$V'(t) = 80e^{0.1t}$$

At this rate, what will the piece of jewelry be worth in 20 years?

Solution To find $V(t)$, we need to integrate $V'(t)$.

$$V(t) = \int V'(t)\, dt$$

$$= \int 80e^{0.1t}\, dt$$

$u = 0.1t$
$du = 0.1\, dt \quad \text{or} \quad \frac{1}{10}\, dt$

$$= 80 \cdot 10 \int e^{0.1t} \cdot \tfrac{1}{10}\, dt$$

$$= 800 \int e^u\, du$$

$$= 800e^u + C = 800e^{0.1t} + C$$

At $t = 0$, $V(t) = 900$, so

$$900 = 800e^{0.1(0)} + C$$
$$900 = 800 + C \quad \text{or} \quad C = 100$$

So $V(t) = 800e^{0.1t} + 100$ and, after 20 years,

$$V(20) = 800e^{0.1(20)} + 100$$
$$\approx 800(7.3891) + 100$$
$$\approx 6011.28$$

The piece of jewelry will be worth approximately \$6011.28 in 20 years. ■

6.2 EXERCISES

SET A

In exercises 1 through 44, evaluate the indicated indefinite integral. (See Examples 1–7.)

1. $\displaystyle\int (x + 3)^5\, dx$

2. $\displaystyle\int (3 - x)^5\, dx$

3. $\displaystyle\int (4x - 7)^3\, dx$

4. $\displaystyle\int (\tfrac{1}{2}x + 3)^4\, dx$

5. $\displaystyle\int 3x^2(x^3 + 1)^5\, dx$

6. $\displaystyle\int 4x^3\sqrt{x^4 - 1}\, dx$

7. $\displaystyle\int 5x^4\sqrt{x^5 - 1}\, dx$

8. $\displaystyle\int 6x^5\sqrt{x^6 - 1}\, dx$

9. $\displaystyle\int 3y^2(y^3 - 1)^{1/3}\, dy$

10. $\displaystyle\int 3y^2(y^3 - 4)^{-2/3}\, dy$

11. $\displaystyle\int \frac{x^2}{\sqrt{x^3 - 1}}\, dx$

12. $\displaystyle\int \frac{x^2}{(x^3 - 1)^2}\, dx$

13. $\displaystyle\int \frac{x^2}{x^3 - 1}\, dx$

14. $\displaystyle\int xe^{x^2}\, dx$

15. $\displaystyle\int x^2 e^{x^3}\, dx$

16. $\displaystyle\int 3x^4 e^{2x^5}\, dx$

17. $\displaystyle\int \frac{t}{3t^2 - 1}\, dt$

18. $\displaystyle\int \frac{z}{1 - 3z^2}\, dz$

19. $\displaystyle\int \frac{x}{x^2 - 4}\, dx$

20. $\displaystyle\int \frac{dx}{(x - 1)^4}$

21. $\displaystyle\int xe^{-x^2}\, dx$

22. $\displaystyle\int \frac{x}{e^{x^2}}\, dx$

23. $\displaystyle\int 5\sqrt{2x - 1}\, dx$

24. $\displaystyle\int x\sqrt{x^2 - 1}\, dx$

25. $\displaystyle\int \frac{3x^2 - 6x}{x^3 - 3x^2 + 5}\, dx$

26. $\displaystyle\int \frac{x^2 - 2x}{(x^3 - 3x^2 + 5)^3}\, dx$

27. $\displaystyle\int \frac{x^2 - \frac{10}{3}x + \frac{7}{3}}{x^3 - 5x^2 + 7x - 1}\, dx$

28. $\displaystyle\int \frac{x^2 - \frac{10}{3}x + \frac{7}{3}}{(x^3 - 5x^2 + 7x - 1)^2}\, dx$

29. $\displaystyle\int xe^{-0.2x^2}\, dx$

30. $\displaystyle\int e^{-0.2x}\, dx$

31. $\displaystyle\int (1 + 0.5x)^4\, dx$

32. $\displaystyle\int \frac{1}{e^{2x}}\, dx$

33. $\displaystyle\int \frac{e^{\sqrt{x}}}{\sqrt{x}}\, dx$

34. $\displaystyle\int \frac{dx}{\sqrt{x} + x}$

(*Hint:* $\sqrt{x} + x = \sqrt{x}(1 + \sqrt{x})$.)

35. $\displaystyle\int (x^3 - 5x)^{2/3}(3x^2 - 5)\, dx$

36. $\displaystyle\int \frac{2x + 7}{x^2 + 7x + 12}\, dx$

37. $\displaystyle\int \frac{6x + 21}{x^2 + 7x + 12}\, dx$

38. $\displaystyle\int \frac{2\, dx}{\sqrt{x}(1 - \sqrt{x})}$

39. $\displaystyle\int \frac{dx}{\sqrt{x}(1 - \sqrt{x})^3}$

40. $\displaystyle\int \frac{dx}{\sqrt{x}(9 - \sqrt{x})^2}$

41. $\displaystyle\int \frac{\ln x}{x}\, dx$

42. $\displaystyle\int \frac{\ln x^2}{x}\, dx$

43. $\displaystyle\int \frac{\ln 2x}{x}\, dx$

44. $\displaystyle\int \ln e^x\, dx$

45. Verify the result of Example 1 by finding $\dfrac{dy}{dx}$ if $y = \dfrac{1}{8}(2x^2 - 1)^8 + C$.

46. Verify the result of Example 2 by finding $\dfrac{dy}{dx}$ if $y = \dfrac{1}{20}(5x + 2)^4 + C$.

47. Verify the result of Example 3 by finding $\dfrac{dy}{dx}$ if $y = \dfrac{2}{9}(x^3 + 1)^{3/2} + C$.

48. Verify the result of Example 5 by finding $\dfrac{dy}{dx}$ if $y = \dfrac{1}{4}e^{x^4} + C$.

49. Verify the result of Example 6 by finding $\dfrac{dy}{dx}$ if $y = \dfrac{-1}{2}\ln |1 - x^2| + C$.

SET B

50. For the integral

$$\int x(x - 2)^7\, dx$$

let $u = x - 2$, $du = dx$, and $x = u + 2$ to obtain

$$\int (u + 2)u^7 \, du$$

Expand the integrand and evaluate.

51. Use exercise 50 as a guide for evaluating each of the following integrals using the substitution technique:

a. $\int x\sqrt{x - 2} \, dx$

b. $\int \dfrac{x}{\sqrt{x + 2}} \, dx$

c. $\int \dfrac{x}{x + 2} \, dx$

d. $\int \dfrac{x}{(x - 2)^4} \, dx$

BUSINESS
PROFIT FUNCTION

52. The marginal profits of a company are given by the equation

$$P'(x) = \frac{2x}{\sqrt{2x^2 - 400}}$$

If the company breaks even $(P(x) = 0)$ when $x = 100$ units are sold, find the specific profit function, $P(x)$.

COST FUNCTION

53. The Sunshine Greeting Card Company has determined that the marginal cost for producing x greeting cards per month is given by

$$C'(x) = \frac{x}{\sqrt{x^2 + 90,000}}$$

If the fixed cost per month is $3500, find the cost of producing 400 greeting cards.

PROFIT FUNCTION

54. The marginal profit for a company is given by

$$P'(x) = \frac{4}{\sqrt[3]{(12x - 200)^2}}$$

where x is the number of units sold and $P'(x)$ is measured in hundreds of dollars. If $P(100) = 0$, which means that the company breaks even at 100 units, find the profit function, $P(x)$.

DEPRECIATION

55. The value of a certain piece of manufacturing equipment is decreasing at a rate given by

$$V'(t) = -560te^{-0.2t^2}$$

where t is measured in years. If the piece of equipment originally sold for $4200, how much will it be worth after 3 years?

BIOLOGY
ZOOLOGY

56. Scientists have projected that the alligator population of a certain region in Florida is increasing at a rate described by the function

$$G'(t) = 32e^{0.25t}$$

where t is measured in years.
a. Find $G(t)$ if $G(t) = 80$ when $t = 0$.
b. Predict how many alligators there will be in 4 years. (Assume that $t = 0$ gives the number of alligators present initially.)

PHYSICAL
SCIENCE

RADIOACTIVE DECAY

57. The rate at which a certain radioactive substance decays is given by

$$R'(t) = -500e^{-0.4t}$$

where t is measured in days. If there were 2000 g of the substance originally, how many grams will there be after 10 days?

WRITING AND CRITICAL THINKING

Given the integral $\int x(2x^2 + 5)^2\,dx$, there are two ways we can perform the integration. Write out in words how the two approaches work.

6.3   DEFINITE INTEGRALS, AREA, AND THE FUNDAMENTAL THEOREM OF CALCULUS

Two of the most important problems in the history of mathematics were

1. Finding the slope of the tangent to a curve
2. Finding the area under a curve

Both of these problems are of a geometrical nature and were solved using calculus, but the ideas involved in their development have had a far-reaching impact on applications in many areas other than mathematics.

Historically, mathematicians attempted to find the area of a plane region by dividing the region into sections of known area. For example, to calculate the area of a circle, the circle could be divided into triangular sections, as shown in Figure 6.2. Because the formula for the area of a triangle was known at that time, the area of the circle was approximated by adding up the areas of triangles. To get a better approximation, more triangles could be used, as in Figure 6.3.

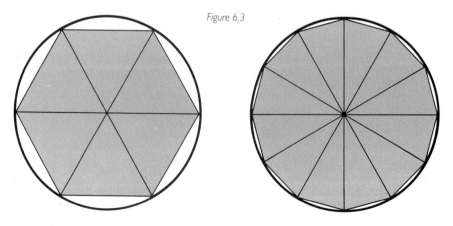

Figure 6.2

Figure 6.3

We can use a similar idea to generate a method for finding the area under a curve, $y = f(x)$, by dividing the region into rectangles and adding up the areas of the rectangles, as follows. Let the function f be nonnegative on the closed interval $[a, b]$. We divide the interval from a to b into n equal subintervals using the numbers x_0, $x_1, \ldots, x_n$ with $a = x_0$ and $b = x_n$, as shown in Figure 6.4. Because each subinterval is of equal length, we denote its length by Δx. Using the right-hand endpoint in each case, we construct rectangles with heights $f(x_1), f(x_2), \ldots, f(x_n)$ and width Δx (see Figure 6.5).

Figure 6.4

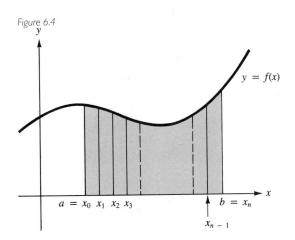

Figure 6.5

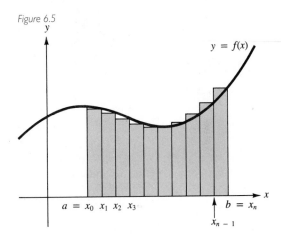

The area of each rectangle is the height times the width, so that the sum of the areas of all n rectangles is

$$
\underset{\substack{\uparrow \\ \text{height of first} \\ \text{rectangle}}}{\overset{\substack{\text{width} \\ \downarrow}}{f(x_1)\, \Delta x}}
\quad + \quad
\underset{\substack{\uparrow \\ \text{height of second} \\ \text{rectangle}}}{\overset{\substack{\text{width} \\ \downarrow}}{f(x_2)\, \Delta x}}
\quad + \cdots + \quad
\underset{\substack{\uparrow \\ \text{height of } n\text{th} \\ \text{rectangle}}}{\overset{\substack{\text{width} \\ \downarrow}}{f(x_n)\, \Delta x}}
$$

This sum can be used to approximate the area under the curve from a to b or, symbolically,

$$\text{area} \approx [\, f(x_1) + f(x_2) + \cdots + f(x_n)\,]\, \Delta x$$

Example 1 Approximate the area under the curve $y = x^2 + 1$ from $x = 0$ to $x = 2$ using four rectangles of width $\Delta x = \frac{1}{2}$ and using the right-hand endpoints to calculate the height of each rectangle.

Solution First we sketch a diagram to help in our calculation (see Figure 6.6).

Figure 6.6

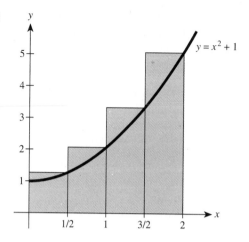

$$\text{area} \approx [f(\tfrac{1}{2}) + f(1) + f(\tfrac{3}{2}) + f(2)](\tfrac{1}{2})$$
$$= [\tfrac{5}{4} + 2 + \tfrac{13}{4} + 5](\tfrac{1}{2})$$
$$= \tfrac{23}{4}$$

The area under the curve is approximately $\tfrac{23}{4}$ square units.   ■

Let's return to our approximation for area under a curve given by

$$\text{area} \approx [f(x_1) + f(x_2) + \cdots + f(x_n)] \, \Delta x$$

and explore what happens if we change the number of rectangles, n, that we use. For small values of n, the expression on the right may not give a very good approximation to the actual area represented. If, however, we increase the value of n, which in turn increases the number of rectangles and decreases their width, we can obtain a better approximation. In fact, if we take the limit as n approaches infinity, we get the exact value for the area under the curve. This limit is called the **definite integral** and is denoted by

$$\int_a^b f(x) \, dx$$

where a is called the **lower limit** of integration and b is called the **upper limit** of integration. From this we can determine the exact area under a curve.

Area under a curve

If f is continuous and nonnegative for all x in the interval $[a, b]$, then the area bounded by $f(x)$ and the x axis from a to b is given by

$$\text{area} = \int_a^b f(x) \, dx$$

In the preceding discussion, we chose our subintervals so the width of each rectangle was the same. We also used the right-hand endpoints to construct the height. Both of these choices were arbitrary and were used to make the process easier. It is not necessary to make these particular choices; widths of subintervals may vary, and any point in the subinterval can be used to determine the height. For a more complete discussion of the theory related to the definite integral, the reader should consult the most recent edition of *Calculus* by James Stewart (Brooks/Cole).

The question now arises as to how we use this information to actually calculate the area under a curve. Must we divide the area into sections, add up the pieces, and take the limit? The answer is no, provided the function is continuous on the interval and has an antiderivative. This idea represents one of the most important results in the study of calculus and is stated in the **Fundamental Theorem of Calculus.**

Theorem 6.4 *The Fundamental* *Theorem of Calculus*	If f is continuous over the closed interval $[a, b]$ and F is an antiderivative of f, then $$\int_a^b f(x)\, dx = F(x) \Big]_a^b = F(b) - F(a)$$

Notice that this theorem links the definite integral (thus, the area under a curve) to the process of finding and evaluating an antiderivative; that is, it establishes a relationship between the processes of differentiation and integration. At the end of this section, we present an informal discussion of how this relationship arises.

Example 2   Find the area of the region bounded by $y = x^2$, $x = 1$, $x = 3$, and the x axis.

Solution   We first graph the function and outline the desired area in Figure 6.7.

Figure 6.7

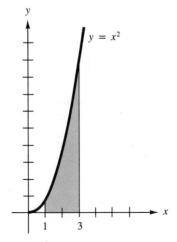

The area is the definite integral of $f(x)$ from 1 to 3.

$$\text{area} = \int_1^3 x^2\, dx = \frac{x^3}{3}\Bigg]_1^3$$

$$= \frac{3^3}{3} - \frac{1}{3} = \frac{26}{3}$$

Thus, the area is $\frac{26}{3}$ square units.   ■

Example 3   Find the area of the region bounded by $f(x) = -x^2 + 8x - 12$, $x = 3$, $x = 6$, and the x axis.

Solution   See Figure 6.8. We must find

$$\int_3^6 (-x^2 + 8x - 12)\, dx = \frac{-x^3}{3} + 4x^2 - 12x\Bigg]_3^6$$

$$= (-72 + 144 - 72) - (-9 + 36 - 36)$$

$$= 9$$

The area is 9 square units.

Figure 6.8

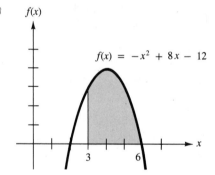

In the next chapter, we describe methods for computing approximations to areas for functions that do not have an antiderivative on the given interval. We would now like, however, to turn our attention away from the area problem to another application of the definite integral.

Example 4   In Example 13 of Section 6.1, we saw that with a marginal cost function $C'(x) = 10 + 0.1x$ and a fixed cost of $3000, the cost function is

$$C(x) = 10x + 0.05x^2 + 3000 \qquad \text{(where } x \text{ is measured in hundreds)}$$

What is the increase in the cost of production if the number of units produced is raised from 1000 ($x = 10$) to 2000 ($x = 20$)?

Solution $C(x) = \int C'(x)\, dx$

The increase in cost from $x = 10$ to $x = 20$ is $C(20) - C(10)$. In other words, we must find the difference in values of $\int (10 + 0.1x)\, dx$ when $x = 10$ and $x = 20$. We denote $C(20) - C(10)$ directly by using the notation

$$C(20) - C(10) = \int_{10}^{20} (10 + 0.1x)\, dx = C(x)]_{10}^{20}$$

With $C(x) = 10x + 0.05x^2 + 3000$, we have

$$C(20) - C(10) = \int_{10}^{20} (10 + 0.1x)\, dx = [10x + 0.05x^2 + 3000]_{10}^{20}$$
$$= 3220 - 3105$$
$$= 115$$

Therefore, the increased cost in raising the level of production from 1000 units to 2000 units is $115. ∎

We can see from this example that the definite integral can be used to find the **total cost** for raising the production level from 1000 to 2000 units. In the next example, we find a similar use of the definite integral for finding **total distance** in a physics application. Total distance is defined by $\int |v(t)|\, dt$. See also exercise 70.

Example 5   If the velocity (in feet per second) of a free-falling object is given by $v(t) = -32t + 80$, where t is measured in seconds, find the total distance (in feet) traveled from $t = 1$ to $t = 2$.

Solution   Total distance $= \int |v(t)|\, dt$

Since $-32t + 80$ is always positive from $t = 1$ to $t = 2$, we have

$$\text{total distance} = \int_1^2 |-32t + 80|\, dt$$
$$= \int_1^2 (-32t + 80)\, dt$$
$$= -16t^2 + 80t]_1^2$$
$$= [-16(2)^2 + 80(2)] - [-16(1)^2 + 80(1)]$$
$$= 96 - 64$$
$$= 32 \text{ ft} \qquad ∎$$

The next three examples provide practice in evaluating the definite integrals and learning how to deal with the limits of integration in the case where an integration by substitution method is used.

Example 6 Evaluate $\int_1^4 (3x^2 - 2x + 3)\, dx$.

Solution $\int_1^4 (3x^2 - 2x + 3)\, dx = [x^3 - x^2 + 3x]_1^4$

$$= [4^3 - 4^2 + 3(4)] - [1^3 - 1^2 + 3(1)]$$
$$= 57 \qquad\blacksquare$$

Example 7 Evaluate $\int_1^2 \dfrac{4}{x}\, dx$.

Solution $\int_1^2 \dfrac{4}{x}\, dx = 4 \ln x]_1^2$

$$= 4 \ln 2 - 4 \ln 1$$
$$\approx 2.7726 - 0$$
$$= 2.7726 \qquad\blacksquare$$

Example 8 Evaluate $\int_0^1 (2x + 1)^4\, dx$.

Solution Using a "u substitution," we have

$$\int_0^1 (2x + 1)^4\, dx = \frac{1}{2}\int_0^1 (2x + 1)^4 \cdot 2\, dx$$

$$= \frac{1}{2}\int u^4\, du \qquad \begin{aligned} u &= 2x + 1 \\ du &= 2\, dx \end{aligned}$$

Notice that we have left off the limits of integration. We have two ways to proceed:

1. Change the limits of integration for the new variable, u (original limits are for x only!).
2. Leave off limits of integration until we resubstitute the original variable, x, and use original limits.

Method 1	Method 2

Method 1

Changing limits:

$x = 0$	$x = 1$
$u = 2x + 1$	$u = 2x + 1$
$u = 2 \cdot 0 + 1$	$u = 2 \cdot 1 + 1$
$u = 1$	$u = 3$

We now have

$$\frac{1}{2} \int_1^3 u^4 \, du = \frac{1}{2} \cdot \frac{u^5}{5} \Big]_1^3$$

$$= \frac{1}{10} [3^5 - 1^5]$$

$$= \frac{121}{5}$$

Method 2

$$\frac{1}{2} \int u^4 \, du = \frac{1}{2} \cdot \frac{u^5}{5}, \text{ which gives}$$

$$\int_0^1 (2x + 1)^4 \, dx = \frac{1}{10} (2x + 1)^5 \Big]_0^1$$

$$= \frac{1}{10} [3^5 - 1^5]$$

$$= \frac{121}{5}$$

Of course, the results are the same, so you may use either method.   ∎

Properties of the definite integral may prove useful for evaluating various definite integrals. They are listed in the following theorem. (Assume for each statement that f and g are continuous on the interval $[a, b]$ and that k represents any real number.)

Theorem 6.5

1. $\displaystyle \int_a^b kf(x) \, dx = k \int_a^b f(x) \, dx$

2. $\displaystyle \int_a^b [f(x) \pm g(x)] \, dx = \int_a^b f(x) \, dx \pm \int_a^b g(x) \, dx$

3. $\displaystyle \int_a^a f(x) \, dx = 0$

4. $\displaystyle \int_a^b f(x) \, dx = -\int_b^a f(x) \, dx$

The first two parts of the theorem are similar to parts 3 and 4 of Theorem 6.1. The third part is self-explanatory. Notice that the fourth part indicates that the limits of integration are interchangeable, but only with an accompanying change in sign.

Returning to the idea of area, the next example involves the area of a region that lies entirely below the x axis. If we use the definite integral $\int_a^b f(x) \, dx$, the result will be negative. In this case, the area can be computed by taking the absolute value of the definite integral.

Example 9   Find the area of the region bounded by $f(x) = x^3$, $x = -3$, $x = -1$, and the x axis.

Solution   The region is pictured in Figure 6.9.

Figure 6.9

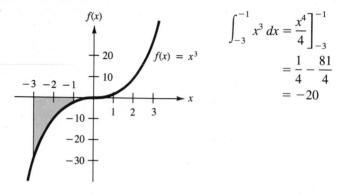

$$\int_{-3}^{-1} x^3\, dx = \frac{x^4}{4}\Bigg]_{-3}^{-1}$$

$$= \frac{1}{4} - \frac{81}{4}$$

$$= -20$$

Since $|-20| = 20$, the area of the region is 20 square units. (See also exercises 49 and 50.) ∎

We conclude this section by giving, as promised, an outline of the relationship of the area under a curve and the antiderivative. We begin by introducing an **area function**, $A(x)$. Think of x as a variable that "moves" from point a to point b; $A(x)$ is the area under the curve $y = f(x)$ between a and x. Thus,

$A(a) = 0$ (there is no area under a *point*)
$A(b) =$ the desired area   (see Figure 6.10)

Figure 6.10

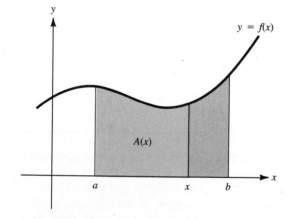

Now we examine a small section of the area between x and $x + h$ where h is a small number. The area of this region is "almost" rectangular, and we, in fact, use a rectangle to approximate its area (see Figure 6.11).

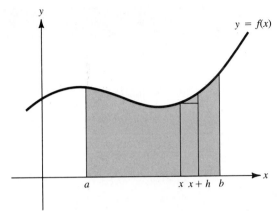

Figure 6.11

Shaded area is $A(x + h) - A(x)$

The area of this region is the area from a to $x + h$ minus the area from a to x or, symbolically,

$$A(x + h) - A(x)$$

Using the rectangle with height $f(x)$ and width h, we can write

$$A(x + h) - A(x) \approx hf(x)$$

Dividing both sides by h gives

$$\frac{A(x + h) - A(x)}{h} \approx f(x)$$

The significance of this is that $\dfrac{A(x + h) - A(x)}{h}$ approaches $f(x)$ as $h \to 0$, but recall from the definition of derivative in Chapter 2 that

$$\lim_{h \to 0} \frac{A(x + h) - A(x)}{h} = A'(x)$$

Thus, we can conclude that A *is an antiderivative of f:*

$$A'(x) = f(x)$$

Furthermore, if we *evaluate*

$$
\begin{aligned}
\int_a^b f(x)\, dx &= \int_a^b A'(x)\, dx \\
&= A(x)]_a^b \\
&= A(b) - A(a) \\
&= A(b) - 0 \\
&= \text{the desired area of the region bounded by } y = f(x), \\
&\quad\ x = a, x = b, \text{ and the } x \text{ axis, provided } f(x) \geq 0
\end{aligned}
$$

6.3 EXERCISES

SET A

In exercises 1 through 4, approximate the area under the curve over the given interval using four rectangles of width $\Delta x = \dfrac{1}{2}$ and using right-hand endpoints to calculate the height of each rectangle. (See Example 1.)

1. $y = x^3 + 1, x = 0$ to $x = 2$

2. $f(x) = \dfrac{1}{2}x + 1, x = 0$ to $x = 2$

3. $f(x) = 2x^2, x = 1$ to $x = 3$

4. $y = x^2 + 2, x = 1$ to $x = 3$

In exercises 5 through 40, evaluate each definite integral. (See Examples 6, 7, and 8.)

5. $\displaystyle\int_1^4 (2x + 3)\, dx$

6. $\displaystyle\int_1^2 (2x + 3)\, dx$

7. $\displaystyle\int_0^4 3x^2\, dx$

8. $\displaystyle\int_0^3 3x^2\, dx$

9. $\displaystyle\int_0^2 3x^2\, dx$

10. $\displaystyle\int_3^5 (x^2 - 5x + 6)\, dx$

11. $\displaystyle\int_1^2 (6x^2 - 2x + 1)\, dx$

12. $\displaystyle\int_0^4 x(x^2 - 1)\, dx$

13. $\displaystyle\int_1^2 (3x^2 - 2x + 3)\, dx$

14. $\displaystyle\int_0^1 e^x\, dx$

15. $\displaystyle\int_0^1 xe^{x^2}\, dx$

16. $\displaystyle\int_1^2 \dfrac{1}{x}\, dx$

17. $\displaystyle\int_1^3 \dfrac{1}{x}\, dx$

18. $\displaystyle\int_0^1 x^3\, dx$

19. $\displaystyle\int_0^2 x^3\, dx$

20. $\displaystyle\int_1^2 (x^3 + x^2 - 1)\, dx$

21. $\displaystyle\int_{-2}^{-1} \dfrac{1}{x^2}\, dx$

22. $\displaystyle\int_4^9 \sqrt{x}\, dx$

23. $\displaystyle\int_1^4 \dfrac{1}{\sqrt{x}}\, dx$

24. $\displaystyle\int_0^1 \dfrac{1}{2x + 1}\, dx$

25. $\displaystyle\int_0^1 \dfrac{1}{(2x + 1)^2}\, dx$

26. $\displaystyle\int_0^1 e^{x+1}\, dx$

27. $\displaystyle\int_2^4 \dfrac{x}{x^2 - 1}\, dx$

28. $\displaystyle\int_1^2 \dfrac{\ln x}{x}\, dx$

29. $\displaystyle\int_{-2}^0 \sqrt{1 - 4x}\, dx$

30. $\displaystyle\int_{-2}^0 \dfrac{1}{\sqrt{1 - 4x}}\, dx$

31. $\displaystyle\int_0^2 x(x^2 - 1)^4\, dx$

32. $\displaystyle\int_2^3 \dfrac{x}{(x^2 - 1)^2}\, dx$

33. $\displaystyle\int_{-1}^1 (1 - 2x)^3\, dx$

34. $\displaystyle\int_{-1}^1 (2 - x)^4\, dx$

35. $\displaystyle\int_0^1 e^x(e^x + 1)\, dx$

36. $\displaystyle\int_0^1 e^x(e^x + 1)^4\, dx$

37. $\displaystyle\int_0^1 (\sqrt{x} - \sqrt[3]{x})\, dx$

38. $\displaystyle\int_0^1 (\sqrt{x} + \sqrt[3]{x})\, dx$

39. $\displaystyle\int_{-2}^1 x\sqrt{x + 3}\, dx$

40. $\displaystyle\int_0^7 \sqrt[3]{x + 1}\, dx$

In exercises 41 through 57, find the area of the region bounded by the given information and sketch a graph of the region. (See Examples 2, 3, and 9.)

41. $y = 4 - x^2$, $x = 0$, $x = 1$, and x axis

42. $y = x^3$, $x = 0$, $x = 2$, and x axis

43. $y = \dfrac{1}{x^2}$, $x = 1$, $x = 2$, and x axis

44. $y = 2\sqrt{x}$, $x = 1$, $x = 4$, and x axis

45. $y = 6 - x - x^2$ and x axis (*Hint:* You need to find x intercepts for the limits of integration.)

46. $y = 3x - x^2$ and x axis

47. $y = x^2 - 1$, $x = 0$, $x = 1$, and x axis

48. $y = x^2 - 1$, $x = -1$, $x = 0$, and x axis

49. $y = x^3$, $x = 1$, $x = 3$, and x axis (Compare with Example 9.)

50. $y = x^3$, $x = -1$, $x = 2$, and x axis (*Hint:* Part of the area is above the x axis and part is below the x axis. Use two integrals.)

51. $y = x^3$, $x = -1$, $x = 1$, and x axis

52. $y = e^x$, $x = 0$, $x = 1$, and x axis (Round answer to two decimal places.)

53. $y = e^{2x}$, $x = 0$, $x = 1$, and x axis (Round answer to two decimal places.)

54. $y = \dfrac{1}{x}$, $x = 1$, $x = 2$, and x axis (Round answer to two decimal places.)

55. $y = \dfrac{1}{x}$, $x = 1$, $x = 100$, and x axis (Round answer to two decimal places.)

56. $y = \dfrac{1}{x}$, $x = 1$, $x = e$, and x axis

57. $y = \dfrac{1}{x}$, $x = e$, $x = e^3$, and x axis

In exercises 58 and 59, find the area of the shaded region for the given piecewise function.

58. $f(x) = \begin{cases} x^2, & x \le 0 \\ x, & x > 0 \end{cases}$

59. $f(x) = \begin{cases} x^2 + 1, & x \le 1 \\ 3 - x, & x > 1 \end{cases}$

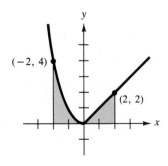

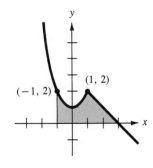

60.   A method has not been presented yet to integrate the natural logarithm function. However, we can approximate the area under this curve by using the idea of the areas of rectangles. Approximate the area bounded by $f(x) = \ln x$ and the x axis from $x = 1$ to $x = 5$ using the sum of the area of the four shaded rectangles. (*Hint:* The height of the first rectangle is $f(2) = \ln 2$.)

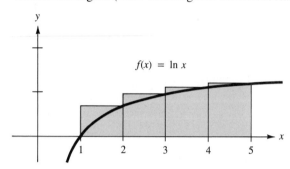

61.   Redo exercise 60 using twice as many rectangles as shown.

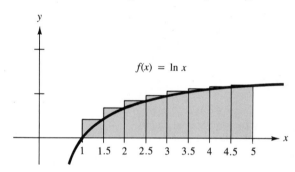

SET B

In exercises 62 through 70, use Examples 4 and 5 as a guide.

62.   The marginal cost function for producing x units of a certain product is

$$C'(x) = 5 + 0.01x$$

Find the total cost incurred by increasing the production level from 200 to 400 units.

63.   A company has determined that its profits are increasing at a rate given by the following marginal profit function

$$P'(x) = 100 + 200x - 12x^2$$

where x represents the number of units sold. Determine the total profit that would be generated by increasing the number of units sold from 10 to 16.

64.   From years of data, a watch manufacturer has determined that the life span of one of his watches is given by

$$f(x) = 0.25e^{-0.25x}$$

where x is measured in years. This function is called a **probability density function** and the graph of $y = f(x)$ is shown here. The definite integral $\int_a^b f(x)\,dx$ gives the probability that the life span of a randomly selected watch will be between a and b years. Determine the probability that the life span of a watch selected at random will be:

a.   Between 2 and 8 years   **b.**   Less than 2 years

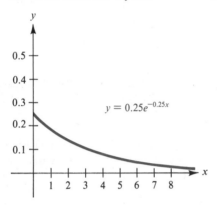

TOTAL SALES

65.   The rate of change for sales is given by

$$S'(x) = 8x + 3$$

where x is measured in months and S is measured in thousands of dollars. Find the total sales from the third month to the eighth month.

PRODUCTIVITY

66.   A company has installed some new equipment and the production level has increased. The rate per day of the number of parts assembled is given by

$$A'(n) = \frac{3}{2\sqrt{3n + 4}}$$

where n is measured in days and A is measured in hundreds of parts. Find the total number of parts assembled during the first week ($n = 0$ to $n = 7$).

BIOLOGY
DISEASE

67.   The rate at which a certain infectious disease spreads is given by

$$N'(t) = 10e^{0.1t}$$

where t is measured in days and N represents the number of people infected. Find the total number of people infected during the first ten days.

REFORESTATION

68.   A company is testing a new kind of fertilizer used to enhance the number of seedlings that survive during a reforestation project. The rate of survival after the application of the fertilizer is given by

$$S'(x) = 3x^2 + x$$

where x is measured in days and S gives the number of seedlings. Find the total number of seedlings that survive over the first two-week period.

69. The velocity of a projectile is given by

$$v(t) = -32t + 108$$

where $v(t)$ is measured in feet per second and t is measured in seconds. Find the distance in feet traveled from $t = 0$ to $t = 5$.

70. The **total distance** traveled by an object moving in a straight line from $t = a$ to $t = b$ is given by

$$\int_a^b |v(t)|\, dt$$

where $v(t)$ gives the velocity at any time t. Find the *total* distance traveled by an object moving in a straight line with

$$v(t) = 4t - t^2$$

where v is measured in feet per second and t is measured in seconds from $t = 1$ to $t = 5$. (*Hint:* This requires two separate integrals.)

WRITING AND CRITICAL THINKING

Explain the difference between indefinite integration and definite integration.

6.4 THE AREA BETWEEN TWO CURVES

In Section 6.3, we found the area of the region bounded above by the curve $y = f(x)$ and below by the x axis between $x = a$ and $x = b$ to be the definite integral

$$\int_a^b f(x)\, dx$$

We now wish to find the area between two curves $y = f(x)$ and $y = g(x)$ on the interval $[a, b]$. Assuming that both f and g are continuous and f lies entirely above g on this interval, we can compute the area between them by subtracting the area under g from the area under f (see Figure 6.12).

Figure 6.12

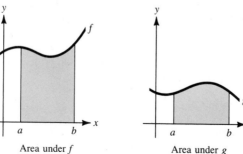

Area under f · Area under g

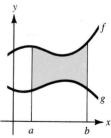

Area between f and g (area under f − area under g)

Both the area under f and the area under g between $x = a$ and $x = b$ can be computed using definite integrals. Thus, we have

$$\text{area between } f \text{ and } g = \int_a^b f(x)\, dx - \int_a^b g(x)\, dx$$

$$= \int_a^b [f(x) - g(x)]\, dx \qquad \text{Theorem 6.5, part 2}$$

This formula will work as long as $f(x) \geq g(x)$ for all x in the interval $[a, b]$. Variations of this formula can be used if such is not the case (see exercise 19). In some problems, it may be your job to figure out which curve is "on top" and which one is "on the bottom." (See Example 4 of this section.)

Example I   Find the area between $f(x) = x^2 + 4$ and $g(x) = 1 - x$ between $x = -2$ and $x = 1$.

Solution   Making a sketch of this region (see Figure 6.13), we have

$$\text{area} = \int_a^b [f(x) - g(x)]\, dx$$

$$= \int_{-2}^1 [(x^2 + 4) - (1 - x)]\, dx$$

$$= \int_{-2}^1 (x^2 + x + 3)\, dx$$

$$= \frac{x^3}{3} + \frac{x^2}{2} + 3x \Big]_{-2}^1$$

$$= \left(\frac{1}{3} + \frac{1}{2} + 3\right) - \left(\frac{-8}{3} + 2 - 6\right)$$

$$= \frac{21}{2}$$

The area between f and g is $\frac{21}{2}$ square units.

Figure 6.13

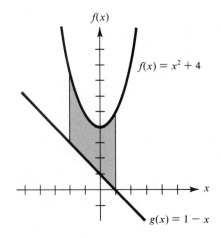

$f(x)$

$f(x) = x^2 + 4$

x

$g(x) = 1 - x$

Example 2   Find the area between $f(x) = x + 2$ and $g(x) = -x^2$ between $x = 0$ and $x = 2$.

Solution   Making a sketch (see Figure 6.14), we see that part of the area is below the x axis. The formula above will still work. (See exercise 25.)

$$
\begin{aligned}
\text{area} &= \int_0^2 [f(x) - g(x)]\, dx \\
&= \int_0^2 [(x + 2) - (-x^2)]\, dx \\
&= \int_0^2 (x^2 + x + 2)\, dx \\
&= \frac{x^3}{3} + \frac{x^2}{2} + 2x \Big]_0^2 = \left(\frac{8}{3} + 2 + 4\right) - (0 + 0 + 0) = \frac{26}{3}
\end{aligned}
$$

Figure 6.14

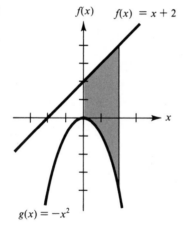

$f(x)$ $f(x) = x + 2$

$g(x) = -x^2$

■

Example 3   If the rate of change in the revenue for a certain product is given by $R'(x) = 2400 - x^2$ and the rate of change of cost of production is given by $C'(x) = 0.5x^2 + 1000$, where x is the number of units measured in hundreds, find the total profit generated by increasing the production level from 400 to 800 units.

Solution   Since, from Chapter 3, $P(x) = R(x) - C(x)$, we can say that

$$
\begin{aligned}
P(x) &= \int R'(x)\, dx - \int C'(x)\, dx \\
&= \int [R'(x) - C'(x)]\, dx
\end{aligned}
$$

This means that we need to find the area between the two curves $R'(x)$ and $C'(x)$ for $x = 4$ to $x = 8$.

$$P(x) = \int_4^8 [(2400 - x^2) - (0.5x^2 + 1000)] \, dx$$

$$= \int_4^8 (1400 - 1.5x^2) \, dx$$

$$= 1400x - 0.5x^3]_4^8$$

$$= (11{,}200 - 256) - (5600 - 32)$$

$$= 5376$$

The total profit is $5376 (see Figure 6.15).

Figure 6.15

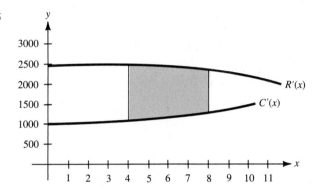

Sometimes it is necessary to find the area between two curves when no interval is given. In such a situation, the area between the two curves is defined to be the area of the region bounded above and below by the curves between the two points where the curves intersect. (The curves may have more than two points of intersection. See exercise 19.) To find the points of intersection that give the limits of integration, the equations for the curves must be solved simultaneously, as Example 4 illustrates.

Example 4 Find the area between the curves $y = x^2 - 2$ and $y = 2x + 1$.

Solution First we solve the equations simultaneously to find the points of intersection and then we make a sketch to see which curve is on top. Solving the equations simultaneously (by substitution) gives

$$x^2 - 2 = 2x + 1$$
$$x^2 - 2x - 3 = 0$$
$$(x - 3)(x + 1) = 0$$
$$x - 3 = 0 \quad \text{or} \quad x + 1 = 0$$
$$x = 3 \qquad\qquad x = -1$$

Figure 6.16 The curves intersect in the points $(-1, -1)$ and $(3, 7)$. Making a sketch, as in Figure 6.16, we see that $y = 2x + 1$ is on top.

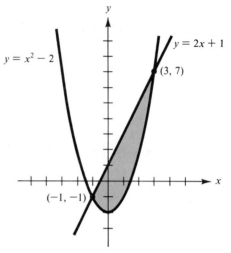

$$\text{area} = \int_{-1}^{3} [(2x + 1) - (x^2 - 2)] \, dx$$

$$= \int_{-1}^{3} (3 + 2x - x^2) \, dx$$

$$= 3x + x^2 - \frac{x^3}{3} \Big]_{-1}^{3}$$

$$= (9 + 9 - 9) - \left(-3 + 1 + \frac{1}{3}\right)$$

$$\text{area} = \frac{32}{3} \text{ square units} \qquad \blacksquare$$

One place in economics where the area between two curves is important is in the calculation of a quantity called **consumer surplus** and a quantity called **producer surplus.** First we define the point where supply and demand curves intersect to be the *equilibrium point* or *price*. The *x* coordinate gives the number of units and the *p* coordinate gives the price per unit (see Figure 6.17).

Figure 6.17

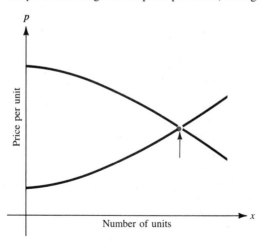

The *consumer surplus* is defined to be the total amount saved by consumers who were willing to pay a price higher than the equilibrium price. Geometrically, this savings is represented by the area under the demand curve and above the horizontal line through the equilibrium point (see Figure 6.18).

Figure 6.18

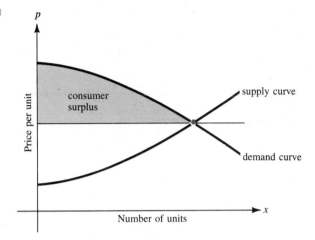

The *producer surplus*, represented geometrically by the area between the same horizontal line and the supply curve, is the increased profit realized by those producers who would have been willing to sell the product at a price lower than the equilibrium price (see Figure 6.19).

Figure 6.19

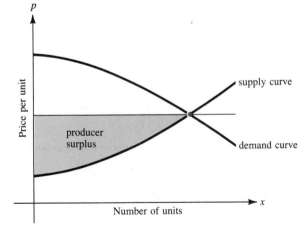

Example 5   Find the consumer surplus and the producer surplus given that the demand equation is $p = 500 - 10x$ and the supply equation is $p = 0.5x^2 + 100$ for a certain product.

Solution   First it is necessary to find the equilibrium point (price) by solving simultaneously

$$0.5x^2 + 100 = 500 - 10x$$
$$0.5x^2 + 10x - 400 = 0$$
$$x^2 + 20x - 800 = 0$$
$$(x + 40)(x - 20) = 0$$
$$x = -40 \quad \text{or} \quad x = 20$$

Because the number of units is positive, we have $x = 20$ and $p = 300$. For the *consumer surplus*, we need to find the area between $p = 500 - 10x$ and $p = 300$ (see Figure 6.20).

$$\text{consumer surplus} = \int_0^{20} [(500 - 10x) - 300]\, dx$$
$$= 200x - 5x^2]_0^{20}$$
$$= \$2000$$

Figure 6.20

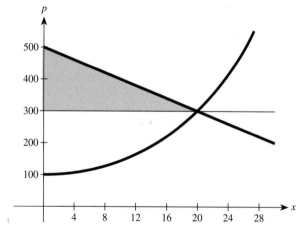

The area between $p = 300$ and $p = 0.5x^2 + 100$ gives the *producer surplus* (see Figure 6.21).

$$\text{producer surplus} = \int_0^{20} [300 - (0.5x^2 + 100)]\, dx$$
$$= 200x - \tfrac{1}{6}x^3]_0^{20}$$
$$\approx \$2667$$

Figure 6.21

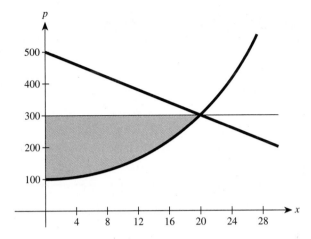

In Example 9 of Section 6.3, we found the area bounded by a curve that fell below the x axis. The integration of that example yielded a negative result and required us to take the absolute value to represent the area. We can now do this type of problem by using the concept of area between two curves if we think of the x axis as $y = 0$.

Example 6 Find the area between $y = x^2 - 3x$ and the x axis.

Solution $\text{area} = \displaystyle\int_0^3 [0 - (x^2 - 3x)]\, dx$

$= \displaystyle\int_0^3 (-x^2 + 3x)\, dx$

$= -\dfrac{x^3}{3} + \dfrac{3x^2}{2} \Bigg]_0^3$

$= (-9 + \tfrac{27}{2}) - 0$

$= \tfrac{9}{2}$ square units

See Figure 6.22.

Figure 6.22

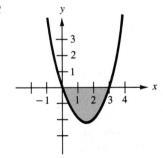

6.4 EXERCISES

SET A

In exercises 1 through 8, find the area of the region bounded above by f(x) and below by g(x) on the given interval. (See Examples 1 and 2.)

1. $f(x) = 3x$ and $g(x) = x$ between $x = 0$ and $x = 4$
2. $f(x) = 2x - x^2$ and $g(x) = -x$ between $x = 0$ and $x = 3$
3. $f(x) = 4 - x^2$ and $g(x) = x^2$ between $x = -1$ and $x = 1$
4. $f(x) = -x^2$ and $g(x) = x^2 - 4$ between $x = -1$ and $x = 1$
5. $f(x) = x^3 + 2$ and $g(x) = -x^2$ between $x = 0$ and $x = 2$
6. $f(x) = \sqrt{x}$ and $g(x) = x^2$ between $x = 0$ and $x = 1$
7. $f(x) = 1 - x^2$ and $g(x) = x^2 - 4x + 1$ between $x = 1$ and $x = 2$
8. $f(x) = 2 + 2x - x^2$ and $g(x) = 2x - 2$ between $x = -2$ and $x = 2$

In exercises 9 through 16, find the area between the two given curves. Make a sketch of the region. (See Example 4.)

9. $y = x^2$ and $y = x + 2$
10. $y = 10x - x^2$ and $y = 10 - x$
11. $y = -x^2 + 5x - 4$ and $y = x - 9$
12. $y = x^2 - 5x + 4$ and $y = -x + 9$
13. $y = x^2 - 5x + 4$ and $y = -x^2 + 5x - 4$
14. $y = x^2$ and $y = x^3$
15. $y = \dfrac{1}{x}$ and $y = -x + \dfrac{5}{2}$ (Express answer to two decimal places.)
16. $y = x^2 - 4x$ and $y = -x^2 + 2x + 8$

In exercises 17 and 18, find the area between the curve $y = f(x)$ and the x axis on the given interval. (See Example 6.)

17. $f(x) = -x^3$ between $x = 1$ and $x = 2$.   18. $f(x) = 1 - \sqrt{x}$ between $x = 1$ and $x = 4$.

SET B

19. The curves $f(x) = \frac{1}{2}x^3$ and $g(x) = 2x$ intersect in three points, as shown in the accompanying figure. Find the area between the two curves (shaded area) by using two separate integrals. The integrand for one section will be $f(x) - g(x)$ and the other will be $g(x) - f(x)$.

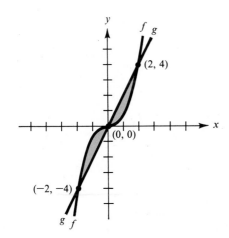

20. Find the area of the shaded region in exercise 19 by using the area between $x = 0$ and $x = 2$ and doubling it. Compare with the answer for exercise 19. Explain why this works in this case.

21. Find the area between $y = x^3 - 3x$ and $y = 2x^2$. (*Hint:* Solve simultaneously to find points of intersection.)

22. Find the area between $y = e^x$ and $y = -x^2$ between $x = 0$ and $x = 1$. Make a sketch of this region.

23. Find the area between $y = e^x$ and $y = e^{-x}$ between $x = -2$ and $x = 2$. Make a sketch of this region.

24. Find the area between $y = \dfrac{1}{x}$ and $y = -2$ between $x = 1$ and $x = 4$. Make a sketch of this region.

25. a. Find the area bounded by $f(x) = x + 2$ and the x axis between $x = 0$ and $x = 2$.
    b. Find the area bounded by $g(x) = -x^2$ and the x axis between $x = 0$ and $x = 2$.
    c. Add the two areas from parts a and b.
    d. Compare the answer to part c with the result in Example 2 for the area between $f(x) = x + 2$ and $g(x) = -x^2$.

BUSINESS   *In exercises 26 through 30, give the coordinates of the equilibrium point and find the **consumer surplus** and the **producer surplus** for the given supply and demand functions. (See Example 5.)*

	Demand function	*Supply function*
26.	$p = 50 - x$	$p = 0.25x$
27.	$p = 50 - x$	$p = 0.05x^2 + 10$
28.	$p = 1000 - 0.1x^2$	$p = 48x$
29.	$p = 500 - 0.2x^2$	$p = 48x$
30.	$p = 450 - 10x$	$p = 0.4x^2 + 210$

Using Example 3 as a guide and letting $R'(x) = 4750 - x^2$ and $C'(x) = 0.5x^2 + 1000$, complete exercises 31 and 32.

31. Find x^*, the value of x for which $R'(x) = C'(x)$. (Note that beyond this level of production it would no longer be profitable to produce the item.)

32. Find the total profit (that is, the total profit generated by increasing the production level from 0 to x^* units).

33. Rework exercise 31 if $R'(x) = 3600 - x^2$ and $C'(x) = 0.5x^2 + 1200$.

34. Rework exercise 32 if $R'(x) = 3600 - x^2$ and $C'(x) = 0.5x^2 + 1200$.

35. Rework exercise 31 if $R'(x) = 3000 - x^2$ and $C'(x) = 2x + 600$.

36. Rework exercise 32 if $R'(x) = 3000 - x^2$ and $C'(x) = 2x + 600$.

PRODUCER SURPLUS

37. Assume that the supply function for a product is given by $p = 3e^{x/3}$, where x is the number of units supplied and equilibrium is reached when $x = 3$. Find the producer surplus.

CONSUMER SURPLUS

38. The demand for a product is given by the equation $p = 100e^{-x}$, where x is the number of units demanded and equilibrium is reached when $x = 10$. Find the consumer surplus.

MANAGEMENT

39. A large corporation ran a study from 1980 to 1985 to determine how much money was being spent on office equipment. With $t = 0$ representing 1980, the amount is given by

$$A(t) = 300\sqrt{t + 4}, \qquad 0 \le t \le 5$$

where t is measured in years and $A(t)$ is measured in thousands of dollars. Beginning in 1985, a new supplier was being used and all old equipment was replaced by newer but cheaper state-of-the-art equipment. From 1985 to 1988, the amount spent can be approximated by

$$B(t) = 1150 - 50t, \qquad 5 \le t \le 8$$

where $B(t)$ is measured in thousands. Determine the total savings realized by the corporation from 1985 to 1988.

SOCIAL SCIENCE
ACCIDENT PREVENTION

40. A city conducted a study to determine where new traffic signals should be installed. In the study, the number of accidents at specified locations was determined to be

$$N(t) = 0.1t^2 + 0.5t + 120, \qquad 0 \le t \le 15$$

where t is measured in months, with $t = 0$ corresponding to March 1986. After the installation of new traffic signals in June 1987, another study was conducted that showed that the number of accidents had decreased and could be approximated by

$$M(t) = 105 + 6t - 0.2t^2, \qquad t \geq 15$$

Approximate the total number of accidents that were prevented with the installation of new traffic signals from June 1987 to December 1987.

WRITING AND CRITICAL THINKING

For the curves $f(x) = x^3$ and $g(x) = x$, if we set up the definite integral $\int_{-1}^{1} (x - x^3)\, dx$ or $\int_{-1}^{1} (x^3 - x)\, dx$ to find the area between the two curves from $x = -1$ to $x = 1$, we get an answer of 0 in either case. Explain in words why that occurs. (*Hint:* A diagram may be helpful.)

USING TECHNOLOGY IN CALCULUS

Consider the two functions $f(x) = 1 - x^2$ and $g(x) = x^2 - 4x + 1$. The first *DERIVE*® screen graphs $y = f(x)$ and $y = g(x)$ and the second graphs $y = f(x) - g(x)$.

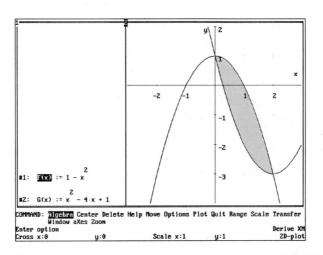

(continued)

(*continued*)

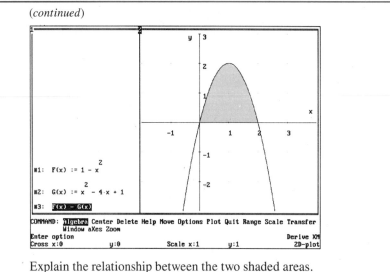

#1: F(x) := 1 - x²

#2: G(x) := x² - 4·x + 1

#3: F(x) - G(x)

```
COMMAND: Algebra Center Delete Help Move Options Plot Quit Range Scale Transfer
         Window aXes Zoom
Enter option                                                        Derive XM
Cross x:0           y:0           Scale x:1      y:1                 2D-plot
```

Explain the relationship between the two shaded areas.

6.5   AVERAGE VALUE OF A FUNCTION AND VOLUMES

Two additional applications of the definite integral are covered in this section: average value of a function and volumes of solids of revolution. The definite integral has application to many other problems that we will not be covering, such as length of a curve and center of mass of a solid.

AVERAGE VALUE OF A FUNCTION

Suppose you are enrolled in a math course in which you will be taking five exams. The instructor indicates that each exam will be weighted equally in computing your final grade. If, at the end of the course, your five grades are 82, 71, 93, 88, and 86, what is your final *average*?

The following simple computation gives the average of the five scores:

$$\frac{82 + 71 + 93 + 88 + 86}{5} = 84$$

We would like to use this idea to find the **average value of a function** over some closed interval. Let f be a function that is continuous over the interval $[a, b]$. Theoretically, we would like to add up all of the y values ($f(x)$'s) for this function over the interval and divide by the total number of them. Obviously this is not possible because there are an infinite number of them, but we can use the concepts of calculus to execute this computation.

We begin by selecting a representative sample of $f(x)$'s equally spaced over the interval $[a, b]$. Choosing n y values, we can get an approximation to the average value of f, which we denote by f_{av}.

$$f_{av} \approx \frac{f(x_1) + f(x_2) + f(x_3) + \cdots + f(x_n)}{n}$$

Because our y values are equally spaced, the distance between any pair of them, denoted by Δx, is

$$\Delta x = \frac{b - a}{n}$$

which gives $n = \dfrac{b - a}{\Delta x}$. Rewriting f_{av}, we have

$$f_{av} \approx \frac{f(x_1) + f(x_2) + \cdots + f(x_n)}{\dfrac{b - a}{\Delta x}}$$

$$= \frac{f(x_1)\,\Delta x + f(x_2)\,\Delta x + \cdots + f(x_n)\,\Delta x}{b - a} \tag{1}$$

Now, if we take a larger sample of y values (let n get larger), we improve our approximation. In fact, if we let $n \to \infty$, we will get the *exact* average value.

Recall from Section 6.3 that the numerator of the right-hand side of equation (1) approaches the definite integral $\int_a^b f(x)\,dx$ as $n \to \infty$. This gives

$$f_{av} = \frac{\displaystyle\int_a^b f(x)\,dx}{b - a}$$

More formally, we have the following definition:

Average value of a function

> For a function f continuous over the interval $[a, b]$, the **average value** is
>
> $$f_{av} = \frac{1}{b - a}\int_a^b f(x)\,dx$$

Example 1   Find the average value of the function $f(x) = x^2 + 1$ over the interval $[0, 2]$.

Solution   From the definition,

$$f_{av} = \frac{1}{b-a} \int_a^b f(x)\, dx$$

$$= \frac{1}{2-0} \int_0^2 (x^2 + 1)\, dx$$

$$= \frac{1}{2} \left[\frac{x^3}{3} + x \right]_0^2$$

$$= \frac{1}{2} \left[\left(\frac{8}{3} + 2 \right) - 0 \right] = \frac{7}{3}$$

The average of all the y values for $0 \le x \le 2$ is $\frac{7}{3}$. See Figure 6.23 for a graph of this function.

Figure 6.23

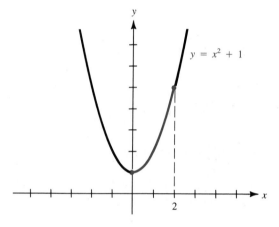

$y = x^2 + 1$

Example 2   The number of workers for a large company is given by

$$N(t) = \tfrac{1}{2}(t + 2)^{2/3}$$

where t is measured in years and $N(t)$ is measured in hundreds. Determine the average number of employees over the last 6 years ($t = 0$ to $t = 6$). A graph of this function appears in Figure 6.24.

Solution   The average number of employees is

$$\frac{1}{6-0} \int_0^6 \frac{1}{2}(t + 2)^{2/3}\, dt = \frac{1}{12} \int_0^6 (t + 2)^{2/3}\, dt$$

$$= \frac{1}{12} \cdot \frac{3}{5} (t + 2)^{5/3} \bigg]_0^6$$

$$\approx \frac{1}{20} (32 - 3.175)$$

$$\approx 1.44$$

Figure 6.24

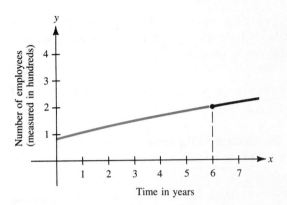

Time in years

The average number of employees over the last 6 years is approximately 144. ∎

VOLUMES OF SOLIDS OF REVOLUTION

Start with the area bounded above by the curve $y = f(x)$ and below by the x axis between $x = a$ and $x = b$ (see Figure 6.25). We revolve this region about the x axis and "sweep out" the volume of a solid called a **solid of revolution** (see Figure 6.26).

Figure 6.25   Figure 6.26

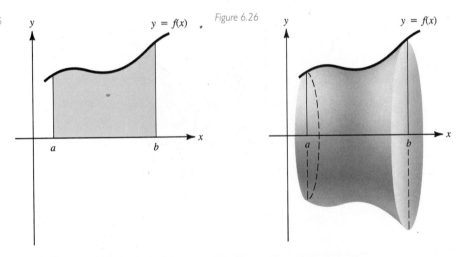

In Section 6.3, we showed how the areas of rectangles could be used to approximate the area under the curve $y = f(x)$, where f is continuous on $[a, b]$. We return to that idea to help us compute the volume of the solid of revolution. Again divide the interval from a to b into n subintervals, each of width Δx. Construct n rectangles using $f(x_1), f(x_2), f(x_3), \ldots, f(x_n)$ as the heights, as before. If we take the ith rectangle and compute its area, we have

area of ith rectangle $= f(x_i)\,\Delta x$

↑ ↑

height   width

Now revolve the area bounded by this rectangle about the x axis and it sweeps out a volume in the shape of a circular disk (see Figure 6.27). The volume of this circular disk can be found as follows:

$$\text{volume of circular disk} = (\text{surface area})(\text{thickness})$$
$$= \pi[f(x_i)]^2 \cdot \Delta x$$

If we do this for each of the rectangles from a to b and add them up, we get an approximation to the desired volume.

$$\text{volume} \approx \pi[f(x_1)]^2 \, \Delta x + \pi[f(x_2)]^2 \, \Delta x + \cdots + \pi[f(x_n)]^2 \, \Delta x$$

Letting $n \to \infty$ for this sum, we get the exact volume that can be evaluated using the definite integral.

$$\text{volume} = \int_a^b \pi[f(x)]^2 \, dx$$

Figure 6.27

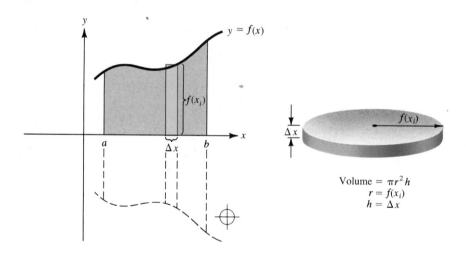

Volume = $\pi r^2 h$
$r = f(x_i)$
$h = \Delta x$

Example 3   Find the volume generated by revolving the region bounded by $y = x^2 + 2$ and the x axis between $x = -1$ and $x = 2$ about the x axis.

Solution   Sketching this region and the volume generated in Figure 6.28, we have

$$\text{volume} = \int_a^b \pi[f(x)]^2 \, dx$$

$$= \int_{-1}^2 \pi(x^2 + 2)^2 \, dx$$

$$= \pi \int_{-1}^2 (x^4 + 4x^2 + 4) \, dx$$

$$= \pi \left(\frac{x^5}{5} + \frac{4x^3}{3} + 4x \right)_{-1}^2$$

$$= \pi \left[\left(\frac{32}{5} + \frac{32}{3} + 8 \right) - \left(\frac{-1}{5} - \frac{4}{3} - 4 \right) \right]$$

$$\text{volume} = \frac{153\pi}{5} \approx 96.13 \text{ cubic units}$$

Figure 6.28

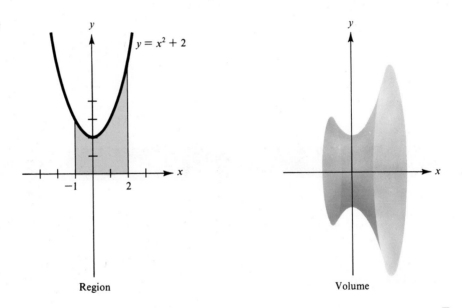

Region

Volume

Example 4 Find the volume of the solid generated by revolving the region bounded by $y = \sqrt{x}$, $x = 4$, and $y = 0$ about the x axis.

Solution A sketch of this region and volume appears in Figure 6.29.

Figure 6.29

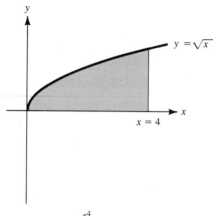

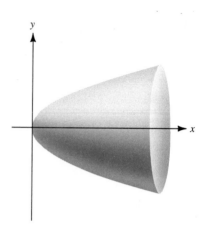

$$\text{volume} = \int_0^4 \pi(\sqrt{x})^2 \, dx$$

$$= \int_0^4 \pi x \, dx$$

$$= \pi \left. \frac{x^2}{2} \right]_0^4$$

$$= \pi(8 - 0)$$

$$\text{volume} = 8\pi \text{ cubic units}$$

6.5 EXERCISES

SET A

In exercises 1 through 8, find the average value of the given function over the given interval. (See Examples 1 and 2.)

1. $f(x) = 2x$, $[0, 2]$
2. $f(x) = 2x + 3$, $[0, 2]$
3. $f(x) = x^3 + 1$, $[0, 1]$
4. $y = 8 - x^3$, $[0, 2]$
5. $y = 1 - x^2$, $[-1, 1]$
6. $S(t) = 4 - t^2$, $[-2, 2]$
7. $f(x) = \sqrt{x}$, $[0, 4]$
8. $f(x) = \sqrt[3]{x}$, $[0, 8]$

In exercises 9 through 20, find the volume of the solid generated by revolving the region bounded by the given information about the x axis. (See Examples 3 and 4.)

9. $y = x + 3$, $y = 0$, $x = 0$, and $x = 1$
10. $y = 5 - x$, $y = 0$, $x = 0$, and $x = 3$
11. $y = 1 - x^2$, $y = 0$, $x = -1$, and $x = 1$
12. $y = 4 - x^2$, $y = 0$, $x = -2$, and $x = 2$
13. $y = x^3$, $y = 0$, $x = 0$, and $x = 1$
14. $y = 1 - x^3$, $y = 0$, $x = 0$, and $x = 1$
15. $y = \sqrt{x + 1}$, $y = 0$, $x = 0$, and $x = 3$
16. $y = \sqrt{x - 1}$, $y = 0$, and $x = 5$
17. $y = e^x$, $y = 0$, $x = 0$, and $x = 1$
    (Round answer to three decimal places.)
18. $y = e^{-x}$, $y = 0$, $x = 0$, and $x = 1$
    (Round answer to three decimal places.)

19. $y = \dfrac{1}{\sqrt{x}}$, $y = 0$, $x = 1$, and $x = 2$

(Round answer to three decimal places.)

20. $y = \dfrac{1}{2\sqrt{x}}$, $y = 0$, $x = 1$, and $x = 4$

(Round answer to three decimal places.)

SET B

21. Find the volume of the solid generated by revolving the region bounded by the horizontal line $y = 2$ and the x axis between $x = 0$ and $x = 4$ about the x axis. Sketch a picture of this volume. Compare your result with the volume of a cylinder of radius 2 and height 4.

22. Find the volume of the solid generated by revolving the region bounded by the line $y = x$, the x axis, and the line $x = 4$ about the x axis. Sketch a picture of this volume. Compare your result with the volume of a cone of radius 4 and height 4.

23. Find the volume of the solid generated by revolving the region bounded by the curves $f(x) = 2x - x^2$ and $g(x) = x$ about the x axis by first finding the points of intersection of the two curves and then using the integral

$$\pi \int ([f(x)]^2 - [g(x)]^2) \, dx$$

Using a diagram, explain why this formula works.

24. Find the volume of the solid generated by revolving the region bounded by the two curves $f(x) = \sqrt{x}$ and $g(x) = x^2$ about the x axis. (*Hint:* See exercise 23.)

BUSINESS
AVERAGE COST

25. The Suds-O-Matic Company, which manufactures washing machines, has determined that the cost of producing its washing machines has been increasing over the years because of the rising costs of materials and labor. If $C(t)$ (in thousands of dollars) is given by

$$C(t) = 1.5t^2 + 0.5$$

where t is measured in years, determine the average cost for the first 10 years.

AVERAGE REVENUE

26. The revenue from the sale of x units of a product is given by $R(x) = 15x + 6x^2 - x^3$. Determine the average revenue earned by the company for the first five units produced.

AVERAGE OUTPUT

27. The weekly output of a product t weeks after the start of a production run is given by $y = 600 - 600e^{-0.3t}$. What is the average output over the first 10 weeks of production?

BIOLOGY
BACTERIAL GROWTH

28. The number of bacteria present at any time, t (in hours), for an experiment is given by

$$N(t) = 80e^{0.5t}$$

Find the average number of bacteria during the first 12 hours of the experiment.

AIR POLLUTION

29. An environmentalist is studying the effects of air pollution from a factory in a small town. The level of air pollution, $P(x)$, is a function of the distance, x (in miles), from the factory and is given by

$$P(x) = e^{-0.2x} + 5$$

Determine the average level of pollution between 5 and 10 miles from the factory.

ZOOLOGY

30. The water temperature in an experiment involving the breeding of a certain type of fish measured over the period of one month is determined to be

$$W(t) = 50 + 2.4t - 0.08t^2$$

where $W(t)$ is the water temperature measured in degrees Fahrenheit and t is measured in days. Find the average temperature for the first 10 days of the experiment.

SOCIAL SCIENCE
SUGAR CONSUMPTION

31. The rate of sugar consumption per person per year by a large sample of the population has been found to be

$$S(t) = 90 + 4.8t - 0.08t^2$$

where $S(t)$ is measured in pounds and t is measured in years. Find the average rate of sugar consumption between 1988 and 1992 if $t = 0$ corresponds to 1982.

32. Find the average value for the function $f(x) = x^3$ over the interval $[-1, 1]$. Sketch a graph of the function and interpret your result.

USING TECHNOLOGY IN CALCULUS

One advantage of computer software packages and supercalculators is their ability to do the most tedious calculations.

1. Using an appropriate technology, find the average value of the function $y = e^{x^2}$ on the interval $[0, 1]$. Call this value v.

2. Use the graphing capabilities of the technology to graph $y = e^{x^2}$ and $y = v$ on the same axes. Can you make an observation?

3. Use the technology to approximate $\int_{0}^{0.616644} (v - e^{x^2})\, dx$ and $\int_{0.616644}^{1} (e^{x^2} - v)\, dx$. What is the significance of these calculations, and what observations can you make about them?

4. Where does the value 0.616644 come from?

CHAPTER 6 REVIEW

VOCABULARY/FORMULA LIST

antiderivative
antidifferentiation
indefinite integral
integration
constant of integration
integrand
Power Rule for Integrals:

$$\int x^n \, dx = \frac{x^{n+1}}{n+1} + C, n \neq -1$$

integration formulas:

$$\int kf(x) \, dx = k \int f(x) \, dx$$

$$\int [f(x) + g(x)] \, dx$$

$$= \int f(x) \, dx + \int g(x) \, dx$$

$$\int g'(x) \, dx = g(x) + C$$

$$\int e^x \, dx = e^x + C$$

$$\int \frac{1}{x} \, dx = \ln |x| + C$$

definite integral
limits of integration
integration by substitution

area under a curve:

$$\text{area} = \int_a^b f(x) \, dx$$

Fundamental Theorem
area between curves:

$$\text{area} = \int_a^b [f(x) - g(x)] \, dx$$

equilibrium point
consumer surplus
producer surplus
average value of a function:

$$f_{av} = \frac{1}{b-a} \int_a^b f(x) \, dx$$

solid of revolution:

$$\text{volume} = \int_a^b \pi [f(x)]^2 \, dx$$

CHAPTER 6 REVIEW EXERCISES

In exercises 1 through 10, evaluate each indefinite integral.

[6.1] **1.** $\int (3x^2 - x + 1) \, dx$

[6.1] **2.** $\int (4x^3 - 4x) \, dx$

[6.2] **3.** $\int x^2 \sqrt{1 - x^3} \, dx$

[6.2] **4.** $\int \frac{x^2}{\sqrt{1 - x^3}} \, dx$

[6.2] **5.** $\int (2x - 1)^5 \, dx$

[6.2] **6.** $\int \sqrt[5]{2x - 1} \, dx$

[6.2] **7.** $\int \frac{1}{2x + 5} \, dx$

[6.2] **8.** $\int \frac{x}{2x^2 + 5} \, dx$

[6.1] **9.** $\int \frac{1}{x\sqrt{x}} \, dx$

[6.1] **10.** $\int \frac{x^2 - 1}{\sqrt{x}} \, dx$

In exercises 11 through 18, evaluate each definite integral. Where appropriate, round answers to three decimal places.

[6.3]   **11.** $\displaystyle\int_0^1 (x^2 - x + 3)\, dx$

[6.3]   **12.** $\displaystyle\int_1^8 \frac{2}{\sqrt[3]{x}}\, dx$

[6.3]   **13.** $\displaystyle\int_0^2 (x - 2)^4\, dx$

[6.3]   **14.** $\displaystyle\int_2^4 (x - 2)^4\, dx$

[6.3]   **15.** $\displaystyle\int_0^1 e^{4x}\, dx$

[6.3]   **16.** $\displaystyle\int_1^2 \frac{3}{(2x - 1)^2}\, dx$

[6.3]   **17.** $\displaystyle\int_{-1}^1 \frac{1}{\sqrt{x + 3}}\, dx$

[6.3]   **18.** $\displaystyle\int_{-2}^{-1} \frac{1}{x^2}\, dx$

In exercises 19 through 24, find the area of the region described.

[6.3]   **19.** Bounded by $f(x) = 1 - x^4$, $x = -1$, $x = 1$, and x axis

[6.3]   **20.** Bounded by $f(x) = x^3 + 2$, $x = -1$, $x = 1$, and x axis

[6.3]   **21.** Bounded by $y = 2x + 1$, $x = 0$, $x = 3$, and x axis

[6.3]   **22.** Bounded by $y = 1 - x$, $x = 0$, $x = 1$, and x axis

[6.4]   **23.** Between $y = x^3$ and $y = x^2$

[6.4]   **24.** Between $y = x$ and $y = 2x - x^2$

[6.5]   **25.** Find the average value for the function $f(x) = \sqrt{x - 2}$ from $x = 2$ to $x = 6$.

[6.5]   **26.** Find the average value for the function $f(x) = \dfrac{1}{x^2}$ from $x = 1$ to $x = 4$.

[6.5]   **27.** Find the volume of the solid generated by revolving the region bounded by $y = \sqrt{x}$, $x = 1$, $x = 9$, and the x axis about the x axis.

[6.5]   **28.** Find the volume of the solid generated by revolving the region bounded by $y = x^{2/3}$, $x = 0$, $x = 1$, and the x axis about the x axis.

[6.1]   **29.** Given $f'(x) = x^2 - \dfrac{1}{x^2}$ and $f(1) = \dfrac{4}{3}$, find $f(x)$.

[6.2]   **30.** Given $f'(x) = \sqrt{1 - 2x}$ and $f(-4) = -3$, find $f(x)$.

BUSINESS
PROFIT
[6.2]   **31.** The Tiny Tot Toy Company has determined that its marginal profit, $P'(x)$, for selling x stuffed animals (measured in hundreds) is given by

$$P'(x) = \frac{200}{\sqrt[3]{(x - 5)^2}}$$

If the company breaks even ($P(x) = 0$) when it produces 500 stuffed animals, determine $P(x)$.

CONSUMER SURPLUS [6.4]   **32.** Find the consumer surplus if the demand equation is $p(x) = 700 - 20x$ and the supply equation is $p(x) = 100 + 0.5x^2$. (See Section 6.4 for the definition of consumer surplus.)

QUALITY CONTROL [6.5]   **33.** The amount of fruit juice concentrate added in the production process for a new dessert, during the experimentation stage, is given by

$$C(t) = \frac{2t}{t^2 + 1} + 3$$

where t is measured in hours. Find the average amount of concentrate (measured in parts per gallon) during the first 6 hours ($t = 0$ to $t = 6$). Round your answer to one decimal place.

PRODUCER SURPLUS [6.4]   **34.** Find the producer surplus in exercise 32.

COST   [6.2]   **35.** The marginal cost for a particular commodity is given by the function

$$C'(x) = 1 + 10x - e^{-x}$$

If fixed cost is 4, find the total cost of 10 units of the commodity.

CHAPTER 6 TEST

PART ONE
Multiple choice. In exercises 1 through 10, put the letter of your correct response in the blank provided.

_____ **1.** The average value for $f(x) = x^3 - x$ between $x = 1$ and $x = 3$ is

    a. 16   **b.** 8

    c. 32   **d.** 6

_____ **2.** Evaluate $\int \left(x^3 - \frac{1}{x^2} \right) dx.$

    a. $\dfrac{x^4}{4} - \dfrac{1}{x} + C$ **b.** $\dfrac{x^4}{4} + \dfrac{2}{x^3} + C$

    c. $\dfrac{x^4}{4} + \dfrac{1}{x} + C$ **d.** $\dfrac{x^4}{4} - \dfrac{1}{2x} + C$

_____ **3.** Evaluate $\int e^{-2x}\, dx.$

    a. $-2e^{-2x} + C$ **b.** $e^{-2x} + C$

    c. $-e^{-2x} + C$ **d.** $\dfrac{-1}{2} e^{-2x} + C$

_____ **4.** The area of the region bounded by $f(x) = x - x^2$, $x = 0$, $x = 1$, and the x axis is

    a. 0 square units   **b.** $\frac{1}{6}$ square unit

    c. $\frac{1}{5}$ square unit   **d.** 1 square unit

_____ **5.** Evaluate $\int_0^2 \sqrt{4x + 1}\, dx.$

    a. $\frac{13}{3}$ **b.** $\frac{9}{2}$

    c. $\frac{52}{3}$ **d.** 3

_____ 6. If $g'(x) = 3x^2 - 2x + 4$ and $g(1) = 3$, find $g(x)$.

  a. $x^3 - x^2 + 4x + 3$ b. $x^3 - x^2 + 4x + 1$

  c. $6x - 2$ d. $x^3 - x^2 + 4x - 1$

_____ 7. Which of the following is **false**?

  a. $\int kf(x)\,dx = k \int f(x)\,dx$

  b. $\int [f(x) + g(x)]\,dx = \int f(x)\,dx + \int g(x)\,dx$

  c. $\int [f(x) \cdot g(x)]\,dx = \int f(x)\,dx \cdot \int g(x)\,dx$

  d. $\int x^{-1}\,dx = \ln |x| + C$

_____ 8. Find the area between the two curves $y = x^2$ and $y = 2 - x^2$.

  a. $\frac{8}{3}$ square units   b. $\frac{4}{3}$ square units

  c.   4 square units   d. $\frac{16}{3}$ square units

_____ 9. Find the volume of the solid generated by revolving the region bounded by $y = x^2$, $x = 0$, $x = 1$, and the x axis about the x axis.

  a. π cubic units   b. $\dfrac{\pi}{3}$ cubic units

  c. $\dfrac{\pi}{4}$ cubic unit   d. $\dfrac{\pi}{5}$ cubic unit

_____ 10. Evaluate $\displaystyle\int \frac{5x}{\sqrt[3]{1 - x^2}}\,dx$.

  a. $-5\sqrt{1 - x^2} + C$ b. $\dfrac{-15}{4}(1 - x^2)^{2/3} + C$

  c. $\dfrac{-15}{8}(1 - x^2)^{4/3} + C$ d. $\dfrac{15}{2}(1 - x^2)^{2/3} + C$

PART TWO

Solve and show all work.

11. Evaluate $\displaystyle\int x^2(x^3 + 1)^5\,dx$.

12. For $f'(x) = 2x - \dfrac{1}{x^2}$, find $f(x)$ given that $f(1) = 2$.

BUSINESS
CONSUMER AND
PRODUCER SURPLUS

13. Find the consumer surplus and the producer surplus given that the demand equation is $p = 150 - x$ and the supply equation is $p = 0.1x^2 + 30$ for a certain product.

PROFIT

14. If the rate of change of the profit is given by

$$P'(x) = 2100 + 0.2x$$

where x is the number of units in hundreds, find the total profit generated by increasing the production level from 500 to 1000 units.

COST

15. If the marginal cost for the production of x products is given by

$$C'(x) = 3 + 4x^2 + \frac{e^{-x}}{2}$$

and the fixed cost is 5, what is the total cost function, $C(x)$?

BIOLOGY
BACTERIOLOGY

16. The number of bacteria present at any time, t (in hours), for an experiment is given by

$$N(t) = 50e^{0.2t}$$

Find the average number of bacteria present during the first 24 hours of the experiment.

7

ADDITIONAL TOPICS IN INTEGRATION

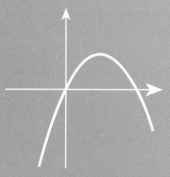

7.1   INTEGRATION BY PARTS

In Chapter 5, we developed the formula for the derivative of the exponential function, $f(x) = e^x$; and in Chapter 6, we developed the integration formula

$$\int e^x \, dx = e^x + C$$

The derivative of the natural logarithm function, $f(x) = \ln x$, was introduced in Chapter 5, but we have not yet seen an integration formula for

$$\int \ln x \, dx$$

This integral, along with many other integration problems, can be integrated using a technique called **integration by parts.**

The technique of integration by parts is especially useful for integrating certain types of products. We can use the Product Rule for differentiation to develop a formula for this technique. From Chapter 3, if $f(x) = u(x)v(x)$, then

$$f'(x) = u(x)v'(x) + v(x)u'(x)$$

Rewriting, we have

$$u(x)v'(x) = f'(x) - v(x)u'(x)$$

Now integrating both sides, we get

$$\int u(x)v'(x) \, dx = \int f'(x) \, dx - \int v(x)u'(x) \, dx$$

$$= f(x) - \int v(x)u'(x) \, dx$$

Replacing $f(x)$ with $u(x)v(x)$ gives the following formula:

Integration by parts formula

$$\int u(x)v'(x) \, dx = u(x)v(x) - \int v(x)u'(x) \, dx$$

To use this formula, we take our original integrand and separate it into two factors, $u(x)$ and $v'(x)$. To get the result on the right side, it is necessary to differentiate our choice for $u(x)$ to produce $u'(x)$ and to integrate our choice for $v'(x)$ to produce $v(x)$. The most important thing to keep in mind when making your choices for $u(x)$ and $v'(x)$ is that the new integral on the right side should be "simpler" than the original integral. Some examples will make this clearer.

Example 1   Evaluate $\int x \ln x \, dx$.

Solution   Because we do not yet know how to integrate ln x, we make the following choices:

$$u(x) = \ln x \xrightarrow{\text{differentiating}} u'(x) = \frac{1}{x}$$

$$v'(x)\, dx = x\, dx \xrightarrow{\text{integrating}} v(x) = \frac{x^2}{2}$$

Now applying the formula, we get

$$\int u(x)v'(x)\, dx = u(x)v(x) - \int v(x)u'(x)\, dx$$

$$\int x \ln x\, dx = (\ln x)\left(\frac{x^2}{2}\right) - \int \frac{x^2}{2} \cdot \frac{1}{x}\, dx$$

$$= \frac{x^2}{2} \ln x - \frac{1}{2} \underbrace{\int x\, dx}_{\text{simpler integral}}$$

$$= \frac{x^2}{2} \ln x - \frac{1}{4} x^2 + C$$

■

In practice, we often use a simpler alternative notation for the formula for integration by parts, as follows:

$$\int u\, dv = uv - \int v\, du$$

We use this version of the formula for the remaining examples in this section.

Example 2   Evaluate $\int xe^{-x}\, dx$.

Solution   Because differentiating or integrating e^{-x} yields the same result, $-e^{-x}$, we make our choice based on the fact that differentiating x gives a simpler result than does integrating x.

$$u = x \qquad dv = e^{-x}\, dx$$
$$du = dx \qquad v = -e^{-x}$$

$$\int xe^{-x}\, dx = -xe^{-x} - \int -e^{-x}\, dx$$

$$= -xe^{-x} + \int e^{-x}\, dx$$

$$= -xe^{-x} - e^{-x} + C$$

We would like to again point out that the answers may be checked using differentiation.   ∎

Returning to the original problem of this section, we now integrate the function $f(x) = \ln x$.

Example 3   Evaluate $\displaystyle\int \ln x \, dx$.

Solution   Letting $u = \ln x$ means that dv must be equal to dx.

$$u = \ln x \qquad dv = 1 \, dx$$
$$du = \frac{1}{x} \, dx \qquad v = x$$

$$\int \ln x \, dx = x \ln x - \int x \cdot \frac{1}{x} \, dx$$
$$= x \ln x - \int dx$$

$$\int \ln x \, dx = x \ln x - x + C$$

∎

Example 4   Evaluate $\displaystyle\int x\sqrt{x-2} \, dx$.

Solution

$$u = x \qquad dv = (x-2)^{1/2} \, dx$$
$$du = dx \qquad v = \tfrac{2}{3}(x-2)^{3/2}$$

$$\int x\sqrt{x-2} \, dx = \tfrac{2}{3}x(x-2)^{3/2} - \int \tfrac{2}{3}(x-2)^{3/2} \, dx$$
$$= \tfrac{2}{3}x(x-2)^{3/2} - \tfrac{4}{15}(x-2)^{5/2} + C$$

Compare this result with the answer to exercise 51a in Section 6.2. Both solutions should be factored to see that the results are equivalent.   ∎

Some problems require that we use this technique more than once in the same problem, as you can see in Example 5.

Example 5   Evaluate $\int \dfrac{x^2 e^x}{2}\, dx$.

Solution

$$u = \frac{x^2}{2} \qquad dv = e^x\, dx$$
$$du = x\, dx \qquad v = e^x$$

$$\int \frac{x^2 e^x}{2}\, dx = \frac{x^2 e^x}{2} - \int x e^x\, dx$$

We must use "parts" again for this new integral, with

$$u = x \qquad dv = e^x\, dx$$
$$du = dx \qquad v = e^x$$

$$\int \frac{x^2 e^x}{2}\, dx = \frac{x^2 e^x}{2} - \left[x e^x - \int e^x\, dx \right]$$
$$= \frac{x^2 e^x}{2} - x e^x + e^x + C$$

■

Example 6   The Doralite Company has determined that, because new equipment was installed, the company's production level has been increasing at the rate

$$n'(t) = 2t \ln t, \qquad t > 0$$

where $n(t)$ is the number of units measured in hundreds and t is measured in days. Find the total number of units produced for the 10-day period from $t = 20$ to $t = 30$.

Solution $n(t) = \displaystyle\int n'(t)\, dt$

$$= \int 2t \ln t\, dt$$

$$u = \ln t \qquad dv = 2t\, dt$$
$$du = \frac{1}{t}\, dt \qquad v = t^2$$

$$n(t) = t^2 \ln t - \int t^2 \cdot \frac{1}{t} \, dt$$

$$= t^2 \ln t - \int t \, dt$$

$$= t^2 \ln t - \frac{t^2}{2} + C$$

$$\int_{20}^{30} 2t \, \ln t \, dt = t^2 \ln t - \frac{t^2}{2} \Bigg]_{20}^{30} \approx 1612$$

The total number of units is 161,200.   ■

7.1   EXERCISES

SET A

In exercises 1 through 22, evaluate each indefinite integral. (See Examples 1–5.)

1. $\displaystyle\int x^2 \ln x \, dx$

2. $\displaystyle\int x e^x \, dx$

3. $\displaystyle\int x e^{x^2} \, dx$

4. $\displaystyle\int x e^{3x} \, dx$

5. $\displaystyle\int x e^{-3x} \, dx$

6. $\displaystyle\int x\sqrt{x+1} \, dx$

7. $\displaystyle\int x^2 e^x \, dx$

8. $\displaystyle\int x^2 e^{2x} \, dx$

9. $\displaystyle\int \ln 2x \, dx$

10. $\displaystyle\int x\sqrt{x-2} \, dx$

11. $\displaystyle\int x e^{5x} \, dx$

12. $\displaystyle\int x^2\sqrt{x+1} \, dx$

13. $\displaystyle\int \frac{x}{e^{2x}} \, dx$

14. $\displaystyle\int x e^{-x^2} \, dx$

15. $\displaystyle\int \frac{\ln x}{x} \, dx$

16. $\displaystyle\int x(e^x - 2) \, dx$

17. $\displaystyle\int \ln \sqrt{x} \, dx$

18. $\displaystyle\int \ln \sqrt{2x} \, dx$

19. $\displaystyle\int \ln (x^2) \, dx$

20. $\displaystyle\int (\ln x)^2 \, dx$

21. $\displaystyle\int x \ln (x^2) \, dx$

22. $\displaystyle\int x^2 \ln (x^2) \, dx$

In exercises 23 through 32, evaluate each definite integral. If necessary, round answers to two decimal places.

23. $\displaystyle\int_1^2 \ln x \, dx$

24. $\displaystyle\int_1^e \ln x \, dx$

25. $\displaystyle\int_1^2 x \ln x \, dx$

26. $\displaystyle\int_1^e x \ln x \, dx$ 　　　　　　**27.** $\displaystyle\int_3^6 x\sqrt{x-2} \, dx$ 　　　　　　**28.** $\displaystyle\int_3^{11} x\sqrt{x-2} \, dx$

29. $\displaystyle\int_0^1 xe^x \, dx$ 　　　　　　**30.** $\displaystyle\int_0^2 xe^x \, dx$ 　　　　　　**31.** $\displaystyle\int_0^1 \frac{x^2 e^x}{2} \, dx$

32. $\displaystyle\int_0^2 \frac{x^2 e^x}{2} \, dx$

SET B

In exercises 33 through 37, find the area of the region bounded by the given curves. Where necessary, round answers to two decimal places.

33. $y = \ln x$, $x = 2$, $y = 0$

34. $y = xe^x$, $x = -1$, $y = 0$ (*Hint:* See the accompanying figure.)

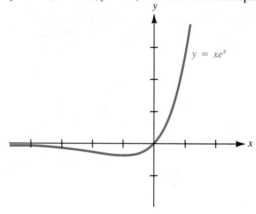

35. $y = xe^x$, $x = 1$, $y = 0$

36. $y = x \ln x$, $x = 2$, $y = 0$ (*Hint:* See the accompanying figure.)

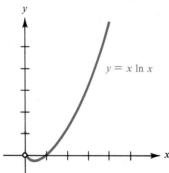

37. $y = x \ln x$, $x = 0.5$, $x = 2$, $y = 0$ (*Hint:* At $x = 1$, $y = 0$ and the function is negative for $0 < x < 1$ and positive for $x > 1$.)

BUSINESS
REVENUE FUNCTION

38. A company finds that the marginal revenue function for one of its products is given by $R'(x) = -120xe^{-0.3x} + 4.00e^{-0.3x}$. Determine the revenue function. Assume that $R(0) = 0$.

COST FUNCTION

39. If the marginal cost function for a product is given by $C'(x) = 2 \ln (x + 1)$, determine the cost function, $C(x)$, if fixed cost is 10.

INFLATION

40. In a certain country, the change in the inflation rate can be described by the equation

$$R'(t) = \ln (t + 2)$$

where t is time measured in years. Determine the equation for the inflation with respect to time, $R(t)$. Assume $R(0) = 0$.

PROFIT

41. The marginal profit function for x items is given by

$$P'(x) = \frac{1000 \ln (x + 1)}{x + 1}$$

Find the profit function if $P(0) = -100$.

BIOLOGY
HABITAT DESTRUCTION

42. The population of a certain species of bird is decreasing due to the destruction of its natural habitat. The rate at which the number of birds, N, is decreasing is given by the model

$$N'(x) = -50xe^{0.1x}$$

where x is measured in years.
a. Find $N(x)$ if there were 3000 birds initially.
b. Predict how many birds will remain after 8 years.

SOCIAL SCIENCE
URBAN PLANNING

43. New office space is increasing at the rate

$$S'(t) = 50 \ln 2t$$

where S is the number of new offices and t is measured in months.
a. Find $S(t)$ if $S(1) = 450$.
b. How many new offices will there be after 6 months?

WRITING AND CRITICAL THINKING

For an integral of the form $\int \frac{f(x)}{g(x)} \, dx$, which of the following might be used to do an integration by parts for this integral? In each case, explain why or why not.

1. Let $u = f(x)$ and $dv = g(x) \, dx$.
2. Let $u = f(x) \, dx$ and $dv = g(x)$.
3. Let $u = f(x)$ and $dv = \dfrac{1}{g(x)}$.
4. Let $u = f(x)$ and $dv = [g(x)]^{-1} \, dx$.

7.2 INTEGRAL TABLES

In our presentation of techniques for evaluating integrals, we have not covered all of the possibilities. Some techniques require use of the trigonometric functions, and others require complicated algebraic manipulations. It is possible to evaluate some of these by using a **table of integrals.**

A table of integrals consists of a list of formulas. To evaluate integrals, we make appropriate substitutions into the formulas. We have included a small list of such formulas in Appendix B.

The integral tables give formulas for *indefinite* integration; for definite integration, we need only substitute the limits of integration. However, even with an extensive list of integration formulas, there are still some definite integrals for which no formula exists and which may be evaluated using approximation techniques. Some of these techniques are covered in Section 7.3.

Example 1 Evaluate $\int \dfrac{1}{\sqrt{x^2 + 9}}\, dx$.

Solution We use formula 17 from the Table of Integrals in Appendix B.

$$\int \frac{1}{\sqrt{x^2 + a^2}}\, dx = \ln |\, x + \sqrt{x^2 + a^2}\,| + C$$

For our integral, $a = 3$, giving

$$\int \frac{1}{\sqrt{x^2 + 9}}\, dx = \ln |\, x + \sqrt{x^2 + 9}\,| + C$$

Example 2 Evaluate $\int \dfrac{1}{\sqrt{16x^2 + 9}}\, dx$.

Solution Using the same formula 17 and a u substitution, we have

$$\int \frac{1}{\sqrt{16x^2 + 9}}\, dx = \int \frac{1}{\sqrt{(4x)^2 + 3^2}}\, dx$$

$$= \frac{1}{4}\int \frac{1}{\sqrt{(4x)^2 + 3^2}}\,(4\, dx) \qquad \begin{array}{l} u = 4x \\ du = 4\, dx \\ a = 3 \end{array}$$

$$= \frac{1}{4}\int \frac{1}{\sqrt{u^2 + a^2}}\, du$$

$$= \frac{1}{4}\ln |\, u + \sqrt{u^2 + a^2}\,| + C$$

$$= \frac{1}{4}\ln |\, 4x + \sqrt{16x^2 + 9}\,| + C$$

Example 3   Evaluate $\int \dfrac{x}{(2x+1)^2}\,dx.$

Solution   Using formula 2, we get

$$\int \frac{x}{(a+bx)^2}\,dx = \frac{1}{b^2}\left[\frac{a}{a+bx} + \ln|a+bx|\right] + C$$

With $a=1$ and $b=2$, we have

$$\int \frac{x}{(2x+1)^2}\,dx = \frac{1}{4}\left[\frac{1}{1+2x} + \ln|1+2x|\right] + C$$ ∎

In the next example, we use a formula that can be referred to as a **recursion formula** or **reduction formula.** In this kind of formula, the right-hand side contains a new integral that is simpler than the original integral. This is similar to what occurred in the last section when using the integration by parts technique. Many of these reduction formulas were derived using integration by parts.

Example 4   Evaluate $\int (\ln x)^3\,dx.$

Solution   Using formula 23, we get

$$\int (\ln x)^n\,dx = x(\ln x)^n - n\int (\ln x)^{n-1}\,dx$$

With $n=3$ we have

$$\int (\ln x)^3\,dx = x(\ln x)^3 - 3\int (\ln x)^2\,dx$$

We can apply the same formula again to this new integral with $n=2$.

$$\int (\ln x)^3\,dx = x(\ln x)^3 - 3\left[x(\ln x)^2 - 2\int \ln x\,dx\right]$$
$$= x(\ln x)^3 - 3x(\ln x)^2 + 6\int \ln x\,dx$$

Now, using formula 22, we have

$$\int (\ln x)^3\,dx = x(\ln x)^3 - 3x(\ln x)^2 + 6x\ln x - 6x + C$$ ∎

Example 5   Under certain conditions the population, $N(t)$, of the cicada in a small area can be modeled by the bounded growth equation

$$N(t) = \frac{10}{6 + 4e^{-0.8t}}$$

where t is measured in days and $t = 0$ is the time of first hatching. Here, $N(t)$ is measured in thousands. Find the average number of cicadas present in the first 10 days by evaluating

$$\frac{1}{10} \int_0^{10} \frac{10}{6 + 4e^{-0.8t}} \, dt$$

Solution   Using formula 30 with $a = 6$, $b = 4$, and $n = -0.8$, we have

$$\frac{1}{10} \int_0^{10} \frac{10}{6 + 4e^{-0.8t}} \, dt = \left[\frac{t}{6} + \frac{1}{4.8} \ln |6 + 4e^{-0.8t}| \right]_0^{10}$$

$$\approx \left(\frac{10}{6} + \frac{1}{4.8} \ln (6.001) \right) - \left(\frac{1}{4.8} \ln 10 \right)$$

$$\approx 2.04 - 0.48$$

$$\approx 1.56$$

Thus, the average number of cicadas for the first 10 days is 1560. ∎

7.2 EXERCISES

SET A

In exercises 1 through 26, use the formulas in the Table of Integrals in Appendix B to evaluate each of the following. (See Examples 1–4.)

1. $\int \dfrac{1}{x(x + 3)} \, dx$

2. $\int \dfrac{2}{x(3x + 1)} \, dx$

3. $\int \dfrac{1}{x^2 - 9} \, dx$

4. $\int \dfrac{1}{9x^2 - 1} \, dx$

5. $\int \dfrac{x}{x^2 + 1} \, dx$

6. $\int \dfrac{1}{x(x + 2)^2} \, dx$

7. $\int \sqrt{x^2 + 9} \, dx$

8. $\int \sqrt{x^2 - 4} \, dx$

9. $\int \dfrac{\sqrt{x^2 + 9}}{x} \, dx$

10. $\int \dfrac{1}{x\sqrt{4 - x^2}} \, dx$

11. $\int \dfrac{x}{2 + 3x^2} \, dx$

12. $\int \dfrac{2x}{1 + 3x^2} \, dx$

13. $\int \dfrac{1}{9 - 16x^2} \, dx$

14. $\int \dfrac{1}{x^2(2x + 1)} \, dx$

15. $\int \dfrac{x}{5 - 2x} \, dx$

16. $\int \dfrac{x}{(5 - 2x)^2} \, dx$

17. $\int \dfrac{x}{(5 - 2x)^3} \, dx$

18. $\int \dfrac{1}{\sqrt{4x^2 + 1}} \, dx$

19. $\int x^3 \ln x \, dx$

20. $\int x(\ln x)^3 \, dx$

21. $\int x^2 e^x \, dx$

22. $\int xe^{4x}\,dx$

23. $\int \frac{1}{1+e^{2x}}\,dx$

24. $\int x^2 e^{2x}\,dx$

25. $\int \frac{\sqrt{x+4}}{x^2}\,dx$

26. $\int \frac{\sqrt{x+4}}{x}\,dx$

In exercises 27 through 36, use the formulas in Appendix B to evaluate the definite integrals. Round answers to three decimal places.

27. $\int_0^1 x^2 e^x\,dx$

28. $\int_0^1 xe^{2x}\,dx$

29. $\int_0^1 \frac{x}{(x+1)^2}\,dx$

30. $\int_0^1 \frac{1}{9-x^2}\,dx$

31. $\int_0^4 \frac{1}{\sqrt{x^2+9}}\,dx$

32. $\int_4^5 \frac{1}{x\sqrt{x^2+9}}\,dx$

33. $\int_{-1}^1 \frac{1}{(x^2-4)^2}\,dx$

34. $\int_{-1}^1 \frac{1}{x^2-4}\,dx$

35. $\int_1^e (\ln x)^2\,dx$

36. $\int_1^2 x^2 \ln x\,dx$

SET B

In exercises 37 through 41, use a u substitution together with the integral tables. (See Example 2.)

37. $\int \frac{x^2}{1-x^6}\,dx$

38. $\int x\sqrt{x^4+16}\,dx$

39. $\int \frac{x^3}{4+x^2}\,dx$

40. $\int \frac{x^2}{9x^6-1}\,dx$

41. $\int \frac{x}{1+e^{x^2}}\,dx$

42. Redo Example 2 by factoring out 16 under the square root in the denominator instead of using a u substitution.

43. Find the area of the region bounded by the curve $y=\sqrt{x^2-9}$ and the x axis between $x=3$ and $x=5$. Round your answer to three decimal places.

44. Find the area of the region between the curves $y=e^x$ and $y=xe^x$ for $-1 \le x \le 0$ (see the figure accompanying exercise 34 in Section 7.1). Round your answer to three decimal places.

45. Find the volume of the solid generated by revolving the region bounded by the curve $y=xe^x$ and the x axis between $x=0$ and $x=1$ about the x axis. Round your answer to three decimal places.

46. Find the volume of the solid generated by revolving the region bounded by the curve $y=x\ln x$ and the x axis between $x=1$ and $x=2$ about the x axis. Round your answer to three decimal places.

BUSINESS
TOTAL PROFIT

47. The marginal profits of a company are given by

$$P'(x) = 650 - \frac{500}{\sqrt{x^2+1}}$$

where x is measured in hundreds of units and $P'(x)$ is measured in thousands of dollars. Find the total profits generated by increasing the number of units from 400 to 500.

TOTAL COST

48. The marginal cost function for the Alpha Can Company, which manufactures aluminum cans, is

$$C'(x) = \frac{\sqrt{x^2 + 9}}{x}$$

where x is the number of cans measured in thousands and $C'(x)$ is measured in hundreds of dollars. Find the total cost, $C(x)$, of increasing production from 10,000 to 12,000 cans.

REVENUE

49. The marginal revenue function for a company is given by

$$R'(x) = -xe^{-x} + e^{-x}$$

Use the Table of Integrals in Appendix B to determine the revenue function for the company. Assume that $R(0) = 0$.

SOCIAL SCIENCE
POPULATION

50. The population of a certain town is given by

$$P(t) = \sqrt{4t^2 + 5}$$

where t is measured in years and $P(t)$ is measured in thousands. Determine the average population over the time from $t = 1$ to $t = 10$.

POPULATION

51. Calculator Exercise
The world population could be approximated using the function

$$N(t) = \frac{82}{4.1 + 15.9e^{-0.0248t}}$$

where $N(t)$ is measured in billions. Using this function, estimate the average world population from 1980 ($t = 0$) to the year 2000.

PHYSICAL
SCIENCE
POSITION EQUATION

52. The velocity of an object moving in a straight line is given by

$$v(t) = \frac{1}{1 + 2t} + 5$$

where $v(t)$ is measured in feet per second and t is measured in seconds. Find the position, $s(t)$, of the object at any time, t, given that $s(0) = 3$.

7.3 NUMERICAL INTEGRATION TECHNIQUES

We can use the Fundamental Theorem to evaluate a definite integral only if we can find an antiderivative for our function. In this section, we learn to evaluate definite integrals for which we can find no antiderivative. The techniques provide us with only approximate answers; however, accuracy can be improved as desired by extended repetition of the processes.

The two techniques we use are called the **Trapezoidal Rule** and **Simpson's Rule.** You can perform the necessary calculations with a calculator.

Because we have determined that a definite integral may be interpreted as the area under a curve, it is with this interpretation that we first develop the Trapezoidal Rule. Let the function $y = f(x)$ be continuous between $x = a$ and $x = b$. Divide the interval from a to b into n subintervals using the numbers $x_0, x_1, x_2, \ldots, x_n$, with $x_0 = a$ and $x_n = b$. Each subinterval is of length Δx, where $\Delta x = \dfrac{b-a}{n}$ (see Figure 7.1).

Figure 7.1

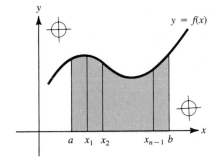

This is the same way we started when we approximated the area under a curve using rectangles. This time, however, instead of using rectangles, we use trapezoids. The formula for the area of a trapezoid is given in Figure 7.2. To form the trapezoids, we connect the points on the curve corresponding to the values $x_0, x_1, x_2, \ldots, x_n$ (see Figure 7.3).

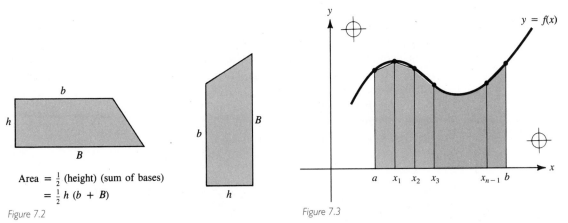

Area $= \frac{1}{2}$ (height) (sum of bases)
$= \frac{1}{2} h (b + B)$

Figure 7.2

Figure 7.3

The height of each trapezoid is Δx and the bases are represented by $f(x_0), f(x_1)$, $f(x_2), \ldots, f(x_n)$. The areas of the trapezoids are as follows:

area of 1st trapezoid $= \frac{1}{2}\Delta x[f(x_0) + f(x_1)]$

area of 2nd trapezoid $= \frac{1}{2}\Delta x[f(x_1) + f(x_2)]$

area of 3rd trapezoid $= \frac{1}{2}\Delta x[f(x_2) + f(x_3)]$

$\vdots$

area of nth trapezoid $= \frac{1}{2}\Delta x[f(x_{n-1}) + f(x_n)]$

By adding all of these areas together, we get an approximation to the area under the curve and thus an approximation to the definite integral.

$$\int_a^b f(x)\, dx \approx \frac{1}{2}\Delta x[f(x_0) + 2f(x_1) + 2f(x_2) + \cdots + 2f(x_{n-1}) + f(x_n)]$$

Replacing Δx with $\dfrac{b - a}{n}$, we have the following formula:

Trapezoidal Rule

For a function f continuous in the interval $[a, b]$,

$$\int_a^b f(x)\, dx \approx \frac{b - a}{2n}[f(x_0) + 2f(x_1) + 2f(x_2) + \cdots + 2f(x_{n-1}) + f(x_n)]$$

Notice in the formula that $f(x_0)$ and $f(x_n)$ are used only once, whereas the rest of the $f(x)$'s are doubled. This is because $f(x_0)$ and $f(x_n)$ act as bases for the trapezoids on the ends. The others are each used for bases for two trapezoids positioned side by side. Also, in most cases, the larger the value of n you choose, the better the approximation will be.

For our first example, we use a function that has an antiderivative, which means that the definite integral may be evaluated directly. We do so merely to give you some feel for the possible accuracy of this method.

Example 1   Use the Trapezoidal Rule to approximate $\int_0^2 (x^2 + 1)\, dx$. First let $n = 4$ and then let $n = 8$. Compare these results to the actual value.

Solution   We start by letting $n = 4$ and making a sketch, as in Figure 7.4.

Figure 7.4

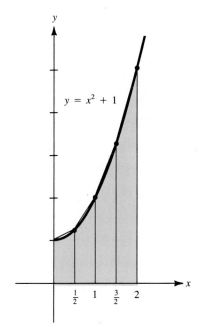

$y = x^2 + 1$

$\frac{1}{2}$ 1 $\frac{3}{2}$ 2

$$\Delta x = \frac{b - a}{n} = \frac{2 - 0}{4} = \frac{1}{2}$$

$a = x_0 = 0 \qquad f(x_0) = f(0) = 1$

$x_1 = \dfrac{1}{2} \qquad f(x_1) = f\left(\dfrac{1}{2}\right) = \dfrac{5}{4}$

$x_2 = 1 \qquad f(x_2) = f(1) = 2$

$x_3 = \dfrac{3}{2} \qquad f(x_3) = f\left(\dfrac{3}{2}\right) = \dfrac{13}{4}$

$b = x_4 = 2 \qquad f(x_4) = f(2) = 5$

Using the formula and substituting, we have

$$\int_0^2 (x^2 + 1)\, dx \approx \frac{2 - 0}{2(4)}\left[1 + 2\left(\frac{5}{4}\right) + 2(2) + 2\left(\frac{13}{4}\right) + 5\right]$$

$$= \frac{1}{4}\left(1 + \frac{5}{2} + 4 + \frac{13}{2} + 5\right)$$

$$= \frac{19}{4} \quad \text{or} \quad 4.75$$

Make a new sketch for $n = 8$ (see Figure 7.5) and compute again.

Figure 7.5

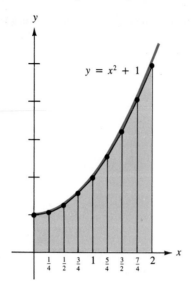

$$\Delta x = \frac{b - a}{n} = \frac{2 - 0}{8} = \frac{1}{4}$$

$$a = x_0 = 0 \qquad f(x_0) = f(0) = 1$$

$$x_1 = \frac{1}{4} \qquad f(x_1) = f\left(\frac{1}{4}\right) = \frac{17}{16}$$

$$x_2 = \frac{1}{2} \qquad f(x_2) = f\left(\frac{1}{2}\right) = \frac{5}{4}$$

$$x_3 = \frac{3}{4} \qquad f(x_3) = f\left(\frac{3}{4}\right) = \frac{25}{16}$$

$$x_4 = 1 \qquad f(x_4) = f(1) = 2$$

$$x_5 = \frac{5}{4} \qquad f(x_5) = f\left(\frac{5}{4}\right) = \frac{41}{16}$$

$$x_6 = \frac{3}{2} \qquad f(x_6) = f\left(\frac{3}{2}\right) = \frac{13}{4}$$

$$x_7 = \frac{7}{4} \qquad f(x_7) = f\left(\frac{7}{4}\right) = \frac{65}{16}$$

$$b = x_8 = 2 \qquad f(x_8) = f(2) = 5$$

Using the Trapezoidal Rule, we have

$$\int_0^2 (x^2 + 1)\, dx \approx \frac{2 - 0}{2(8)}\left[1 + 2\left(\frac{17}{16}\right) + 2\left(\frac{5}{4}\right) + 2\left(\frac{25}{16}\right)\right.$$

$$\left. + 2(2) + 2\left(\frac{41}{16}\right) + 2\left(\frac{13}{4}\right) + 2\left(\frac{65}{16}\right) + 5\right]$$

$$= \frac{1}{8}\left[\frac{600}{16}\right] = \frac{75}{16} \quad \text{or} \quad 4.6875$$

The actual value is computed by performing the integration.

$$\int_0^2 (x^2 + 1)\, dx = \left[\frac{x^3}{3} + x\right]_0^2 = \left(\frac{8}{3} + 2\right) - (0) = \frac{14}{3} \approx 4.6667$$

Notice that the second approximation ($n = 8$) gives a better estimate of the actual value. ∎

Example 2   Use the Trapezoidal Rule with $n = 5$ to approximate $\int_0^1 e^{-x^2}\, dx$.

Solution   This time, no antiderivative exists for the function $f(x) = e^{-x^2}$. We use decimals in all of our calculations for this example. A sketch of the function appears in Figure 7.6.

Figure 7.6

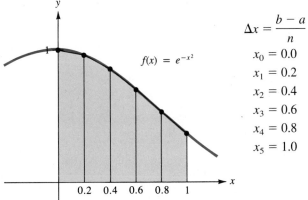

$$\Delta x = \frac{b-a}{n} = \frac{1-0}{5} = 0.2$$

$x_0 = 0.0 \qquad f(x_0) = f(0) = e^0$

$x_1 = 0.2 \qquad f(x_1) = f(0.2) = e^{-0.04}$

$x_2 = 0.4 \qquad f(x_2) = f(0.4) = e^{-0.16}$

$x_3 = 0.6 \qquad f(x_3) = f(0.6) = e^{-0.36}$

$x_4 = 0.8 \qquad f(x_4) = f(0.8) = e^{-0.64}$

$x_5 = 1.0 \qquad f(x_5) = f(1.0) = e^{-1.0}$

$f(x) = e^{-x^2}$

$$\int_0^1 e^{-x^2}\, dx \approx \frac{1-0}{2(5)}\, [e^0 + 2e^{-0.04} + 2e^{-0.16} + 2e^{-0.36} + 2e^{-0.64} + e^{-1.0}]$$

Using a calculator and rounding to four decimal places gives

$$\int_0^1 e^{-x^2}\, dx \approx 0.1[1 + 1.9216 + 1.7043 + 1.3954 + 1.0546 + 0.3679]$$

$$\approx 0.7444 \qquad \blacksquare$$

The second method of approximating a definite integral, *Simpson's Rule*, is presented without verification. This process, too, is based on using n subintervals of width $\Delta x = \dfrac{b-a}{n}$. However, in order to apply the formula, we must choose only even numbers for n. The process involves taking three of the points on the curve determined by the subintervals and connecting them with a parabola.

Simpson's Rule

For a function f continuous on the interval $[a, b]$,

$$\int_a^b f(x)\, dx \approx \frac{b-a}{3n}\, [f(x_0) + 4f(x_1) + 2f(x_2) + 4f(x_3)$$
$$+ \cdots + 4f(x_{n-1}) + f(x_n)]$$

where n is an even integer.

Note: The multiples 4 and 2 alternate ending with a "4."

Example 3   Use Simpson's Rule with $n = 4$ to approximate

$$\int_1^2 \frac{1}{x}\, dx$$

Solution   With $n = 4$, $\Delta x = \dfrac{b - a}{n} = \dfrac{2 - 1}{4} = \dfrac{1}{4}$. This gives

$$a = x_0 = 1 \qquad f(x_0) = f(1) = 1$$

$$x_1 = \frac{5}{4} \qquad f(x_1) = f\left(\frac{5}{4}\right) = \frac{4}{5}$$

$$x_2 = \frac{3}{2} \qquad f(x_2) = f\left(\frac{3}{2}\right) = \frac{2}{3}$$

$$x_3 = \frac{7}{4} \qquad f(x_3) = f\left(\frac{7}{4}\right) = \frac{4}{7}$$

$$b = x_4 = 2 \qquad f(x_4) = f(2) = \frac{1}{2}$$

Substituting into the formula gives

$$\int_1^2 \frac{1}{x}\,dx \approx \frac{2 - 1}{3(4)}\left[1 + 4\left(\frac{4}{5}\right) + 2\left(\frac{2}{3}\right) + 4\left(\frac{4}{7}\right) + \frac{1}{2}\right]$$

$$= \frac{1}{12}\left[\frac{1747}{210}\right] \approx 0.693254$$

Our original function $f(x) = \dfrac{1}{x}$ has an antiderivative, so we may calculate the exact value.

$$\int_1^2 \frac{1}{x}\,dx = [\ln x]_1^2 = \ln 2 - \ln 1 = \ln 2$$

But, using a calculator, $\ln 2 \approx 0.6931471$.   ∎

We conclude this section by mentioning that, for both the Trapezoidal Rule and Simpson's Rule, there are formulas for estimating how accurate an approximation will be for a particular value of n. We state these formulas (without proof) and then, in the remaining example of this section, we examine one of the formulas with help from technology.

Error using the Trapezoidal Rule

The error, T_E, in the approximation of a definite integral using the Trapezoidal Rule is given by

$$0 \le T_E \le \frac{(b - a)^3 M}{12n^2}$$

where M is the maximum value for $|f''(x)|$ on the interval $[a, b]$.

<table>
<tr><td>*Error using*
Simpson's Rule</td><td>The error, S_E, in the approximation of a definite integral using Simpson's Rule is given by

$$0 \le S_E \le \frac{(b-a)^5 M}{180n^4}$$

where M is the maximum value for $|f^{(4)}(x)|$ on the interval $[a, b]$.</td></tr>
</table>

Example 4   The real application of the error function in Simpson's Rule is to determine the value of n needed to ensure a given degree of precision. For the function $f(x) = e^{-x^2}$, find the value of n necessary to ensure that the Simpson's Rule calculation of $\displaystyle\int_0^1 e^{-x^2}\, dx$ is within 10^{-6} of the actual value.

Solution   We use a symbolic algebra program called *DERIVE®* to find $f^{(4)}(x) = 4e^{-x^2}(4x^4 - 12x^2 + 3)$. But what is M, the maximum value of $|f^{(4)}(x)|$ on $[0, 1]$? One way to answer this question is to graph $y = |4e^{-x^2}(4x^4 - 12x^2 + 3)|$ and see where it is greatest on $[0, 1]$. The answer is 12 and can be observed in the *DERIVE®* screen graph in Figure 7.7.

Figure 7.7

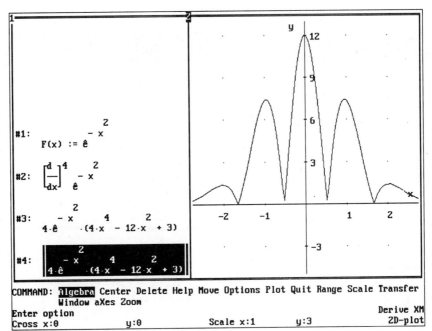

Thus, we have

$$S_E \le \frac{(b-a)^5 \cdot 12}{180n^4}$$

Now, if we want to find the value of n such that $S_E \le 10^{-6}$, we have the following:

$$S_E \le \frac{(b-a)^5 \cdot 12}{180n^4} \le 10^{-6}$$

$$\frac{1}{180n^4} \le \frac{10^{-6}}{12}$$

$$\frac{1}{n^4} \le \frac{10^{-6} \cdot 180}{12}$$

$$n^4 \ge 66666.67$$

$$n \ge 16.06$$

We round 16.06 up to the next even integer, so $n = 18$. In other words, the smallest (even) value of n that will ensure that Simpson's Rule is accurate to within 0.000001 for the given function on [0, 1] is $n = 18$. ∎

You should verify that $\int_0^1 e^{-x^2}\, dx \approx 0.7468241$ (see exercise 34); that approximation is precise to six decimal places.

You will probably find it extremely helpful to use a calculator for the exercises of this section.

7.3 EXERCISES

SET A

In exercises 1 through 12, using the given n, approximate the value of the definite integral using (a) the Trapezoidal Rule and (b) Simpson's Rule. And (c) calculate the exact value by evaluating the integral. (See Examples 1, 2, and 3.)

1. $\int_0^4 x^2\, dx,\ n = 4$

2. $\int_2^4 x^2\, dx,\ n = 4$

3. $\int_0^1 (1 - x^2)\, dx,\ n = 4$

4. $\int_0^2 (4 - x^2)\, dx,\ n = 4$

5. $\int_0^1 (1 - x^2)\, dx,\ n = 8$

6. $\int_0^2 (4 - x^2)\, dx,\ n = 8$

7. $\int_0^1 e^x\, dx,\ n = 6$

8. $\int_0^1 e^{-x}\, dx,\ n = 6$

9. $\int_1^4 \frac{1}{x}\, dx,\ n = 6$

10. $\int_1^4 \frac{1}{x^2}\, dx,\ n = 6$

11. $\int_0^4 \sqrt{x}\, dx,\ n = 4$

12. $\int_0^4 \sqrt{x}\, dx,\ n = 8$

In exercises 13 through 18, let $n = 4$ and approximate each definite integral using (a) the Trapezoidal Rule and (b) Simpson's Rule. (See Examples 1, 2, and 3.)

13. $\int_0^1 e^{x^2}\, dx$

14. $\int_1^2 e^{x^2}\, dx$

15. $\int_1^2 \sqrt{x^3 - 1}\, dx$

16. $\int_0^1 \sqrt{1 - x^3}\, dx$ **17.** $\int_0^1 \dfrac{1}{\sqrt{x^3 + 1}}\, dx$ **18.** $\int_0^2 \dfrac{1}{\sqrt{x^3 + 1}}\, dx$

SET B

19. The Trapezoidal Rule (or Simpson's Rule) can be used to approximate the area under a curve when no equation for the curve is known. It is possible to use a set of points whose x values are equally spaced over some interval to find the area under the curve. Use the following table of values and the Trapezoidal Rule to approximate the area under the curve (see the accompanying figure). (*Hint:* $a = 0$, $b = 6$, and $n = 6$.)

x	0	1	2	3	4	5	6
$f(x)$	2	4	6	5	2	3	2

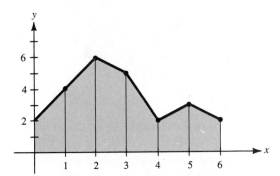

20. Use the Trapezoidal Rule to approximate the area under the curve through the points given in the table. See exercise 19.

x	1.0	1.2	1.4	1.6	1.8	2.0
$f(x)$	12	15	16	15	12	10

BUSINESS
TOTAL PROFIT

21. A company has been keeping a monthly record of its rate of increase in profits, $P'(t)$, over a 10-month period. The following table gives the data, where t is measured in months and $P'(t)$ is measured in thousands of dollars. Use the Trapezoidal Rule to approximate the total profit over the 10-month period.

t	1	2	3	4	5	6	7	8	9	10
$P'(t)$	12.4	15.1	15.4	17.0	20.5	20.5	21.0	24.0	24.2	25

COST

22. A company has determined the marginal cost, $C'(x)$, for the following levels of production. Using these data, and the Trapezoidal Rule, determine the total cost of producing eight items.

x	0	1	2	3	4	5	6	7	8
$C'(x)$	10	12	16	18	21	24	27	32	36

HOUSING STARTS

23. The following graph gives the rate of change in housing starts, $H'(t)$, for a city for a 6-month period.

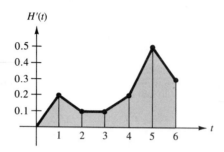

(H(t) is measured in hundreds of new homes.) Use the Trapezoidal Rule to find the total number of new homes for the 6-month period.

SALES

24. The sales management team for a company prepared a year-end report that included the following chart outlining the rate of change in sales, $S'(t)$, for the previous year (t is measured in months; sales is measured in thousands of dollars).

t	1	2	3	4	5	6	7	8	9	10	11	12
$S'(t)$	0.5	1.2	1.1	1.7	0.3	1.2	1.4	1.8	2	1.5	1.6	0.7

a. Someone on the team suggested that a graph would give a better overall view. Make a graph and, from it, decide during which month the rate of change in sales saw the biggest increase.

b. Use the Trapezoidal Rule to find total sales for the year.

PRODUCER SURPLUS

25. A company is willing to supply its product according to the equation

$$p = \sqrt{x^2 + x + 1}$$

where x is the number of units supplied and equilibrium is reached when $x = 5$. Use Simpson's Rule to determine the producer surplus, letting $n = 4$.

BIOLOGY
BACTERIA

26. The rate of growth of bacteria in a certain culture is given by

$$f(x) = 500xe^{0.1x^2}$$

where x is measured in hours. Find the total number of bacteria over the first 4 hours using Simpson's Rule with $n = 6$.

SOCIAL SCIENCE
DRUG USE

27. The rate of decline in birth defects in a certain neighborhood was believed to be due to an increased awareness of the effects of drug use by mothers through a new educational program. The rate of change in the number of babies born with birth defects is given by

$$B'(t) = e^{-0.2t^2}$$

where t is measured in months. Use Simpson's Rule with $n = 6$ to find the total number of babies born with birth defects for the first 6 months.

PHYSICAL SCIENCE
DISTANCE TRAVELED

28. The velocity of an object was measured at 10-sec intervals for a period of 1 minute. If $v(t)$ represents velocity measured in feet per second and t is time measured in seconds, use the Trapezoidal Rule to approximate the distance traveled over this time period.

t	0	10	20	30	40	50	60
$v(t)$	12	12.5	13	12.5	13	15	16

ELECTRICITY

29. The following graph gives the discharge curve through a fixed resistance of an electrical flashlight cell. The area under the curve gives the quantity of electricity consumed. Use Simpson's Rule with $n = 10$ to find the quantity of electricity consumed.

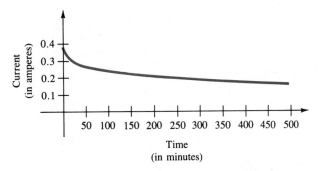

Time	0	50	100	150	200	250	300	350	400	450	500
Current	0.39	0.27	0.25	0.23	0.22	0.21	0.20	0.19	0.18	0.17	0.16

30. Find the upper bound for the error in the approximation for $\int_1^2 x^3 \, dx$ using the Trapezoidal Rule with $n = 4$.

31. Determine the value of n needed in using the Trapezoidal Rule to estimate $\int_1^2 \sqrt{x} \, dx$ so that the error is less than 0.001.

32. Find the upper bound for the error in the approximation for $\int_0^1 x^4 \, dx$ using Simpson's Rule with $n = 6$.

33. Determine the value of n needed in using Simpson's Rule to compute $\int_1^2 \frac{1}{x^2} \, dx$ so that the error is less than 0.002.

34. Use an appropriate form of technology to verify the value of 0.7468241 as an approximation to $\int_0^1 e^{-x^2} \, dx$ using Simpson's Rule with $n = 18$. (See Example 4.)

USING TECHNOLOGY IN CALCULUS

1. Use a spreadsheet to calculate the area given in exercise 19. Establish three columns; place x values in column A, $f(x)$ values in column B, and the values in the terms (either $f(x_i)$ or $2f(x_i)$ depending on i) in column C. Finally, sum the values in column C.

2. Some software packages—Converge®, for example—offer numerical approximations. The following screens show a Converge® graph and approximations using rectangles, the Trapezoidal Rule, and Simpson's Rule for the area under the curve $y = e^{x^2}$ in [0, 1]:

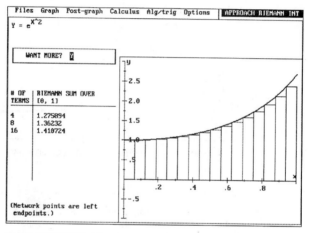

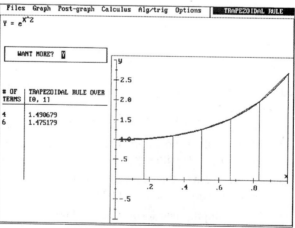

(*continued*)

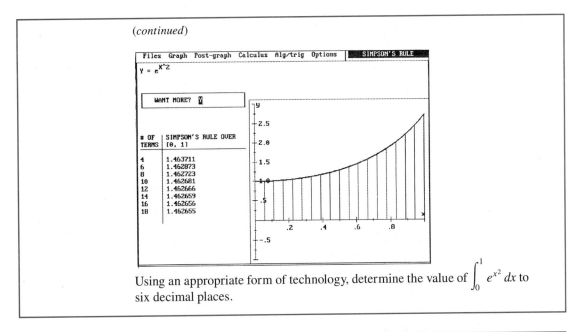

(continued)

Using an appropriate form of technology, determine the value of $\int_0^1 e^{x^2}\,dx$ to six decimal places.

7.3A   SIMPSON'S RULE ON A GRAPHICS CALCULATOR

Many methods are available for approximating the area under a curve and definite integrals. In this section, we examine **Simpson's Rule**, the method that forms the basis for many computer and calculator programs with such a capability.

Suppose we have a graph $y = f(x)$ over the interval $[a, b]$. To approximate the area between the curve and the x axis, we first break up the interval $[a, b]$ into an even number of subintervals: $x_0 = a, x_1, x_2, \ldots, x_{n-1}, x_n = b$. (See Figure 7.3A.1.)

Figure 7.3A.1 (a) $y = f(x)$. (b) The same function broken into an even number (eight) of subintervals.

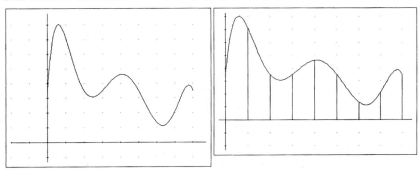

With Simpson's Rule, we use "pieces" of parabolas to approximate the area under portions of the curve. We do this using three x subinterval values at a time, starting at x_0. For example, we use x_0, x_1, and x_2 to determine the parabola that goes through the three points (x_0, y_0), (x_1, y_1), and (x_2, y_2), which approximates the area under the portion of the curve from x_0 to x_2. (See Figure 7.3A.2.)

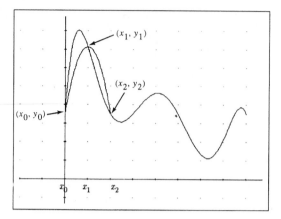

The area under that parabola approximates the area under the curve for the interval $[x_0, x_2]$. We continue the "parabola-finding" procedure for the interval $[x_2, x_4]$, then for $[x_4, x_6]$, and so on. (See Figure 7.3A.3.) The total area under the curve on $[a, b]$ is found by adding up the areas under the parabolas.

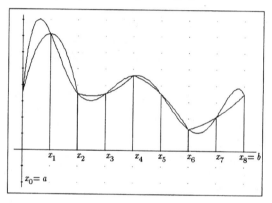

This can be shown by the following approximation:

Simpson's Rule

For a function f continuous on the interval $[a, b]$,

$$\int_a^b f(x)\, dx \approx \frac{h}{3}[f(x_0) + 4f(x_1) + 2f(x_2) + 4f(x_3) + 2f(x_4)$$
$$+ \cdots + 4f(x_{n-1}) + f(x_n)]$$

where h is the width of each subinterval, $\dfrac{b-a}{n}$, and n is an even integer.

In this section, we use the programming feature of a graphics calculator to compute areas using Simpson's Rule. Before we do, however, it would be helpful to rewrite the approximation by moving the $f(x_n)$ term to a location just after $f(x_0)$:

$$\frac{h}{3}[f(x_0) + f(x_n) + 4f(x_1) + 2f(x_2) + 4f(x_3) + 2f(x_4) + \cdots + 4f(x_{n-1})] \qquad (1)$$

Notice that once we calculate $f(x_0) = f(a)$ and $f(x_n) = f(b)$, all other y values are multiplied by either 4 (if the x subscript index is odd) or 2 (if the x subscript index is even). The sum of those y values (with appropriate coefficients) is then multiplied by $\frac{h}{3}$.

The SIMPSON program follows. Notice that we use the value of K in the program as a counter; think of K as the subscript, or index, of the x's. Also, we use the value of S as the accumulator of the terms within brackets in equation (1).

SIMPSON'S RULE PROGRAM

```
PROGRAM:SIMPSON
:ClrHome
::ClrDraw
:Disp "LOWER LIM
IT":Input A
:Disp "UPPER LIM
IT":Input B
:
```

1. For clarity, we clear the home (text) and graphics screens.

2. We prompt and have the user input the values of A (lower limit) and B (upper limit).

```
PROGRAM:SIMPSON
:Disp "N INTERVA
LS"
:Input N
:ClrHome
:Disp "WAIT..."
:0→S
:1→K
```

3. The number of intervals, N, is input by the user. N must be an even integer.

4. The variable S is an accumulator and stores the sum; here it is initialized to 0. K is a counter and is initialized to 1.

```
PROGRAM:SIMPSON
:(B-A)/N→H
:A→X
:Y₁→S
:B→X
:S+Y₁→S
:
:
```

5. The value of H is the constant (B-A)/N.

6. The values of $f(A)$ and $f(B)$ are accumulated in the variable S.

```
PROGRAM:SIMPSON
:Lbl 1
:A+K*H→X
:If iPart (K/2)=
K/2:2→M
:If iPart (K/2)≠
K/2:4→M
:
```

7. LBL 1 is a loop that is executed as long as K is less than or equal to N−1.
8. The values of X are increased by H each time through the loop.
9. The values of $f(X)$ are multiplied either by M=2 (if the index K is even) or by M=4 (if the index K is odd).

```
PROGRAM:SIMPSON
:M*Y₁+S→S
:IS>(K,N-1)
:Goto 1
:(H/3)*S→S
:Shade(0,Y₁,1,A,
B)
:
```

10. S in increased by the value M times $f(X)$.
11. K is incremented and the loop LBL 1 is repeated provided K is less than or equal to N−1.
12. When we are done accumulating (and out of the loop), the sum is multiplied by H/3.
13. The area is shaded in the graphics screen.

```
PROGRAM:SIMPSON
:Pause
:ClrHome
:Disp "AREA ="
:Disp S
```

14. The final approximation is displayed on the home screen.

Let's execute this program to estimate the value of $\int_0^1 e^{-x^2}\,dx$; we will use 20 intervals. To see how the program works, it is important to realize that the function must first be entered as "Y1" (see Step 1).

Step 1: Before we can execute the program, we enter the function as Y1 and choose a suitable graphing window.

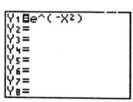

Step 2: Upon executing the program, we input the values of 0 (for *a*), 1 (for *b*), and 20 (for *n*).

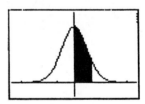

Step 3: The area between the function and the *x* axis is shaded between *a* and *b*.

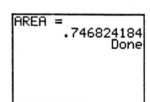

Step 4: The approximation of the area is displayed as 0.746824184.

```
AREA =
         .746824184
             Done
```

Thus, we can write $\int_0^1 e^{-x^2}\,dx \approx 0.746824184$.

The advantage of entering the function as Y1 and not directly into the program is that the program does not have to be rewritten for different functions. For example, to find $\int_1^2 \frac{1}{x}\,dx$, we need only change the function Y1. Next, we run the program for a variety of different values of *n*:

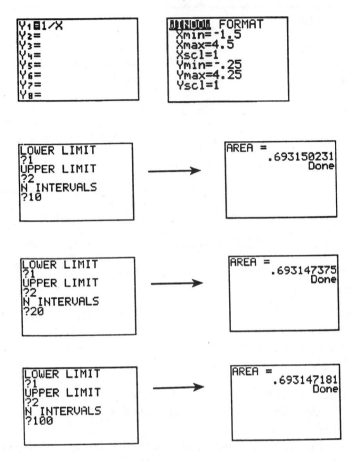

In this case, we know the *exact* value of the integral: $\int_{1}^{2} \frac{1}{x} \, dx = \ln 2 \approx 0.69314718056$.

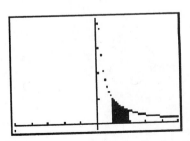

7.3A   EXERCISES

In exercises 1 through 8, use the program presented in this section to approximate the requested definite integral with the specified value of n. Then, use calculus to find the exact value of the integral.

1. $\int_1^3 \dfrac{1}{x}\, dx,\ n = 10$

2. $\int_0^1 e^x\, dx,\ n = 10$

3. $\int_1^3 \dfrac{1}{x}\, dx,\ n = 50$

4. $\int_0^1 e^x\, dx,\ n = 50$

5. $\int_1^4 \dfrac{1}{x^2}\, dx,\ n = 4$

6. $\int_1^4 \dfrac{1}{x^2}\, dx,\ n = 20$

7. $\int_0^4 \sqrt{x}\, dx,\ n = 4$

8. $\int_0^4 \sqrt{x}\, dx,\ n = 100$

In exercises 9 through 16, use n = 10 and approximate each definite integral using Simpson's Rule.

9. $\int_1^3 e^{x^2}\, dx$

10. $\int_1^3 e^{-x^2}\, dx$

11. $\int_1^4 \sqrt{x^3 - 1}\, dx$

12. $\int_0^1 \sqrt{1 - x^3}\, dx$

13. $\int_0^4 \dfrac{1}{\sqrt[3]{x^2 + 1}}\, dx$

14. $\int_0^4 \dfrac{1}{\sqrt{x^3 + 1}}\, dx$

15. $\int_0^8 |4 - x|\, dx$

16. $\int_0^4 |4 - x^2|\, dx$

17. The actual function graphed in Figure 7.3A.1 on the interval [0, 8] is

$$f(x) = -2x^6 + 51x^5 - 492x^4 + 2221x^3 - 4650x^2 + 3600x + 700$$

   If we use eight intervals, we have $x_0 = 0$, $x_1 = 1$, and so on. Find the equation of the actual parabola in Figure 7.3A.2. (*Hint:* You must find A, B, and C in $y = Ax^2 + Bx + C$, and each of the points satisfies that equation.)

18. Suppose we generalize exercise 17 and let the three points through which the first parabola passes be represented by (x_0, y_0), (x_1, y_1), and (x_2, y_2). If we write the equation of the parabola as $y = Ax^2 + Bx + C$, we can write:

$$y_0 = Ax_0^2 + Bx_0 + C$$
$$y_1 = Ax_1^2 + Bx_1 + C$$
$$y_2 = Ax_2^2 + Bx_2 + C$$

   a. Show that $y_1 - y_0 = A(x_1^2 - x_0^2) + B(x_1 - x_0)$ and

   $$y_2 - y_1 = A(x_2^2 - x_1^2) + B(x_2 - x_1)$$

   b. Show that $\dfrac{y_1 - y_0}{h} = A(x_1 + x_0) + B$ and that

   $$\dfrac{y_2 - y_1}{h} = A(x_2 + x_1) + B$$

   (*Hint:* Divide the first equation in part a by $(x_1 - x_0)$. Since $(x_1 - x_0)$ represents the width of one interval on the x axis, it equals h.)

c.   Subtract the two equations in part b to show that we can write $A = -\dfrac{y_0 - 2y_1 + y_2}{2h^2}$.

d.   Show that $Bx_1 + C = y_1 - Ax_1^2$.

19.   **a.**   Why can we write

$$\int_{x_0}^{x_2} (Ax^2 + Bx + C)\, dx = \int_{x_1 - h}^{x_1 + h} (Ax^2 + Bx + C)\, dx?$$

b.   Show that $\displaystyle\int_{x_1 - h}^{x_1 + h} (Ax^2 + Bx + C)\, dx$ can be rewritten as

$$\frac{A}{3}(6hx_1^2 + 2h^3) + 2h(Bx_1 + C)$$

c.   Using the substitution from exercise 18d, rewrite the area under the first parabola as $2h\left(\dfrac{Ah^2}{3} + y_1\right)$ and then, using the substitution from exercise 18c, show that this area can be written as $\dfrac{h}{3}(y_0 + 4y_1 + y_2)$.

20.   **a.**   If the procedures of exercises 18 and 19 are repeated for the next parabola using x_2, x_3, and x_4, what would we find the area to be?

b.   If the areas are found under the four parabolas, show that they add up to:

$$\frac{h}{3}(y_0 + 4y_1 + 2y_2 + 4y_3 + 2y_4 + 4y_5 + 2y_6 + 4y_7 + y_8)$$

21.   For the function in Figure 7.3A.1,

$$f(x) = -2x^6 + 51x^5 - 492x^4 + 2221x^3 - 4650x^2 + 3600x + 700$$

it can be shown that the equations of the four parabolas are

On $[0, 2]$: $y = -728x^2 + 1456x + 700$
On $[2, 4]$: $y = 112x^2 - 560x + 1372$
On $[4, 6]$: $y = -104x^2 + 712x - 260$
On $[6, 8]$: $y = 64x^2 - 680x + 2044$

Show that the quantity

$$\int_0^2 (-728x^2 + 1456x + 700)\, dx + \int_2^4 (112x^2 - 560x + 1372)\, dx$$

$$+ \int_4^6 (-104x^2 + 712x - 260)\, dx + \int_6^8 (64x^2 - 680x + 2044)\, dx$$

equals the Simpson's Rule approximation of

$$\int_0^8 (-2x^6 + 51x^5 - 492x^4 + 2221x^3 - 4650x^2 + 3600x + 700)\, dx$$

with $n = 8$.

7.4 IMPROPER INTEGRALS

In Chapter 6, we learned to compute the area under a continuous curve on a closed interval. In this section, we examine how to find the area under a continuous curve on an interval that extends infinitely in one or two directions. Such calculations are important in applications, especially in the field of probability and statistics. The integrals that represent these areas are called **improper integrals.**

Let's begin with the area bounded by the curve $y = \dfrac{1}{x^2}$ from $x = 1$ to $x = 2$. This area can be computed using the integral:

$$\int_1^2 \frac{1}{x^2}\,dx = \int_1^2 x^{-2}\,dx$$
$$= -\frac{1}{x}\Bigg]_1^2$$
$$= \frac{1}{2}$$

This area is shown in Figure 7.8.

Figure 7.8

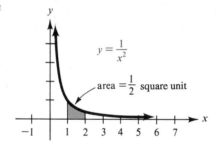

Now suppose we want to determine the area bounded by the curve from $x = 1$ to $x = b$, where b is some finite number greater than 1. The integral representing this area is given by

$$\int_1^b \frac{1}{x^2}\,dx$$

We make a table of values for some representative values of b, with b steadily increasing in value. See Table 7.1.

Table 7.1

Value of b	Area in square units
2	$1/2 = 0.5$
5	$4/5 = 0.8$
10	$9/10 = 0.9$
100	$99/100 = 0.99$
1000	$999/1000 = 0.999$
10,000	$9999/10,000 = 0.9999$

From the table, it appears that as b increases, the area gets close to 1. This brings to mind the notion of a limit. We can symbolically represent the idea that as b grows without bound, the integral value approaches 1 as

$$\lim_{b \to \infty} \left[\int_1^b \frac{1}{x^2}\,dx \right] = 1$$

We are now in a position to identify our first type of *improper integral* and see how to evaluate such integrals. The integral represented by

$$\int_1^\infty \frac{1}{x^2}\,dx$$

is considered to be an *improper integral* because the upper limit of integration is ∞. In Example 1, we demonstrate the step-by-step procedure for evaluating this integral; in Example 2, we also see how to interpret our findings.

Example 1   Determine the numerical value, if it exists, of

$$\int_1^\infty \frac{1}{x^2}\,dx$$

Solution   We begin by rewriting the improper integral as a limit and then solve.

$$\int_1^\infty \frac{1}{x^2}\,dx = \lim_{b \to \infty} \left[\int_1^b \frac{1}{x^2}\,dx \right]$$

$$= \lim_{b \to \infty} \left[\int_1^b x^{-2}\,dx \right]$$

$$= \lim_{b \to \infty} \left[-\frac{1}{x} \right]_1^b$$

$$= \lim_{b \to \infty} \left[-\frac{1}{b} - (-1) \right]$$

$$= \lim_{b \to \infty} \left[1 - \frac{1}{b} \right]$$

$$= 1 - 0$$

$$= 1$$ ∎

Example 2   Determine the numerical value, if it exists, for

$$\int_2^\infty \frac{1}{(x + 1)^4}\,dx$$

Solution   Notice that we are representing the area bounded by the curve $y = \dfrac{1}{(x+1)^4}$ from $x = 2$ to ∞. See Figure 7.9.

$$\int_2^\infty \frac{1}{(x+1)^4}\,dx = \lim_{b \to \infty}\left[\int_2^b (x+1)^{-4}\,dx\right]$$

$$= \lim_{b \to \infty}\left[\frac{(x+1)^{-3}}{-3}\right]_2^b$$

$$= \lim_{b \to \infty}\left[\frac{-1}{3(b+1)^3} + \frac{1}{81}\right]$$

$$= \frac{1}{81}$$

Figure 7.9

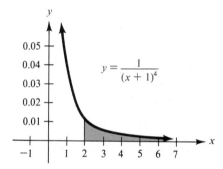

$$y = \frac{1}{(x+1)^4}$$

The next example shows another type of improper integral—this time involving $-\infty$. Notice again the use of the limit and a constant to represent the concept of approaching $-\infty$.

Example 3   Determine the numerical value, if it exists, for

$$\int_{-\infty}^{-1} \frac{5}{x^3}\,dx$$

Solution
$$\int_{-\infty}^{-1} \frac{5}{x^3}\,dx = \lim_{a \to -\infty}\left[\int_a^{-1} 5x^{-3}\,dx\right]$$

$$= \lim_{a \to -\infty}\left[\frac{5x^{-2}}{-2}\right]_a^{-1}$$

$$= \lim_{a \to -\infty}\left[\frac{-5}{2(-1)^2} - \frac{-5}{2a^2}\right]$$

$$= \lim_{a \to -\infty}\left[\frac{5}{2a^2} - \frac{5}{2}\right]$$

$$= \frac{-5}{2}$$

So far, all three of our examples of improper integrals have resulted in finite numerical values. If such is the case, then the improper integral is said to **converge** to that value. The next example illustrates how an improper integral can **diverge.**

Example 4 Determine the numerical value, if it exists, for

$$\int_1^\infty \frac{1}{\sqrt{x+3}}\, dx$$

Solution $\displaystyle \int_1^\infty \frac{1}{\sqrt{x+3}}\, dx = \lim_{b \to \infty}\left[\int_1^b (x+3)^{-1/2}\, dx\right]$

$$= \lim_{b \to \infty} [2(x+3)^{1/2}]_1^b$$

$$= \lim_{b \to \infty} [2(b+3)^{1/2} - 2(4)^{1/2}]$$

$$= \lim_{b \to \infty} [2(b+3)^{1/2} - 4]$$

Because this limit approaches infinity as b approaches infinity, the improper integral is said to *diverge*. ∎

We are most interested in three types of improper integrals. They are outlined here.

Improper integrals

Suppose f is a continuous curve and a and b are arbitrary real numbers. Then

1. $\displaystyle \int_a^\infty f(x)\, dx = \lim_{b \to \infty} \int_a^b f(x)\, dx$

2. $\displaystyle \int_{-\infty}^b f(x)\, dx = \lim_{a \to -\infty} \int_a^b f(x)\, dx$

3. $\displaystyle \int_{-\infty}^\infty f(x)\, dx = \int_{-\infty}^c f(x)\, dx + \int_c^\infty f(x)\, dx,$ where c is any real number

We have not yet seen an example of type 3. In the next example, we see that an improper integral with limits of $-\infty$ and ∞ can be split into two separate improper integrals.

Example 5 Determine whether the improper integral converges or diverges. If it converges, find the value to which it converges. See Figure 7.10.

Figure 7.10

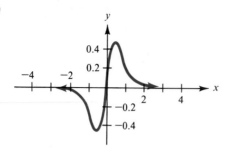

$$\int_{-\infty}^{\infty} \frac{2x}{(x^2 + 1)^4} \, dx$$

Solution   We split the integral using $c = 0$.

$$\int_{-\infty}^{\infty} \frac{2x}{(x^2 + 1)^4} \, dx = \int_{-\infty}^{0} 2x(x^2 + 1)^{-4} \, dx + \int_{0}^{\infty} 2x(x^2 + 1)^{-4} \, dx$$

$$= \lim_{a \to -\infty} \int_{a}^{0} 2x(x^2 + 1)^{-4} \, dx + \lim_{b \to \infty} \int_{0}^{b} 2x(x^2 + 1)^{-4} \, dx$$

Each integral requires a u substitution with $u = x^2 + 1$ and $du = 2x \, dx$.

$$= \lim_{a \to -\infty} \left[\frac{-(x^2 + 1)^{-3}}{3} \right]_{a}^{0} + \lim_{b \to \infty} \left[\frac{-(x^2 + 1)^{-3}}{3} \right]_{0}^{b}$$

$$= \lim_{a \to -\infty} \left[-\frac{1}{3} - \frac{-1}{3(a^2 + 1)^3} \right] + \lim_{b \to \infty} \left[\frac{-1}{3(b^2 + 1)^3} - \frac{-1}{3} \right]$$

$$= \frac{-1}{3} + \frac{1}{3}$$

$$= 0$$

This improper integral converges to zero.   ∎

The *probability density function* was introduced in exercise 64 in Section 6.3. At that time, given the function $f(x) = 0.25e^{-0.25x}$, the integral

$$\int_{a}^{b} f(x) \, dx$$

was used to represent the probability that the life span of a randomly selected watch would be between a and b years. In general, a probability density function must satisfy two conditions: (1) The function must be nonnegative on a given interval and (2) the integral of the function over that interval must be equal to 1.

Example 6   **a.**   Verify that $f(x) = 0.25e^{-0.25x}$ satisfies the conditions for a probability density function on $[0, \infty)$.

b.   Determine the probability that the life span of a randomly selected watch will be longer than 2 years.

Solution   **a.**   In Chapter 5, we learned that a number of the form e^{-n} is always positive; thus, $0.25e^{-0.25x}$ is also positive. For the second condition to be met, we must evaluate the improper integral.

$$\int_0^\infty 0.25e^{-0.25x} \, dx$$

$$\int_0^\infty 0.25e^{-0.25x} \, dx = \lim_{b \to \infty} \int_0^b 0.25e^{-0.25x} \, dx$$

$$= \lim_{b \to \infty} \, [-e^{-0.25x}]_0^b$$

$$= 0 - (-1)$$

$$= 1$$

b.   We must find $\int_2^\infty 0.25e^{-0.25x} \, dx$.

$$\int_2^\infty 0.25e^{-0.25x} \, dx = \lim_{b \to \infty} \, [-e^{-0.25x}]_2^b$$

$$= 0 - (-e^{-0.5})$$

$$\approx 0.61 \qquad \blacksquare$$

7.4 EXERCISES

SET A

In exercises 1 through 24, determine whether the improper integrals converge or diverge, and for those that converge, find their value. (See Examples 1 through 5.)

1. $\displaystyle\int_1^\infty \frac{4}{x^2} \, dx$

2. $\displaystyle\int_2^\infty \frac{7}{x^3} \, dx$

3. $\displaystyle\int_8^\infty \frac{6}{\sqrt[3]{x}} \, dx$

4. $\displaystyle\int_1^\infty \frac{1}{\sqrt[4]{x}} \, dx$

5. $\displaystyle\int_2^\infty \frac{4}{\sqrt{x-1}} \, dx$

6. $\displaystyle\int_1^\infty \frac{3}{(x+2)^2} \, dx$

7. $\displaystyle\int_3^\infty \frac{1}{(x-2)^2} \, dx$

8. $\displaystyle\int_5^\infty \frac{2}{\sqrt[3]{x+3}} \, dx$

9. $\displaystyle\int_{-\infty}^3 \frac{1}{\sqrt{4-x}} \, dx$

10. $\displaystyle\int_{-\infty}^0 \frac{1}{(1-x)^2} \, dx$

11. $\displaystyle\int_0^\infty \frac{2x}{\sqrt{x^2+1}} \, dx$

12. $\displaystyle\int_1^\infty \frac{2x}{(x^2+1)^2} \, dx$

13. $\displaystyle\int_{-\infty}^{-2} \frac{5}{(x+1)^2} \, dx$

14. $\displaystyle\int_{-\infty}^{-1} \frac{6}{\sqrt[3]{(x+2)^2}} \, dx$

15. $\displaystyle\int_1^\infty \frac{1}{(2x-1)^3} \, dx$

16. $\displaystyle\int_4^\infty \frac{1}{x\sqrt{x}} \, dx$

17. $\displaystyle\int_1^\infty \frac{1}{\sqrt{4x-3}} \, dx$

18. $\displaystyle\int_2^\infty \frac{4}{\sqrt{2x-3}} \, dx$

19. $\displaystyle\int_{-\infty}^\infty \frac{8x}{\sqrt[3]{x^2+1}} \, dx$

20. $\displaystyle\int_{-\infty}^\infty \frac{4x}{(x^2+1)^5} \, dx$

21. $\displaystyle\int_{-\infty}^0 e^{3x} \, dx$

22. $\displaystyle\int_0^\infty e^{-4x}\,dx$

23. $\displaystyle\int_0^\infty 2xe^{-x^2}\,dx$

24. $\displaystyle\int_{-\infty}^\infty xe^{-x^2}\,dx$

SET B

In exercises 25 through 28, use the fact that as $N \to \infty$, $\ln N \to \infty$ to determine whether the improper integrals converge or diverge.

25. $\displaystyle\int_1^\infty \frac{1}{x+3}\,dx$

26. $\displaystyle\int_5^\infty \frac{3}{x-4}\,dx$

27. $\displaystyle\int_0^\infty \frac{4x}{x^2+1}\,dx$

28. $\displaystyle\int_{-\infty}^0 \frac{x}{x^2+1}\,dx$

29. Find the area of the region bounded by the curve $y = \dfrac{2}{x^2}$ and the x axis for $x \ge 1$. Sketch the region.

30. Find the area of the region bounded by the curve $y = e^{-x}$ and the x axis for $x \ge 0$. Sketch the region.

31. Sketch the graphs of $y = e^{-x}$ and $y = -e^{-x}$ on the same set of axes. Find the area between these two curves for $x \ge 1$. Round the answer to three decimal places.

32. Use integration by parts to determine whether the improper integrals converge or diverge:

    a. $\displaystyle\int_{-\infty}^0 xe^x\,dx$ **b.** $\displaystyle\int_0^\infty x^2 e^{-x}\,dx$

    (*Hint:* Use the fact that $\lim\limits_{a \to -\infty} ae^a = 0$.)

33. Use the results of Example 5 to show that the area bounded by the curve $y = \dfrac{2x}{(x^2+1)^4}$ and the x axis is $\dfrac{2}{3}$.

34. A trust fund has been set up to provide $10,000 a year starting a year from now and continuing indefinitely. The amount of money that must be invested now at 8% compounded continuously is given by

$$\int_0^\infty 10{,}000e^{-0.08x}\,dx$$

Find the amount that must be invested now.

35. Redo exercise 34 if the trust fund is to provide $15,000 a year.

36. A local industry has been found to be exceeding EPA standards for the release of pollutants into the air in a certain city. The rate at which the pollutants are being released is given by

$$P(t) = \frac{1}{(t+0.01)^2}$$

where t is measured in months and P is measured in parts per square mile. Find the total number of pollutants in parts per square mile if the industry is allowed to continue polluting at the current rate into the future. $\left(\text{Hint: Evaluate the improper integral } \int_0^\infty P(t)\,dt.\right)$

37. Suppose the rate for exercise 36 had been

$$P(t) = \frac{1}{t + 0.01}$$

What would that mean for the level of pollutants into the future?

WRITING AND CRITICAL THINKING

In addition to the types of improper integrals mentioned in this section, a definite integral on [a, b] is also called "improper" if it is undefined at any point on [a, b].

Describe in complete sentences why the following *is not correct* and explain what a logical answer to the integral might be:

$$\int_{-1}^{1} \frac{1}{x^2}\,dx = -2$$

USING TECHNOLOGY IN CALCULUS

Here we consider the *Horn of Gabriel.** We start with the function $y = \dfrac{1}{x}$. It can be shown that if this function is revolved about the x axis from $x = 1$ to ∞, the improper integral representing the volume of that solid of revolution is $\int_1^\infty \pi \left(\dfrac{1}{x}\right)^2 dx$. This solid of revolution is sometimes referred to as the *Horn of Gabriel*. The calculation of that volume is shown in the following *DERIVE®* screen; it is π, which is a finite volume. The surface area of the horn, however, is *infinite*. To substantiate that fact, we can show that the surface area, S, is given by the formula:

$$S = 2\pi \int_1^\infty f(x)\sqrt{1 + [f'(x)]^2}\,dx = 2\pi \int_1^\infty \frac{1}{x}\sqrt{1 + \frac{1}{x^4}}\,dx$$

(See lines 1 through 3.)

(continued)

---

*This exercise is adapted from *Calculus and the DERIVE Program: Experiments with the Computer*, 2d ed., by Lawrence G. Gilligan and James F. Marquardt (Cincinnati: Gilmar Publishing, 1991), pp. 93–94.

(continued)

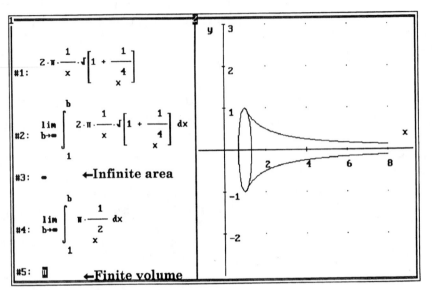

#1: $2 \cdot \pi \cdot \dfrac{1}{x} \cdot \sqrt{\left[1 + \dfrac{1}{x^4}\right]}$

#2: $\lim\limits_{b \to \infty} \displaystyle\int_{1}^{b} 2 \cdot \pi \cdot \dfrac{1}{x} \cdot \sqrt{\left[1 + \dfrac{1}{x^4}\right]}\, dx$

#3: ∎   ←**Infinite area**

#4: $\lim\limits_{b \to \infty} \displaystyle\int_{1}^{b} \pi \cdot \dfrac{1}{x^2}\, dx$

#5: ⊡   ←**Finite volume**

Thus, there is no amount of paint that could paint the *Horn of Gabriel*, yet it holds π cubic units of paint!

Use an appropriate technology to verify the calculations in the screen.

CHAPTER 7 REVIEW

VOCABULARY/FORMULA LIST

integration by parts:

$$\int u(x)v'(x)\, dx = u(x)v(x) - \int v(x)u'(x)\, dx$$

Trapezoidal Rule
Simpson's Rule
improper integral

converge
diverge

CHAPTER 7 REVIEW EXERCISES

In exercises 1 through 8, evaluate each indefinite integral. Use integral tables where necessary.

[7.1]   **1.** $\displaystyle\int \ln 4x\, dx$ [7.1]   **2.** $\displaystyle\int \ln \sqrt{x}\, dx$

[7.2]   **3.** $\displaystyle\int \frac{3x}{(1 + 2x)^2} \, dx$

[7.2]   **4.** $\displaystyle\int \frac{1}{9x^2 - 16} \, dx$

[7.1]   **5.** $\displaystyle\int xe^{4x} \, dx$

[7.1]   **6.** $\displaystyle\int 2xe^{-x} \, dx$

[7.2]   **7.** $\displaystyle\int \frac{1}{1 + e^{2x}} \, dx$

[7.2]   **8.** $\displaystyle\int \frac{\sqrt{x^2 + 9}}{x} \, dx$

In exercises 9 through 12, evaluate each definite integral. Use integral tables where necessary. Where appropriate, round answers to three decimal places.

[7.1]   **9.** $\displaystyle\int_0^1 xe^{2x} \, dx$

[7.1]   **10.** $\displaystyle\int_1^3 x \ln x \, dx$

[7.2]   **11.** $\displaystyle\int_0^1 \frac{1}{9 - x^2} \, dx$

[7.1]   **12.** $\displaystyle\int_0^3 x\sqrt{x + 1} \, dx$

In exercises 13 through 16, using the given n, approximate the value of the definite integral using (a) the Trapezoidal Rule and (b) Simpson's Rule. Round answers to four decimal places.

[7.3]   **13.** $\displaystyle\int_1^3 \frac{1}{x} \, dx, \, n = 8$

[7.3]   **14.** $\displaystyle\int_0^1 x^4 \, dx, \, n = 4$

[7.3]   **15.** $\displaystyle\int_0^3 \sqrt[3]{x} \, dx, \, n = 6$

[7.3]   **16.** $\displaystyle\int_0^1 \frac{1}{\sqrt{4 - x^3}} \, dx, \, n = 4$

In exercises 17 through 22, determine whether the improper integrals converge or diverge, and for those that converge, find their value.

[7.4]   **17.** $\displaystyle\int_1^\infty \frac{3}{x^2} \, dx$

[7.4]   **18.** $\displaystyle\int_1^\infty \frac{1}{\sqrt[3]{x^2}} \, dx$

[7.4]   **19.** $\displaystyle\int_{-\infty}^0 e^{4x} \, dx$

[7.4]   **20.** $\displaystyle\int_1^\infty \frac{1}{x^3} \, dx$

[7.4]   **21.*** $\displaystyle\int_0^2 \frac{1}{(x - 2)^2} \, dx$

[7.4]   **22.** $\displaystyle\int_{-\infty}^0 xe^x \, dx$

[7.2]   **23.** Find the area of the region bounded by the curve $y = \sqrt{x^2 + 16}$ and the x axis between $x = 0$ and $x = 3$. Round your answer to three decimal places.

[7.1]   **24.** If the marginal revenue function for a product is given by

$$R'(x) = -200xe^{-0.2x} + 5e^{-0.2x}$$

determine the revenue function given that $R(0) = 0$.

---

*This is an improper integral because the function becomes infinite at the upper limit. For another variation, see the Writing and Critical Thinking exercise on page 515.

CHAPTER 7 TEST

PART ONE

Multiple choice. In exercises 1 through 7, put the letter of your correct response in the blank provided.

_____ **1.** Evaluate $\int x \ln 3x \, dx$.

   a. $\dfrac{x^2}{2} \ln 3x - \dfrac{1}{12} x^2 + C$

   b. $\dfrac{x^2}{2} \ln 3x - \dfrac{1}{4} x^2 + C$

   c. $x^2 \ln 3x - x \ln 3x + C$

   d. $\dfrac{x^2}{2} (x \ln 3x - x) + C$

_____ **2.** In using the Trapezoidal Rule to approximate $\int_1^2 x^3 \, dx$ with $n = 4$, $x_0 = 1$, and $x_4 = 2$, the value for x_2 is

   a. $\frac{3}{2}$ **b.** $\frac{5}{4}$ **c.** $\frac{7}{4}$ **d.** $\frac{5}{2}$

_____ **3.** Evaluate $\int xe^{-x} \, dx$.

   a. $xe^{-x} - e^{-x} + C$

   b. $\dfrac{-x^2 e^{-x}}{2} + C$

   c. $-xe^{-x} + e^{-x} + C$

   d. $-xe^{-x} - e^{-x} + C$

_____ **4.** Evaluate $\int (\ln x)^2 \, dx$ using the integral tables.

   a. $x(\ln x)^2 - x \ln x + C$

   b. $x(\ln x)^2 - 2x \ln x + 2x + C$

   c. $x(\ln x)^2 - \ln x + C$

   d. $x(\ln x)^2 - 2x \ln x - 2x + C$

_____ **5.** Which of the following improper integrals will converge?

   a. $\displaystyle\int_1^\infty \dfrac{1}{x^2} \, dx$

   b. $\displaystyle\int_1^\infty \dfrac{1}{\sqrt[3]{x}} \, dx$

   c. $\displaystyle\int_1^\infty \dfrac{1}{\sqrt{x}} \, dx$

   d. $\displaystyle\int_1^\infty \dfrac{1}{x} \, dx$

_____ **6.** Evaluate $\int_1^2 \ln x \, dx$.

   a. $-\frac{1}{2}$ **b.** $2 \ln 2$ **c.** $2 \ln 2 - 1$ **d.** $2 \ln 2 - 3$

_____ **7.** Evaluate $\displaystyle\int_3^\infty \dfrac{1}{x^3} \, dx$.

   a. $\frac{1}{18}$ **b.** $\frac{1}{12}$ **c.** $\frac{1}{27}$ **d.** $\frac{1}{9}$

PART TWO

Solve and show all work.

8. Evaluate $\int x^2 \ln 3x \, dx$.

9. Approximate $\int_0^2 (x^3 + 1) \, dx$ using the Trapezoidal Rule with $n = 4$.

10. Find the area bounded by the curve $y = \dfrac{4}{x^2}$ and the x axis for $x \geq 1$.

11. If the marginal cost for a certain product is given by

$$C'(x) = 5e^x - xe^{-x}$$

and the fixed cost is $800, find the cost function.

8

DIFFERENTIAL EQUATIONS

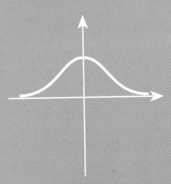

8.1   **Introduction**

8.2   **Separation of Variables**

8.3   **Applications**

        Chapter Review

             Vocabulary/Formula List

             Review Exercises

             Test

8.1   INTRODUCTION

Is $y = 4$ a solution of the following equation?

$$y^2 + 3y - 14\sqrt{y} = 0$$

One way to answer the question is by solving the equation. An easier way would be to merely substitute 4 for y to see if it satisfies the equation. To do so, it would be necessary to square y, multiply y by 3, multiply -14 times the square root of y, and finally, add these results together, giving

$$4^2 + 3(4) - 14\sqrt{4} = 0$$
$$0 = 0$$

So $y = 4$ is indeed a solution.

Is $y = 2x^3$ a solution of the following equation?

$$x\frac{dy}{dx} - 3y = 0$$

To answer this question we again use substitution rather than trying to solve this equation. To accomplish this, we need to calculate $\frac{dy}{dx}$ for $y = 2x^3$. Using the Power Rule, we get

$$\frac{dy}{dx} = 6x^2$$

Substituting $6x^2$ for $\frac{dy}{dx}$ and $2x^3$ for y in the equation gives

$$x\frac{dy}{dx} - 3y = 0$$
$$x(6x^2) - 3(2x^3) = 0$$
$$0 = 0$$

This verifies that $y = 2x^3$ is a **solution.** The preceding equation, $x\frac{dy}{dx} - 3y = 0$, is called a **differential equation.** Following are more examples of *differential equations*, equations that involve derivatives.

$$\frac{dy}{dx} = 2x - 7 \qquad \frac{dy}{dx} - x + 4 = 0 \qquad \frac{d^2y}{dx^2} + 2\frac{dy}{dx} - 8y = 0$$

$$2x\frac{dy}{dx} - 3y = 0 \qquad \frac{d^2y}{dx^2} = x^2 - x \qquad \frac{d^3y}{dx^3} = 5 - 3x$$

Notice that a differential equation may contain a first derivative, a second derivative, or higher-order derivatives; in fact, it may contain a combination of derivatives. Before we learn to solve a differential equation, we present two examples that involve verifying that a given solution satisfies a given differential equation.

Example 1   Verify that $y = \dfrac{1}{x}$ is a solution of the differential equation $x\dfrac{dy}{dx} + y = 0$.

Solution   If $y = \dfrac{1}{x}$, then $\dfrac{dy}{dx} = \dfrac{-1}{x^2}$. Substituting into the differential equation gives

$$x\frac{dy}{dx} + y = 0$$

$$x\left(\frac{-1}{x^2}\right) + \frac{1}{x} = 0$$

$$\frac{-1}{x} + \frac{1}{x} = 0$$

$$0 = 0$$

Therefore, $y = \dfrac{1}{x}$ is a solution.   ∎

Example 2   Verify that $y = e^{2x}$ is a solution to the differential equation

$$\frac{d^2y}{dx^2} + 2\frac{dy}{dx} - 8y = 0$$

Solution   For $y = e^{2x}$, we have $\dfrac{dy}{dx} = 2e^{2x}$. Differentiating again to produce the second deriva-

tive gives $\dfrac{d^2y}{dx^2} = 4e^{2x}$. Substituting yields

$$\frac{d^2y}{dx^2} + 2\frac{dy}{dx} - 8y = 0$$

$$4e^{2x} + 2(2e^{2x}) - 8(e^{2x}) = 0$$

$$4e^{2x} + 4e^{2x} - 8e^{2x} = 0$$

$$0 = 0$$

This verifies that $y = e^{2x}$ is a solution.   ∎

We now direct our attention to the process of *solving* a differential equation. In Chapter 6, we saw that if

$$\frac{dy}{dx} = 3x^2 + 2x + 1, \quad \text{then } y = x^3 + x^2 + x + C$$

is referred to as the *antiderivative* and can be found by integration. The same concept and process are used to solve differential equations, which can be written in the form

$$\frac{dy}{dx} = f(x)$$

Using the differential form for a derivative gives

$$dy = f(x)\,dx$$

Integrating both sides, we have

$$\int dy = \int f(x)\, dx$$

or $y = \displaystyle\int f(x)\, dx = F(x) + C$

where $F(x)$ is an antiderivative of $f(x)$. Performing the integration on the right side produces what can be referred to as a **general solution** to the differential equation. The constant of integration makes it a *general* rather than a *particular* solution. A **particular solution** is found when it is possible to evaluate the constant of integration subject to some **initial conditions** (or **boundary conditions**), as we did in Chapter 6.

Example 3   Find the general solution for the differential equation $\dfrac{dy}{dx} = 2x$.

Solution   Writing this derivative in the differential form gives

$$dy = 2x\, dx$$

Integrating, we have

$$\int dy = \int 2x\, dx$$ (*Note:* It is only necessary to have a constant of integration on one side.)

$$y = x^2 + C$$

This means that $y = x^2 + C$ is a general solution.   ∎

In Example 3, the general solution $y = x^2 + C$ represents a *family* of curves. Three curves from this family are sketched in Figure 8.1.

Figure 8.1

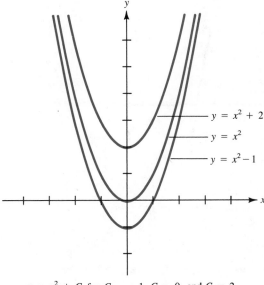

$$y = x^2 + C \text{ for } C = -1, C = 0, \text{ and } C = 2$$

Example 4   Find the general solution for $\dfrac{dy}{dx} = 3x - e^x$.

Solution $dy = (3x - e^x)\, dx$

$\displaystyle\int dy = \int (3x - e^x)\, dx$

$y = \frac{3}{2}x^2 - e^x + C$ ■

In the next two examples, we find a particular solution for a differential equation.

Example 5   Find the particular solution for $\dfrac{dy}{dx} = 5 - 2x$ if $y = 3$ when $x = 0$.

Solution $dy = (5 - 2x)\, dx$

$\displaystyle\int dy = \int (5 - 2x)\, dx$

$y = 5x - x^2 + C$

Substituting $x = 0$ and $y = 3$ gives

$3 = 5 \cdot 0 - 0^2 + C$

$3 = C$

Therefore, the particular solution is $y = 3 + 5x - x^2$ (see Figure 8.2 for its graph).

Figure 8.2

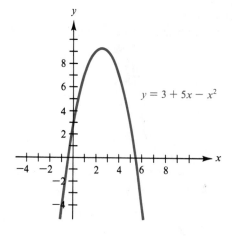

$y = 3 + 5x - x^2$

     ■

Example 6   Find the particular solution for $\dfrac{dy}{dx} - 4x + 3 = 0$ if $y = 7$ when $x = 2$.

Solution   First solve for $\dfrac{dy}{dx}$.

$$\frac{dy}{dx} = 4x - 3$$
$$dy = (4x - 3)\,dx$$
$$\int dy = \int (4x - 3)\,dx$$
$$y = 2x^2 - 3x + C \qquad \text{Now substitute 7 for } y \text{ and 2 for } x.$$
$$7 = 2(2)^2 - 3(2) + C$$
$$7 = 2 + C$$
$$5 = C$$

The particular solution is $y = 2x^2 - 3x + 5$ (see Figure 8.3 for a graph of this function).

Figure 8.3

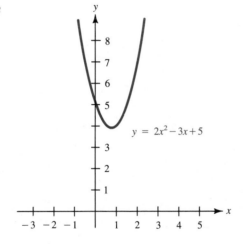

$$y = 2x^2 - 3x + 5$$

Example 7   Find the general solution for $\dfrac{d^2y}{dx^2} = 2x$.

Solution   We put $\dfrac{d^2y}{dx^2} = 2x$ in the form $d^2y = 2x\,dx^2$. Now, rewriting and setting up the integrals, we have

$$\int d(dy) = \left[\int (2x\,dx)\right]dx$$
$$dy = (x^2 + C_1)\,dx$$
$$\int dy = \int (x^2 + C_1)\,dx$$
$$y = \frac{x^3}{3} + C_1 x + C_2$$

8.1 EXERCISES

SET A

In exercises 1 through 12, verify that the given function is a solution to the given differential equation. (See Examples 1 and 2.)

1. $\dfrac{dy}{dx} = 5,\ y = 5x + 3$

2. $\dfrac{dy}{dx} = 2x,\ y = x^2 - 7$

3. $\dfrac{dy}{dx} = 5x^4,\ y = x^5 + 4$

4. $\dfrac{dy}{dx} = 6x^5 + 1,\ y = x^6 + x - 4$

5. $3x\dfrac{dy}{dx} - 6y = 0,\ y = x^2$

6. $3x\dfrac{dy}{dx} - 6y + 6 = 0,\ y = x^2 + 1$

7. $x\dfrac{dy}{dx} - 1 = 0,\ y = \ln x$

8. $2x\dfrac{dy}{dx} - 1 = 0,\ y = \ln \sqrt{x}$

9. $\dfrac{dy}{dx} - 3y + 15 = 0,\ y = 5 + e^{3x}$

10. $\dfrac{dy}{dx} + 4y = 0,\ y = 3e^{-4x}$

11. $x\dfrac{dy}{dx} - xy - y = 0,\ y = xe^x$

12. $\dfrac{d^2y}{dx^2} - 4\dfrac{dy}{dx} + 4y = 0,\ y = e^{2x} + xe^{2x}$

In exercises 13 through 30, find the general solution for the differential equation. (See Examples 3, 4, and 7.)

13. $\dfrac{dy}{dx} = 3$

14. $\dfrac{dy}{dx} = -4$

15. $\dfrac{dy}{dx} = 2x + 1$

16. $\dfrac{dy}{dx} = 1 - 2x$

17. $\dfrac{dy}{dx} = 6x^3 + x - 2$

18. $\dfrac{dy}{dx} = 8x^4 - x^2 + 3$

19. $\dfrac{dy}{dx} = \sqrt{x} + 2$

20. $\dfrac{dy}{dx} = \sqrt{x + 2}$

21. $\dfrac{dy}{dx} - 4x = 0$

22. $\dfrac{dy}{dx} + 1 = 5x$

23. $2\dfrac{dy}{dx} = 3x^2$

24. $3\dfrac{dy}{dx} + 2x - 1 = 0$

25. $x^2\dfrac{dy}{dx} + 4 = 0$

26. $x^2\dfrac{dy}{dx} + 1 = x^4$

27. $\dfrac{d^2y}{dx^2} = 6x$

(*Hint:* Integrate twice.)

28. $\dfrac{d^2y}{dx^2} = \dfrac{-1}{x^2}$

29. $\dfrac{d^2y}{dx^2} = 8e^{2x}$

30. $\dfrac{d^2y}{dx^2} = x - e^x$

In exercises 31 through 40, find the particular solution subject to the given conditions. (See Examples 5 and 6.)

31. $\dfrac{dy}{dx} = 12x,\ y = 8$ when $x = 1$

32. $\dfrac{dy}{dx} = 12x,\ y = -1$ when $x = 1$

33. $\dfrac{dy}{dx} = 7 - 4x,\ y = 3$ when $x = 0$

34. $\dfrac{dy}{dx} = 7 - 4x,\ y = 3$ when $x = -1$

35. $\dfrac{dy}{dx} + x^2 - x = 0$, $y = 1$ when $x = 1$

36. $\dfrac{dy}{dx} + 2x = x^3$, $y = -2$ when $x = 0$

37. $3\dfrac{dy}{dx} - 4x = 0$, $y = -2$ when $x = 0$

38. $5\dfrac{dy}{dx} - 10x = 0$, $y = 4$ when $x = 2$

39. $x^2\dfrac{dy}{dx} - x^2 + 1 = 0$, $y = 1$ when $x = 1$

40. $x\dfrac{dy}{dx} - 2x^2 + x = 0$, $y = 3$ when $x = 0$

SET B

In exercises 41 through 44, find the particular solution for the differential equation subject to the given conditions.

41. $x\dfrac{dy}{dx} + 3 = 0$, $y = 4$ when $x = 1$

42. $2x\dfrac{dy}{dx} - x^2 + 1 = 0$, $y = \dfrac{1}{4}$ when $x = 1$

43. $\dfrac{dy}{dx} = \sqrt[3]{x}$, curve passes through the point $(8, 12)$

44. $\dfrac{dy}{dx} = \sqrt[3]{x} + 1$, curve passes through the point $(8, 16)$

45. Find the particular solution for the differential equation $\dfrac{d^2y}{dx^2} - 2 = 0$ given that $\dfrac{dy}{dx} = 3$ when $x = 0$, and $y = 5$ when $x = 0$.

46. Find the particular solution for the differential equation $\dfrac{d^2y}{dx^2} - 2x = 0$ given that $\dfrac{dy}{dx} = 4$ when $x = 0$, and $y = -3$ when $x = 0$.

BUSINESS
COST FUNCTION

47. The marginal cost of producing x units of an item is given by

$$\dfrac{dC}{dx} = 6 - 0.04x$$

where C is measured in dollars. Determine the particular cost function if it has been determined that the cost of producing 100 items is $400.

MARKETING

48. A new cookie company is conducting a marketing experiment in a chain of grocery stores in a certain county. It has been determined that the demand for its cookies is changing at a rate described by the following:

$$\dfrac{dx}{dp} = \dfrac{-2}{\sqrt[3]{p}}$$

where x gives the number of boxes of cookies and p is the price per box measured in cents. If 200 boxes were sold when the price was $1.25, determine the number of boxes that would be sold if the price were dropped to $1.00.

BIOLOGY
BACTERIOLOGY

49. The rate of growth of a certain bacteria in an experiment is given by the differential equation

$$\frac{dy}{dt} = 600e^{3t}$$

where t is measured in hours.
  a. Find the general solution for the differential equation.
  b. Find the particular solution if there were 200 present initially ($t = 0$).
  c. How many bacteria are present after 5 hours?

SOCIAL SCIENCE
UNDERDEVELOPED
COUNTRIES

50. A new piece of farm machinery has been brought into an underdeveloped country. Resistance to its usage has been great, but the rate of acceptance has been increasing with time. The rate is given by

$$\frac{dA}{dt} = t + 2$$

where A gives the number of people who have accepted it and t is measured in months. If the number of people who accepted it originally is 20, determine the number who have accepted it after 1 year.

WRITING AND CRITICAL THINKING

Explain in words what you think a differential equation is. Also, explain the difference between a general solution and a particular solution to a differential equation.

8.2 SEPARATION OF VARIABLES

In the previous section, we solved differential equations that could be written in the form

$$\frac{dy}{dx} = f(x) \quad \text{or} \quad dy = f(x)\,dx$$

In this section, we expand our capabilities by learning a technique for solving differential equations that can be written in the form

$$g(y)\,dy = f(x)\,dx$$

This technique is called **separation of variables.** The name comes from the fact that the left side is expressed entirely in terms of y and the right side in terms of x. Thus, the variables are *separated* by the equal sign. To obtain the solution, we integrate both sides, just as before. This type of differential equation is called a **separable differential equation.**

Example 1   Find the general solution for $2y\dfrac{dy}{dx} - 3x = 0$.

Solution   We need to separate the variables as follows:

$$2y\frac{dy}{dx} = 3x$$

$$2y\,dy = 3x\,dx$$

Now, integrating both sides gives

$$\int 2y\,dy = \int 3x\,dx$$

$$y^2 = \frac{3x^2}{2} + C$$

As before, it is not necessary to have a constant of integration on both sides. We could solve this solution for y so that we have y expressed explicitly in terms of x, but, as a general rule, this is not necessary. ∎

Example 2   Find the general solution for $x^2y\dfrac{dy}{dx} - 3x = 0,\; x \neq 0$.

Solution   Separating the variables gives

$$x^2y\frac{dy}{dx} = 3x$$

$$y\frac{dy}{dx} = \frac{3x}{x^2} \qquad \text{Dividing both sides by } x^2$$

$$y\,dy = \frac{3}{x}\,dx$$

Integrating both sides, we get

$$\int y\,dy = \int \frac{3}{x}\,dx$$

$$\frac{y^2}{2} = 3\ln|x| + C_1 \qquad \text{Using } C_1 \text{ temporarily until final solution}$$

$$y^2 = 6\ln|x| + 2C_1 \qquad \text{Multiplying both sides by 2}$$

Finally,

$$y^2 = 6\ln|x| + C \qquad\qquad ∎$$

In the next example, we find a particular solution for a differential equation.

Example 3   Find the particular solution for $3y^2\dfrac{dy}{dx} - 2x = 3$ if $y = 2$ when $x = 1$.

Solution $3y^2 \dfrac{dy}{dx} = 2x + 3$

$3y^2 \, dy = (2x + 3) \, dx$

$\displaystyle\int 3y^2 \, dy = \int (2x + 3) \, dx$

$y^3 = x^2 + 3x + C$

Now substituting $x = 1$ and $y = 2$ gives

$2^3 = 1^2 + 3 \cdot 1 + C$

$4 = C$

The particular solution is, therefore,

$y^3 = x^2 + 3x + 4$ ∎

Example 4   Find the general solution of $\dfrac{dy}{dt} = ky$ where k is a constant.

Solution   Separating the variables gives

$\dfrac{1}{y} \, dy = k \, dt$

Integrating both sides, we have

$\displaystyle\int \dfrac{1}{y} \, dy = k \int dt$

$\ln |y| = kt + C_1$

Writing this in exponential form gives

$|y| = e^{kt + C_1}$

$|y| = e^{kt} \cdot e^{C_1}$ Using a law of exponents

$y = \pm e^{C_1} \cdot e^{kt}$

or, more simply,

$y = Ce^{kt}$ Since $\pm e^{C_1}$ is a constant C ∎

8.2 EXERCISES

SET A

In exercises 1 through 14, find the general solution for the given differential equation. (See Examples 1, 2, and 4.)

1. $y \dfrac{dy}{dx} - 2x = 0$

2. $y^2 \dfrac{dy}{dx} - x = 0$

3. $\dfrac{dy}{dx} = 2xy^2$

4. $\dfrac{dy}{dx} - xy^4 = 0$ 　　　　**5.** $x^2 \dfrac{dy}{dx} - y^2 = 0$ 　　　　**6.** $xy \dfrac{dy}{dx} - 3x = 0$

7. $\dfrac{dy}{dx} - 2y = 0$ 　　　　**8.** $\dfrac{dy}{dx} - y + 1 = 0$ 　　　　**9.** $\dfrac{dy}{dx} = x^2 e^y$

10. $\dfrac{dy}{dx} = y^2 e^x$ 　　　　**11.** $\dfrac{dy}{dx} = 2x(y - 1)$ 　　　　**12.** $y \dfrac{dy}{dx} = 2x(y^2 - 1)$

13. $\dfrac{dy}{dx} = 2xy^2 - y^2$ 　　　　**14.** $\dfrac{dy}{dx} = 4x\sqrt{y} + \sqrt{y}$

In exercises 15 through 22, find the particular solution for the given differential equation subject to the given conditions. (See Example 3.)

15. $y^2 \dfrac{dy}{dx} - 2x = 0$, $y = 3$ when $x = 0$ 　　　　**16.** $2y \dfrac{dy}{dx} - x^2 = 0$, $y = 2$ when $x = 0$

17. $x^2 \dfrac{dy}{dx} + y^2 = 0$, $y = 1$ when $x = 1$ 　　　　**18.** $x^2 \dfrac{dy}{dx} = 4y^3$, $y = 1$ when $x = 1$

19. $y \dfrac{dy}{dx} = x - 4$, $y = -2$ when $x = 0$ 　　　　**20.** $x \dfrac{dy}{dx} - xy = 0$, $y = 1$ when $x = 1$

21. $2x \dfrac{dy}{dx} - x^2 y = 0$, $y = 1$ when $x = 2$ 　　　　**22.** $x^2 \dfrac{dy}{dx} = y + 3$, $y = -2$ when $x = 1$

SET B

In exercises 23 through 30, find the general solution for the given differential equation.

23. $y \dfrac{dy}{dx} - 4x = 0$ 　　　　**24.** $y^2 \dfrac{dy}{dx} - e^x = 0$

25. $y^2 \dfrac{dy}{dx} + e^{2x} = 0$ 　　　　**26.** $x^4 \dfrac{dy}{dx} = y^2$

27. $\dfrac{dy}{dx} = x^3 e^{2y}$ 　　　　**28.** $\dfrac{dy}{dx} = ye^{2x}$

29. $\dfrac{dy}{dx} = xye^{x^2}$ 　　　　**30.** $y \dfrac{dy}{dx} = 4x(y^2 - 5)$

In exercises 31 through 35, find the particular solution for the given differential equation subject to the given conditions.

31. $y \dfrac{dy}{dx} = x^2 - 4$, $y = 4$ when $x = 0$ 　　　　**32.** $x \dfrac{dy}{dx} = x^4 y$, $y = 10$ when $x = 3$

33. $x^3 \dfrac{dy}{dx} = y + 3$, $y = -2$ when $x = 1$ 　　　　**34.** $3y^2 \dfrac{dy}{dx} - 4x = 1$, $y = \dfrac{2}{3}$ when $x = 0$

35. $\dfrac{dy}{dx} = xye^{x^2}$, $y = e$ when $x = 0$

BUSINESS
SALES

36. The rate of change in sales for a large company is given by

$$\frac{dS}{dx} = \sqrt{S}(6x + 1)$$

where S is measured in thousands of dollars and x is measured in years.
a. Find S as a function of x given that initial sales were $25,000.
b. Find the sales figure at $x = 10$ years.

SOCIAL SCIENCE
POPULATION

37. The population of a city is decreasing at a rate given by

$$\frac{dP}{dt} = \frac{-200}{P^{0.1}e^{0.4t}}$$

where t is measured in months. If $P = 38,000$ when $t = 0$, find the population after 10 months.

PHYSICAL SCIENCE
MOTION

38. A particle is moving so that the rate of change of its position in feet at any time, t (in seconds), is given by

$$\frac{ds}{dt} = \frac{2t + 1}{s^2}$$

Find s as a function of t if its initial position is given as $s = 3$ ft.

BIMOLECULAR REACTION

39. The velocity of a bimolecular reaction, when the concentrations of the two reacting materials are the same, is given by

$$\frac{dy}{dt} = k(a - y)^2$$

Find the general solution to this differential equation.

WRITING AND CRITICAL THINKING

Examine these differential equations:

$$(3x + y)\frac{dy}{dx} = x^2 \quad \text{and} \quad (3xy)\frac{dy}{dx} = x^2$$

We can "separate" the variables in one but not in the other. Decide which one is separable and explain in words why one is and one is not.
Now consider

$$(3x + xy)\frac{dy}{dx} = x^2 \quad \text{and} \quad 3x + y\frac{dy}{dx} = x^2$$

Both are separable. Why?
Can you draw any conclusions about the structure of the four differential equations that allows or does not allow the variables to be separated?

8.3 APPLICATIONS

Recall from Chapter 2 that a derivative is an instantaneous *rate of change.* It is just this idea that gives the differential equation such a wide range of applications in business and in the natural and social sciences. Many of these applications involve a rate of change of some quantity with respect to time. Consider, for example, the separable differential equation introduced in Example 4 of Section 8.2:

$$\frac{dy}{dt} = ky \qquad \text{where } k \text{ is a constant}$$

This may be interpreted verbally by the following:

The rate of change of y with respect to time t is proportional to y.

The constant k is referred to as the **constant of proportionality.**

In Example 4 of Section 8.2, we found the general solution to this differential equation to be

$$y = Ce^{kt}$$

If k is positive, recall from Chapter 5 that this function represents **exponential growth,** and if k is negative, it represents **exponential decay.** Notice, also, that if $t = 0$, then $y = C$. For this reason, C is referred to as the *initial value* (value of y at time $t = 0$). The next example gives an application of this function to continuous compounding.

Example I   Suppose that $1000 is deposited into an account. The account balance increases as it earns interest at 8% per year, compounded *continuously.* The differential equation describing this rate of change is given by

$$\frac{dA}{dt} = 0.08A$$

where A is the amount of money in the account at any time t. Solve this differential equation for A.

Solution   From Example 4 of Section 8.2, we know that this differential equation has a solution of the form

$$A = Ce^{0.08t} \qquad \text{where } k = 0.08$$

We can determine C, because we know that initially, at time $t = 0$, there was $1000 in the account.

$$1000 = Ce^{(0.08)(0)}$$
$$1000 = Ce^0$$
$$1000 = C$$

Therefore,

$$A = 1000e^{0.08t}$$

The graph of $A = 1000e^{0.08t}$ appears in Figure 8.4.

Figure 8.4

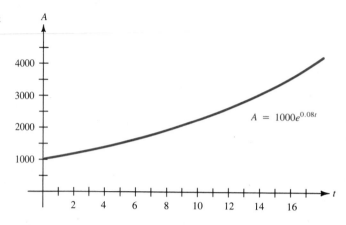

This was an example of an *unbounded growth* model. In Chapter 5, we saw examples of both bounded and unbounded growth. Our next two examples involve differential equations that exhibit bounded growth.

Recall the **learning curve** introduced in Chapter 5. It is an example of a *bounded growth* model and has applications in business as well as in psychology. The general form of a learning curve is

$$y = B(1 - e^{-kt})$$

The number B is its upper bound. The equation can be generated from the separable differential equation

$$\frac{dy}{dt} = k(B - y)$$

as we see in the next example.

Example 2   On an assembly line, the maximum number of units that an employee can produce per day is 50. If y represents the number of units produced per day, then any *new* employee can produce at a *rate* that is proportional to the difference, $50 - y$.

   a.   Determine the differential equation that represents this situation and find its general solution.

   b.   If, after 10 days, the production level is 20 units per day for Jean, a new employee, how many units can she be expected to produce per day after 25 days?

Solution   **a.**   With y representing the number of units produced per day and t the number of days worked, the differential equation is

$$\frac{dy}{dt} = k(50 - y)$$

because $\frac{dy}{dt}$ represents the *rate* at which y is increasing with respect to t. Separating the variables gives

$$\frac{1}{50 - y} dy = k \, dt$$

Integrating both sides, we have

$$\int \frac{1}{50 - y} dy = \int k \, dt$$

$-\ln(50 - y) = kt + C_1$	Since $50 - y$ is positive, we do not need absolute value symbols.
$\ln(50 - y) = -kt - C_1$	Multiply both sides by -1.
$50 - y = e^{-kt - C_1}$	Write in exponential form.
$50 - y = e^{-C_1} \cdot e^{-kt}$	Using a law of exponents
$50 - y = Ce^{-kt}$	Replace constant e^{-C_1} with constant C.

Finally, the general solution is

$$y = 50 - Ce^{-kt}$$

b. Now we need to determine the values for C and k. Since $y = 0$ when $t = 0$, we have

$$0 = 50 - C$$
$$\text{or} \quad C = 50$$

Thus,

$$y = 50 - 50e^{-kt}$$

We use the fact that after 10 days ($t = 10$) the employee was producing 20 units ($y = 20$).

$$20 = 50 - 50e^{-10k}$$
$$-30 = -50e^{-10k}$$
$$0.6 = e^{-10k}$$
$$\ln 0.6 = -10k$$
$$k = -0.1(\ln 0.6) \approx 0.0511$$

To determine the number of units after 25 days, we need to substitute 25 for t in the equation

$$y \approx 50 - 50e^{-0.0511t}$$

For $t = 25$,

$$y \approx 50 - 50e^{(-0.0511)(25)}$$
$$\approx 50 - 14$$
$$\approx 36$$

The new employee, Jean, can be expected to produce 36 units a day after 25 days. A table for this relationship and an accompanying graph are shown in Figure 8.5.

Figure 8.5

ASSEMBLY LEARNING CURVE
Employee: Jean

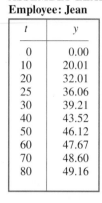

t	y
0	0.00
10	20.01
20	32.01
25	36.06
30	39.21
40	43.52
50	46.12
60	47.67
70	48.60
80	49.16

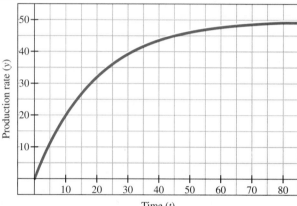

■

Example 3   The *rate* at which information spreads through a population is proportional to both the number of people who have heard the information and the number who haven't. Let y represent the number who have heard the information at any time, t, and let N be the total population. Determine the differential equation describing this situation.

Solution   Using the following notation to set up the differential equation:

$$\frac{dy}{dt} = \text{rate at which the information spreads}$$

$$k = \text{constant of proportionality}$$

$$N - y = \text{number of people who have not heard the information}$$

we have

$$\frac{dy}{dt} = ky(N - y)$$ ■

It can be shown that after separating the variables, the solution of the differential equation in Example 3 is of the form

$$y = \frac{y_0 N}{y_0 + (N - y_0)e^{-Nkt}}$$

where y_0 is the number of people at time $t = 0$ who have heard the information. This curve exhibits bounded growth and is called a **logistics curve.** It was first introduced in exercise 57 of Section 5.2. See Figure 8.6 for a graph of this function.

Figure 8.6

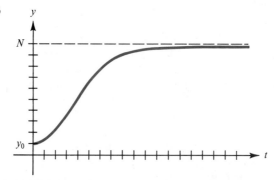

We conclude this section with Table 8.1, which summarizes the types of functions examined here.

Table 8.1

Differential equation	General solution
$\dfrac{dy}{dt} = ky$	$y = Ce^{kt}$ (unbounded growth curve)
$\dfrac{dy}{dt} = k(M - y)$	$y = M + Ce^{-kt}$ (bounded growth, learning curve)
$\dfrac{dy}{dt} = ky(M - y)$	$y = \dfrac{CM}{C + (M - C)e^{-Mkt}}$ (bounded growth, logistics curve)

8.3 EXERCISES

SET A

Exercises 1 through 22 involve applications of differential equations. (Use Examples 1, 2, and 3 as a guide.)

BUSINESS
BANKING

1. The amount of money, A, in a savings account grows at a rate that is proportional to the amount present. Construct the differential equation for this and find the general solution.

FINANCE

2. Suppose that an investment is increasing at a rate of 6% per year, compounded continuously. If the initial investment was $2000, what will the investment be worth in 10 years?

FINANCE

3. Redo exercise 2 for an initial investment of $4000.

REAL ESTATE

4. The value of a certain piece of real estate, which originally sold for $3000, is increasing at a rate that is proportional to its value at any time, t. If, after 5 years, the property is worth $15,000, what will it be worth after 10 years?

REAL ESTATE

5. Redo exercise 4 assuming that after 5 years the property is worth $10,000.

LEARNING CURVE

6. A factory has just installed a new piece of machinery in an area of the plant having 800 employees. Initially, only 20 employees knew how to use the machinery. The number of people, x, who learn how to use the machinery increases at a rate that is proportional to the difference between the total number of employees and x. If, after 2 weeks, 230 employees can use the machinery, how many employees should know how to use the machinery after 4 weeks?

MARKETING

7. A company has introduced a new shampoo into the marketplace. Initially, free samples were sent out to 1000 people in a town of 20,000 people. The number of people who hear about the new shampoo is proportional to both the number of people who have heard about it and the number who have not yet heard about it. If, 2 weeks after the free samples were sent out, 8000 people knew about the shampoo, determine the equation for the number who have heard about the shampoo at any time, t, measured in weeks.

MARKETING

8. Using the equation from exercise 7, approximate the number of people who have heard about the new shampoo after 4 weeks.

SOCIAL SCIENCE
SPREAD OF A RUMOR

9. The rate at which a rumor spreads in a population of 4000 is proportional to the number of people who have not heard the rumor. Let x be the number who have heard the rumor. Find the differential equation that represents this.

SPREAD OF INFORMATION

10. The rate at which a town of population P hears about the proposed freeway going through the town is proportional to the number who have not heard about the new freeway. Let x be the number of people who have heard. Find the differential equation that represents this situation.

ELECTIONS

11. The rate at which a city hears about a new political candidate is proportional to the number of people who have not heard about the candidate. In a city with a population of 50,000 people, it was found that 10 days after the candidate declared his candidacy, 12,000 people had heard the news. After 30 days, how many people should the candidate expect to have heard the news? (*Hint:* At time $t = 0$, assume that no one knew about the candidate.)

SET B

BIOLOGY
ANIMAL POPULATIONS

12. The population of a certain type of animal cannot exceed the value N due to the amount of food available. The rate of change of the number of animals is proportional to both the number of animals, P, and the difference between the upper limit, N, and the number of animals. Represent this with a differential equation.

MEDICINE

13. In an experiment involving laboratory mice, it was found that the rate of change of the blood sugar level was increasing at a rate proportional to the square of the amount of sugar in the blood, S. Represent this as a differential equation.

BACTERIOLOGY

14. The number of bacteria in a certain culture is increasing at a rate that is proportional to the number of bacteria present. Express this as a differential equation and find its general solution.

BACTERIOLOGY

15. A biologist is conducting an experiment involving bacteria that are increasing by a number proportional to the number present, Q. If, after 3 hours, there are 600 bacteria present, and there were 200 present initially, determine the equation for Q in terms of t and find the number of bacteria present after 5 hours.

ECOLOGY

16. The fish population in a certain river is growing at a rate proportional to the number of fish in the river. During one count, there were approximately 8500 fish, and for a count conducted 2 months later, there were approximately 11,000. At this rate how many fish will there be in 6 months?

PHYSICAL SCIENCE
RADIOACTIVE DECAY

17. The rate at which the amount of radioactive material, R, decays is proportional to the amount present. Find the differential equation that represents this and find its general solution.

CHEMISTRY

18. In a certain type of chemical reaction, the rate at which an old substance (initial amount is A) is converted into a new substance is proportional to both the amount of new substance, x, and the difference between the amounts of old and new substances. Express this as a differential equation.

NEWTON'S LAW OF COOLING

19. According to *Newton's law of cooling*, the rate of change of the temperature of a mass is proportional to the difference between its temperature and the temperature of the surrounding medium. A small piece of metal has been heated to 400°F and submerged in a large tank of water having a constant temperature of 70°F. If the temperature of the metal is 300°F after 2 minutes, determine the temperature of the metal at any time, t, measured in minutes.

20. Determine the temperature of the piece of metal in exercise 19 after 5 minutes.

NEWTON'S LAW OF COOLING

21. The rate of change in the temperature, T, of a small object placed in a large body of water with a temperature of 70°F is proportional to the difference between the temperature of the object and the temperature of the water. Find the differential equation that represents this situation and find its general solution.

NEWTON'S LAW OF COOLING

22. A side of beef of temperature B is placed in a large freezer in a meat-packing warehouse. The rate of change in the temperature of the beef is proportional to the difference between its temperature and the temperature of the freezer (which is 28°F). Find the differential equation that represents this and find its general solution.

> **USING TECHNOLOGY IN CALCULUS**
>
> 1. **a.** Establish a two-column spreadsheet for the information and associated equation of exercise 2. The first column should represent time, t, and the second column should be the amount of investment, A. Let t be integer formatted and range from 0 to 12 years. Let A be formatted as "$" with two decimal places.
>   **b.** Use the graphing feature of your spreadsheet and graph A as a function of t.
> 2. **a.** Establish a spreadsheet for the information in exercise 6. The first column should represent time, t (in weeks), and the second column should represent the number of employees who know how to use the machinery, x. Let t be integer formatted and range from 0 to 15, and let x also be an integer.
>   **b.** Use the graphing feature of your spreadsheet and graph x as a function of t.

CHAPTER 8 REVIEW

VOCABULARY/FORMULA LIST

differential equation
general solution
initial conditions
boundary conditions

exponential decay
logistics curve
separation of variables
separable differential equation

constant of proportionality
exponential growth
learning curve
particular solution

CHAPTER 8 REVIEW EXERCISES

In exercises 1 through 10, find the general solution to the given differential equation.

[8.1] **1.** $\dfrac{dy}{dx} = 4$

[8.1] **2.** $\dfrac{dy}{dx} = 3 - 4x$

[8.1] **3.** $\dfrac{dy}{dx} = 5x^2 + 3x - 2$

[8.1] **4.** $\dfrac{dy}{dx} = 8x^4 - x^3 + 5$

[8.1] **5.** $x^2\dfrac{dy}{dx} + 6 = 0$

[8.1] **6.** $\dfrac{dy}{dx} = 4x + e^x$

[8.2] **7.** $\dfrac{dy}{dx} = \dfrac{x}{y}$

[8.2] **8.** $\dfrac{dy}{dx} = xy^3$

[8.2]   **9.** $\dfrac{dy}{dx} = xe^y$ [8.2]   **10.** $\dfrac{dy}{dx} = y^2e^x$

In exercises 11 through 18, find the particular solution for the given differential equation subject to the given conditions.

[8.1]   **11.** $\dfrac{dy}{dx} = 5 - 3x,\ y = 0$ when $x = 2$

[8.1]   **12.** $2\dfrac{dy}{dx} - 3x = 0,\ y = 7$ when $x = -2$

[8.1]   **13.** $x^2\dfrac{dy}{dx} - x^2 + 2 = 0,\ y = 4$ when $x = 1$

[8.2]   **14.** $y^2\dfrac{dy}{dx} = 2x,\ y = 9$ when $x = 0$

[8.2]   **15.** $x^2\dfrac{dy}{dx} = 3y^2,\ y = \dfrac{1}{6}$ when $x = 1$

[8.2]   **16.** $3y^2\dfrac{dy}{dx} - 2x = 3,\ y = 8$ when $x = 0$

[8.2]   **17.** $x^2\dfrac{dy}{dx} = y + 1,\ y = 0$ when $x = -\dfrac{1}{2}$

[8.2]   **18.** $\dfrac{dy}{dx} = xye^{x^2},\ y = 1$ when $x = 0$

BUSINESS
BANKING
   [8.3]   **19.** Suppose that $2000 is invested in a bank account earning 10% per year, compounded continuously. The differential equation describing this rate of change is

$$\frac{dA}{dt} = 0.10A$$

where A is the amount of money in the account at any time, t. Solve this differential equation for A.

LEARNING MODEL   [8.3]   **20.** On a certain assembly line, the maximum number of units an employee can produce per day is 50. Any new employee can produce units at a rate that is proportional to the difference $(50 - y)$ between the maximum number of units (50) and the number of units (y) produced daily.

   a. Determine the differential equation that represents this situation.

   b. Find the general solution to your answer to part a assuming that, for $t = 0$, the number of units is zero.

   c. Suppose that, after 10 days, the production level is 30 units per day for Randy, a new employee. How many units can he be expected to produce per day after 25 days?

CHAPTER 8 TEST

PART ONE

Multiple choice. Put the letter of your correct response in the blank provided.

_____ 1.   The general solution for $\dfrac{dy}{dx} = 6x^2 + 8x$ is

   a. $y = 6x^3 + 8x^2 + C$ **b.** $y = 2x^3 + 4x^2 + C$

   c. $y = 3x^3 + 4x^2 + C$ **d.** $y = 12x + 8$

_____ 2.   The general solution to $\dfrac{dy}{dx} = y(2x + 1)$ is

   a. $\ln|2x + 1| = \dfrac{y^2}{2} + C$ **b.** $y = \dfrac{y^2}{2}(x^2 + x) + C$

   c. $\dfrac{-y^2}{2} = x^2 + x + C$ **d.** $\ln|y| = x^2 + x + C$

_____ 3.   The particular solution for $\dfrac{dy}{dx} = 5 - 4x$ given that $y = 10$ when $x = 1$ is

   a. $y = -2x^2 + 5x + 1$ **b.** $y = -2x^2 + 5x + 10$

   c. $y = -4x^2 + 5x + 9$ **d.** $y = -2x^2 + 5x + 7$

_____ 4.   Given $x^3 y \dfrac{dy}{dx} + 1 = 0$, find the particular solution given that the curve passes through the
   point $(-1, 2)$.

   a. $\dfrac{x^4 y^2}{8} + x = \dfrac{3}{2}$ **b.** $y^2 = \dfrac{1}{x^2} + 3$

   c. $\dfrac{x^4}{4} = -\ln\left|\dfrac{y}{2}\right| + \dfrac{1}{4}$ **d.** $x^2 + y^2 = 5$

_____ 5.   The rate at which a rumor spreads in a town of 8000 is proportional to the number who have
   not heard the rumor. Letting x be the number who have heard the rumor and using k as the
   constant of proportionality, the differential equation describing this situation is

   a. $\dfrac{dx}{dt} = 8000kx$ **b.** $\dfrac{dx}{dt} = k(8000 - x)$

   c. $\dfrac{dx}{dt} = 8000 - kx$ **d.** $\dfrac{dx}{dt} = k(x - 8000)$

PART TWO

Solve and show all work.

6.   Find the general solution for $\dfrac{dy}{dx} = 2x - e^{-x}$.

7.   Find the general solution for $y^2 \dfrac{dy}{dx} - x = 2$.

8. Find the particular solution for $\dfrac{dy}{dx} = \dfrac{1}{2\sqrt{x+1}}$ given that $y = 6$ when $x = 0$.

BUSINESS
CONTINUOUS
COMPOUNDING

9. Suppose that an investment is increasing at 9% per year compounded continuously. If the initial investment was $5000, what will the investment be worth in 6 years?

BIOLOGY
BACTERIOLOGY

10. The rate of growth of bacteria in an experiment is given by

$$\frac{dQ}{dt} = 800e^{0.2t}$$

where t is measured in hours. Find the number of bacteria present at any time, t, given that there were 4000 present initially.

9

MULTIVARIABLE CALCULUS

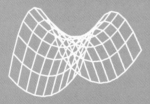

9.1 FUNCTIONS OF SEVERAL VARIABLES

The functions we have discussed so far in this text have all involved two variables—
one independent variable and **one dependent variable.** Some examples are the
following:

Equation for function	Independent variable	Dependent variable
$y = 1 - 0.5x^2$	x	y
$s = -16t^2$	t	s

Solutions to these equations took the form of ordered pairs, and their graphs were
created in a two-dimensional rectangular coordinate system.

For the remainder of this chapter, we are concerned with functions of one
dependent variable and several (usually two) independent variables. For example,
the area of a rectangle (A) is the product of the length (l) and width (w).

$$A = lw$$

Here, A is the dependent variable and l and w are the independent variables. Such
functions arise naturally in business. The cost of shipping a particular product may
depend on three variables: weight, size, and shelf life. In biology, the amount of
DDT a carnivorous animal ingests may depend on levels of DDT in its prey as well
as DDT in the vegetation the prey feeds on. Examples also arise in other disciplines.

The *solutions* to equations involving one dependent and two independent variables take the form of **ordered triples** of real numbers. For example, if x and y are
independent variables and $z = 2xy^2 - 3xy$, then by choosing an x value and a y value
we can find a corresponding z value. If $x = 1$ and $y = -2$ for this function, then

$$\begin{aligned} z &= 2xy^2 - 3xy \\ &= 2(1)(-2)^2 - 3(1)(-2) \\ &= 14 \end{aligned}$$

Ordered triples of the form (x, y, z) are solutions to $z = 2xy^2 - 3xy$. In particular,
$(1, -2, 14)$ is a solution.

Using function notation, we may write

$$z = f(x, y) = 2xy^2 - 3xy$$

to indicate that z is a function of two variables. The ordered triple $(1, -2, 14)$ can be
represented by

$$f(1, -2) = 14$$

where the independent variables x and y are 1 and -2, respectively; the dependent
variable $z = 14$.

Example 1   Suppose $g(x, y) = \dfrac{x^2}{x + y}$. Find:

    a. $g(1, 2)$ **b.** $g(-3, 4)$

Solution   **a.** $g(1, 2) = \dfrac{1^2}{1 + 2} = \dfrac{1}{3}$

        b. $g(-3, 4) = \dfrac{(-3)^2}{-3 + 4} = 9$ ■

Example 2   A caterer has established the cost for large parties to be a function of the number of waiters needed and the number of hours the banquet hall is used. The charge per waiter is $50 and the fee for the use of the facility is $300 per hour. Cost (C) is a function of the number of waiters (x) and hours of use of the facility (y) according to the formula

$$C(x, y) = 50x + 300y$$

    a.   Find the cost if six waiters are needed and the facility is used for 5 hours.
    b.   Find $C(3, 10)$.

Solution   **a.**   For six waiters, we have $x = 6$, and 5 hours means $y = 5$.

$$\begin{aligned} C(x, y) &= 50x + 300y \\ C(6, 5) &= 50(6) + 300(5) \\ &= 300 + 1500 \\ &= \$1800 \end{aligned}$$

        b. $\begin{aligned}[t] C(x, y) &= 50x + 300y \\ C(3, 10) &= 50(3) + 300(10) \\ &= \$3150 \end{aligned}$ ■

Example 3   The cost (C) of renting a Buick depends on the number of days used (x) and the number of miles driven (y) according to the formula

$$C(x, y) = 20x + 0.25y + 10$$

assuming a charge of $20 per day, 25¢ per mile, and a $10 insurance fee.
    a.   What is the charge for renting the car for three days and 300 miles?
    b.   Find $C(2, 350)$.

Solution   **a.** $\begin{aligned}[t] C(x, y) &= 20x + 0.25y + 10 \\ C(3, 300) &= 20(3) + 0.25(300) + 10 \\ &= 60 + 75 + 10 \\ &= \$145 \end{aligned}$

        b. $\begin{aligned}[t] C(x, y) &= 20x + 0.25y + 10 \\ C(2, 350) &= 20(2) + 0.25(350) + 10 \\ &= 40 + 87.50 + 10 \\ &= \$137.50 \end{aligned}$ ■

Although most of the functions studied throughout the remainder of this chapter involve only *two* independent variables, the notation can easily be extended to three or more independent variables. We conclude this section with one such example.

Example 4   Let $g(x, y, t) = x^2 + y^2 - 2t^2$. Find:
  a. $g(1, 2, 3)$ **b.** $g(0, 5, -2)$ **c.** $g(1, 2, -8)$

Solution   **a.** $g(x, y, t) = x^2 + y^2 - 2t^2$
$g(1, 2, 3) = 1^2 + 2^2 - 2(3)^2$
$= 1 + 4 - 18$
$= -13$

b. $g(0, 5, -2) = 0^2 + 5^2 - 2(-2)^2$
$= 0 + 25 - 8$
$= 17$

c. $g(1, 2, -8) = 1^2 + 2^2 - 2(-8)^2$
$= 1 + 4 - 128$
$= -123$ ∎

9.1 EXERCISES

SET A

In exercises 1 through 20, find the indicated value. Use

$f(x, y) = 2x^2 + 3y^2 - xy$
$g(x, y) = \sqrt{100 - x^2 - y^2}$

(See Example 1.)

1. $f(0, 1)$	**2.** $f(1, 2)$	**3.** $f(-1, 2)$	**4.** $f(2, -1)$
5. $f(0, 0)$	**6.** $f(2, 1)$	**7.** $f(-5, 0)$	**8.** $f(-5, 5)$
9. $f(-5, -5)$	**10.** $g(0, 0)$	**11.** $g(3, 4)$	**12.** $g(6, 8)$
13. $g(-6, 8)$	**14.** $g(5, -5)$	**15.** $g(0, 2)$	**16.** $g(1, 2)$
17. $f(3, 4)$	**18.** $f(1, 2) + g(6, 8)$	**19.** $f(1, 2) - g(6, 8)$	**20.** $f(g(0, 0), 3)$

In exercises 21 through 30, use

$f(x, y, t) = -2x^2 + 3y^2 + xyt$

and find the indicated values. (See Example 4.)

21. $f(0, 0, 1)$	**22.** $f(1, 0, 0)$	**23.** $f(0, 1, 0)$	**24.** $f(1, 1, 1)$
25. $f(-2, 3, -1)$	**26.** $f(2, -3, 1)$	**27.** $f(10, 3, -1)$	**28.** $f(3, 10, -1)$
29. $f(2, 4, 6)$	**30.** $f(2, 1, \frac{1}{2})$		

SET B

In exercises 31 through 36, use Examples 2 and 3 as models.

BUSINESS
REAL ESTATE

31. The Ridgefill Construction Company owns an apartment complex with one- and two-bedroom apartments. Their monthly revenue (R) is a function of the number of one-bedroom apartments (x) rented and the number of two-bedroom apartments (y) rented that month.

$$R(x, y) = 300x + 420y$$

because each one-bedroom apartment rents for $300 monthly and each two-bedroom apartment rents for $420.

   a. Find the income from rentals in June when 50 one-bedroom units and 60 two-bedroom units were rented.

   b. Find $R(45, 65)$.

MARKETING

32. The cost (C) of renting an economy car depends on the number of miles it is driven (x) and the number of days it is rented (y). The cost is given by the formula

$$C(x, y) = 0.15x + 20y$$

   a. How much does it cost to rent a car for three days if it is driven 300 miles?

   b. Find $C(500, 5)$.

BIOLOGY
MEDICINE

33. The dose (D) of a drug for a child between the ages of 3 and 12 is a function of the child's age (x) and the adult dose (y) for the drug. This is often referred to as **Young's Rule.**

$$D(x, y) = \frac{xy}{x + 12}, \qquad 3 \le x \le 12$$

   a. Find $D(6, 30)$.

   b. Find $D(8, 200)$.

ECOSYSTEMS

34. The amount (D) of DDT taken in by a carnivorous animal is a function of the DDT levels (x) in its prey and the DDT levels (y) in the vegetation consumed by its prey according to the formula

$$D(x, y) = x^2 + 0.5y + 0.5xy$$

   a. Find $D(10, 20)$.

   b. Find $D(5, 4)$.

**PHYSICAL
SCIENCE**
TIDAL ENERGY

35. The amount of energy (E, in foot-pounds) from sea waves is a function of the wave's height (x), length (y), and width (z)—each measured in feet—and is given by

$$E(x, y, z) = 8xyz$$

How much energy is created by a wave 50 ft long, 8 ft high, and 125 ft wide?

HEALTH

36. Arm strength (*A*) in an adolescent is a function of the number of dips on parallel bars (*x*), the number of pull-ups (*y*), the adolescent's weight (*w*) in pounds, and his/her height (*h*) in inches according to the formula*

$$A(x, y, w, h) = (x + y)\left(\frac{w}{10} + h - 60\right), \qquad h \geq 60$$

a. Find *A*(5, 10, 100, 65).
b. Find *A*(12, 20, 150, 65).

WRITING AND CRITICAL THINKING

Using the definition for a function (of one independent variable) from Section 2.1 as a guide, write a definition for a function with two independent variables.

9.2 GRAPHING IN THREE DIMENSIONS

Just as ordered pairs (represented graphically by points in a plane) are solutions to a function of one independent variable, we use **ordered triples** of numbers as solutions to a function of two independent variables. These ordered triples are also represented graphically by points, but in a three-dimensional coordinate system. This system with three axes—an *x* axis, *y* axis, and *z* axis—is used to graph points that represent solutions to *z* = *f*(*x*, *y*), a function of two independent variables.

Although the choice of labels for each of the mutually perpendicular axes may vary, in this text we use a "right-handed" system as depicted in Figure 9.1. Notice that to label a point (*x*, *y*, *z*), we start at the origin and first travel along the *x* axis ("out" for positive, "in" for negative), second, travel along the *y* axis (right for positive, left for negative), and finally, travel vertically for the *z* value (up for positive, down for negative). Notice that the three planes (*xy*, *yz*, and *xz*) formed by the axes divide three-dimensional space into eight parts, or **octants.**

---

*Donald K. Matthews, *Measurement in Physical Education* (Philadelphia: W. B. Saunders Co., 1973), p. 88.

Figure 9.1

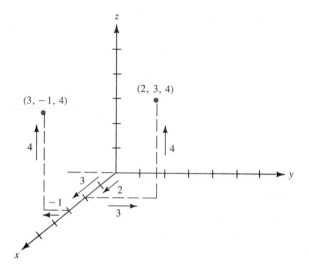

Example 1   Plot the point $(-1, 5, -2)$.

Solution   The -1 means travel "inward" one unit for x. Then we move five units to the right for y and two units down for z, as shown in Figure 9.2.

Figure 9.2

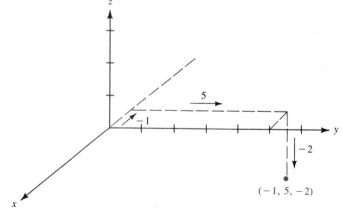

Example 2   Plot the following points:

$$A(3, 2, 0) \qquad B(0, 0, 3) \qquad C(4, 0, 0) \qquad D(2, -4, 1)$$

Solution   Verify the points' locations in Figure 9.3.

Figure 9.3

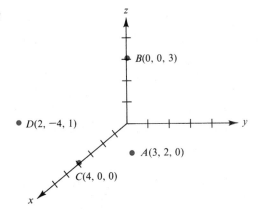

For the remainder of this section we consider the three-dimensional graphs of functions of two types: planes and spheres.

PLANES

Recall that the graph of $ax + by + c = 0$ is a straight line in two dimensions. In three dimensions, the graph of $ax + by + cz + d = 0$ is a **plane** (where not all of a, b, and c can be zero). To graph a line (in two dimensions) requires a minimum of two points. To graph a plane (in three dimensions) requires a minimum of three noncollinear points.

Example 3   Graph $x + 2y + z = 8$.

Solution   We graph the plane that represents the collection of ordered triples that satisfy the given equation by creating a table of values and plotting the points in three dimensions. For our table, we arbitrarily choose values for x and y and calculate the corresponding z values. Notice that we used zero for several of our choices, which makes both the calculations and plotting the points easier. Also note that we used a fourth point as a safety measure. The points are now plotted and the desired plane appears in Figure 9.4. (Notice that only the portion of the plane in the first octant is depicted.)

x	y	$z = 8 - x - 2y$
4	2	0
8	0	0
0	4	0
0	0	8

Figure 9.4

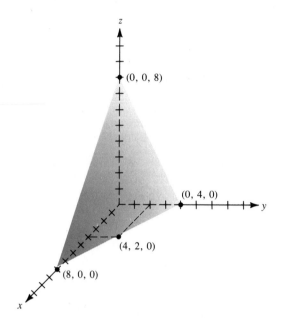

Example 4   Graph $x + 2y = 4$.

Solution   Note that this equation does not define z as a function of x and y. The plane will be parallel to (that is, never intersect) the z axis because no z is present in $x + 2y = 4$. This means that any value of z can be associated with each ordered pair (x, y). In the table, we choose values for x and z (arbitrary) and compute the corresponding y values. The graph appears in Figure 9.5.

x	y	z
4	0	0
4	0	3
0	2	0
0	2	3
2	1	0
2	1	3

Figure 9.5

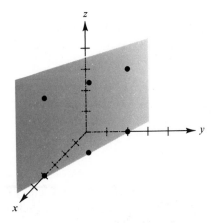

Example 5   Graph $y = 3$.

Solution   The plane is parallel to the xz plane, as seen in Figure 9.6.

Figure 9.6

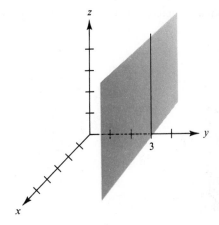

In two dimensions, the distance between two points (x_1, y_1) and (x_2, y_2) is given by

$$d = \sqrt{(x_2 - x_1)^2 + (y_2 - y_1)^2}$$

In three dimensions, the distance between the two points (x_1, y_1, z_1) and (x_2, y_2, z_2) is given by

$$d = \sqrt{(x_2 - x_1)^2 + (y_2 - y_1)^2 + (z_2 - z_1)^2} \qquad (1)$$

Example 6   Find the distance between $(2, 5, -1)$ and $(-3, 3, 5)$.

Solution   Using equation (1), we have

$$d = \sqrt{(2 + 3)^2 + (5 - 3)^2 + (-1 - 5)^2}$$
$$= \sqrt{25 + 4 + 36}$$
$$= \sqrt{65}$$

∎

SPHERES

In two dimensions, the graph of $x^2 + y^2 = r^2$ is a **circle** of radius r centered at the origin. The graph of $(x - h)^2 + (y - k)^2 = r^2$ is a circle of radius r centered at the point (h, k). In three dimensions, the graph of $x^2 + y^2 + z^2 = r^2$ is a **sphere** of radius r centered at the origin, and the graph of $(x - h)^2 + (y - k)^2 + (z - l)^2 = r^2$ is a sphere of radius r centered at the point (h, k, l).

Example 7   Graph $(x - 1)^2 + (y + 2)^2 + (z - 3)^2 = 9$.

Solution   This is a sphere of radius 3 centered at $(1, -2, 3)$. The graph appears in Figure 9.7.

Figure 9.7

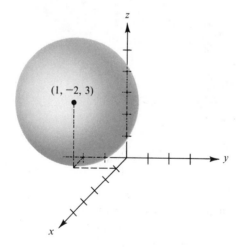

$(1, -2, 3)$

∎

We stress here that the points *on the surface* of the sphere (not inside the sphere) make up the graph of $(x - h)^2 + (y - k)^2 + (z - l)^2 = r^2$. Also, as an aid to graphing in three dimensions, we may investigate where a surface intersects the xy plane $(z = 0)$, the xz plane $(y = 0)$, and the yz plane $(x = 0)$. These intersections are called **traces.**

For the surface of Example 7, the xy trace is

$$(x - 1)^2 + (y + 2)^2 + (0 - 3)^2 = 9$$
$$(x - 1)^2 + (y + 2)^2 = 0$$

which is merely the point $(1, -2, 0)$.

The yz trace is

$$(0 - 1)^2 + (y + 2)^2 + (z - 3)^2 = 9$$
$$(y + 2)^2 + (z - 3)^2 = 8$$

which is a circle of radius $\sqrt{8}$ centered at $(0, -2, 3)$.

The xz trace is

$$(x - 1)^2 + (0 + 2)^2 + (z - 3)^2 = 9$$
$$(x - 1)^2 + (z - 3)^2 = 5$$

This is a circle of radius $\sqrt{5}$ centered at $(1, 0, 3)$; its graph appears in Figure 9.8.

Figure 9.8

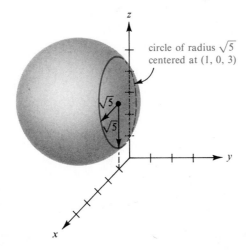

We conclude this section by mentioning that many other types of surfaces in space parallel two-dimensional curves. We focused our attention on the two most basic, the plane and the sphere. Our purpose in examining the general nature of three-dimensional graphs via these two surfaces is to give you a better perspective on the differential calculus of functions of several variables, which takes up the remainder of this chapter.

9.2 EXERCISES

SET A

In exercises 1 through 6, plot the given points. Use one set of axes for all six points. (See Examples 1 and 2.)

1. $(0, 0, 5)$ **2.** $(3, 0, 2)$ **3.** $(-2, 0, 3)$

4. $(0, -3, 0)$ **5.** $(1, 5, -2)$ **6.** $(2, 4, -2)$

In exercises 7 through 16, graph the plane whose equation is given. (See Examples 3, 4, and 5.)

7. $x + y + 2z = 8$ **8.** $2x + y + z = 8$

9. $x + y + z = 4$ **10.** $x + y - z = 4$

11. $2x - 3y + 4z = 12$ **12.** $3x + y = 8$

13. $3x + z = 8$ **14.** $x = 4$

15. $y = 5$ **16.** $z = -1$

In exercises 17 through 20, find the distance between the given points. (See Example 6.)

17. $(3, 1, 2), (0, 1, 4)$ **18.** $(3, -2, 0), (5, -1, 5)$

19. $(2, 8, 5), (-1, 8, 9)$ **20.** $(6, 1, 2), (-3, 4, 7)$

In exercises 21 through 25, graph the sphere whose equation is given. (See Example 7.)

21. $x^2 + y^2 + z^2 = 25$ **22.** $(x - 3)^2 + (y + 1)^2 + z^2 = 25$

23. $x^2 + (y - 1)^2 + (z + 1)^2 = 16$ **24.** $(x - 1)^2 + y^2 + (z - 2)^2 = 25$

25. $x^2 + y^2 + (z + 2)^2 = 16$

SET B

26. Recall that the midpoint of the line segment joining two points (x_1, y_1) and (x_2, y_2) in two dimensions is given by $\left(\dfrac{x_1 + x_2}{2}, \dfrac{y_1 + y_2}{2}\right)$. In three dimensions, the midpoint of the line segment joining (x_1, y_1, z_1) and (x_2, y_2, z_2) is given by $\left(\dfrac{x_1 + x_2}{2}, \dfrac{y_1 + y_2}{2}, \dfrac{z_1 + z_2}{2}\right)$. Find the coordinates of the midpoint of the line segment connecting each pair of points.

a. $(1, 5, -3)$ and $(3, 7, -7)$

b. $(2, 4, -6)$ and $(0, 4, 5)$

In exercises 27 through 30, determine the xy trace.

27. $(x - 2)^2 + y^2 + (z - 3)^2 = 25$ **28.** $5x^2 + y^2 + z^2 = 1$
(*Hint:* the xy trace is an ellipse.)

29. $5x^2 - y^2 + z^2 = 1$ **30.** $x^2 + 2y^2 + z^2 = 1$

USING TECHNOLOGY IN CALCULUS

Graphing software offers speed and versatility when it comes to surface plots of functions of two independent variables. Consider the graph of $z = 10e^{-x^2 - 0.5y^2}$. We use *DERIVE*® to graph four different viewpoints of the graph.

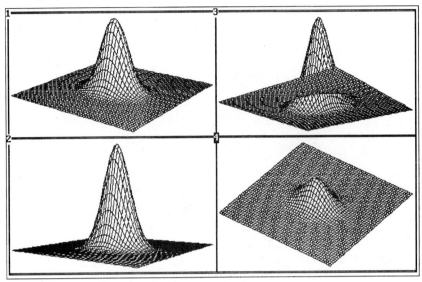

1. From the graph, determine the coordinates of the peak (which lies on the z axis).

2. Graph the function $z = \dfrac{50}{50 + (\sqrt{x^2 + y^2} - 2)^2} - 1$, and include sketches from four different viewpoints.

9.3   PARTIAL DERIVATIVES

Our first look at derivatives and differentiation occurred in Chapter 2, where all functions were functions of a *single* independent variable. In this case, a derivative gives the rate of change of the function with respect to that particular variable.

For functions of *several* independent variables, we can still find a rate of change, or derivative, for the function by considering the independent variables one at a time. This type of derivative is called a **partial derivative** and gives the rate of change of the function with respect to *one* of its independent variables. It is calculated by treating the remaining independent variables as constants. In this section, we restrict ourselves to functions of two independent variables.

Partial derivatives

Suppose that $z = f(x, y)$. The **first partial derivative of f with respect to x is the derivative of f considering y as a constant and x as a variable.** It is denoted by $\dfrac{\partial f}{\partial x}$, where

$$\frac{\partial f}{\partial x} = \lim_{h \to 0} \frac{f(x + h, y) - f(x, y)}{h}$$

We may also denote the partial derivative of f with respect to x by $\dfrac{\partial z}{\partial x}$, $f_x(x, y)$, or merely f_x.

Similarly, $f_y = \dfrac{\partial f}{\partial y}$ is given by the following:

$$\frac{\partial f}{\partial y} = \lim_{h \to 0} \frac{f(x, y + h) - f(x, y)}{h}$$

This, of course, is called the **partial derivative of f with respect to y.** In this case, we hold x constant and treat y as the variable. Some examples follow.

Example 1   Suppose that $f(x, y) = 5x - 6x^2y^3 + 7y$. Find:

    a. $\dfrac{\partial f}{\partial x}$ **b.** $\dfrac{\partial f}{\partial y}$

Solution   **a.** $\dfrac{\partial f}{\partial x}$ is found by considering y as a constant.

$$f(x, y) = 5x - 6y^3x^2 + 7y \qquad \text{Think of } -6y^3 \text{ as a numerical coefficient of } x^2.$$

$$\frac{\partial f}{\partial x} = 5 - 6y^3(2x) + 0$$

$$\frac{\partial f}{\partial x} = 5 - 12xy^3$$

b.   Treating x as a constant, we have

$$f(x, y) = 5x - 6x^2y^3 + 7y \qquad \text{Think of } -6x^2 \text{ as a numerical coefficient of } y^3.$$

$$\frac{\partial f}{\partial y} = 0 - 6x^2(3y^2) + 7$$

$$\frac{\partial f}{\partial y} = -18x^2y^2 + 7 \qquad\qquad\qquad\qquad\qquad ■$$

We denote partial derivatives evaluated at a point, say $(2, 3)$, by $f_x(2, 3)$ or $\left. \dfrac{\partial f}{\partial x} \right|_{(2, 3)}$.

Example 2 Suppose that $f(x, y) = e^{xy} + 5xy^2$. Find:
  a. $f_x(0, 1)$ **b.** $f_y(1, 0)$

Solution **a.** Treating y as a constant, we have

$$f_x = ye^{xy} + 5y^2 \qquad \text{Since the derivative of } e^{ax} \text{ is } ae^{ax}$$

Substituting $x = 0$ and $y = 1$ gives

$$f_x(0, 1) = 1e^{0(1)} + 5(1)^2$$
$$= 6$$

b. $$f_y = xe^{xy} + 10xy$$
$$f_y(1, 0) = 1e^{1(0)} + 10(1)(0)$$
$$= 1 \qquad\qquad\blacksquare$$

Example 3 The cost (C) of producing a color video monitor depends on the cost of material (x) and the cost of labor (y) and is approximated by

$$C(x, y) = 2x + 5y^2 - 3xy + 6$$

where C, x, and y are measured in dollars.
  a. Find $C_y(3, 8)$.
  b. What interpretation can be given to $C_y(3, 8)$?

Solution **a.** $C_y = 10y - 3x$
    $C_y(3, 8) = 80 - 9 = 71$

  b. C_y represents the rate of change of cost with respect to labor costs. Thus, $C_y(3, 8) = \$71$ represents the approximate change in cost of producing a color monitor per unit change in labor cost (that is, as y changes from $7 to $8 or from $8 to $9). $\blacksquare$

We calculated second (and higher-order) derivatives in Chapter 3. It is possible to find second (and higher-order) *partial* derivatives as well. *Four* such **second partial derivatives** can be found for $z = f(x, y)$.

1. The partial derivative with respect to x of the partial derivative with respect to x is denoted by

$$\frac{\partial}{\partial x}\left(\frac{\partial f}{\partial x}\right) \quad \text{or} \quad \frac{\partial^2 f}{\partial x^2} \quad \text{or} \quad f_{xx}$$

2. The partial derivative with respect to x of the partial derivative with respect to y is denoted by

$$\frac{\partial}{\partial x}\left(\frac{\partial f}{\partial y}\right) \quad \text{or} \quad \frac{\partial^2 f}{\partial x\,\partial y} \quad \text{or} \quad f_{yx}$$

3. The partial derivative with respect to y of the partial derivative with respect to x is denoted by

$$\frac{\partial}{\partial y}\left(\frac{\partial f}{\partial x}\right) \quad \text{or} \quad \frac{\partial^2 f}{\partial y\,\partial x} \quad \text{or} \quad f_{xy}$$

4. The partial derivative with respect to y of the partial derivative with respect to y is denoted by

$$\frac{\partial}{\partial y}\left(\frac{\partial f}{\partial y}\right) \quad \text{or} \quad \frac{\partial^2 f}{\partial y^2} \quad \text{or} \quad f_{yy}$$

Caution: The notation f_{yx} means to *first* find f_y and then find the partial derivative of that result with respect to x. The notation $\dfrac{\partial^2 f}{\partial x\,\partial y}$ represents the same second partial derivative, but the order of the x and y has been reversed.

Example 4 Find all four second partial derivatives of

$$f(x, y) = x^3 + 5x^2y^3 - y^4$$

Solution First calculating $\dfrac{\partial f}{\partial x}$ and $\dfrac{\partial f}{\partial y}$, we have

$$\frac{\partial f}{\partial x} = 3x^2 + 10xy^3 \qquad \frac{\partial f}{\partial y} = 15x^2y^2 - 4y^3$$

Using these results, we can compute the four second partial derivatives:

1. $\dfrac{\partial}{\partial x}\left(\dfrac{\partial f}{\partial x}\right) = 6x + 10y^3$ or $f_{xx} = 6x + 10y^3$

2. $\dfrac{\partial}{\partial x}\left(\dfrac{\partial f}{\partial y}\right) = 30xy^2$ or $f_{yx} = 30xy^2$

3. $\dfrac{\partial}{\partial y}\left(\dfrac{\partial f}{\partial x}\right) = 30xy^2$ or $f_{xy} = 30xy^2$

4. $\dfrac{\partial}{\partial y}\left(\dfrac{\partial f}{\partial y}\right) = 30x^2y - 12y^2$ or $f_{yy} = 30x^2y - 12y^2$

Notice that $f_{xy} = f_{yx}$. This is true for many, but not all, functions of two variables. ∎

Example 5 Let $f(x, y) = 5x^2y - 3y + e^{xy}$. Find $f_{xy}(-1, 2)$.

Solution Be careful of the notation! For f_{xy}, *first* find f_x and *then* differentiate with respect to y.

$$f_x = 10xy + ye^{xy}$$

Now differentiating this result with respect to y, we need to treat the second term as a product and use the Product Rule.

$$f_{xy} = 10x + [y(xe^{xy}) + e^{xy}]$$
$$= 10x + xye^{xy} + e^{xy}$$

Evaluating this for $x = -1$ and $y = 2$ gives

$$f_{xy}(-1, 2) = -10 - 2e^{-2} + e^{-2}$$
$$= -10 - e^{-2}$$
$$\approx -10.135335$$

■

We conclude this section with an application of partial differentiation to economics, the *Cobb-Douglas production formula*.

Example 6 The **Cobb-Douglas production formula**, used by economists to set production levels (P) based on available units of labor (x) and capital (y), is given by

$$P(x, y) = kx^a y^{1-a}$$

where k and a are constants and $0 < a < 1$. Suppose $P(x, y) = kx^{1/2}y^{1/2}$.

a. Find $\dfrac{\partial P}{\partial x}$, the **marginal productivity of labor.**

b. Find $\dfrac{\partial P}{\partial y}$, the **marginal productivity of capital.**

Solution **a.** $P(x, y) = kx^{1/2}y^{1/2}$

$$\frac{\partial P}{\partial x} = ky^{1/2} \frac{1}{2} x^{-1/2}$$

$$= \frac{k}{2} \sqrt{\frac{y}{x}}$$

b. $\dfrac{\partial P}{\partial y} = kx^{1/2} \dfrac{1}{2} y^{-1/2}$

$$= \frac{k}{2} \sqrt{\frac{x}{y}}$$

■

9.3 EXERCISES

SET A

In exercises 1 through 12, find (a) f_x and (b) f_y. (See Example 1.)

1. $f(x, y) = 2x + 3y + 4$

2. $f(x, y) = 2x + 3y + 4xy$

3. $f(x, y) = x - y + 6x^2y$

4. $f(x, y) = 4x^3 - 3xy^2$

5. $f(x, y) = e^{2x + 3y}$

6. $f(x, y) = xe^{2y}$

7. $f(x, y) = x^2\sqrt{y}$

8. $f(x, y) = y^3\sqrt{x}$

9. $f(x, y) = \ln(4x + 3y)$

10. $f(x, y) = \ln(5x - 2y)$

11. $f(x, y) = \ln(x^2 + 2y^2)$

12. $f(x, y) = \ln(x^2 + 3xy)$

In exercises 13 through 22, evaluate each function. (See Example 2.)

13. $f_x(2, 4)$ if $f(x, y) = x^2y$

14. $f_y(2, 4)$ if $f(x, y) = x^2y$

15. $\left.\dfrac{\partial f}{\partial x}\right|_{(3, 4)}$ if $f(x, y) = \sqrt{x^2 + y^2}$

16. $\left.\dfrac{\partial f}{\partial y}\right|_{(3, 4)}$ if $f(x, y) = \sqrt{x^2 + y^2}$

17. $\left.\dfrac{\partial z}{\partial x}\right|_{(1, 2)}$ if $z = \ln|xy + 3y^2|$

18. $\left.\dfrac{\partial z}{\partial y}\right|_{(1,2)}$ if $z = \ln|xy + 3y^2|$

19. $f_x(-2, 1)$ if $f(x, y) = e^{2x + 4y}$

20. $f_y(-2, 1)$ if $f(x, y) = e^{2x + 4y}$

21. $f_x(3, \frac{1}{4})$ if $f(x, y) = 3x\sqrt{y}$

22. $f_y(3, \frac{1}{4})$ if $f(x, y) = 3x\sqrt{y}$

In exercises 23 through 34, find the indicated second partial derivative. (See Examples 4 and 5.)

Let $f(x, y) = 5x^2y^3 - 6x^2 + 7y^2$.

23. Find f_{xx}.   **24.** Find f_{xy}.   **25.** Find f_{yx}.   **26.** Find f_{yy}.

Let $g(x, y) = e^{3x^2 + 2y}$.

27. Find g_{xx}.   **28.** Find g_{xy}.   **29.** Find g_{yx}.   **30.** Find g_{yy}.

Let $z = 2x^2y^3 - 5xy^2$.

31. Find $\left.\dfrac{\partial^2 z}{\partial x^2}\right|_{(1, -2)}$

32. Find $\left.\dfrac{\partial^2 z}{\partial x\, \partial y}\right|_{(1, -2)}$

33. Find $\left.\dfrac{\partial^2 z}{\partial y^2}\right|_{(1, -2)}$

34. Find $\left.\dfrac{\partial^2 z}{\partial y\, \partial x}\right|_{(1, -2)}$

In exercises 35 through 37, find all four second partial derivatives for the stated function. (See Example 4.)

35. $f(x, y) = 6xe^{y^2}$

36. $g(x, y) = \ln(x^2 - y)$

37. $z = 5x^2y - ye^x$

SET B

BUSINESS
COST FUNCTIONS

38. The cost (C) of producing one box of greeting cards is a function of the cost of the paper (x) and the cost of labor (y) and is approximated by

$$C(x, y) = 2x + 3y^2 - xy + 2.5$$

where C, x, and y are measured in dollars.

     a. Find C_x.

     b. Find $C_y(1, 3)$.

     c. Interpret $C_y(1, 5)$.

PROFIT

39. The profit from the sale of x hundred electric stoves and y hundred gas stoves for an appliance manufacturer is given by

$$P(x, y) = 3x^2 + 2y^2$$

where P is measured in thousands of dollars.

   a. Find the profit if 300 electric stoves and 400 gas stoves are sold.

   b. Find $\dfrac{\partial P}{\partial x}$.

   c. Find $\dfrac{\partial P}{\partial x}\bigg|_{(3, 4)}$ and interpret.

SALES

40. A company has determined that its sales depend on the amount spent on TV ads (T), the amount spent on magazine ads (M), and the amount spent on billboard ads (B). The approximate yearly sales are given by

$$S(T, M, B) = 0.1T^2 + 2M + B$$

where T, M, B, and S are measured in thousands of dollars. What is the yearly sales figure if \$75,000 is spent on TV ads, \$130,000 on magazine ads, and \$40,000 on billboard ads?

PRODUCTION

41. In Example 6 we saw an illustration of the *Cobb-Douglas production formula*. Suppose $P(x, y) = 1000x^{0.5}y^{0.5}$.

   a. Find the marginal productivity of labor.

   b. Find the marginal productivity of capital.

PRODUCTION

42. Consider the Cobb-Douglas model

$$P(x, y) = 2000x^{0.25}y^{0.75}$$

   a. Find $\dfrac{\partial P}{\partial x}$.

   b. Find $\dfrac{\partial P}{\partial y}$.

43. Use the model of exercise 42 to find:

   a. $\dfrac{\partial^2 P}{\partial x^2}$

   b. $\dfrac{\partial^2 P}{\partial y\, \partial x}$

DEMAND EQUATIONS

44. The Electrodisk Company produces two items: deluxe diskettes and standard diskettes. The selling prices of the two items are p and q, respectively. The demand, x, for deluxe diskettes is given by the equation

$$x = f(p, q) = -3p + 4q + 6$$

and the demand, y, for standard diskettes is

$$y = g(p, q) = p - 5q + 5$$

where x and y are measured in hundreds, and p and q are measured in dollars.

a. Find $\dfrac{\partial x}{\partial p}$, the **partial marginal demand of x with respect to p.**

b. Find $\dfrac{\partial x}{\partial q}$, the **partial marginal demand of x with respect to q.**

c. Find $\dfrac{\partial y}{\partial p}$, the **partial marginal demand of y with respect to p.**

d. $\dfrac{\partial y}{\partial q} = -5$ and can be interpreted as: "If the price of deluxe diskettes (p) remains constant (fixed), then a \$1 increase in the price of standard diskettes will cause a *decrease* of 500 units in the demand of standard diskettes."

Interpret parts a through c.

DEMAND

45. A toy manufacturer produces two kinds of doll houses. The one-story ranch model sells for p dollars and the two-story colonial model sells for q dollars. The demand, x, for the ranch model is given by

$$x = -2p + 5q + 3$$

and the demand, y, for the colonial model is given by

$$y = 3p - 2q + 4$$

where x and y are measured in hundreds.

a. Find $\dfrac{\partial x}{\partial q}$ and interpret.

b. Find $\dfrac{\partial y}{\partial p}$ and interpret.

REVENUE

46. The monthly revenue (R) for an insurance company is a function of the number of policies sold (x) and the total number of hours (h) put in by the sales force. It can be approximated by

$$R(x, h) = x^2 + 2h$$

where x and h are measured in hundreds and R is measured in thousands of dollars. Find $R_h(x, h)$.

ADVERTISING

47. The total sales (S) for a particular product for a certain corporation depends on the amount of money (x) spent on magazine ads and the amount (y) spent on TV commercials. This function is given by

$$S(x, y) = 5 + 4xy - x^2$$

where S, x, and y are measured in millions of dollars.

a. Find $\dfrac{\partial S}{\partial x}\Big|_{(2, 5)}$.

b.   The corporation is currently spending $2 million on magazine ads and $5 million on TV commercials. They are considering increasing the magazine ad expenditure to $3 million and keeping the TV commercial expenditure the same. Do you think this is a good idea? Explain your answer.

PROFIT

48.   A toy manufacturer that makes dolls has found that its profit (P) in any particular year is affected by the number of shopping days (x) left until Christmas and the number of dolls (y) available. This function is approximated by

$$P(x, y) = y^2 + 5xy - 0.2x^2$$

where P is measured in thousands of dollars and y is measured in thousands. If there are now 10 shopping days left until Christmas and 5000 dolls are available, find the marginal profit if the number of dolls remains fixed and we move one shopping day closer to Christmas. (*Hint:* Calculate $\left.\dfrac{\partial P}{\partial x}\right|_{(10,\,5)}$ and interpret your results.)

BIOLOGY
ECOLOGY

49.   A large factory is dumping waste into a nearby stream, which is affecting the fish population. The population (P, in hundreds) at a certain point in the stream is given by

$$P(d, x) = -3x - 2d - xd + 100$$

where d is the distance from the waste outlet to the point in the stream and x is the number of units produced daily at the factory.
a.   Find P_d.
b.   Find P_x.

TREE GROWTH

50.   The growth per year (G, in feet) for a certain tree depends on the number of inches of rainfall (r) and the number of days (d) when the temperature is above 65°F and is given by

$$G(r, d) = \ln (2r^2 + 5d)$$

Find $G_d(15, 120)$ and interpret the result.

SOCIAL SCIENCE
TRAFFIC ENGINEERING

51.   A study was conducted to determine the need for new traffic lights in a city. An index (I) was established to prioritize locations. Those with a higher I value were put at the top of the list. The index was established as a function of the average number of cars (n) per hour through the intersection and the number of traffic accidents (a) per month at the intersection. Given that

$$I(n, a) = 0.03n + 4a + 0.5an$$

find the rate of change in the index if the number of accidents per month is assumed to be a constant.

WRITING AND CRITICAL THINKING

Explain in words the difference between calculating the partial derivative with respect to x, $\dfrac{\partial z}{\partial x}$, compared with finding the partial derivative with respect to y, $\dfrac{\partial z}{\partial y}$, for some function $z = f(x, y)$.

9.4 EXTREMA OF FUNCTIONS OF TWO VARIABLES

In Chapter 4, we studied how the derivative could be used to find extreme values for a function of one variable. In this section, we show how partial derivatives can be used to locate extreme values for a function of two variables. Although the process can also be applied to functions of three or more variables, we restrict our discussion to two independent variables.

Recall that the two basic types of extreme values are called **relative maximum** and **relative minimum.** It is necessary that we redefine them for functions of two variables.

Relative extrema

Let $z = f(x, y)$ be defined at (a, b).

1. If $f(x, y) \leq f(a, b)$ for all (x, y) in some circular region around (a, b), then f has a **relative maximum** at (a, b).
2. If $f(x, y) \geq f(a, b)$ for all (x, y) in some circular region around (a, b), then f has a **relative minimum** at (a, b).

If a function has a relative maximum at (a, b) and $f(a, b) = c$, then the number c is the **relative maximum** value. In Figure 9.9, this is depicted geometrically by a point in three-dimensional space with coordinates (a, b, c).

Figure 9.9

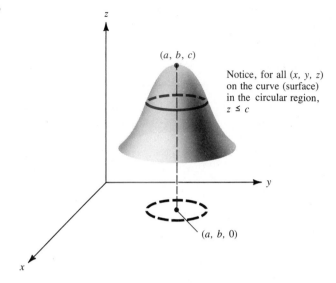

A relative minimum is illustrated in Figure 9.10.

Figure 9.10

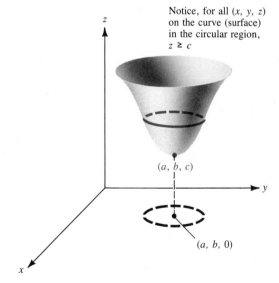

The process of finding relative extrema for functions of two variables is similar to the process developed in Chapter 4. There, we first found any critical values that served as candidates for extrema by finding values for which the first derivative was zero. Similarly, *critical values* for a function of two variables can be found by finding values for which the first *partial* derivatives are zero.

Theorem 9.1

> If $z = f(x, y)$ has a relative extremum at (a, b) and both $f_x(a, b)$ and $f_y(a, b)$ exist, then
>
> $$f_x(a, b) = f_y(a, b) = 0$$

A pair (a, b) that has the property $f_x(a, b) = 0$ and $f_y(a, b) = 0$ is called a **critical value** of f. The associated ordered triple (a, b, c) is called a **critical point,** where $c = f(a, b)$. Although we do not investigate them here, points for which f_x and f_y do not exist are also called critical points of f.

Example 1   For $f(x, y) = x^2 + y^2 - 12x + 6y - 7$, find any critical values.

Solution   First calculate f_x and f_y.

$$f_x = 2x - 12 \qquad f_y = 2y + 6$$

Now we must find a point (x, y) for which $f_x = 0$ and $f_y = 0$, so we set each equal to zero.

$$\begin{aligned} 2x - 12 &= 0 & 2y + 6 &= 0 \\ x &= 6 & y &= -3 \end{aligned}$$

The only critical value for f is $(6, -3)$.   ■

Example 2   Find the critical points for

$$f(x, y) = 3x^2 + 3y^2 - 2xy - 12x + 4y - 10$$

Solution   First, we find f_x and f_y.

$$f_x = 6x - 2y - 12 \qquad f_y = 6y - 2x + 4$$

Now, setting both equal to zero gives

$$\begin{aligned} 6x - 2y - 12 &= 0 & 6y - 2x + 4 &= 0 \\ \text{or} \qquad 3x - y - 6 &= 0 & 3y - x + 2 &= 0 \end{aligned}$$

Solve simultaneously by multiplying the equation on the left by 3 and adding the equation on the right.

$$\begin{aligned} 9x - 3y - 18 &= 0 \\ -x + 3y + 2 &= 0 \\ \hline 8x - 16 &= 0 \\ 8x &= 16 \\ x &= 2 \end{aligned}$$

Substituting $x = 2$ gives

$$\begin{aligned} 3(2) - y - 6 &= 0 \\ y &= 0 \end{aligned}$$

Thus, (2, 0) is a critical value of f, and the ordered triple (2, 0, −22) is a critical point. ∎

We caution the reader that although (2, 0, −22) is a critical point of the function f in Example 2, we do not know whether it is a relative maximum, a relative minimum, or *neither*. Similar problems arose in studying functions of a single variable in Chapter 4. Recall that for the function $y = x^3$, $x = 0$ is a critical value; however, it is not an extreme value. A similar situation for three dimensions is depicted in Figure 9.11 by a **saddle point.**

Figure 9.11

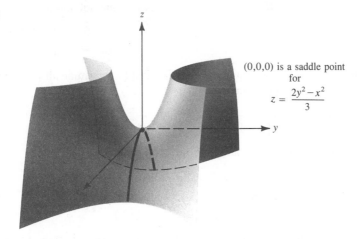

(0,0,0) is a saddle point
for
$$z = \frac{2y^2 - x^2}{3}$$

The point (0, 0, 0) is called a saddle point because (0, 0, 0) is considered a minimum relative to the points in one plane (the yz plane) and a maximum relative to points in another plane (the xz plane).

We are now ready to determine which, if any, of the critical points for a function of two variables produces a relative extremum. This can be determined using something called the **Second Partial Derivative Test.**

Second Partial Derivative Test

Consider the function $z = f(x, y)$ and suppose (a, b) is a pair such that

$$f_x(a, b) = 0 \quad \text{and} \quad f_y(a, b) = 0$$

Let $K = f_{xx}(a, b) \cdot f_{yy}(a, b) - [f_{xy}(a, b)]^2$.

1. If $K > 0$ and $f_{xx}(a, b) > 0$, then $f(a, b)$ is a **relative minimum.**
2. If $K > 0$ and $f_{xx}(a, b) < 0$, then $f(a, b)$ is a **relative maximum.**
3. If $K < 0$, then (a, b, c) is a **saddle point** of f, where $c = f(a, b)$.

Example 3   Return to the function of Example 2 and its critical value (2, 0). Is $f(2, 0)$ a relative maximum, relative minimum, or neither?

Solution We must calculate all second partials of f.

$$f(x, y) = 3x^2 + 3y^2 - 2xy - 12x + 4y - 10$$
$$f_x = 6x - 2y - 12 \qquad f_y = 6y - 2x + 4$$
$$f_{xx} = 6 \qquad f_{yx} = -2$$
$$f_{xy} = -2 \qquad f_{yy} = 6$$
$$K = f_{xx}(2, 0) \cdot f_{yy}(2, 0) - [f_{xy}(2, 0)]^2$$
$$K = 6 \cdot 6 - (-2)^2$$
$$K = 36 - 4 = 32$$

Now, since $K > 0$ and $f_{xx}(2, 0) > 0$, the function has a relative minimum at $(2, 0)$. That minimum value is $f(2, 0) = -22$. ∎

Example 4 Use the Second Partial Derivative Test to determine all possible relative maximum and minimum points of

$$f(x, y) = 3x^2 + y^3 - 6xy - 9y + 2$$

Solution Calculate f_x and f_y and set both equal to zero.

$$f_x = 6x - 6y \qquad f_y = 3y^2 - 6x - 9$$
$$6x - 6y = 0 \qquad 3y^2 - 6x - 9 = 0$$

Solve the equation on the left for x and substitute into the equation on the right.

$$6x - 6y = 0 \qquad\qquad 3y^2 - 6x - 9 = 0$$
$$6x = 6y \qquad\qquad 3y^2 - 6y - 9 = 0$$
$$x = y \qquad\qquad 3(y - 3)(y + 1) = 0$$
$$y = 3 \quad \text{or} \quad y = -1$$

Since $x = y$, the critical values are $(3, 3)$ and $(-1, -1)$.
Now we calculate the four second partial derivatives.

$$f_{xx} = 6 \qquad f_{xy} = -6 \qquad f_{yy} = 6y \qquad f_{yx} = -6$$

For the pair $(3, 3)$ we have

$$K = f_{xx}(3, 3) \cdot f_{yy}(3, 3) - [f_{xy}(3, 3)]^2$$
$$= 6 \cdot 18 - 36$$
$$= 72$$

Since $K > 0$ and $f_{xx} > 0$, f has a relative minimum at $(3, 3, -25)$.
For the pair $(-1, -1)$:

$$K = f_{xx}(-1, -1) \cdot f_{yy}(-1, -1) - [f_{xy}(-1, -1)]^2$$
$$= 6(-6) - 36 = -72$$

Since $K < 0$, the point $(-1, -1, 7)$ is a saddle point.   ■

We conclude this section with an applied problem.

Example 5   The cost of manufacturing a space telescope (C) is a function of the labor (x) in grinding the lens and the raw materials (y) and is given by

$$C(x, y) = \tfrac{1}{3}x^3 - 6xy + y^2 + 200$$

where x and y are measured in \$100 units. What combination of x and y results in minimum cost?

Solution   Computing C_x and C_y and setting equal to zero give

$$C_x = x^2 - 6y \qquad\qquad C_y = -6x + 2y$$
$$x^2 - 6y = 0 \qquad\qquad -6x + 2y = 0$$
$$y = \tfrac{1}{6}x^2 \qquad\qquad y = 3x$$

Solving simultaneously, we have

$$\tfrac{1}{6}x^2 = 3x$$
$$x^2 - 18x = 0$$
$$x(x - 18) = 0$$
$$x = 0 \;\rightarrow y = 0$$
or $\quad x = 18 \rightarrow y = 54$

Calculate all four second partial derivatives.

$$C_{xx} = 2x \qquad\qquad C_{yx} = -6$$
$$C_{xy} = -6 \qquad\qquad C_{yy} = 2$$

For $(18, 54)$:

$$K = C_{xx}(18, 54) \cdot C_{yy}(18, 54) - [C_{xy}(18, 54)]^2$$
$$K = 36 \cdot 2 - (-6)^2 = 36$$

Now, $K > 0$ and $C_{xx}(18, 54) > 0$ and the cost is minimized when 18 units of labor and 54 units of raw materials are used. We leave it to the reader (in exercise 15) to verify that $(0, 0)$ does not yield a relative minimum.   ■

9.4 EXERCISES

SET A

In exercises 1 through 12, find the critical points and then determine any relative maximum, relative minimum, or saddle points for the given function. (See Examples 1, 2, and 3.)

1. $f(x, y) = 3x^2 + 3y^2 - 2xy + 4x - 12y + 8$
2. $f(x, y) = 3x^2 + 3y^2 - 2xy + 6x - 18y + 15$

3. $f(x, y) = 3x^2 + 3y^2 - 2xy - 20x - 4y + 48$

4. $f(x, y) = 12x^2 + 3y^2 + 4xy - 8x + 4y + 2$

5. $f(x, y) = -3x^2 - 3y^2 - 2xy + 4x + 12y - 100$

6. $f(x, y) = 3x^2 - 3y^2 - 2xy + 4x + 12y - 100$

7. $f(x, y) = -12x^2 - 3y^2 + 4xy + 16x - 8y + 60$

8. $f(x, y) = y^3 - 6y^2 - x^2 + 8x + 20$

9. $f(x, y) = y^3 - 3y^2 - x^2 - 9y + 10x - 60$

10. $f(x, y) = \dfrac{4}{x} - \dfrac{32}{y} + \dfrac{xy}{8}$

11. $f(x, y) = 12xy - x^3 - 36y^2$

12. $f(x, y) = xy - x - x^3 - y^2 + 2y + 10$

SET B

BUSINESS
REVENUE EQUATIONS

13. Suppose a company produces two products: electric pencil sharpeners and electric staplers. The total revenue, R (in thousands of dollars), from selling x pencil sharpeners (in hundreds) and y staplers (in hundreds) is

$$R(x, y) = -5x^2 - 8y^2 + 39x + 39y - 2xy + 20$$

    a. Find x and y so that R will be maximized.

    b. What is that maximum revenue?

PROFIT EQUATIONS

14. The cost of producing two of its products has been determined by a company to be

$$C = 0.04(3x^2 + 2xy + 4y^2) + 4x - 2y - 100$$

    where x is the number of units produced of the first product and y is the number of units produced of the second product.

    a. If the revenue from selling x units of the first product and y units of the second product is given by $5x + 2y$, determine the profit function.

    b. Find the values of x and y that will maximize the company's profit.

COST

15. a. Verify that $(0, 0)$ in Example 5 is a critical value.

    b. Show that $(0, 0, 200)$ is a saddle point for the cost function.

COST

16. A sporting goods company makes two types of tennis racquets, the regular model and a pro model. The cost of producing x hundred regulars and y hundred pros is given by

$$C(x, y) = \tfrac{1}{3}x^3 + y^2 - 4xy + 100$$

    where C is measured in thousands of dollars. Find the number of racquets of each type that should be made to minimize cost.

PROFIT

17. For a small appliance company, the monthly profit from the sale of x mixers and y blenders is given by

$$P(x, y) = -2x^2 - 3y^3 + 6xy + 3$$

where x and y are measured in hundreds and P is measured in thousands of dollars. Find the number of mixers and blenders that should be sold to maximize profit. What is the maximum monthly profit?

MANUFACTURING COSTS

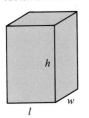

18. Boxes (as shown in the accompanying picture) are to be constructed to hold 100 in.^{3}. The cost of the material for the sides is 1.5¢ per square inch, and the cost of the top and bottom material is 2¢ per square inch. The cost per box is given by

$$C = 1.5(2lh) + 1.5(2wh) + 2(2lw)$$

Use the fact that the volume $lwh = 100$ to eliminate a variable, and then determine what dimensions the boxes should have to minimize cost.

MANUFACTURING COSTS

19. Repeat exercise 18 if the boxes are to hold 200 in.^{3}.

PROFIT

20. A pet food manufacturer sells x hundred pounds of dog food and y hundred pounds of cat food per week. The total profit in thousands of dollars is given by

$$P(x, y) = -3x^2 - 2y^2 + 4xy + 50x$$

How many pounds of dog food and cat food should be sold to maximize the profit?

WRITING AND CRITICAL THINKING

Write out in words the procedure for finding a relative maximum or a relative minimum for a function of two variables.

9.5 THE METHOD OF LAGRANGE MULTIPLIERS

In this section, we continue the process of maximizing or minimizing a function of two variables, but with an added condition. For example, a certain company that produces two styles of dishwashers may want to minimize the cost of production given by

$$C(x, y) = x^3 - xy + y^3$$

with the added condition that the total number of dishwashers produced is 500, or

$$x + y = 500$$

This last equation is called a **constraint equation.** For our work in this section, we need to set the constraint equation equal to zero, or in this case

$$x + y - 500 = 0$$

To solve this problem, observe that one possibility would be to solve the constraint equation for one variable, substitute into $C(x, y)$, and minimize the resulting function of *one* variable using the methods from Chapter 4. However, for some constraint equations it would be extremely complicated, and sometimes impossible, to solve for one variable. Another method was developed by mathematician Joseph Louis Lagrange (1736–1813), involving a five-step procedure outlined as follows.

The method of
Lagrange multipliers

Problem: Minimize (or maximize) $f(x, y)$, subject to the constraint $g(x, y) = 0$.

1. Introduce a new variable λ (Greek letter lambda) to create a new function, F, as follows:

$$F(x, y, \lambda) = f(x, y) + \lambda g(x, y)$$

2. Find F_x, F_y, and F_λ.
3. Set each of F_x, F_y, F_λ equal to zero.
4. Solve the system of the three equations in three unknowns from Step 3.
5. The values of x and y from Step 4 are the solutions that optimize (minimize or maximize) f.

Before proceeding with examples, we point out that (1) the method just described can be generalized to functions, f, of three or more independent variables; (2) the proof and theory related to the procedure are beyond the scope of this text; and (3) the resulting values are really just critical points, and the method does not guarantee that these points will produce a maximum or minimum. Our exercises and examples are restricted to those that do produce the desired maximum or minimum.

We begin the set of examples by returning to an application from Chapter 4 (Example 1 from Section 4.6).

Example I   Find the maximum rectangular area that can be enclosed using 24 ft of fencing if fencing is required on only three of the four sides.

Solution   We must maximize the area:
$f(x, y) = xy$, subject to the perimeter constraint: $2x + y = 24$.

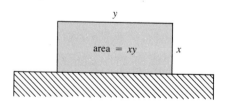

$$f(x, y) = xy$$
$$g(x, y) = 24 - 2x - y = 0$$

The five steps to solve by Lagrange multipliers are

1. Form $F(x, y, \lambda) = f(x, y) + \lambda g(x, y)$
$$= xy + \lambda(24 - 2x - y)$$
$$= xy + 24\lambda - 2x\lambda - y\lambda$$

2. Find F_x, F_y, F_λ.

$$F_x = y - 2\lambda$$
$$F_y = x - \lambda$$
$$F_\lambda = 24 - 2x - y$$

3. Set $F_x = 0$, $F_y = 0$, and $F_\lambda = 0$.

$$y - 2\lambda = 0 \tag{1}$$
$$x - \lambda = 0 \tag{2}$$
$$24 - 2x - y = 0 \tag{3}$$

4. Solve equations (1), (2), and (3).

From equation (2), $x = \lambda$; substitute λ for x in equation (3) and add equation (1).

$$24 - y - 2\lambda = 0$$
$$\underline{y - 2\lambda = 0}$$
$$24 \qquad - 4\lambda = 0$$
$$4\lambda = 24$$
$$\lambda = 6$$

Substitution into equation (1) gives $y = 12$ and substitution into equation (2) gives $x = 6$.

5. The values of $x = 6$ and $y = 12$ maximize the area subject to $2x + y = 24$. The maximum area is 72 ft².

The solution, 72 ft², is consistent with the solution found by substitution in Chapter 4. ∎

We conclude this section with two additional examples.

Example 2   Maximize $4x^2 + 2xy - 3y^2$ subject to the constraint $x + y - 1 = 0$.

Solution $f(x, y) = 4x^2 + 2xy - 3y^2$ and $g(x, y) = x + y - 1$

$$F(x, y, \lambda) = 4x^2 + 2xy - 3y^2 + \lambda(x + y - 1)$$
$$= 4x^2 + 2xy - 3y^2 + x\lambda + y\lambda - \lambda$$

Now, we find all three first partial derivatives of F and set equal to zero.

$$F_x = 8x + 2y + \lambda = 0 \tag{1}$$
$$F_y = 2x - 6y + \lambda = 0 \tag{2}$$
$$F_\lambda = x + y - 1 = 0 \tag{3}$$

Solving the system by subtracting equation (2) from equation (1) gives

$$8x + 2y + \lambda = 0 \tag{1}$$
$$\underline{2x - 6y + \lambda = 0} \tag{2}$$
$$6x + 8y \qquad = 0$$
$$x = \frac{-4}{3}y$$

Substitute $\dfrac{-4}{3}y$ for x in equation (3) and solve for y.

$$\frac{-4}{3}y + y - 1 = 0$$

$$\frac{-1}{3}y = 1$$

$$y = -3$$

Now, since $x = \dfrac{-4}{3}y$, we have

$$x = \frac{-4}{3}(-3) = 4$$

Thus, f is maximized at $(4, -3)$ and that maximum value is

$$f(4, -3) = 4(4)^2 + 2(4)(-3) - 3(-3)^2 = 13$$ ∎

Example 3   A firm produces air filters and oil filters and has estimated its monthly cost function for producing x units of air filters and y units of oil filters to be

$$C(x, y) = 8x^2 + 6y^2 - 20y + 8xy - 8x + 10$$

If the firm must produce a total of 90 units (due to consumers' demand), how many of each should it produce in order to minimize cost?

Solution   We must maximize $C(x, y)$ subject to $x + y = 90$ or $x + y - 90 = 0$.

$$F(x, y, \lambda) = C(x, y) + \lambda(x + y - 90)$$
$$F(x, y, \lambda) = 8x^2 + 6y^2 - 20y + 8xy - 8x + 10 + \lambda x + \lambda y - 90\lambda$$

Finding the first partial derivatives and setting equal to zero gives

$$F_x = 16x + 8y - 8 + \lambda = 0 \tag{1}$$
$$F_y = 12y - 20 + 8x + \lambda = 0 \tag{2}$$
$$F_\lambda = x + y - 90 = 0 \tag{3}$$

Now, we eliminate λ from equations (1) and (2) by subtracting

$$\begin{array}{r} 16x + 8y - 8 + \lambda = 0 \\ 8x + 12y - 20 + \lambda = 0 \\ \hline 8x - 4y + 12 \phantom{{}+\lambda} = 0 \\ y = 2x + 3 \end{array}$$

Substituting $2x + 3$ for y in equation (3) yields $x + 2x + 3 - 90 = 0$ or $x = 29$. In equation (3), we see that with $x = 29$, $y = 61$.

Finally, the cost is minimized when 29 units of air filters and 61 units of oil filters are produced. That minimum cost is $41,764. ∎

9.5 EXERCISES

SET A

In exercises 1 through 14, solve the stated optimization problem. (See Examples 1, 2, and 3.)

1. Maximize $f(x, y) = 2xy$ subject to $4x + y = 8$.
2. Maximize $f(x, y) = 3xy + 4x - 2y - 4$ subject to $4x + y = 10$.
3. Minimize $f(x, y) = 4x^2 + 9y^2$ subject to $4x + 3y = 10$.
4. Minimize $f(x, y) = x^2 + y^2 - 32y + 256$ subject to $2x + y = 26$.
5. Minimize $f(x, y) = 4x^2 + y^2 - 32y + 256$ subject to $4x + y = 26$.
6. Maximize $f(x, y) = 4x^2 + 16x - 10y^2 + 40y - 24$ subject to $2x - y = 12$.
7. Maximize $f(x, y) = xy - 2x^2$ subject to $2x + 3y = 14$.
8. Minimize $f(x, y) = 2y^2 - 6x^2 - 64y + 12x + 506$ subject to $2x + y = 22$.
9. Maximize $f(x, y) = 4x^2 - 12y^2 + 4xy - 10x + 60y - 75$ subject to $x + 2y = 6$.
10. Minimize $f(x, y) = 2xy$ subject to $x^2 + 4y^2 = 4$.
11. Find two numbers whose sum is 124 and whose product is as large as possible. What is that product?
12. Find two numbers whose sum is 160 and whose product is as large as possible. (Compare your result with the result of exercise 1 in Section 4.6.)
13. A rectangle is to be constructed so that the perimeter will be 80 cm. What is the maximum area for the rectangle?
14. A rectangle is to be constructed so that the area will be 64 cm². What should the dimensions of the rectangle be so that the perimeter is a minimum?

SET B

BUSINESS
SHIPPING

15. Suppose that the postal service has decided to restrict the size of packages to be mailed so that the combined length and girth (perimeter of a cross section) can be no more than 90 in. A local manufacturer needs to determine dimensions for its shipping cartons so that volume is maximized. Also, the cross sections must be twice as long as they are wide (see the figure). Find the dimensions that maximize the manufacturer's volume subject to the postal service's restrictions.

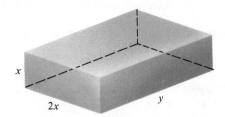

ECONOMICS

16. A company has been selling (on the average) 3500 boxes of its cologne per month at a price of $5.00 each but has found that for each $0.15 reduction in price, sales go up 150 boxes each month.

    a. Show that the demand equation of x units based on price, p, is given by

$$x = 8500 - 1000p$$

    b. Maximize $R(x, p) = xp$ subject to the constraint of the demand equation.

PACKAGING

17. A company is changing the shape of its containers for a certain product. Previously, the product was packaged in cans that held 64 in.^{3}. The company would now like to use cardboard boxes with a square base but that still hold 64 in.^{3}. What dimensions should the box have so as to minimize the amount of cardboard used?

ECONOMICS

18. An electronics firm has determined that the number of units produced per worker, y, is related to the number of its workers according to the "supply" equation

$$y = 960 - 30x$$

Maximize the company's total output

$$f(x, y) = xy$$

PACKAGING

19. Sholtz Beer wants its 12-oz beer cans to be reconstructed so that the cylindrical container will have minimum surface area. Solve Scholtz's problem using the fact that the volume $\pi x^2 y$ is to remain at 12 oz or 345 cm^3. Also, the total surface area is $2\pi x^2 + 2\pi xy$. (See the figure.)

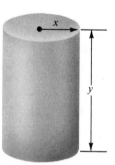

COST

20. The monthly cost of producing x gas stoves and y electric stoves is given by

$$C(x, y) = 2x^2 + 3y^2$$

Find the number of gas stoves and electric stoves that will minimize the cost if a total of 400 stoves can be produced per month.

CONSTRUCTION

21. A company is planning to fence in a rectangular area at the back of one of its warehouses to use as a storage area. Fencing will be needed on three sides only, as pictured.

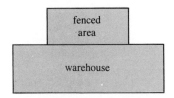

They plan to use 200 feet of leftover fencing material from another project. Find the dimensions of the maximum area that can be enclosed subject to the amount of fencing available.

PRODUCTION COST

22. Suppose the production formula for a certain company is given by

$$P(x, y) = 100x^{0.6}y^{0.4}$$

where x is the units of labor at \$200 each and y is the units of capital at \$300 each. The total expenditure for labor and capital is not to exceed \$50,000. Find the maximum production level subject to the constraint. (*Hint:* $200x + 300y = 50,000$.)

PACKAGING

23. Boxes with square bases are to be constructed to hold 96 in.[3]. The cost of the material for the sides and bottom is \$0.05 per square inch and for the top is \$0.10 per square inch. Find the dimensions for the most economical boxes.

9.6   THE METHOD OF LEAST SQUARES

We have often been in situations where, given the equation of a line, we must find whether or not a point lies on that line. Now consider that process in reverse: Given many points, what line (or equation) best "fits" the collection of points?

Table 9.1

Year	Electrical energy used (in hundred billion kWh)
1940	1.82
1950	3.96
1960	8.49
1970	16.42

As an illustration, consider the information for the total use of electrical energy in the United States for the years 1940, 1950, 1960, and 1970 given in Table 9.1. We want to find the *best* linear equation that describes the relationship between the year, x, and electrical energy used (in units of hundred billion kilowatt-hours). Using as the unit of x, 10 years, and 1940 as our first year, we can rewrite the points, called **data points**, as follows:

x	y	
1	1.82	(1940)
2	3.96	(1950)
3	8.49	(1960)
4	16.42	(1970)

The approach we use to find the best linear equation is referred to as **the method of least squares.** The line that we find, $y = mx + b$, is referred to as the **regression line** and is of great importance in the study of statistics.

We begin by defining what is meant by "best" fit. If $y = mx + b$ is a straight line and (x_1, y_1), (x_2, y_2), (x_3, y_3), . . . represent data points, then $y = mx + b$ will be a "best" line if the sum of squares of the distance each data point lies from the line is a minimum. That is, if d_i represents the (vertical) distance each data point (x_i, y_i) is from the line, we want the following to be a minimum:

$$\sum_{i=1}^{n} d_i^2 = d_1^2 + d_2^2 + \cdots + d_n^2$$

By minimizing $\sum d_i^2$, we can reach our goal of finding m and b.

We return to the electrical energy illustration and its graph in Figure 9.12.

Figure 9.12

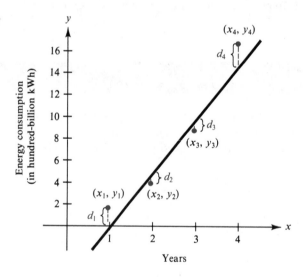

We find each d_i in the following table, noting that each $d_i = y_i - y = y_i - (mx_i + b)$.

x_i	y_i	$d_i = y_i - (mx_i + b)$
1	1.82	$d_1 = 1.82 - (m + b) = 1.82 - m - b$
2	3.96	$d_2 = 3.96 - (2m + b) = 3.96 - 2m - b$
3	8.49	$d_3 = 8.49 - (3m + b) = 8.49 - 3m - b$
4	16.42	$d_4 = 16.42 - (4m + b) = 16.42 - 4m - b$

Now,

$$\sum_{i=1}^{n} d_i^2 = (1.82 - m - b)^2 + (3.96 - 2m - b)^2$$
$$+ (8.49 - 3m - b)^2 + (16.42 - 4m - b)^2$$

is a function of the two variables m and b, or $f(m, b)$. To minimize f, we use the Second Partial Derivative Test and find $f_m, f_b, f_{mm}, f_{mb}, f_{bm},$ and f_{bb} as follows:

$$f(m, b) = (1.82 - m - b)^2 + (3.96 - 2m - b)^2 + (8.49 - 3m - b)^2$$
$$+ (16.42 - 4m - b)^2$$
$$f_m(m, b) = 2(1.82 - m - b)(-1) + 2(3.96 - 2m - b)(-2)$$
$$+ 2(8.49 - 3m - b)(-3) + 2(16.42 - 4m - b)(-4)$$
$$= -201.78 + 60m + 20b$$
$$f_b(m, b) = 2(1.82 - m - b)(-1) + 2(3.96 - 2m - b)(-1)$$
$$+ 2(8.49 - 3m - b)(-1) + 2(16.42 - 4m - b)(-1)$$
$$= -61.38 + 20m + 8b$$
$$f_{mm} = 60 \qquad f_{mb} = 20 \qquad f_{bm} = 20 \qquad f_{bb} = 8$$

Recall from Section 9.3 that f has a relative minimum at (m, b) provided that $f_m(m, b) = f_b(m, b) = 0$, and both

$$f_{mm}(m, b) \cdot f_{bb}(m, b) - [f_{mb}(m, b)]^2 > 0 \qquad \text{and} \qquad f_{mm}(m, b) > 0$$

Since $60 \cdot 8 - (20)^2 = 80 > 0$ and $60 > 0$, we set $f_m = 0$ and $f_b = 0$ to find the values of m and b.

$$f_m = -201.78 + 60m + 20b = 0 \qquad \text{or} \qquad b + 3m = 10.089$$
$$f_b = -61.38 + 20m + 8b = 0 \qquad \text{or} \qquad 2b + 5m = 15.345$$

Solving this system of two equations in m and b yields

$$m = 4.833$$
and $b = -4.41$

Finally, the linear equation of best fit (that is, the line that minimizes the sum of the distances data points are from it) in Figure 9.12 is

$$y = 4.833x - 4.41$$

Example 1   The per capita income, y (in thousands of dollars), in the United States for several years, x, is given in the following table:

	x	y
(1960)	1	2.2
(1964)	2	2.6
(1968)	3	3.4
(1972)	4	4.5
(1976)	5	6.4
(1980)	6	9.5

Determine the equation of the regression line.

Solution We first find $f(m, b)$ by finding each $d_i = y_i - (mx_i + b)$.

$$d_1 = y_1 - (mx_1 + b) = 2.2 - m - b$$
$$d_2 = y_2 - (mx_2 + b) = 2.6 - 2m - b$$
$$d_3 = y_3 - (mx_3 + b) = 3.4 - 3m - b$$
$$d_4 = y_4 - (mx_4 + b) = 4.5 - 4m - b$$
$$d_5 = y_5 - (mx_5 + b) = 6.4 - 5m - b$$
$$d_6 = y_6 - (mx_6 + b) = 9.5 - 6m - b$$
$$f(m, b) = \sum_{i=1}^{6} d_i^2$$
$$f(m, b) = (2.2 - m - b)^2 + (2.6 - 2m - b)^2 + (3.4 - 3m - b)^2$$
$$+ (4.5 - 4m - b)^2 + (6.4 - 5m - b)^2 + (9.5 - 6m - b)^2$$

Now, we take the partial derivatives.

$$f_m(m, b) = 2(2.2 - m - b)(-1) + 2(2.6 - 2m - b)(-2)$$
$$+ 2(3.4 - 3m - b)(-3) + 2(4.5 - 4m - b)(-4)$$
$$+ 2(6.4 - 5m - b)(-5) + 2(9.5 - 6m - b)(-6)$$
$$= -249.2 + 182m + 42b \qquad (1)$$
$$f_b(m, b) = 2(2.2 - m - b)(-1) + 2(2.6 - 2m - b)(-1)$$
$$+ 2(3.4 - 3m - b)(-1) + 2(4.5 - 4m - b)(-1)$$
$$+ 2(6.4 - 5m - b)(-1) + 2(9.5 - 6m - b)(-1)$$
$$= -57.2 + 42m + 12b \qquad (2)$$
$$f_{mm} = 182 \qquad f_{mb} = 42 \qquad f_{bb} = 12 \qquad f_{mm}f_{bb} - (f_{mb})^2 = 420$$

So, by setting $f_m = 0$ and $f_b = 0$, we have the desired minimum.

$$182m + 42b = 249.2 \qquad (1)$$
$$42m + 12b = 57.2 \qquad (2)$$

We leave it to the reader (in exercise 29) to verify that

$$m = 1.4$$
$$b = -\tfrac{2}{15} \approx -0.1333$$

The line of best fit, $y = 1.4x - \tfrac{2}{15}$, and the original data points are graphed in Figure 9.13.

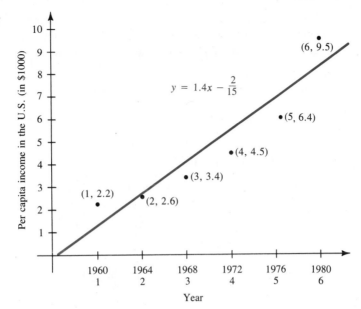

Figure 9.13

$y = 1.4x - \frac{2}{15}$

(6, 9.5)

(5, 6.4)

(4, 4.5)

(3, 3.4)

(1, 2.2)

(2, 2.6)

Per capita income in the U.S. (in $1000)

Year					
1960	1964	1968	1972	1976	1980
1	2	3	4	5	6

Example 2 Use the regression line of Example 1 to *predict* the per capita income for $x = 10$ (the year 1996).

Solution We find the *predicted* y value by finding the ordinate of the point on the line of best fit for $x = 10$.

$y = 1.4x - \frac{2}{15}$

$= 1.4(10) - \frac{2}{15}$

$= 13.867$

The predicted per capita income in 1996 is 13.867 thousand dollars, or $13,867. ■

Example 3 In the illustration preceding Example 1, we found that the relationship between electrical energy consumption, y (in units of hundred billion kilowatt-hours), and year, x, was given by

$y = 4.833x - 4.41$

What amount of electrical energy could be expected based on this model in the year 2000 (use $x = 7$)?

Solution $y = 4.833(7) - 4.41$

$= 29.421$ hundred billion kilowatt-hours

or 2,942,100,000,000 kWh. ■

Before concluding this section with a final example of the method of least squares, we point out that the method described thus far can be *generalized*. In general, consider a collection of (at least three) data points (x_1, y_1), (x_2, y_2), $\ldots$, (x_n, y_n). Our objective is to minimize $f(m, b) = \sum\limits_{i=1}^{n} d_i^2$, where each $d_i = y_i - (mx_i + b)$.

$$f(m, b) = (y_1 - mx_1 - b)^2 + (y_2 - mx_2 - b)^2 + (y_3 - mx_3 - b)^2 + \cdots + (y_n - mx_n - b)^2$$

Notice that $f_m(m, b)$ is given as

$$
\begin{aligned}
f_m(m, b) &= 2(y_1 - mx_1 - b)(-x_1) + 2(y_2 - mx_2 - b)(-x_2) \\
&\quad + 2(y_3 - mx_3 - b)(-x_3) + \cdots + 2(y_n - mx_n - b)(-x_n) \\
&= 2[(-x_1y_1 - x_2y_2 - x_3y_3 - \cdots - x_ny_n) \\
&\quad + m(x_1^2 + x_2^2 + \cdots + x_n^2) + b(x_1 + x_2 + x_3 + \cdots + x_n)] \\
&= 2\left[-\sum_{i=1}^{n} x_iy_i + m \sum_{i=1}^{n} x_i^2 + b \sum_{i=1}^{n} x_i \right] \quad (1)
\end{aligned}
$$

$$
\begin{aligned}
f_b(m, b) &= 2(y_1 - mx_1 - b)(-1) + 2(y_2 - mx_2 - b)(-1) \\
&\quad + 2(y_3 - mx_3 - b)(-1) + \cdots + 2(y_n - mx_n - b)(-1) \\
&= 2[(-y_1 - y_2 - y_3 - \cdots - y_n) + m(x_1 + x_2 + x_3 + \cdots + x_n) \\
&\quad + b(1 + 1 + 1 + \cdots + 1)] \\
&= 2\left[-\sum_{i=1}^{n} y_i + m \sum_{i=1}^{n} x_i + bn \right] \quad (2)
\end{aligned}
$$

Now, we set $f_m = 0$ and $f_b = 0$ and solve that system of two equations for m and b to obtain (see exercise 30)

$$m = \frac{n \sum\limits_{i=1}^{n} x_iy_i - \sum\limits_{i=1}^{n} x_i \sum\limits_{i=1}^{n} y_i}{n \sum\limits_{i=1}^{n} x_i^2 - \left(\sum\limits_{i=1}^{n} x_i \right)^2}$$

$$b = \frac{\sum\limits_{i=1}^{n} y_i \sum\limits_{i=1}^{n} x_i^2 - \sum\limits_{i=1}^{n} x_i \sum\limits_{i=1}^{n} x_iy_i}{n \sum\limits_{i=1}^{n} x_i^2 - \left(\sum\limits_{i=1}^{n} x_i \right)^2}$$

We now conclude this section with some applications of the preceding two formulas. Although the *method* of least squares was used to obtain the formulas, we dispense with the method now and address our attention to *using* the formulas.

Example 4   Find the line of best fit for the following data:

x	y
0	-1
2	2.5
3	5.0
5	10.0

Solution Here, $n = 4$. We see, by the formula on page 584, that we need to find the following quantities: $\sum x, \sum x^2, \sum xy, \sum y.$* A table is the most appropriate way to find these. We exhibit such a table here.

x	y	x^2	xy
0	-1	0	0
2	2.5	4	5
3	5	9	15
5	10	25	50
Sum 10	16.5	38	70

Thus, $\sum x = 10$, $\sum y = 16.5$, $\sum x^2 = 38$, $\sum xy = 70$, and $\left(\sum x\right)^2 = 100$.

$$m = \frac{n\sum xy - \sum x \sum y}{n\sum x^2 - \left(\sum x\right)^2}$$

$$= \frac{4(70) - 10(16.5)}{4(38) - 100}$$

$$= \frac{115}{52} \approx 2.212$$

$$b = \frac{\sum y \sum x^2 - \sum x \sum xy}{n\sum x^2 - \left(\sum x\right)^2}$$

$$= \frac{(16.5)(38) - 10(70)}{4(38) - 100}$$

$$= \frac{-73}{52} \approx -1.404$$

Finally, the regression line is

$$y = 2.212x - 1.404$$ ∎

All quantities in the next example were obtained using a calculator.

---

*We drop the index $i = 1$ to n for the remainder of the chapter.

Example 5 The Consumer Price Index (CPI) for three years is given in the table.

Year	CPI
1965	94.5
1970	116.3
1975	161.2

   a. Find the line of best fit that describes these data.
   b. Estimate the CPI for 1980 using the line.

Solution   **a.** We must assign x values to each evenly spaced year. Although we *could* use 1965, 1970, and 1975, it lessens the burden of calculation if we assign -1 to 1965, 0 to 1970, and 1 to 1975. The reason: $\sum x = 0$. We complete the calculation table as follows:

	x	y	x^2	xy
(1965)	-1	94.5	1	-94.5
(1970)	0	116.3	0	0
(1975)	1	161.2	1	161.2
Sum $\rightarrow$	0	372	2	66.7

$$m = \frac{n \sum xy - \sum x \sum y}{n \sum x^2 - \left(\sum x\right)^2} = \frac{(3)(66.7) - 0}{3(2)} = 33.35$$

$$b = \frac{\sum y \sum x^2 - \sum x \sum xy}{n \sum x^2 - \left(\sum x\right)^2} = \frac{(372)(2) - 0}{3(2)} = 124$$

Thus, $y = 33.35x + 124$ is the desired line.

   b. The 1980 projection is found by substituting the value of 2 for x (why 2?) to find y.

$$y = 33.35x + 124$$
$$y = 33.35(2) + 124$$
$$= 190.7$$

(*Note:* The *actual* 1980 CPI was 247.0.) ∎

9.6 EXERCISES

SET A

In exercises 1 through 8, find the linear equation that best fits the data by (1) finding f(m, b); (2) finding $f_m(m, b)$ and $f_b(m, b)$; and (3) finding m and b by solving the resulting system, $f_m = 0$ and $f_b = 0$. (See Example 1 and the discussion preceding it.)

1.

x	y
-2	4
0	10
2	12

2.

x	y
-2	4
0	10
1	11
2	12

3.

x	y
0	10
2	6
3	5

4.

x	y
0	10
2	6
3	5
5	1

5.

x	y
-4	-11
0	-3
4	6

6.

x	y
-4	-11
-2	-6
0	-3
2	0
4	6

7.

x	y
0	5
1	6
2	7
3	8
4	10

8.

x	y
0	5
1	6
2	7
3	8
4	9

In exercises 9 through 16, find the projected y value for the given x value and data. (See Examples 2 and 3.)

9. $x = 3$ and data from exercise 1

10. $x = 3$ and data from exercise 2

11. $x = 4$ and data from exercise 3

12. $x = 4$ and data from exercise 4

13. $x = 6$ and data from exercise 5

14. $x = 6$ and data from exercise 6

15. $x = 10$ and data from exercise 7

16. $x = 10$ and data from exercise 8

In exercises 17 through 22, find the line of best fit, $y = mx + b$, by using the formulas for m and b developed in this section. (See Examples 4 and 5.)

17.

x	y
-2	7
0	11
2	16

18.

x	y
-1	5
0	6.5
1	7

19.

x	y
-2	14
-1	11
0	10
1	9
2	6

20.

x	y
-4	0
-1	5
0	12
1	20
4	24

21.

x	y
1	2
2	7
3	8
4	10
5	14

22.

x	y
0	-1
1	4
3	4
5	20
8	35

SET B

BUSINESS
ECONOMICS

23. The amount expended for new factories and equipment (excluding agriculture) in the United States for various years is given in the table.

Year	Amount (in tens of billions of dollars)
1970	10.6
1975	15.8
1980	23.1

a. Find the line of best fit that represents the data.

b. Use your answer to project the amount that will be expended in 1995.

ACCOUNTING

24. A researcher suspects that the amount of money paid in personal income tax may be linearly related to the number of IRS employees. The following data apply:

Number of IRS employees (in thousands)		Personal income tax (in hundred billion dollars)
(1970)	68.0	1.03
(1979)	86.6	2.52
(1980)	87.5	2.88

a. Let x represent the number of IRS employees and y the personal income tax taken in by the IRS that year. Find the line of best fit that represents these data.

b. In 1978 the IRS employed 85.3 thousand employees. Use your answer from part a to predict the personal income tax collected by the IRS in 1978.

INVESTMENT STRATEGIES

25. A stamp collector is trying to decide whether his hobby of philately may be worth considering as an investment venture. He uses for his research a collection of 28 commemorative stamps he purchased for face value in 1948 at 86¢. These stamps were appraised over the years as follows:

Year	Value
1948	$0.86
1963	1.65
1981	4.95
1982	4.95

a. Determine the line of best fit.

b. Project the value of the 28 stamps in 1993, 45 years after their purchase.

c. If the $0.86 were invested at 5% compounded quarterly, it would amount to $8.05 by 1993. Do you think these 28 stamps were a good investment?

BIOLOGY
PATHOLOGY

26. A research physician suspects that the death by cancer rate, per thousand, is linearly related to the emission of sulfur oxide annually in a city, shown as follows:

x (sulfur oxide pollutant in thousands of tons)		y (death by cancer rate per thousand)
(1970)	32.8	190
(1975)	28.8	175
(1980)	29.8	185

a. Find the line of best fit.

b. Predict the rate of death by cancer in the city if the sulfur oxide emitted in a given year is 50,000 tons. Use your answer to part a.

SOCIAL SCIENCE
POPULATION PROFILES

27. A researcher suspects that the number of television sets produced annually is linearly related to the school enrollment for that year. Her figures appear in the table.

x (school enrollment in millions)	y (number of TV sets produced in millions)
(1950) 28	7.5
(1960) 42	5.7
(1970) 51	4.9

Find the line of best fit for these data.

DIVORCE RATES

28. Consider the following data:

State	x per capita personal income 1979 (in thousands)	y divorce rate 1979
Alabama	7.0	7.0
California	10.0	6.1
Connecticut	10.0	4.5
Florida	8.5	7.9
Kansas	9.2	5.4
Massachusetts	8.9	3.0
New Jersey	9.7	3.2
Texas	8.8	6.9

Find the line of best fit.

29. Find m and b in Example 1 by solving the system

$$182m + 42b = 249.2$$
$$42m + 12b = 57.2$$

30. The value of m on page 584 is found by solving the following system:

$$m \sum x^2 + b \sum x = \sum xy \tag{1}$$
$$m \sum x + bn = \sum y \tag{2}$$

a. Find m by multiplying equation (1) by $-n$, multiplying equation (2) by $\sum x$, and adding the resulting equations.

b. Use a similar approach to solve for b.

c. Show that an alternate formula for b is

$$b = \frac{1}{n}\left(\sum y - m \sum x\right)$$

WRITING AND CRITICAL THINKING

Using the data points from Example 4, compute $\bar{x}$, the average of the x coordinates, and $\bar{y}$, the average of the y coordinates. Now show that the point $(\bar{x}, \bar{y})$ lies on the line of best fit. Explain in words why you think this happens.

USING TECHNOLOGY IN CALCULUS

Many calculators have statistical functions, like the line of best fit, built in. On the TI-82 graphics calculator, for example, data are entered in *list variables*. For the data in Example 1, we have entered the x data in list variable L_1 and the y data in list variable L_2.

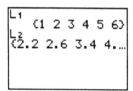

Using the STAT PLOT option, we can get a scatterplot of the data points:

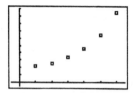

The "2-Var Stats" under the STAT menu allow us to display a variety of statistics regarding the paired data L_1 and L_2:

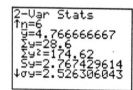

The linear regression ("LinReg") option calculates the line of best fit, $y = 1.4x - 0.13$:

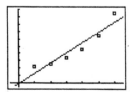

Finally, we display the line along with the scatterplot:

Rework Example 4 using a similar approach.

CHAPTER 9 REVIEW

VOCABULARY/FORMULA LIST

ordered triples

octants

plane

$$d = \sqrt{(x_2 - x_1)^2 + (y_2 - y_1)^2 + (z_2 - z_1)^2}$$

traces

first partial derivative of f with respect to x, f_x

second partial derivatives: $f_{xx}, f_{xy}, f_{yx}, f_{yy}$

$$m = \frac{n \sum_{i=1}^{n} x_i y_i - \sum_{i=1}^{n} x_i \sum_{i=1}^{n} y_i}{n \sum_{i=1}^{n} x_i^2 - \left(\sum_{i=1}^{n} x_i \right)^2}$$

relative maximum

relative minimum

saddle point

Second Partial Derivative Test

constraint equation

Lagrange multipliers

regression line

method of least squares

$$b = \frac{\sum_{i=1}^{n} y_i \sum_{i=1}^{n} x_i^2 - \sum_{i=1}^{n} x_i \sum_{i=1}^{n} x_i y_i}{n \sum_{i=1}^{n} x_i^2 - \left(\sum_{i=1}^{n} x_i \right)^2}$$

CHAPTER 9 REVIEW EXERCISES

[9.1] **1.** Suppose $f(x, y) = 3x^2 + 2x^2 y^3 - xy^2$; find the value of $f(2, -1)$.

[9.2] **2.** On one set of axes, plot the following points: $(0, 1, 3)$, $(2, -3, 1)$, $(0, 0, 2)$, and $(-1, 5, 0)$.

[9.2] **3.** Graph the plane whose equation is $x + 2y - z = 4$.

[9.2] **4.** Find the distance between the two points $(3, 1, 4)$ and $(0, 1, 6)$.

[9.2] **5.** Graph the sphere whose equation is $x^2 + y^2 + z^2 = 64$.

In exercises 6 through 11, use $f(x, y) = e^{xy} + 4x^3 y^2 - 3y^4$. *Find:*

[9.3] **6.** f_x **7.** f_y **8.** f_{xx}

        9. f_{xy} **10.** f_{yx} **11.** f_{yy}

[9.4] **12.** Consider the function $f(x, y) = 3x^2 + 3y^2 + 4xy - 8x + 4y + 2$. Find the critical points and then determine any relative maximum, relative minimum, or saddle points for the function.

BUSINESS

REVENUE

[9.4] **13.** An electronics company manufactures two items: microprocessing chips and circuit boards. The total revenue, R (in thousands of dollars), from selling x chips (in thousands) and y boards (in thousands) is

$$R(x, y) = -8x^2 - 5y^2 + 39x + 39y - 2xy + 20$$

     a. Find the value of x and y that will maximize R.

     b. What is the maximum revenue?

[9.5] **14.** Maximize $f(x, y) = 4x^2 + y^2 - 32y + 256$ subject to $4x + y = 26$.

[9.6]   **15.**   **a.**   Graph the following points:

x	y
2	5
4	10
6	12

b.   Determine $f(m, b)$.
c.   Find $f_m(m, b)$ and $f_b(m, b)$.
d.   Find m and b.
e.   Use the results of part d to graph the line of best fit on your answer to part a.

CHAPTER 9 TEST

PART ONE

Multiple choice. Put the letter of your correct response in the blank provided.

———   **1.**   If $f(x, y) = x^2 - 4y^2 - 2xy$, then $f(-1, -3)$ is

   a. -4 **b.** -41
   c. 29 **d.** -43

———   **2.**   The distance between $(1, 4, 5)$ and $(-3, 7, 5)$ is

   a. $\sqrt{51}$ **b.** $\sqrt{13}$
   c. $\sqrt{38}$ **d.** 5

———   **3.**   The graph that represents $x^2 + y^2 + z^2 = 16$ is

   a.   a plane   **b.**   a straight line
   c.   a sphere of radius 16   **d.**   a sphere of radius 4

———   **4.**   Suppose that $f(x, y) = 6x - 3x^2y^3 - 7y$. Then $\dfrac{\partial f}{\partial y} =$

   a. $-9x^2y^2 - 7$ **b.** $6 - 6xy^2$
   c. $6 - 3x^2$ **d.** $-9x^2y^2 - 6xy^3 - 7$

———   **5.**   Let $f(x, y) = 6x^4 + 5y^2 - 3x^2y^3$. Find $f_{xx}(1, 2)$.

   a. -42 **b.** 34
   c. 24 **d.** 2

———   **6.**   The line of best fit for the following data points is

x	y
0	4
-2	0
1	5

a. $y = \frac{12}{7}x + \frac{25}{7}$

b. $y = 2x + 4$

c. $y = \frac{25}{7}x + \frac{12}{7}$

d. $y = \frac{12}{7}x - \frac{25}{7}$

PART TWO

Solve and show all work.

7. Graph $2x + y + z = 7$.

8. Graph $(x - 2)^2 + (y - 1)^2 + z^2 = 9$.

In exercises 9 through 15, suppose that $f(x, y) = 2x^4 - 5x^2y^3 + 3y^2$. Find:

9. $f(0, -1)$

10. f_x

11. f_y

12. $f_x(0, -1)$

13. f_{xy}

14. f_{yy}

15. $f_{xx}(0, -1)$

16. Use the Second Partial Derivative Test to determine all possible relative maximum and minimum points of $f(x, y) = x^3 + 3y^2 - 6xy - 9x + 2$.

17. Maximize $-3x^2 + 2xy + 4y^2$ subject to $x + y - 1 = 0$.

BUSINESS
MAXIMUM REVENUE

18. A company produces two products, and its revenue, R, from selling x hundred units of one item and y hundred units of the other is

$$R(x, y) = -8x^2 - 5y^2 + 136x + 36y - 2xy + 2000$$

a. How many units of x and y must be sold to maximize the revenue?

b. What is that maximum revenue?

In exercises 19 and 20, use the following data:

x	y
-2	-3
-1	-1
0	2
1	0
2	2

19. **a.** Find $f(m, b)$.   **b.** Find $f_m(m, b)$.   **c.** Find $f_b(m, b)$.

20. Find the equation of the straight line that best fits the data.

THE TI-82 GRAPHICS CALCULATOR

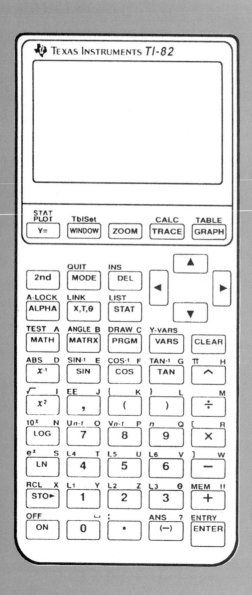

NUMERICAL COMPUTATION ON THE TI-82 GRAPHICS CALCULATOR

We begin by examining how to use the calculator to calculate! Two keys must be mastered early, the $\boxed{2^{nd}}$ key and the $\boxed{\text{ALPHA}}$ key. These keys (like the SHIFT key on a typewriter) are used with other keys to increase each key's functionality. The $\boxed{2^{nd}}$ key is blue, and it accesses the blue-colored feature above the key pressed. Similarly, the $\boxed{\text{ALPHA}}$ key is used with other keys to access the (uppercase) 26 letters of the alphabet and some additional special characters, such as θ, space, quote, and comma.

The $\boxed{\text{MODE}}$ key also needs early mentioning. Pressing the $\boxed{\text{MODE}}$ key will display one of the TI-82's menus. From this menu, we choose things such as the way numbers will be displayed, whether angles are in degrees or radians, and certain graphing features. The mode screen is displayed in Figure A.1.

Figure A.1
The mode screen

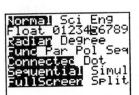

The first two lines indicate that numbers will be expressed with five fixed decimal places. "Sci" stands for scientific notation, and "Eng" stands for engineering notation.

Line 3 indicates that angles will be expressed in radian mode. Lines 4, 5, and 6 refer to graphing details covered later. The bottom line allows the user to split the screen into text and graphics display simultaneously.

Let's begin computing by attempting to find the value of $-5 + 11 - (-3)$. Notice the following two keys on the calculator: $\boxed{(-)}$ and $\boxed{-}$. They are considerably different! To enter negative numbers, we use the $\boxed{(-)}$ key; for subtraction, we use the $\boxed{-}$ key. Thus, to find the value of $-5 + 11 - (-3)$, we would enter the following keystrokes:*

$\boxed{(-)}$ 5 $\boxed{+}$ 1 1 $\boxed{-}$ $\boxed{(-)}$ 3 $\boxed{\text{ENTER}}$

The screen displays:

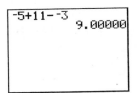

---

*In this appendix we use the following notation convention: Keystrokes are placed in boxes so that when we write $\boxed{\text{MODE}}$, for example, we mean "press the key labeled MODE." We do not, however, place numbers in boxes, and it is assumed that when we write 5.89, the four obvious keys are pressed: $\boxed{5}$ $\boxed{.}$ $\boxed{8}$ $\boxed{9}$. Also, to access the square root symbol, $\sqrt{\ }$, we need to use the 2nd function of the x^2 key. To emphasize that we are selecting the square root, we choose to write that as $\boxed{2^{nd}}$ $\boxed{\sqrt{\ }}$ rather than $\boxed{2^{nd}}$ $\boxed{x^2}$. Likewise, to access the alphabetic character "P," for example, we write $\boxed{\text{ALPHA}}$ $\boxed{\text{P}}$ rather than $\boxed{\text{ALPHA}}$ $\boxed{8}$.

For a more complicated calculation, like $\sqrt{7^2 - 2^2}$, we need to use the left parenthesis key, $\boxed{(}$, and the right parenthesis key, $\boxed{)}$. The keystrokes necessary to enter $\sqrt{7^2 - 2^2}$ are:

$$\boxed{2^{nd}} \quad \boxed{\sqrt{}} \quad \boxed{(} \quad 7 \quad \boxed{x^2} \quad - \quad 2 \quad \boxed{x^2} \quad \boxed{)} \quad \boxed{\text{ENTER}}$$

Now, the screen should resemble the following:

```
-5+11--3
         9.00000
√(7²-2²)
         6.70820



```

To calculate $\sqrt{7^2 + 2^2}$, there is no need to reenter the new expression; instead, we can *edit* the previous expression. To do this, press the editing key, $\boxed{2^{nd}}$ $\boxed{\text{ENTRY}}$. Now, press the left arrow key, $\boxed{\triangleleft}$, four times until the cursor is over the subtraction symbol. Press the $\boxed{+}$ key followed by the $\boxed{\text{ENTER}}$ key. The screen display should look like this:

```
-5+11--3
         9.00000
√(7²-2²)
         6.70820
√(7²+2²)
         7.28011
```

The most recently computed value, in this case 7.28011, is stored in the "Answer" memory and is accessible by pressing the $\boxed{\text{ANS}}$ key. Hence, to find $(7.28011)^2$, we merely have to enter $\boxed{2^{nd}}$ $\boxed{\text{ANS}}$ $\boxed{x^2}$:

```
         9.00000
√(7²-2²)
         6.70820
√(7²+2²)
         7.28011
Ans²
        53.00000
```

Exponents and roots in general can be calculated using the $\boxed{\wedge}$ key. Several examples follow showing the exact keystrokes to enter to obtain a specified answer. Note the use of parentheses and verify the answers given.

$\dfrac{3}{\sqrt{12} - \sqrt{7}}$

$3 \boxed{\div} \boxed{(} \boxed{2^{nd}} \boxed{\sqrt{}} 12 - \boxed{2^{nd}} \boxed{\sqrt{}} 7 \boxed{)} \boxed{\text{ENTER}}$
Answer: 3.66591

$\sqrt[5]{133^2}$ or $(133^2)^{1/5}$

$\boxed{(} 133 \boxed{x^2} \boxed{)} \boxed{\wedge} \boxed{(} 1 \boxed{\div} 5 \boxed{)} \boxed{\text{ENTER}}$
Answer: 7.07197

$\dfrac{-3 - \sqrt{3^2 - 4(2)(-3)}}{2 \cdot 2}$

$\boxed{(} \boxed{(-)} 3 - \boxed{2^{nd}} \boxed{\sqrt{}} \boxed{(} 3 \boxed{x^2} - 4 \boxed{\times} 2 \boxed{\times}$
$\boxed{(-)} 3 \boxed{)} \boxed{)} \boxed{\div} \boxed{(} 2 \boxed{\times} 2 \boxed{)} \boxed{\text{ENTER}}$
Answer: -2.18614

The next two calculations involve the number pi.

$$\frac{\pi + 1}{\sqrt{3}}$$

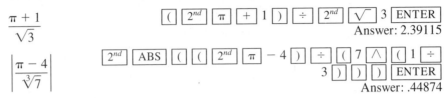

Answer: 2.39115

$$\left|\frac{\pi - 4}{\sqrt[3]{7}}\right|$$

Answer: .44874

Note that the special powers of cubing and cube rooting are available through the [MATH] menu options. The last calculation above can also be done with the following sequence of keystrokes:

[2nd] [ABS] [(] [(] 4 − [2nd] [π] [)] [÷] [(] [MATH] 4 7 [)] [)] [ENTER]

Numbers in scientific notation may be entered directly by using the [2nd] [EE] key. For example, to enter the number 3.89 × 10⁵, enter 3.89 [2nd] [EE] 5 [ENTER]. The display should read:

```
3.89E5
     389000.0000
```

Displaying numbers in scientific notation mode occurs by pressing the [MODE] key and then pressing the right arrow key, [▷], followed by [ENTER]. Then choose the number of digits of precision with the down arrow key [▽], say 5. Press [ENTER] and the following would be displayed:*

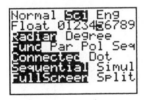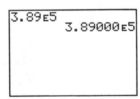

Similarly, a number can be converted to *engineering notation* (where the power of 10 is always a multiple of 3) by choosing ENG in the mode menu.

DATA STORAGE

When a certain number needs to be "remembered" or is going to be used in many subsequent calculations, it can be stored by using the [STO ▷] ("assign a value to a memory location") key in conjunction with a memory location name (the letters

*Leave the [MODE] menu by pressing [2nd] [QUIT].

A through Z and θ). For example, to store the number −42.31 in memory D, we enter the following keystrokes:

$$\boxed{(-)}\ 42.31\ \boxed{\text{STO} \triangleright}\ \boxed{\text{ALPHA}}\ \boxed{\text{D}}\ \boxed{\text{ENTER}}$$

To display the contents of memory location D, we enter $\boxed{\text{ALPHA}}\ \boxed{\text{D}}\ \boxed{\text{ENTER}}$. To increase the value in memory location D by 32.5, we would enter:

$$\boxed{\text{ALPHA}}\ \boxed{\text{D}}\ \boxed{+}\ 32.5\ \boxed{\text{STO} \triangleright}\ \boxed{\text{ALPHA}}\ \boxed{\text{D}}\ \boxed{\text{ENTER}}$$

To "zero" (or clear) the contents of a memory, we store the value of 0 in it. To ensure that memory location D holds nothing, we enter the keystrokes

$$0\ \boxed{\text{STO} \triangleright}\ \boxed{\text{ALPHA}}\ \boxed{\text{D}}\ \boxed{\text{ENTER}}$$

THE GRAPHING KEYS

What makes this calculator so powerful is its ability to graph points, lines, and functions. Before we attempt to graph something, we first must observe a very important graphing-related key, the $\boxed{\text{WINDOW}}$ key (see Figure A.2). When we press this key, six values are displayed: Xmin (the smallest x value that will appear in the graphing window), Xmax (the largest x value in the graphing window), Xscl (the scale or distance between tick marks on the x axis), Ymin (the smallest y value that will appear in the graphing window), Ymax (the largest y value in the graphing window), and Yscl (the distance between tick marks on the y axis). In addition to these six values, there are FORMAT options (see Figure A.2). The FORMAT options can be seen by pressing the right arrow key.

Figure A.2
The WINDOW and
FORMAT settings

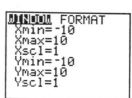 

Notice that the graphics screen (and the text screen, too) is wider than it is high. For that reason, when we begin to plot functions, we need to be aware that there are more plottable "points" (actually called *pixels*) horizontally than there are vertically. In fact, the ratio of width to height is about 3:2 or 1.5:1. For plotting situations where you want about the same distance between x axis hatch marks as between y axis hatch marks, the difference between Xmax and Xmin should be about 1.5 times the difference between Ymax and Ymin.

To graph $y = x^3 - 5x^2$ using the WINDOW settings, as in Figure A.3, first press the $\boxed{\text{Y} =}$ key to enter the function (we assume no other functions have been entered and that the following is entered on the Y1 = line):

$$\boxed{\text{Y} =}\ \boxed{\text{X, T, } \theta}\ \boxed{\wedge}\ 3\ \boxed{-}\ 5\ \boxed{\text{X, T, } \theta}\ \boxed{x^2}\ \boxed{\text{ENTER}}$$

Then, enter the WINDOW values.* To graph the function, simply press the
GRAPH key. Notice, in Figure A.3, that we do not get a complete picture of the
graph. In Figures A.4 and A.5, we present two other views of the same function.

Figure A.3

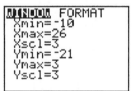

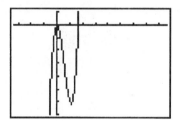

With the given WINDOW settings, the graph does
not appear "complete." Notice that the ratio of X
distance to Y distance is 1.5 to 1: $(6 - (-6)):(4 - (-4))$
$= 12:8 = 1.5:1$. Also notice that "GridOn" is selected.

Figure A.4

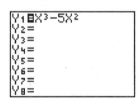

Each tick mark now represents 3 points
(Xscl = Yscl = 3) and the graph appears
slightly more complete. Also notice that
"GridOff" is now selected.

Figure A.5

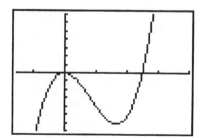

The x and y scales are now different; the
graph displays some distortion, but we
get to see a more "complete" graph.

---

*To change the WINDOW values, press the WINDOW key followed by ENTER. Then enter each
new value followed by the ENTER key. To obtain the values shown in Figure A.3, for example,
press the following keys: ENTER (−) 6 ENTER 6 ENTER 1 ENTER (−) 4 ENTER
4 ENTER 1 ENTER.

TABLES

Related to the graphing keys are the two keys that generate tables of values. To illustrate how to see a table of values displayed, we begin by entering a function for Y_1. We choose $y = 4 - x^2$:

```
Y₁〓4-X²
Y₂=
Y₃=
Y₄=
Y₅=
Y₆=
Y₇=
Y₈=
```

Now, press the $\boxed{2^{nd}}$ $\boxed{\text{TblSet}}$ key. There you will be able to choose the first x value in the table (called *TblMin*) and the increment of the x values (called ΔTbl). We have chosen -4 for the first x value and 0.5 for the increment. With the independent (x) and dependent (Y_1) variables in "Auto" mode, we get the table on the right below when we hit the $\boxed{2^{nd}}$ $\boxed{\text{TABLE}}$ key:

The table can be scrolled down to see other xy pairs:

If you would like the calculator to return a y value for a specified x value, set "Indpnt" to "Ask." We entered the value of 5, and the TI-82 returned the value of -21; it then allows the user to input additional x values, if desired:

THE TI-82 MENU SYSTEM

Probably more than anything else, what has contributed to the popularity of the TI-81/82 calculators is their "user-friendly" system of menus. A summary of the keys: MODE , MATH , Y = , WINDOW , ZOOM , 2nd DRAW , VARS , 2nd TEST , CALC , and PRGM follows.

When moving about from menu to menu, we can always leave and cancel the current menu by pressing CLEAR . To leave the menu screen, you can also type 2nd QUIT (which usually returns you to the last screen you were in) or you can enter another menu (such as Y = or GRAPH or MATRX , and so on).

THE MODE MENU

We have already seen this menu screen (Figure A.6). Recall that it is composed of seven types of settings, all of which are "remembered" when the calculator is turned off. The first two categories relate to the way numerical results are displayed; the third category controls the argument of trigonometric functions; categories four, five, and six address graphing features; and category seven determines whether the screen is split. When you turn the calculator on for the first time, the first entry in each line appears as the default setting.

Figure A.6
The MODE menu

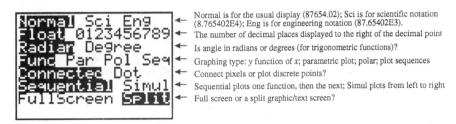

Normal is for the usual display (87654.02); Sci is for scientific notation (8.765402E4); Eng is for engineering notation (87.65402E3).
The number of decimal places displayed to the right of the decimal point
Is angle in radians or degrees (for trigonometric functions)?
Graphing type: *y* function of *x*; parametric plot; polar; plot sequences
Connect pixels or plot discrete points?
Sequential plots one function, then the next; Simul plots from left to right
Full screen or a split graphic/text screen?

THE MATH MENU OPTIONS

When the MATH key is pressed, four submenu options appear: MATH, NUM, HYP (hyperbolic functions), and PRB (probability functions). We will concentrate on the MATH (Figure A.7) and NUM (Figure A.8) options here.

Figure A.7 The MATH menu options under the MATH key

MATH NUM HYP PRB	MATH NUM HYP PRB
1:▶Frac	4↑³√
2:▶Dec	5:ˣ√
3:³	6:fMin(
4:³√	7:fMax(
5:ˣ√	8:nDeriv(
6:fMin(	9:fnInt(
7↓fMax(	0:solve(

1: ▷ Frac Converts display or calculation to fraction
2: ▷ Dec Converts fraction to decimal

3: x^3 Raises x to the third power

4: $\sqrt[3]{Z}$ Takes the cube root of Z

5: $n\sqrt[x]{R}$ Takes the nth root of R

6: fMin(Y1,X,a,b)   Approximates the minimum value of Y1 (which is a function of X) between a and b

7: fMax(Similar to fMin(

8: nDeriv(f,x,a,ϵ) Calculates the slope of the secant line going through the points $(a - \epsilon, f(a - \epsilon))$ and $(a + \epsilon, f(a + \epsilon))$ as an approximation to the slope of the tangent line at $(x, f(x))$; if ϵ is not given, it defaults to 0.001.

9: fnInt(f,x,a,b,tol)   Approximates $\int_b^a f(x)\,dx$ to within an accuracy of tol. If tol is not specified, it defaults to 0.00001.

0: solve(f,x,c)   Approximates a solution to $f(x) = 0$ using an initial guess of $x = c$

Figure A.8
The NUM, HYP, and
PRB menu options under
the MATH *key*

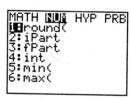
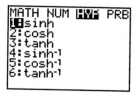
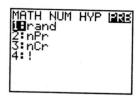

The NUM functions are as follows: The function round(A, n) returns the value of A rounded off n decimal places, where n is a positive integer less than or equal to 9. The function iPart(U) returns the integer part of U and fPart(U) returns the fractional part of U, as in the following example:

```
round(π,4)
            3.1416
iPart -58.12
               -58
fPart -58.12
              -.12
```

Examples of int, min, and max appear as follows:

```
int -58.12
               -59
min(π,4)
        3.141592654
max(π,4)
                 4
```

THE Y = MENU

Depending on whether "Func" or "Par" or "Pol" or "Seq" is chosen on the MODE menu, this screen will appear differently. Here we discuss only the "Function" option. Up to ten functions can be graphed on one set of axes on the graphics screen; they are Y1, Y2, Y3, . . . , Y0. Even if ten functions are listed in this menu, any of them can be turned on or turned off. When you turn a function on (default), the equals sign is highlighted:

```
Y1∎X³-5X²
Y2=5-6X
Y3∎abs X
Y4=
Y5=
Y6=
Y7=
Y8=
```

Three functions have been entered but only Y1 and Y3 will be graphed because Y2 was turned off. The on/off selection is toggled by placing the cursor over the = sign and hitting the ENTER key.

THE WINDOW MENU

Here, we choose the minimum and maximum x values, the minimum and maximum y values, the x scaling factor, and the y scaling factor. The default settings, called the *standard setting,* are shown in Figure A.9.

Figure A.9
The WINDOW *menu*

```
WINDOW FORMAT
Xmin=-10
Xmax=10
Xscl=1
Ymin=-10
Ymax=10
Yscl=1
```

If Xscl = 0, there will be no tick marks on the x axis; if Yscl = 0, there will be no tick marks on the y axis.

THE ZOOM MENU

There are nine options for zooming; seven are depicted in Figure A.10. Note that the eighth and ninth options, Integer and ZoomStat, appear on the menu screen only when the down arrow is pressed seven times. The first eight options are as follows.

Figure A.10
The ZOOM *menu*

```
ZOOM MEMORY
1∎ZBox
2:Zoom In
3:Zoom Out
4:ZDecimal
5:ZSquare
6:ZStandard
7↓ZTrig
```

1. ZBox   This option has the user locate two diagonal corners of a box and then redraws the function(s) using that box as a new viewing rectangle. After selecting ZBox from the menu, either by pressing 1

or using the arrow keys, move the cursor that appears on your graph screen to the desired location of the upper left-hand corner of the box and press ENTER . Then move the cursor to the lower right-hand corner location and press ENTER again. Be aware that the WINDOW values have automatically been adjusted to account for this new view. See Figures A.11 to A.14.

Figure A.11

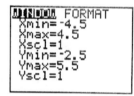

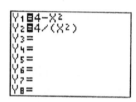

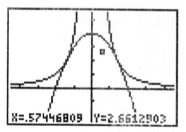

We begin by graphing two functions. We use a ZBox zoom to try to approximate the point of intersection in the first quadrant.

Figure A.12

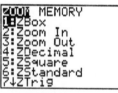

Choose ZBox, position the flashing pixel, and press ENTER. In this example, the cursor is positioned at (0.57446809, 2.6612903).

Figure A.13

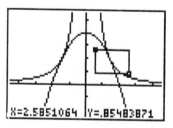

The flashing pixel is moved to the lower right diagonal corner. In this case, the coordinates are (2.5851064, 0.85483871).

Figure A.14 The graph and WINDOW values resulting from the Box Zoom

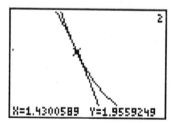

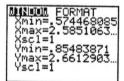

The result of the Box Zoom. Notice how the WINDOW settings have changed automatically.

2. Zoom In
3. Zoom Out

After you choose either Zoom In or Zoom Out, move the cursor to the location of a desired point. That point becomes the center of a redrawn graphics screen when you press ENTER . The Zoom In and Zoom Out features automatically rescale the viewing rectangle based on how the zoom factors are set. (Zoom factors are set under the Memory option of the ZOOM key.)

4. ZDecimal

The decimal option resets the decimal window so that x values displayed increment 0.1 as x increases.

5. ZSquare

The square option causes the TI-82 to draw circles in the correct perspective.

6. ZStandard

The standard option automatically changes the WINDOW settings to the default values mentioned earlier. Compare the graphs of $x^2 + y^2 = 100$ in standard and then square modes in Figures A.15 and A.16.*

Figure A.15

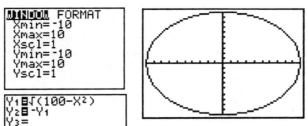

The circle is distorted when we use the default (Standard) settings.

Figure A.16

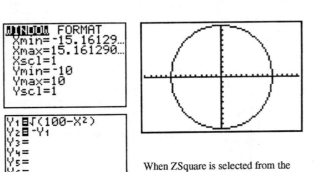

When ZSquare is selected from the ZOOM menu, the circle looks circular!

7. ZTrig

This is a collection of convenient WINDOW settings for graphing many trigonometric functions. The TI-82 assigns the following preset values: $\text{Xmin} = -(\frac{47}{24})\pi$, $\text{Xmax} = (\frac{47}{24})\pi$, $\text{Xscl} = \pi/2$, $\text{Ymin} = -3$, $\text{Ymax} = 3$, $\text{Yscl} = .25$.

*Actually, we graph the two *functions*, $Y1 = \sqrt{100 - X^2}$ and $Y2 = -\sqrt{100 - X^2}$.

8. ZInteger   When ZInteger is selected from the $\boxed{\text{ZOOM}}$ menu, the x (and y) values increment by an integer value as you TRACE to the right (or up).

THE $\boxed{2^{nd}}$ $\boxed{\text{DRAW}}$ MENU

The $\boxed{2^{nd}}$ $\boxed{\text{DRAW}}$ menu (Figure A.17) has eleven options. None of these options affects the WINDOW settings. Usually, before using these options, you turn off (or $\boxed{\text{CLEAR}}$) the functions in the $\boxed{Y =}$ menu.

Figure A.17 The $\boxed{2^{nd}}$ $\boxed{\text{DRAW}}$ *menu*

```
DRAW POINTS STO        DRAW POINTS STO
1:ClrDraw              5↑Tangent(
2:Line(               6:DrawF
3:Horizontal          7:Shade(
4:Vertical            8:DrawInv
5:Tangent(            9:Circle(
6:DrawF               0:Text(
7↓Shade(              A:Pen
```

1: ClrDraw   Clears previous drawings on the graphics screen
2: Line(X1,Y1,X2,Y2)   Draws a straight line segment connecting (x_1, y_1) to (x_2, y_2); can also be used from the graphics screen by anchoring segment's endpoints.
3: Horizontal(n)   Draws the horizontal line $y = n$
4: Vertical(m)   Draws the vertical line $x = m$
5: Tangent(Y1,n)   Draws the curve $y = $ Y1 and the line tangent to $y = $ Y1 at $x = n$.
6: DrawF(expression)   Draws $y = $ expression
7: Shade(Y1,Y2,s,a,b)   Shades the area below Y2 and above Y1 between a and b with shade factor s
8: DrawInv Y1   Draws the function $y = $ Y1 and its inverse
9: Circle(X,Y,r)   Draws a circle of radius r at center (X, Y)
0: Text(Will draw text directly on a graphics screen. This option is best if selected from the graphics screen. Then just position the cursor and type your text. From home screen or programming mode, Text(a,b,"text") will write the word "text" at pixel location (a, b); $0 \leq a \leq 57$ and $0 \leq b \leq 94$
A: Pen   Press $\boxed{\text{ENTER}}$ to turn on a specified point, then draw, then press $\boxed{\text{ENTER}}$ to turn off drawing pen.

The menu options are executed differently depending on which screen the option was selected from. For example, the "Line(" option has these two modes of selection:

1. *To draw a line directly from the graphics screen:* After selecting "Line(" from the $\boxed{\text{DRAW}}$ menu, move the cursor to the location where you want the line segment to start and press $\boxed{\text{ENTER}}$. Then move the cursor to the other end of your line segment and press $\boxed{\text{ENTER}}$ again.

2. *To draw a line from a program or from the home (text) screen:* Enter the four coordinates from the two points after the left parenthesis in "Line(". For

example, to draw a line connecting the point (3, 5) to the point (7, 8), enter the following keystrokes:

2nd   DRAW   2 3 , 5 , 7 , 8)   ENTER

Two additional menu options under the 2nd   DRAW   key are POINTS and STO (Figure A.18). For example, under the POINTS menu, the "PT-On(," "PT-Off(," and "PT-Chg(" options have similar syntax and modes to the "Line(" option just described.

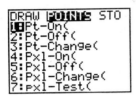

Figure A.18   POINTS and STO options under the 2nd   DRAW   menu

THE VARS AND Y-VARS MENUS

Choose the VARS option (Figure A.19) (usually in program edit mode) to access any of the TI-82's predefined variables (such as Xmin, Ymax, and so on). There are six VARS submenu options: Window, Zoom, GDB, Picture, Statistics, and Table.

Figure A.19 The VARS menu

The WINDOW variables (Xmin, Xmax, Xscl, Ymin, Ymax, Yscl, ΔX, ΔY, XFact, and YFact) are accessed here. The Y-VARS menu is shown in Figure A.20.

Figure A.20 The Y-VARS menu

THE 2nd TEST MENU

When we may want to compare two variables to see which is the larger of the two, often as part of a program, we use this menu (Figure A.21). The symbols are self-explanatory.

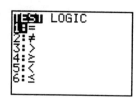

THE CALC MENU OPTIONS

The CALC menu is shown in Figure A.22. Here, we demonstrate some of the CALC options—namely, option 5 (to find the intersection of two curves), option 6 (to calculate a first derivative of a function at a point), and option 7 (to approximate a definite integral).

Figure A.22
The CALC menu

CALCULATE
1:value
2:root
3:minimum
4:maximum
5:intersect
6:dy/dx
7:∫f(x)dx

To find the intersection of $y = 4 - x^2$ and $y = 3x - 1$, we enter these two functions using the Y= key:

Y₁◼4-X²
Y₂◼3X-1
Y₃=
Y₄=
Y₅=
Y₆=
Y₇=
Y₈=

Next, we graph these functions and choose the 2nd CALC key's fifth option (intersect). There will be three prompts: one for "First curve?", one for "Second curve?", and one for "Guess?". For this example, simply press the ENTER key after each prompt. You should see these three screens:

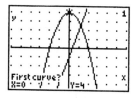 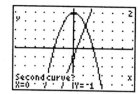 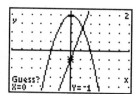

The approximation of the intersection is then displayed:

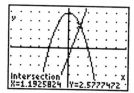

The TI-82 calculates the first derivative at a point. To calculate $f'(2)$ if $f(x) = 4 - x^2$, we enter the function and a suitable collection of window values:

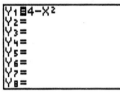 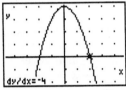

Now, choose the sixth option (dy/dx) under the $\boxed{2^{nd}}$ $\boxed{\text{CALC}}$ menu, move the cursor to $x = 2$, and press $\boxed{\text{ENTER}}$ to see $f'(2) = -4$:

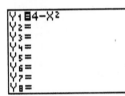

Finally, to evaluate $\int_{-1}^{2} (4 - x^2)dx$. Follow these steps:

Step 1. Begin by graphing $y = 4 - x^2$:

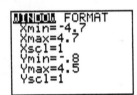 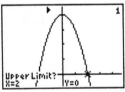

Step 2. After choosing "7: $\int$ f(x)dx" from the CALC menu, move the cursor to the left endpoint $(x = -1)$ and press $\boxed{\text{ENTER}}$. Then move to $x = 2$ and press $\boxed{\text{ENTER}}$ again:

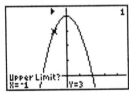 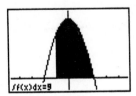

Step 3. The final result, $\int_{-1}^{2} (4 - x^2)dx = 9$, is displayed and shaded:

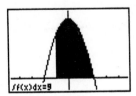

THE $\boxed{\text{PRGM}}$ KEY: PROGRAMMING

The $\boxed{\text{PRGM}}$ key, when selected from nonprogramming mode, offers three options: EXECute a program, EDIT a program, or create a NEW program. Program names are labeled with user-given names and coded with up to 8 characters made up of the 10 digits (1 through 9 and 0), the 26 letters of the alphabet (A through Z), and θ. The menu looks like this when you press the PRGM key from nonprogramming mode:

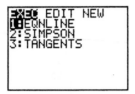

(This example was from a calculator that had three programs in it and their names were EQNLINE, SIMPSON, and TANGENTS.)

The PRGM key produces totally different options when pressed while EDIT-ing a program. In this case, the user has the option of choosing 16 different control (CTL) commands (Figure A.23), 12 different input/output (I/O) commands (Figure A.24), or the ability to run a program from within a program (as a subroutine, EXEC).

Figure A.23   The 16 control commands

Figure A.24   The 12 I/O commands

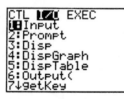 

APPENDIX B
TABLES

TABLE OF INTEGRALS

1. $\displaystyle\int \frac{x}{a+bx}\,dx = \frac{x}{b} - \frac{a}{b^2}\ln|a+bx| + C$

2. $\displaystyle\int \frac{x}{(a+bx)^2}\,dx = \frac{1}{b^2}\left[\frac{a}{a+bx} + \ln|a+bx|\right] + C$

3. $\displaystyle\int \frac{x}{(a+bx)^n}\,dx = \frac{1}{b^2}\left[\frac{-1}{(n-2)(a+bx)^{n-2}} + \frac{a}{(n-1)(a+bx)^{n-1}}\right] + C, \quad n \neq 1, 2$

4. $\displaystyle\int \frac{1}{x(a+bx)}\,dx = \frac{-1}{a}\ln\left|\frac{a+bx}{x}\right| + C \quad \text{or} \quad \frac{1}{a}\ln\left|\frac{x}{a+bx}\right| + C$

5. $\displaystyle\int \frac{1}{x(a+bx)^2}\,dx = \frac{1}{a(a+bx)} - \frac{1}{a^2}\ln\left|\frac{a+bx}{x}\right| + C$

6. $\displaystyle\int \frac{1}{x^2(a+bx)}\,dx = \frac{-1}{ax} + \frac{b}{a^2}\ln\left|\frac{a+bx}{x}\right| + C$

7. $\displaystyle\int \frac{1}{x^2-a^2}\,dx = \frac{1}{2a}\ln\left|\frac{x-a}{x+a}\right| + C$

8. $\displaystyle\int \frac{1}{(x^2-a^2)^2}\,dx = \frac{-x}{2a^2(x^2-a^2)} + \frac{1}{4a^3}\ln\left|\frac{a+x}{a-x}\right| + C$

9. $\displaystyle\int \frac{x}{a+bx^2}\,dx = \frac{1}{2b}\ln|a+bx^2| + C$

10. $\displaystyle\int \frac{1}{x(a+bx^2)}\,dx = \frac{1}{2a}\ln\left|\frac{x^2}{a+bx^2}\right| + C$

11. $\displaystyle\int \frac{x^2}{a+bx^2}\,dx = \frac{x}{b} - \frac{a}{b}\int \frac{1}{a+bx^2}\,dx$

12. $\displaystyle\int \frac{1}{a^2-b^2x^2}\,dx = \frac{1}{2ab}\ln\left|\frac{a+bx}{a-bx}\right| + C$

13. $\displaystyle\int \frac{1}{x\sqrt{a+bx}}\,dx = \frac{1}{\sqrt{a}}\ln\left|\frac{\sqrt{a+bx}-\sqrt{a}}{\sqrt{a+bx}+\sqrt{a}}\right| + C$

14. $\displaystyle\int \frac{\sqrt{a+bx}}{x}\,dx = 2\sqrt{a+bx} + a\int \frac{1}{x\sqrt{a+bx}}\,dx$

15. $\displaystyle\int \frac{\sqrt{a+bx}}{x^2}\,dx = \frac{-\sqrt{a+bx}}{x} + \frac{b}{2}\int \frac{1}{x\sqrt{a+bx}}\,dx$

TABLE OF INTEGRALS (continued)

16. $\int \sqrt{x^2 \pm a^2}\, dx = \frac{1}{2}[x\sqrt{x^2 \pm a^2} \pm a^2 \ln |x + \sqrt{x^2 \pm a^2}|] + C$

17. $\int \dfrac{1}{\sqrt{x^2 \pm a^2}}\, dx = \ln |x + \sqrt{x^2 \pm a^2}| + C$

18. $\int \dfrac{1}{x\sqrt{x^2 + a^2}}\, dx = \dfrac{-1}{a} \ln \left| \dfrac{a + \sqrt{x^2 + a^2}}{x} \right| + C$

19. $\int \dfrac{\sqrt{x^2 + a^2}}{x}\, dx = \sqrt{x^2 + a^2} - a \ln \left| \dfrac{a + \sqrt{x^2 + a^2}}{x} \right| + C$

20. $\int \dfrac{1}{x\sqrt{a^2 - x^2}}\, dx = \dfrac{-1}{a} \ln \left| \dfrac{a + \sqrt{a^2 - x^2}}{x} \right| + C$

21. $\int \dfrac{\sqrt{a^2 - x^2}}{x}\, dx = \sqrt{a^2 - x^2} - a \ln \left| \dfrac{a + \sqrt{a^2 - x^2}}{x} \right| + C$

22. $\int \ln x\, dx = x \ln x - x + C$

23. $\int (\ln x)^n\, dx = x(\ln x)^n - n \int (\ln x)^{n-1}\, dx, n \neq -1$

24. $\int x^n \ln x\, dx = x^{n+1}\left[\dfrac{\ln x}{n+1} - \dfrac{1}{(n+1)^2} \right] + C, n \neq -1$

25. $\int x^n(\ln x)^m\, dx = \dfrac{x^{n+1}(\ln x)^m}{n+1} - \dfrac{m}{n+1}\int x^n(\ln x)^{m-1}\, dx, n \neq -1$

26. $\int x^n e^{ax}\, dx = \dfrac{x^n e^{ax}}{a} - \dfrac{n}{a}\int x^{n-1} e^{ax}\, dx$

27. $\int \dfrac{1}{1 + e^x}\, dx = x - \ln (1 + e^x) + C$

28. $\int \dfrac{1}{a + be^x}\, dx = \dfrac{x}{a} - \dfrac{1}{a} \ln |a + be^x| + C$

29. $\int \dfrac{1}{1 + e^{nx}}\, dx = x - \dfrac{1}{n} \ln |1 + e^{nx}| + C$

30. $\int \dfrac{1}{a + be^{nx}}\, dx = \dfrac{x}{a} - \dfrac{1}{an} \ln |a + be^{nx}| + C$

POWERS OF *e*

x	e^x	e^{-x}	x	e^x	e^{-x}
0.00	1.0000	1.0000	1.5	4.4817	0.2231
0.01	1.0101	0.9901	1.6	4.9530	0.2019
0.02	1.0202	0.9802	1.7	5.4739	0.1827
0.03	1.0305	0.9705	1.8	6.0496	0.1653
0.04	1.0408	0.9608	1.9	6.6859	0.1496
0.05	1.0513	0.9512	2.0	7.3891	0.1353
0.06	1.0618	0.9418	2.1	8.1662	0.1225
0.07	1.0725	0.9324	2.2	9.0250	0.1108
0.08	1.0833	0.9231	2.3	9.9742	0.1003
0.09	1.0942	0.9139	2.4	11.023	0.0907
0.10	1.1052	0.9048	2.5	12.182	0.0821
0.11	1.1163	0.8958	2.6	13.464	0.0743
0.12	1.1275	0.8869	2.7	14.880	0.0672
0.13	1.1388	0.8781	2.8	16.445	0.0608
0.14	1.1503	0.8694	2.9	18.174	0.0550
0.15	1.1618	0.8607	3.0	20.086	0.0498
0.16	1.1735	0.8521	3.1	22.198	0.0450
0.17	1.1853	0.8437	3.2	24.533	0.0408
0.18	1.1972	0.8353	3.3	27.113	0.0369
0.19	1.2092	0.8270	3.4	29.964	0.0334
0.20	1.2214	0.8187	3.5	33.115	0.0302
0.21	1.2337	0.8106	3.6	36.598	0.0273
0.22	1.2461	0.8025	3.7	40.447	0.0247
0.23	1.2586	0.7945	3.8	44.701	0.0224
0.24	1.2712	0.7866	3.9	49.402	0.0202
0.25	1.2840	0.7788	4.0	54.598	0.0183
0.30	1.3499	0.7408	4.1	60.340	0.0166
0.35	1.4191	0.7047	4.2	66.686	0.0150
0.40	1.4918	0.6703	4.3	73.700	0.0136
0.45	1.5683	0.6376	4.4	81.451	0.0123
0.50	1.6487	0.6065	4.5	90.017	0.0111
0.55	1.7333	0.5769	4.6	99.484	0.0101
0.60	1.8221	0.5488	4.7	109.95	0.0091
0.65	1.9155	0.5220	4.8	121.51	0.0082
0.70	2.0138	0.4966	4.9	134.29	0.0074
0.75	2.1170	0.4724	5.0	148.41	0.0067
0.80	2.2255	0.4493	5.5	244.69	0.0041
0.85	2.3396	0.4274	6.0	403.43	0.0025
0.90	2.4596	0.4066	6.5	665.14	0.0015
0.95	2.5857	0.3867	7.0	1096.6	0.0009
1.0	2.7183	0.3679	7.5	1808.0	0.0006
1.1	3.0042	0.3329	8.0	2981.0	0.0003
1.2	3.3201	0.3012	8.5	4914.8	0.0002
1.3	3.6693	0.2725	9.0	8103.1	0.0001
1.4	4.0552	0.2466	10.0	22,026	0.00005

COMMON LOGARITHMS

N	0	1	2	3	4	5	6	7	8	9
10	0000	0043	0086	0128	0170	0212	0253	0294	0334	0374
11	0414	0453	0492	0531	0569	0607	0645	0682	0719	0755
12	0792	0828	0864	0899	0934	0969	1004	1038	1072	1106
13	1139	1173	1206	1239	1271	1303	1335	1367	1399	1430
14	1461	1492	1523	1553	1584	1614	1644	1673	1703	1732
15	1761	1790	1818	1847	1875	1903	1931	1959	1987	2014
16	2041	2068	2095	2122	2148	2175	2201	2227	2253	2279
17	2304	2330	2355	2380	2405	2430	2455	2480	2504	2529
18	2553	2577	2601	2625	2648	2672	2695	2718	2742	2765
19	2788	2810	2833	2856	2878	2900	2923	2945	2967	2989
20	3010	3032	3054	3075	3096	3118	3139	3160	3181	3201
21	3222	3243	3263	3284	3304	3324	3345	3365	3385	3404
22	3424	3444	3464	3483	3502	3522	3541	3560	3579	3598
23	3617	3636	3655	3674	3692	3711	3729	3747	3766	3784
24	3802	3820	3838	3856	3874	3892	3909	3927	3945	3962
25	3979	3997	4014	4031	4048	4065	4082	4099	4116	4133
26	4150	4166	4183	4200	4216	4232	4249	4265	4281	4298
27	4314	4330	4346	4362	4378	4393	4409	4425	4440	4456
28	4472	4487	4502	4518	4533	4548	4564	4579	4594	4609
29	4624	4639	4654	4669	4683	4698	4713	4728	4742	4757
30	4771	4786	4800	4814	4829	4843	4857	4871	4886	4900
31	4914	4928	4942	4955	4969	4983	4997	5011	5024	5038
32	5051	5065	5079	5092	5105	5119	5132	5145	5159	5172
33	5185	5198	5211	5224	5237	5250	5263	5276	5289	5302
34	5315	5328	5340	5353	5366	5378	5391	5403	5416	5428
35	5441	5453	5465	5478	5490	5502	5514	5527	5539	5551
36	5563	5575	5587	5599	5611	5623	5635	5647	5658	5670
37	5682	5694	5705	5717	5729	5740	5752	5763	5775	5786
38	5798	5809	5821	5832	5843	5855	5866	5877	5888	5899
39	5911	5922	5933	5944	5955	5966	5977	5988	5999	6010
40	6021	6031	6042	6053	6064	6075	6085	6096	6107	6117
41	6128	6138	6149	6160	6170	6180	6191	6201	6212	6222
42	6232	6243	6253	6263	6274	6284	6294	6304	6314	6325
43	6335	6345	6355	6365	6375	6385	6395	6405	6415	6425
44	6435	6444	6454	6464	6474	6484	6493	6503	6513	6522
45	6532	6542	6551	6561	6571	6580	6590	6599	6609	6618
46	6628	6637	6646	6656	6665	6675	6684	6693	6702	6712
47	6721	6730	6739	6749	6758	6767	6776	6785	6794	6803
48	6812	6821	6830	6839	6848	6857	6866	6875	6884	6893
49	6902	6911	6920	6928	6937	6946	6955	6964	6972	6981
50	6990	6998	7007	7016	7024	7033	7042	7050	7059	7067
51	7076	7084	7093	7101	7110	7118	7126	7135	7143	7152
52	7160	7168	7177	7185	7193	7202	7210	7218	7226	7235
53	7243	7251	7259	7267	7275	7284	7292	7300	7308	7316
54	7324	7332	7340	7348	7356	7364	7372	7380	7388	7396

COMMON LOGARITHMS (continued)

N	0	1	2	3	4	5	6	7	8	9
55	7404	7412	7419	7427	7435	7443	7451	7459	7466	7474
56	7482	7490	7497	7505	7513	7520	7528	7536	7543	7551
57	7559	7566	7574	7582	7589	7597	7604	7612	7619	7627
58	7634	7642	7649	7657	7664	7672	7679	7686	7694	7701
59	7709	7716	7723	7731	7738	7745	7752	7760	7767	7774
60	7782	7789	7796	7803	7810	7818	7825	7832	7839	7846
61	7853	7860	7868	7875	7882	7889	7896	7903	7910	7917
62	7924	7931	7938	7945	7952	7959	7966	7973	7980	7987
63	7993	8000	8007	8014	8021	8028	8035	8041	8048	8055
64	8062	8069	8075	8082	8089	8096	8102	8109	8116	8122
65	8129	8136	8142	8149	8156	8162	8169	8176	8182	8189
66	8195	8202	8209	8215	8222	8228	8235	8241	8248	8254
67	8261	8267	8274	8280	8287	8293	8299	8306	8312	8319
68	8325	8331	8338	8344	8351	8357	8363	8370	8376	8382
69	8388	8395	8401	8407	8414	8420	8426	8432	8439	8445
70	8451	8457	8463	8470	8476	8482	8488	8494	8500	8506
71	8513	8519	8525	8531	8537	8543	8549	8555	8561	8567
72	8573	8579	8585	8591	8597	8603	8609	8615	8621	8627
73	8633	8639	8645	8651	8657	8663	8669	8675	8681	8686
74	8692	8698	8704	8710	8716	8722	8727	8733	8739	8745
75	8751	8756	8762	8768	8774	8779	8785	8791	8797	8802
76	8808	8814	8820	8825	8831	8837	8842	8848	8854	8859
77	8865	8871	8876	8882	8887	8893	8899	8904	8910	8915
78	8921	8927	8932	8938	8943	8949	8954	8960	8965	8971
79	8976	8982	8987	8993	8998	9004	9009	9015	9020	9025
80	9031	9036	9042	9047	9053	9058	9063	9069	9074	9079
81	9085	9090	9096	9101	9106	9112	9117	9122	9128	9133
82	9138	9143	9149	9154	9159	9165	9170	9175	9180	9186
83	9191	9196	9201	9206	9212	9217	9222	9227	9232	9238
84	9243	9248	9253	9258	9263	9269	9274	9279	9284	9289
85	9294	9299	9304	9309	9315	9320	9325	9330	9335	9340
86	9345	9350	9355	9360	9365	9370	9375	9380	9385	9390
87	9395	9400	9405	9410	9415	9420	9425	9430	9435	9440
88	9445	9450	9455	9460	9465	9469	9474	9479	9484	9489
89	9494	9499	9504	9509	9513	9518	9523	9528	9533	9538
90	9542	9547	9552	9557	9562	9566	9571	9576	9581	9586
91	9590	9595	9600	9605	9609	9614	9619	9624	9628	9633
92	9638	9643	9647	9652	9657	9661	9666	9671	9675	9680
93	9685	9689	9694	9699	9703	9708	9713	9717	9722	9727
94	9731	9736	9741	9745	9750	9754	9759	9763	9768	9773
95	9777	9782	9786	9791	9795	9800	9805	9809	9814	9818
96	9823	9827	9832	9836	9841	9845	9850	9845	9859	9863
97	9868	9872	9877	9881	9886	9890	9894	9899	9903	9908
98	9912	9917	9921	9926	9930	9934	9939	9943	9948	9952
99	9956	9961	9965	9969	9974	9978	9983	9987	9991	9996

NATURAL LOGARITHMS

x	$\ln x$		x	$\ln x$		x	$\ln x$
			4.5	1.5041		9.0	2.1972
0.1	−2.3026		4.6	1.5261		9.1	2.2083
0.2	−1.6094		4.7	1.5476		9.2	2.2192
0.3	−1.2040		4.8	1.5686		9.3	2.2300
0.4	−0.9163		4.9	1.5892		9.4	2.2407
0.5	−0.6931		5.0	1.6094		9.5	2.2513
0.6	−0.5108		5.1	1.6292		9.6	2.2618
0.7	−0.3567		5.2	1.6487		9.7	2.2721
0.8	−0.2231		5.3	1.6677		9.8	2.2824
0.9	−0.1054		5.4	1.6864		9.9	2.2925
1.0	0.0000		5.5	1.7047		10	2.3026
1.1	0.0953		5.6	1.7228		11	2.3979
1.2	0.1823		5.7	1.7405		12	2.4849
1.3	0.2624		5.8	1.7579		13	2.5649
1.4	0.3365		5.9	1.7750		14	2.6391
1.5	0.4055		6.0	1.7918		15	2.7081
1.6	0.4700		6.1	1.8083		16	2.7726
1.7	0.5306		6.2	1.8245		17	2.8332
1.8	0.5878		6.3	1.8405		18	2.8904
1.9	0.6419		6.4	1.8563		19	2.9444
2.0	0.6931		6.5	1.8718		20	2.9957
2.1	0.7419		6.6	1.8871		25	3.2189
2.2	0.7885		6.7	1.9021		30	3.4012
2.3	0.8329		6.8	1.9169		35	3.5553
2.4	0.8755		6.9	1.9315		40	3.6889
2.5	0.9163		7.0	1.9459		45	3.8067
2.6	0.9555		7.1	1.9601		50	3.9120
2.7	0.9933		7.2	1.9741		55	4.0073
2.8	1.0296		7.3	1.9879		60	4.0943
2.9	1.0647		7.4	2.0015		65	4.1744
3.0	1.0986		7.5	2.0149		70	4.2485
3.1	1.1314		7.6	2.0281		75	4.3175
3.2	1.1632		7.7	2.0412		80	4.3820
3.3	1.1939		7.8	2.0541		85	4.4427
3.4	1.2238		7.9	2.0669		90	4.4998
3.5	1.2528		8.0	2.0794		100	4.6052
3.6	1.2809		8.1	2.0919		110	4.7005
3.7	1.3083		8.2	2.1041		120	4.7875
3.8	1.3350		8.3	2.1163		130	4.8676
3.9	1.3610		8.4	2.1282		140	4.9416
4.0	1.3863		8.5	2.1401		150	5.0106
4.1	1.4110		8.6	2.1518		160	5.0752
4.2	1.4351		8.7	2.1633		170	5.1358
4.3	1.4586		8.8	2.1748		180	5.1930
4.4	1.4816		8.9	2.1861		190	5.2470

ANSWERS TO
SELECTED EXERCISES

1.1 EXERCISES (page 7)

1. $3x^2 - 7x + 3$ **3.** $-4x^2 + 2x - 5$ **5.** $-4x - 14$ **7.** $6x - 22$ **9.** $2x^4 - x^3 + 4x^2$ **11.** $2x^2 - 11x - 21$
13. $2x - x^2$ **15.** -5 **17.** 56 **19.** 114 **21.** 13 **23.** $\sqrt[4]{x^3}$ **25.** $4\sqrt{x}$ **27.** $\dfrac{3}{\sqrt{x+2}}$
29. $x^{1/3}$ **31.** $-2x^{5/4}$ **33.** $2x^{2/3}$ **35.** 4 **37.** -3 **39.** 16 **41.** 40 **43.** $\dfrac{-1}{2}$ **45.** $\dfrac{5\sqrt{6}}{6}$
47. $\dfrac{\sqrt{x}-x}{1-x}$ **49.** $\dfrac{6\sqrt{x}+24}{x-16}$ **51.** $\dfrac{1}{\sqrt{x}}$ **53.** $\dfrac{x-1}{2\sqrt{x}-2}$ **55.** $\dfrac{-1}{2\sqrt{x}-2\sqrt{x+1}}$ **57.** $x^{1/3}+2$
59. $x^{3/2}+5x^{1/2}$ **61.** $16x - 18$ **63.** $2x^3 + 5x^2 + x - 3$ **65.** $2x(x^2+1)^{-2/3}$ **67.** 1 **69.** $-\sqrt{3}-\sqrt{5}$
71. $x^2 + 11x + 18$

	x	9
x	x^2	$9x$
2	$2x$	18

1.2 EXERCISES (page 14)

1. $2x^3(3x+1)$ **3.** $(x+3)(x+4)$ **5.** $(3x-2)(x+2)$ **7.** $(4-x)(2+x)$ **9.** $x^2(x+3)(x-3)$
11. $(x-3)(x^2+3x+9)$ **13.** $3x(2x-1)(x+1)$ **15.** $(x-3)(5x+2)$ **17.** $\dfrac{2x^2(2x+3)}{(2x+1)^3}$
19. $\dfrac{(x+1)(x-14)}{(x-4)^3}$ **21.** $\dfrac{x^2(5x+3)}{(2x+1)^{2/3}}$ or $\dfrac{x^2(5x+3)}{\sqrt[3]{(2x+1)^2}}$ **23.** $\dfrac{(1-x)(1-5x)}{x^{1/2}}$ or $\dfrac{(x-1)(5x-1)}{\sqrt{x}}$ **25.** $x+3$
27. $-x-3$ **29.** $\dfrac{x^2(x+2)(x-2)}{x^2+1}$ **31.** $(x+1)(x-3)^{1/2}$ or $(x+1)\sqrt{x-3}$ **33.** $x^{1/2}(x+1)$ or $\sqrt{x}(x+1)$
35. $\dfrac{2x+1}{x^2}$ **37.** $\dfrac{-x^2-5x}{(x+1)(x-3)}$ **39.** $\dfrac{5y^3-1}{y^2}$ **41.** $\dfrac{9x^2-34x+29}{(x-1)(x-2)(x-3)}$ **43.** $\dfrac{6x^3(7x-2)}{(3x-1)^3}$
45. $\dfrac{x^2(8-5x)}{(2-x)^{2/3}}$ or $\dfrac{x^2(8-5x)}{\sqrt[3]{(2-x)^2}}$ **47.** $2(x+3)(x+2)^2(x+4)^4$ **49.** $\dfrac{0.02(x^5-200)}{x^3}$

1.3 EXERCISES (page 22)

1. $x=-3$ **3.** $x=1$ **5.** $t=9$ **7.** $y=\frac{5}{3}$ **9.** $x=0, x=6$ **11.** $S=-3, S=4$
13. $x=\dfrac{-2\pm\sqrt{2}}{2}$ **15.** $x=\pm\sqrt{5}$ **17.** $x=\pm2$ **19.** $x=-5\pm\sqrt{7}$ **21.** $t=-1\pm\sqrt{3}$
23. No real solution **25.** $x=\pm1, x=\pm2$ **27.** $x=64$ **29.** $x=1$ **31.** $x=4$ **33.** $x=\frac{4}{9}$
35. $x=7$ **37.** $x=\pm2$ **39.** $x=0, x=1$ **41.** $x=\pm\sqrt{p}$ **43.** $x=1$ **45.** $x=26$ cm
47. $t=5, t=14$

1.4 EXERCISES (page 32)

1. $x\le-2; (-\infty,-2]$ **3.** $x>\frac{1}{2}; (\frac{1}{2},\infty)$ **5.** $x>-2; (-2,\infty)$ **7.** $x\ge-4; [-4,\infty)$ **9.** $x<\frac{-1}{2}; (-\infty,\frac{-1}{2})$
11. $x>1; (1,\infty)$ **13.** $x\ge-6; [-6,\infty)$ **15.** $-3<x<3; (-3,3)$ **17.** $x\le-4$ or $x\ge-3; (-\infty,-4]\cup[-3,\infty)$
19. $-4<x<2; (-4,2)$ **21.** $\frac{-2}{3}\le x\le4; [\frac{-2}{3},4]$ **23.** $x\ne\frac{-1}{2}; (-\infty,\frac{-1}{2})\cup(\frac{-1}{2},\infty)$
25. $x\le-\sqrt{7}$ or $x\ge\sqrt{7}; (-\infty,-\sqrt{7}]\cup[\sqrt{7},\infty)$ **27.** $\varnothing$ **29.** $\varnothing$ **31.** $x>0; (0,\infty)$ **33.** $0\le x<1; [0,1)$
35. $x\le-4$ or $x>-2; (-\infty,-4]\cup(-2,\infty)$ **37.** $\frac{1}{2}<x<3; (\frac{1}{2},3)$ **39.** $x<-1$ or $0<x<1; (-\infty,-1)\cup(0,1)$
41. $x<3$ or $x>6; (-\infty,3)\cup(6,\infty)$ **43.** $x>0$ **45.** $x=2$ **47.** $x<-4$ **49.** $x>\frac{1}{2}$ **51.** $x>0$

53. $x \neq 1$ **55.** $0 < x \leq 5$ **57.** $\emptyset$ **59.** $x > 0$ **61.** $-3 < x < -1$ or $x > 2$ **63.** $x \leq 0$ or $x = 5$
65. $-2 \leq x \leq 0$ or $x \geq 3$ **67. a.** 100 boxes ($x = 1$)   **b.** More than 100 boxes, but less than 1000 boxes ($1 < x < 10$)

1.5   EXERCISES   (page 39)

1. 4   **3.** 7   **5.** 5   **7.**

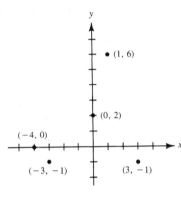

9.

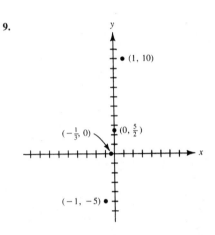

11. 5   **13.** $3\sqrt{2}$ **15.** $\sqrt{5}$ **17.** 4.4   **19.** Quadrants II and III
21. a. Quadrants I and III   **b.** Quadrants II and IV   **c.** Quadrants I and III
23. $(\sqrt{10})^2 + (\sqrt{10})^2 = (\sqrt{20})^2$ **25.** $(-1, -2)$ and $(7, -2)$

1.6   EXERCISES   (page 50)

1. -3 **3.** -1 **5.** $\frac{3}{2}$ **7.** $\frac{-1}{4}$ **9.** 0   **11.** Undefined   **13.** $y = 6x - 14$ **15.** $y = -3x + 1$
17. $y = \frac{1}{2}x + \frac{3}{2}$ **19.** $y = \frac{-2}{3}x + \frac{19}{12}$ **21.** $y = 5$ **23.** $x = -2$ **25.** $x + y - 4 = 0$ **27.** $y - 5 = 0$
29. $5x + 2y + 4 = 0$ **31.** $m = -4, b = 6$ **33.** $m = \frac{1}{2}, b = 3$ **35.** $m = \frac{-5}{4}, b = 0$
37. $m = 0, b = 6$ **39.**

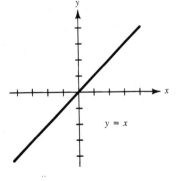

41.

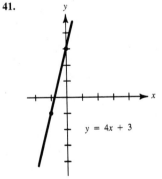

43.

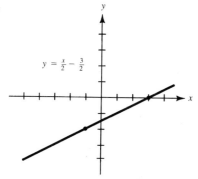

$y = \frac{x}{2} - \frac{3}{2}$

45.

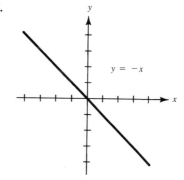

$y = -x$

47.

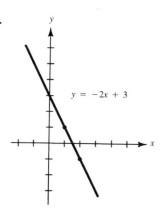

$y = -2x + 3$

49.

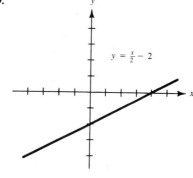

$y = \frac{x}{2} - 2$

51.

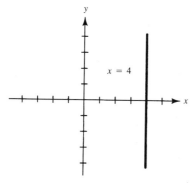

$x = 4$

53. $4x - y + 5 = 0$ **55.** $x + 2y - 10 = 0$

57. $x - 5 = 0$ **59.** $x + 5y - 6 = 0$ **61.** $3x - y - 1 = 0$ **63.** $(5, 9)$ **65.** $(-2, -9)$
67. $(-1, 3)$ **69.** 990 **71.** 60 **73.** 75 **75.** 50 **77.** $x = 20, p = 44$
79. $x = 40, p = 360$ **81.** $p = -0.005x + 9.5$

12. $\dfrac{-1}{4}$ **13.** $x \le 0$ or $x \ge 4$ **14.** $\dfrac{2x(3x-1)}{(x-1)^3}$ **15.** $x + 2y + 6 = 0$ **16.** $x = 0, x = 2$ **17.** 900

18. Loses money: $0 < x < 50$; makes money: $x > 50$

PROBLEM-SOLVING EXERCISES (page 74)

1. 196 cm^2 − 49π cm^2 ≈ 42.1 cm^2 **3.** $A = 2r^2$ **5.** 8 units **7.** $A = 4r^2 + \frac{1}{2}\pi r^2$

9. $A = x^2 + \dfrac{64}{x}$

11. $C = 8x^2 + \dfrac{288}{x}$ where cost, C, is measured in cents.

13. $P = 9750 + 25x - 10x^2$ where profit, P, is measured in dollars.

2.1 EXERCISES (page 85)

1. Yes **3.** Yes **5.** Yes **7.** Yes **9.** Yes **11.** Domain = $\{-3, -2, -1\}$; range = $\{1, 2, 3\}$

13. Domain: all real numbers; range: all real numbers **15.** Domain: $x \ge 0$; range: $y \ge 0$

17. Domain: $x \ge 0$; range: $y \ge 0$ **19. a.** 5 **b.** 5 **21. a.** 23 **b.** 3 **c.** 7 **d.** $7 - 4h$

23. a. -2 **b.** -4 **c.** $3x^2 - x - 4$ **d.** $3x^2 + 6xh + 3h^2 + x + h - 4$

25. a. $4x + 4h + 6$ **b.** $4h$ **c.** 4

27. a. $x^2 + 2xh + h^2 + x + h$ **b.** $2xh + h^2 + h$ **c.** $2x + 1 + h$

29. a. $\dfrac{1}{x + h}$ **b.** $\dfrac{-h}{x(x + h)}$ **c.** $\dfrac{-1}{x(x + h)}$

31.

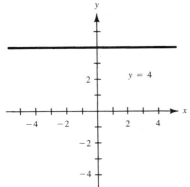

33.

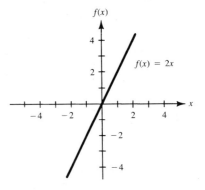

35.

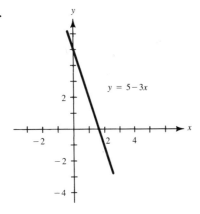

37.

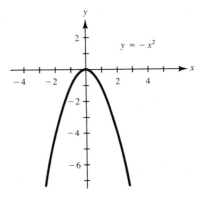

39.

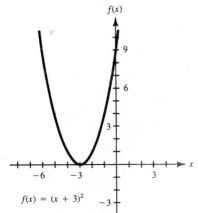

$f(x) = (x + 3)^2$

41.

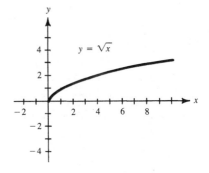

$y = \sqrt{x}$

43.

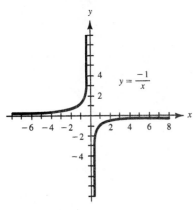

$y = \dfrac{-1}{x}$

45.

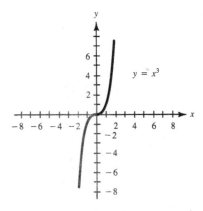

$y = x^3$

47. Because $f(x)$ is equal to zero for those values of x **49.** $-2 \le x \le 2$ **51.** No   **53.** Yes   **55.** $1800
57. $801,000   **59.** $2000   **61.** $100,000   **63.** 200

2.2   **EXERCISES**   (page 98)

1. $2x^2 + 2x + 4$ **3.** $6x^3 + 5x^2 - 4x$ **5.** $\frac{1}{4}$ **7.** 10   **9.** $f[g(x)] = 2 - 3x,\ g[f(x)] = 6 - 3x$
11. $f[g(x)] = x^2 + 3x + 2,\ g[f(x)] = x^2 + x + 1$
13.

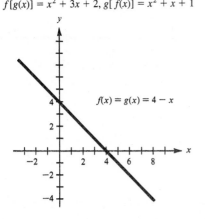

$f(x) = g(x) = 4 - x$

15.

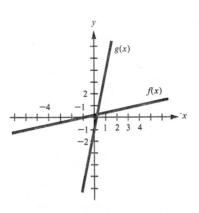

$g(x)$

$f(x)$

17.

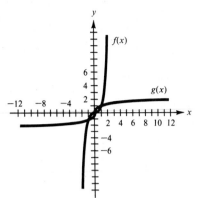

19.

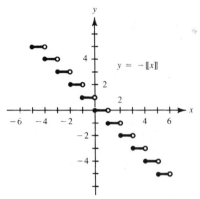

$y = -[\![x]\!]$

21.

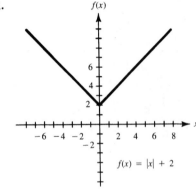

$f(x) = |x| + 2$

23.

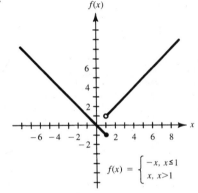

$f(x) = \begin{cases} -x, & x \le 1 \\ x, & x > 1 \end{cases}$

25.

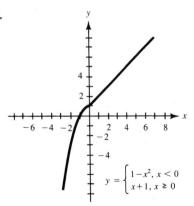

$y = \begin{cases} 1 - x^2, & x < 0 \\ x + 1, & x \ge 0 \end{cases}$

27.

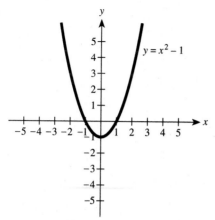

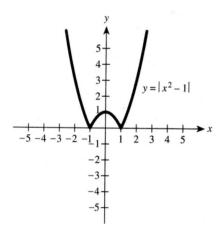

29. Because x^2 is already nonnegative, the absolute value symbols will not change it.

31. $P(w) = 2w + \dfrac{216}{w}$

33.

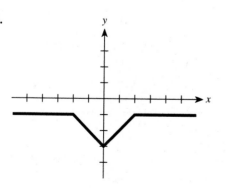

35.

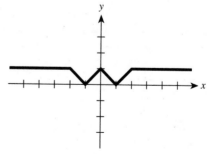

37.

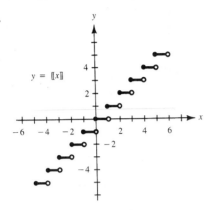

39. $P(x) = 28x - x^2 - 100$ **41.** $P(x) = 1.5x^2 - 0.5x^3 - 100$ **43.** \$4000   **45.** $S(x) = 2x^2 + \dfrac{256}{x}$

47. 608; 500; five days   **49.** c   **51.** a   **53.** $(-2.4, 2)$

2.3   EXERCISES   (page 115)

1.

x	2.5	2.9	2.99	2.999	3.001	3.01	3.1	3.5
$f(x)$	-1	-0.2	-0.02	-0.002	0.002	0.02	0.2	1

Limit: 0

3.

x	-2.5	-2.1	-2.01	-2.001	-1.999	-1.99	-1.9	-1.5
$f(x)$	2	1.1111	1.0101	1.0010	0.9990	0.9901	0.9091	0.6667

Limit: 1

5. 5   **7.** -1 **9.** 0   **11.** $\frac{-3}{2}$ **13.** Does not exist   **15.** 3   **17.** Does not exist
19. $15\frac{3}{4}$ **21.** $\frac{-1}{8}$ **23.** 4   **25.** 13   **27.** 0   **29** 5
31. a. 0   **b.** -1 **c.** Does not exist   **d.** Does not exist   **33. a.** 5   **b.** 3   **c.** 1   **d.** 0

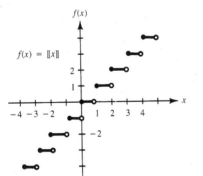

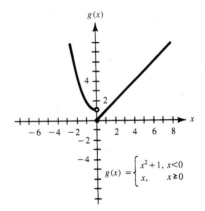

$$g(x) = \begin{cases} x^2 + 1, & x < 0 \\ x, & x \geq 0 \end{cases}$$

35. a. 1   **b.** -1 **c.** Does not exist   **d.** 1

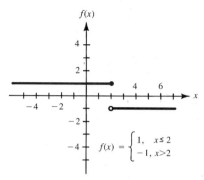

$$f(x) = \begin{cases} 1, & x \le 2 \\ -1, & x > 2 \end{cases}$$

37. 6   **39.** $-\frac{1}{2}$ **41.** -1 **43.** $\frac{1}{6}$ **45.** 0   **47.** 1   **49.** $\frac{3}{2}$

51. a.

x	-0.5	-0.1	-0.01	-0.001	-0.0001	0.0001	0.001	0.01	0.1	0.5
$f(x)$	-3.5	-3.1	-3.01	-3.001	-3.0001	-2.9999	-2.999	-2.99	-2.9	-2.5

   b. -3 **c.** Checks   **53. a.** 300   **b.** 200
55. a. \$5527   **b.** \$26,522   **c.** \$12,107   **d.** \$12,107   **57.** 0

2.4   EXERCISES   (page 125)

1. Continuous   **3.** 0   **5.** -2 and 2   **7.** 1
9.

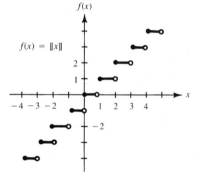

$$f(x) = [\![x]\!]$$

11.

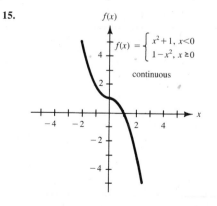

$$g(x) = \begin{cases} x^2 + 1, & x < 0 \\ x, & x \ge 0 \end{cases}$$

13.

$$f(x) = \begin{cases} -2x, & x \le -1 \\ x, & x > -1 \end{cases}$$

discontinuous
at $x = -1$

15.

$$f(x) = \begin{cases} x^2 + 1, & x < 0 \\ 1 - x^2, & x \ge 0 \end{cases}$$

continuous

17.

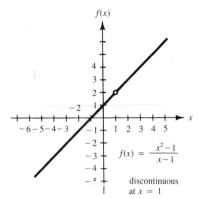

$$f(x) = \frac{x^2 - 1}{x - 1}$$

discontinuous
at $x = 1$

19. Yes **21.** No **23.** Yes **25.** $a = 1$

27. Continuous at $x = -2$ **29.** $P(x) = \begin{cases} 29, & 0 < x \le 1 \\ 23[\![x]\!] + 29, & x > 1 \end{cases}$. Function is not continuous at $x = 1$, $x = 2$, or $x = 3$.
Function is continuous at $x = 6.5$.

31. The limit does not exist as $r \to n$ and the function is not defined.

2.5 EXERCISES (page 137)

1. $x = 0$, $\displaystyle\lim_{x \to 0^-} f(x) = -\infty$, $\displaystyle\lim_{x \to 0^+} f(x) = +\infty$ **3.** $x = -3$, $\displaystyle\lim_{x \to -3^-} \frac{1}{x + 3} = -\infty$, $\displaystyle\lim_{x \to -3^+} \frac{1}{x + 3} = +\infty$

5. $x = -1$, $\displaystyle\lim_{x \to -1^-} \frac{-2}{x + 1} = +\infty$, $\displaystyle\lim_{x \to -1^+} \frac{-2}{x + 1} = -\infty$

7. $x = 0$ and $x = -3$, $\displaystyle\lim_{x \to 0^-} \frac{1}{x^2 + 3x} = -\infty$, $\displaystyle\lim_{x \to 0^+} \frac{1}{x^2 + 3x} = +\infty$, $\displaystyle\lim_{x \to -3^-} \frac{1}{x^2 + 3x} = +\infty$, $\displaystyle\lim_{x \to -3^+} \frac{1}{x^2 + 3x} = -\infty$

9. $y = 0$ **11.** $y = 1$ **13.** $y = \frac{1}{2}$ **15.** $y = 0$ **17.** $y = \frac{1}{4}$

19.

x	0	0.1	0.5	0.9	0.99	0.999	0.9999
y	0	-0.2020	-1.333	-9.474	-99.497	-999.5	-9999.5

21. 60,000 **23.** 10 **25. a.** 366 **b.** 544 **27.** 9200

29.

x	5	10	100	200	1000
$f(x)$	10.2	20.1	200.01	400.005	2000.001
$g(x)$	10	20	200	400	2000

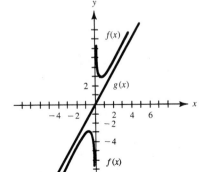

31.

x	5	10	100	200	1000
$f(x)$	12.5	22.22222	202.0202	402.01005	2002.002
$g(x)$	12	22	202	402	2002

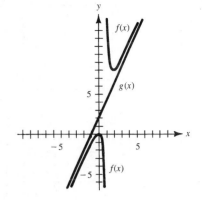

2.5A   EXERCISES   (page 146)

1. Horizontal asymptote: $y = 0$;
vertical asymptote: $x = 5$

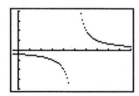

3. Horizontal asymptote: $y = 0$;
vertical asymptotes: $x = \pm 1$

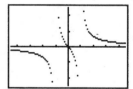

5. Horizontal asymptote: $y = 0$;
vertical asymptote: $x = 1$

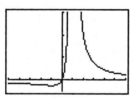

7. Horizontal asymptote: $y = 0$;
vertical asymptote: none

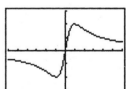

9. Horizontal asymptote: $y = 3$; vertical asymptote: none

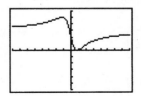

11. No. The left-side and right-side limits do not agree.   **13.** Yes, it is $\frac{1}{2}$.

15. a.

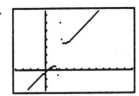

b.

c.

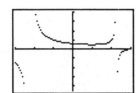

17. The graph is equivalent to the graph of $y = \dfrac{x-4}{x+3}$ except for a puncture hole at $x = 3$.

19. a. The choice of WINDOW values misses both vertical asymptotes, $x = \pm\sqrt{5}$.

b.
```
MINDOW FORMAT
Xmin=-3
Xmax=3
Xscl=1
Ymin=-4.5
Ymax=4
Yscl=1
```

2.6 EXERCISES (page 155)

1. 6   **3.** 2   **5.** −4   **7.** −1   **9.** 7   **11.** 3   **13.** $y = 2x + 1$ **15.** $y = -6x - 1$

17. $y = 6x + 4$ **19.** 2   **21.** 10   **23.** −1   **25.** 29   **27.** 6   **29.** 2   **31.** −1   **33.** 1

35. $-\frac{1}{4}$ **37. a.** −30   **b.** −40   **39.** 16   **41. a.** −180   **b.** −200

43. a. \$2.5 million   **b.** Approx. \$833,333   **c.** Slope is approx. $\frac{1}{2}$

d. Increasing for $0 < t < 4$; decreasing for $t > 4$ **45. a.** 5 lb   **b.** 10 lb   **47.** −10, decreasing

49. a. 240 ft   **b.** 224 ft   **c.** 176 ft   **d.** −32 ft/sec   **e.** −32 ft/sec   **51.** 1.6   **53.** 3

2.6A EXERCISES (page 165)

1. 3   **3.** 7   **5.** 3   **7.** $y = 3x - 4$ **9.** $y = -4x + 16$ **11.** $y = 5x - 2$ **13.** 0   **15.** 4

17. Because the function is not defined at $x = 0$

19. a.
```
MINDOW FORMAT
Xmin=-3.5
Xmax=3.5
Xscl=1
Ymin=-.5
Ymax=3.5
Yscl=1
```

b. The tangent line is vertical; its slope is undefined.

c. The secant lines are not good approximations to the tangent line for such "steep" curves.

2.7   EXERCISES   (page 174)

1. 3   **3.** -1 **5.** $-2x$ **7.** $2x + 1$ **9.** $4x + 3$ **11.** $3x^2$ **13.** **a.** 6   **b.** 18   **c.** -6
15. **a.** 1   **b.** 7   **c.** 9   **17.** $y = 8x - 4$ **19.** $y = 5x - 15$ **21.** $y = -x$

23. **a.** $f'(x) = -\dfrac{1}{x^2}$ **b.** Because $f(0)$ does not exist   **25.** **a.** $y' = m$

27. **a.**

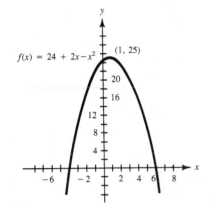

$f(x) = 24 + 2x - x^2$

b. $(1, 25)$ **c.** $x > 1$ **d.** $x < 1$

29. **a.** $y = -2x + 5$ **b.** $y = \frac{1}{2}x + \frac{5}{2}$ **31.** **a.** 1   **b.** 1   **c.** $\frac{-3}{2}$ **d.** 0
33. **a.** $1.6x$ **b.** 8   **c.** 920   **35.** 20   **37.** $\frac{5}{6}$ ft
39. Increasing at a decreasing rate since the slopes of the tangent lines are decreasing as t increases
41. 0   **43.** The function is increasing so $f'(x) > 0$.

CHAPTER 2 REVIEW EXERCISES   (page 181)

1. **a.** $x \geq 1$ **b.** $x \neq -4$ **2.** **a.** -1 **b.** 9   **3.**

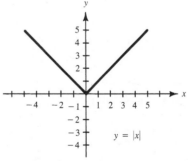

$y = |x|$

4. -35 **5.** 0   **6.** 2   **7.** 0   **8.** Does not exist   **9.** Yes   **10.** No   **11.** Yes   **12.** No
13. **a.** $x = 3$ **b.** $y = 1$ **14.** **a.** $x = 3$ **b.** $y = 0$ **15.** **a.** $x = \pm 3$ **b.** $y = 0$
16. **a.** $x = \pm\sqrt{2}$ **b.** $y = 2$ **17.** $\$160{,}000$ **18.** **a.** 2   **b.** $y = 2x + 5$ **19.** 15   **20.** -4
21. **a.** -9 **b.** -6 **22.** **a.** $\$100$ **b.** -1.4 **c.** $3 - 0.04x$ **d.** -1.8

23. $1200

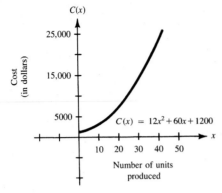

C(x)

Cost (in dollars)

25,000

15,000

5000

$C(x) = 12x^2 + 60x + 1200$

10 20 30 40 50

Number of units produced

CHAPTER 2 TEST (page 183)

1. b **2.** a **3.** d **4.** c **5.** c **6.** b **7.** d **8.** b **9.** b **10.** c **11.** 13
12. $2x + 4 + h$ **13. a.** $-4x - 2h$ **b.** $-4x$ **14.** 15 **15.** $(\frac{3}{2}, -\frac{9}{4})$

16.

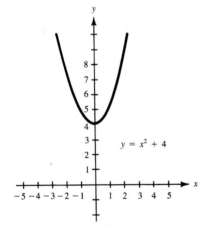

y

8
7
6
5
4
3
2
1

$y = x^2 + 4$

$-5 -4 -3 -2 -1$ 1 2 3 4 5

x

17. $5450 **18. a.** $0.02x$ **b.** 2 **19.** 2100 **20.** -464 ft/sec

---

3.1 EXERCISES (page 191)

1. $2x$ **3.** 0 **5.** 1 **7.** -6 **9.** $6t$ **11.** $6x + 9$ **13.** $\dfrac{3}{2\sqrt{x}}$ **15.** $\dfrac{1}{2\sqrt{x}}$ **17.** $\dfrac{1}{3\sqrt[3]{x^2}}$

19. $\dfrac{1}{7} + \dfrac{10}{7}x$ **21.** $\dfrac{-2}{x^3} - \dfrac{3}{x^4}$ **23.** $x + x^2$ **25.** $-3x^{-5/2} - 2x^{-1/2}$ **27.** $8x - 28$ **29.** $\dfrac{2}{3\sqrt[3]{x}} + \dfrac{5}{2\sqrt{x^3}}$

31. -6 **33.** -2 **35.** -1 **37.** $\dfrac{-1}{48}$ **39.** 5 **41.** $y = 2x - 1$ **43.** $y = -3x - 1$

45. a. $8x + 4$ **b.** 1 **c.** 4 **d.** $\dfrac{-1}{2}$ **e.** 0

47. a.

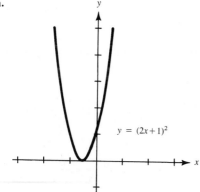

$y = (2x+1)^2$

b. $y = 0$

49. $f'(x) = \dfrac{2}{3}x^{-1/3}$ or $\dfrac{2}{3\sqrt[3]{x}}$

x	-1	-0.1	-0.01	-0.0001	0.0001	0.01	0.1	1
$f(x)$	1	0.21544	0.046416	0.01	0.01	0.046416	0.21544	1
$f'(x)$	-0.667	-1.4363	-3.09439	-6.66667	6.66667	3.09439	1.4363	0.667

a.

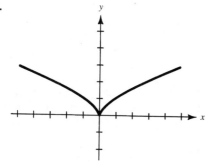

b.   Undefined   **c.** $x = 0$

51. a. $400x^{-1/3}$ **b.** 40 **53.** $\$14$ **55.** -120

57. a. $v(t) = -32t + 128$ **b.** 128 ft/sec   **c.** 32 ft/sec

---

3.2   EXERCISES   (page 201)

1. $12x^3 - 2x$ **3.** $-4x^3$ **5.** $\dfrac{2-x}{x^3}$ **7.** $8t^3 - 3t^2 + 10t - 1$ **9.** $\dfrac{-6}{x^4}$ **11.** $\dfrac{2}{(x+1)^2}$

13. $\dfrac{-5x^2 - 2x - 5}{(x^2-1)^2}$ **15.** $\dfrac{8x - x^4}{(x^3+4)^2}$ **17.** $\dfrac{1-x}{2\sqrt{x}(x+1)^2}$ **19.** $\dfrac{2x}{(x+1)^3}$ **21.** -2

23. a. $0.6x^2 + 1200x$ **b.** $126{,}000$ **25.** $\dfrac{dp}{dx} = \dfrac{-40}{(2x+1)^2}$ **27. a.** $\dfrac{32 - 8x^2}{(4+x^2)^2}$ **b.** 0

29. $\dfrac{-2x^3 - 1500x^2 - 499}{(x^3 + x + 1)^2}$

3.3 EXERCISES (page 209)

1. $\dfrac{dy}{du} = 2u, \dfrac{du}{dx} = 2, \dfrac{dy}{dx} = 8x - 20$ **3.** $\dfrac{dy}{du} = 3, \dfrac{du}{dx} = \dfrac{-1}{x^2}, \dfrac{dy}{dx} = \dfrac{-3}{x^2}$ **5.** $\dfrac{-1}{2\sqrt{x}(\sqrt{x}+1)^2}$ **7.** $6x + 24x^3 + 12x^5$

9. $\dfrac{1}{x^3}$ **11.** $10(x+2)^9$ **13.** $60x^3(x^4-1)^2$ **15.** $\dfrac{-1}{\sqrt[3]{(5-3x)^2}}$ **17.** $-2x(3x+1)^4(21x+2)$

19. $(x+1)^2(2x-3)^3(14x-1)$ **21.** $\dfrac{2(3x+1)}{(1-x)^5}$ **23.** $\dfrac{-5x}{\sqrt{(x^2+1)^3}}$ **25.** $\dfrac{-3(3x+4)}{4\sqrt[4]{(x+1)^5}}$ **27.** $\dfrac{8(x-1)^3}{(x+1)^5}$

29. $\dfrac{-2}{3}$ **31.** -12 **33.** 0 **35. a.** $\dfrac{1}{6t}$ **b.** $\dfrac{\sqrt{2t}}{\sqrt{2t+1}}$ **c.** $\dfrac{-t^2}{(2-t)^2}$ **37.** -5.568 **39.** -2

41. a. \$40 **b.** \$30 **c.** 300 lawnmowers **43. a.** 2450 **b.** 12.5 **45. a.** $\dfrac{-200}{\sqrt{101-x}}$ **b.** -40

3.4 EXERCISES (page 215)

1. $12x^2 + 4$ **3.** $\dfrac{3}{\sqrt{(2x+1)^5}}$ **5.** $108(3x-2)^2$ **7.** 80 **9.** -20 **11.** $\dfrac{3}{128}$ **13.** $108(3x+1)^2$

15. $\dfrac{120}{x^6}$ **17.** $\dfrac{4}{(x-2)^3}$ **19.** 0 **21.** $\dfrac{105}{16\sqrt{x^9}}$ **23.** $\dfrac{4(3-x)}{9\sqrt[3]{(x-1)^7}}$ **25.** $x = 0$ **27.** $x = 3$

3.5 EXERCISES (page 225)

1. a. $C'(x) = 40 - x$ **b.** \$10. The increase in cost for one additional pencil sharpener at a production level of 30 is \$10.
c. \$9.50 **3. a.** $P'(x) = 0.06x^2 + 0.04$ **b.** \$1540. The increase in profit for 100 ($x = 1$) additional televisions at a production level of 500 is \$1540. **c.** \$1860
5. a. $R(x) = 40x - 0.05x^2$ **b.** $C'(x) = 20$ **c.** $P(x) = 20x - 0.05x^2 - 3500$ **d.** $P'(x) = 20 - 0.1x$

7. a. $P(201) - P(200) = \$4.98$ **b.** **c.** 325

x	$P(x)$	$P'(x)$
200	1600	5
250	1800	3
300	1900	1
325	1912.5	0
350	1900	-1

9. a. $\dfrac{200}{x^2} + 5.5$ **b.** $\dfrac{-200}{x^2} + 5.5$ **c.** $\dfrac{-400}{x^3}$ **11. a.** $2 - 6x + 3x^2$ **b.** $2 - 3x + x^2$ **c.** $-3 + 2x$

13. a. $8 - 2x$ **b.** \$4 **c.** $x = 4$ **15.** 25 **17.** $x = 3, x = 25$
19. a. 100 units or 600 units **b.** 400 units **c.** 300 units **d.** 100 to 600 units
21. a. 15 **b.** $E'(1) = 17, E'(2) = 16, E'(3) = 9, E'(4) = -4, E'(5) = -23, E'(6) = -48$ **c.** At approx. 3.7 hours
23. a. Change is minimal **b.** Very small (close to zero) **c.** Increases rapidly and then levels off
25. $N'(t) = \dfrac{550}{(t+2)^2}$ **27.** 25 **29. a.** 48 ft/sec **b.** 32 ft/sec **c.** 4 **d.** $0 < t < 4$ **e.** $4 < t < 10$
31. a. 4.75 ft/sec **b.** 6 ft/sec^2
33. a. -32 ft/sec represents velocity after 1 sec **b.** $\sqrt{30} \approx 5.48$ sec **c.** $-32\sqrt{30} \approx -175.3$ ft/sec
35. a. 2 sec **b.** 176 ft **c.** $2 + \sqrt{11} \approx 5.32$ sec **d.** 64 ft/sec

3.6   EXERCISES   (page 240)

1. $\dfrac{1}{3}$ **3.** $\dfrac{1}{y}$ **5.** $\dfrac{x}{4y}$ **7.** $\dfrac{-x^2}{y^2}$ **9.** $\dfrac{2xy^2 + 3}{12y^2 - 2x^2y}$ **11.** $\dfrac{4y^2 - 2y - 2xy}{x^2 + 2x - 8xy}$ **13.** $\dfrac{-y}{x}$ **15.** -1

17. For $x = 2, y = \pm 1$; for $y = 1, \dfrac{dy}{dx} = \dfrac{-1}{4}$; for $y = -1, \dfrac{dy}{dx} = \dfrac{1}{4}$

19. a. $\dfrac{-9x}{16y}$ **b.** $\dfrac{\pm\sqrt{3}}{4}$ **c.** $\sqrt{3}x + 4y - 8\sqrt{3} = 0$ **d.** $\dfrac{-81}{16y^3}$ **21.** -18 **23.** $\frac{3}{2}$ **25.** 0.124 ft/hr

27. $\dfrac{-p - 2}{x}$ **29.** Increasing at a rate of $1080 per day   **31.** $\dfrac{dC}{dp} = \dfrac{-p}{x}$

33. $\dfrac{dC}{dt} = 3$. The number of clients, C, increases at a rate of 3 per hour.   **35.** $\dfrac{dp}{dt} = -1.25$ lb/in.^{2} per sec

37. $\dfrac{dA}{dt} = 24\pi$ in.^{2}/min ≈ 75.4 in.^{2}/min

3.7   EXERCISES   (page 248)

1. $dy = -4\,dx$ **3.** $dy = (3x^2 + 1)\,dx$ **5.** $dy = \dfrac{dx}{(2 - x)^2}$ **7.** 5.1   **9.** 2.00833   **11.** 0.33148

13. a. 1.261   **b.** 1.2   **15. a.** -2.95 **b.** -3 **17. a.** -0.00249844 **b.** -0.0025

19. a. 0.002518876   **b.** 0.0025   **21.** $dy = (1 - 2x)\,dx$ **23.** $dy = \dfrac{1}{2\sqrt{x} - 3}\,dx$ **25.** 7.5 cm^3

27. -2.4π in.$^2 \approx -7.54$ in.^{2} **29.** $0.36\pi \approx 1.131$ cm^3 **31.** 2.4π cm$^2 \approx 7.54$ cm^2 **33.** $41.06
35. $-0.4\pi \approx -1.26$ mm^2 **37.** 3 in.^{2}

CHAPTER 3 REVIEW EXERCISES   (page 252)

1. $10x - 7$ **2.** $-3x^2$ **3.** $\dfrac{2 - 2x - x^2}{(x^2 + 2)^2}$ **4.** $\dfrac{x(2 - 3x^2)}{\sqrt{1 - x^2}}$ **5.** $2x + \dfrac{1}{2\sqrt{x^3}}$ **6.** $\dfrac{2x - 5}{6y}$

7. a. $\dfrac{-9}{4}$ **b.** $9x + 4y - 3 = 0$ **8.** 11

9. a. $\dfrac{dy}{du} = \dfrac{-1}{(u + 2)^2}, \dfrac{du}{dx} = \dfrac{1}{2\sqrt{x}}, \dfrac{dy}{dx} = \dfrac{-1}{2\sqrt{x}(\sqrt{x} + 2)^2}$ **b.** $y(x) = \dfrac{1}{\sqrt{x} + 2}, y'(x) = \dfrac{-1}{2\sqrt{x}(\sqrt{x} + 2)^2}$

10. $\dfrac{2x(x^2 + 3)}{(x^2 - 1)^3}$ **11.** 192   **12. a.** $1600   **b.** $C'(x) = 25 + 0.2x$ **c.** $35

13. a. $16x - 0.02x^2$ **b.** $16 - 0.04x$ **c.** $16x - 0.024x^2 - 1000$ **d.** $16 - 0.048x$ **14.** $-$16.80

15. -0.5 **16.** $\dfrac{1}{5\pi}$ ft/sec   **17. a.** -16 ft/sec   **b.** $t = \sqrt{2} \approx 1.4$ sec   **18.** 2.025

CHAPTER 3 TEST   (page 253)

1. a   **2.** d   **3.** c   **4.** b   **5.** b   **6.** a   **7.** d   **8.** b   **9.** b   **10.** a   **11.** $\dfrac{1}{27}$

12. $1 + \dfrac{2}{x^3}$ **13.** $x = \pm 1$ **14.** 0   **15. a.** $\dfrac{-1}{\sqrt[3]{x^4}}$ **b.** $\dfrac{4}{3\sqrt[3]{x^7}}$ **c.** $\dfrac{-28}{9\sqrt[3]{x^{10}}}$ **16.** $3750

17. a. $\dfrac{15}{\sqrt{10x - 100}}$ **b.** $\frac{1}{2}$ **18.** -0.5 mg/min   **19.** -240 (decreasing at a rate of 240)

20. a. 36 ft/sec   **b.** 50 ft/sec^2

4.1   **EXERCISES**   (page 267)

1. Increasing: $(-\infty, 2)$; decreasing: $(2, \infty)$ **3.** Increasing for all x **5.** Increasing for all x
7. Increasing: $(-\infty, 0)$; decreasing: $(0, \infty)$ **9.** Increasing for all x
11. Increasing for all x **13.** Increasing: never; decreasing: $(-\infty, 4)$, $(4, \infty)$
15. Increasing: never; decreasing: $(-2, \infty)$
17. $x = 2$ is a critical value; f is increasing for all x

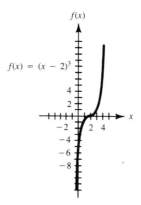

19. a. Increasing: $(-\infty, 0)$; decreasing: $(0, \infty)$ **21. a.** Increasing for all x

b.

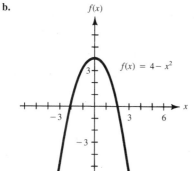

b.

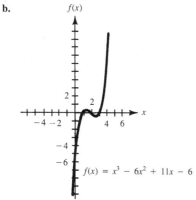

23. a. Increasing: $\left(-\infty, \dfrac{6 - \sqrt{3}}{3}\right)$ or $\left(\dfrac{6 + \sqrt{3}}{3}, \infty\right)$; decreasing: $\left(\dfrac{6 - \sqrt{3}}{3}, \dfrac{6 + \sqrt{3}}{3}\right)$

b.

25. a. Increasing: $(-\infty, -1)$ or $(1, \infty)$;
decreasing: $(-1, 1)$

b.

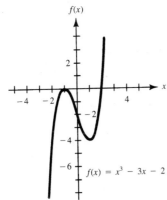

$f(x) = x^3 - 3x - 2$

29. a. $\dfrac{2}{(3-x)^2}$ **b.** No

c.

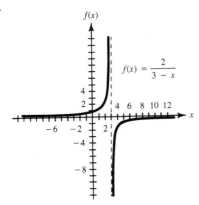

$f(x) = \dfrac{2}{3-x}$

27. a. Increasing: $(\frac{4}{3}, 2)$;
decreasing: $(-\infty, \frac{4}{3})$ or $(2, \infty)$

b.

$f(x) = (x - 2)^2 (1 - x)$

31. g is the function and f is the derivative.
33. f is the function and g is the derivative.
35. The function is decreasing so $f'(x) < 0$.

37. Undefined   **39.** $x = 6$; $30,000   **41. a.** $C'(x) = 0.8x - 1.6$ **b.** Increasing: $x > 2$; decreasing: $0 < x < 2$
43. a. $P(x) = -0.5x^2 + 500x - 500$ **b.** Increasing: $0 < x < 500$; decreasing: $x > 500$ **45.** Between 1970 and 1972

4.1A EXERCISES (page 277)

1. Increasing: $x < 0$;
decreasing: $x > 0$

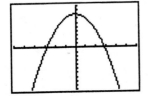

3. Increasing: $-2 < x < 0$ and $x > 1$;
decreasing: $x < -2$ and $0 < x < 1$

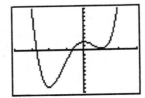

5. Increasing: $-\sqrt{1.5} < x < 0$ and $x > \sqrt{1.5}$; decreasing: $x < -\sqrt{1.5}$ and $0 < x < \sqrt{1.5}$

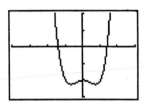

7. Increasing: for all x; decreasing: nowhere

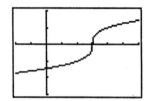

9. Increasing: nowhere; decreasing: $x > -1$

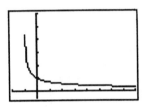

11. **a.** $f'(x) = \dfrac{-3}{(x-4)^2}$ **b.** No

c.

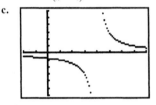

13. g is the function; f is the derivative

15. f is the function; g is the derivative

4.2 EXERCISES (page 282)

1. Concave up for all x **3.** Concave down for all x **5.** Concave up for all x
7. Concave up: $(-1, \infty)$; concave down: $(-\infty, -1)$ **9.** Concave up for all x
11. Concave up: $(-\infty, -1)$ or $(0, \infty)$; concave down: $(-1, 0)$ **13.** $(-\frac{1}{3}, \frac{131}{27})$ **15.** $(0, 0)$ **17.** $(0, 0)$
19. $(-\frac{7}{3}, \frac{128}{27})$ **21.** **a.** Increasing: $(-1, \infty)$; decreasing: $(-\infty, -1)$ **b.** Concave up for all x **c.** None
23. **a.** Increasing: $(-\infty, -4)$ or $(2, \infty)$; decreasing: $(-4, 2)$
b. Concave up: $(-1, \infty)$; concave down: $(-\infty, -1)$ **c.** $(-1, \frac{29}{3})$
25. **a.** Increasing: $(0, \infty)$; decreasing: $(-\infty, 0)$ **b.** Concave down for all x **c.** None
27. It is neither concave up nor concave down for all x.

29.

$$y = \frac{2x}{x^2 - 1}$$

$(0, 0)$ is a point of inflection

31. **a.** $x = 2$ **b.** $8000

33. $C''(x) = \dfrac{-1}{4x^{3/2}}$; $C''(x)$ is negative for all $x > 0$; thus, $C(x)$ is concave down for $x > 0$.
35. **a.** $0 < x < \frac{2}{3}$ **b.** $x > \frac{2}{3}$

4.3   EXERCISES   (page 292)

1. Local minimum: $(1, -1)$ **3.** Local maximum: $(0, 0)$; local minimum: $(\frac{4}{3}, -\frac{32}{27})$ **5.** Local minimum: $(2, -14)$
7. Local maximum: $(0, 2)$; local minima: $(-\sqrt{3}, -7)$, $(\sqrt{3}, -7)$ **9.** Local minimum: $(-3, 0)$
11. Local maximum: $(-1, \frac{1}{2})$; local minimum: $(1, -\frac{1}{2})$ **13.** Local maximum: $(0, 0)$; local minimum: $(2, 4)$
15. Local maximum: $(-2, \frac{50}{3})$; local minimum: $(3, -25)$ **17.** Local maximum: $(-5, 164.5)$; local minimum: $(\frac{1}{2}, -1.875)$
19. Local maximum: $(-2, 64)$; local minimum: $(2, -64)$

21. **a.** $20x - 0.2x^2$ **b.**   **c.** 50   **d.** $10   **e.** $500

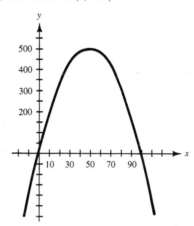

23. $11,000   **25.** $\frac{1}{2}$ **27.** A maximum revenue of $64 occurs when two units are sold at a price of $32 per unit.
29. $60,000   **31.** In about 6 years, 8 months   **33.** 64 ft   **35.** At $t = 1$ sec   **37.** f_6 **39.** f_5 **41.** f_3

4.4   EXERCISES   (page 307)

1. 4   **3.** 1   **5.** 18   **7.** 205   **9.** -9 **11.** $\frac{-1}{4}$ **13.** 0   **15.** $\frac{-9}{4}$
17. Absolute minimum: $f(5) = -49$; absolute maximum: $f(2) = 5$
19. Absolute minimum: $f(3) = -9.5$; absolute maximum: $f(-2) = 11\frac{1}{3}$
21. Absolute minimum: $f(0) = 0$; absolute maximum: $f(-1) = f(1) = 9$
23. Absolute minimum: $f(3) = -135$; absolute maximum: $f(-\sqrt{2}) = f(\sqrt{2}) = 12$
25. Absolute minimum: $g(-2) = -16$; absolute maximum: $g(3) = 14$
27. Absolute minimum: $g(-3) = -46$; absolute maximum: $g(4) = 38$
29. There is no absolute minimum or absolute maximum because of the vertical asymptote at $x = 2$. That is,
$$\lim_{x \to 2^-} f(x) = -\infty$$
$$\lim_{x \to 2^+} f(x) = +\infty$$

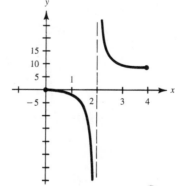

31. **a.** A weekly promotional budget of $10,000 produces a maximum profit of $9000.   **b.** $8888.54   **33.** 3

4.4A EXERCISES (page 313)

1. Absolute minimum of -49 at $x = 5$; absolute maximum of 21 at $x = -2$
3. Absolute minimum of 1 at $x = 0$; absolute maximum of 5 at $x = 2$
5. Absolute minimum of -9.5 at $x = 3$; absolute maximum of 11.3333 at $x = -2$
7. Absolute minimum of 0 at $x = 0$ and $x = \pm 2$; absolute maximum of 12 at $x = \pm\sqrt{2}$
9. Absolute minimum of -31 at $t = -2$; absolute maximum of -3 at $t = 2$
11. Absolute minimum of -2 at $w = -1$; absolute maximum of 2.375 at $w = 1.5$
13. Absolute minimum of 0.875 at $w = -0.5$; absolute maximum of 7.625 at $w = 2.5$
15. 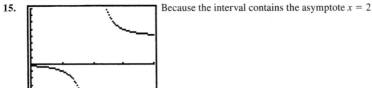 Because the interval contains the asymptote $x = 2$

4.5 EXERCISES (page 322)

1.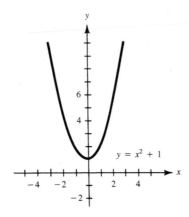

$y = x^2 + 1$

3.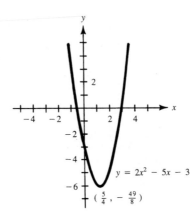

$y = 2x^2 - 5x - 3$

$\left(\frac{5}{4}, -\frac{49}{8}\right)$

5.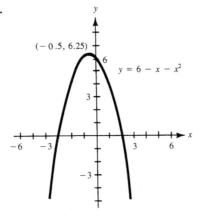

$(-0.5, 6.25)$

$y = 6 - x - x^2$

7.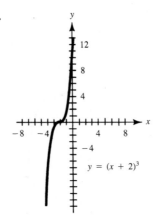

$y = (x + 2)^3$

9.

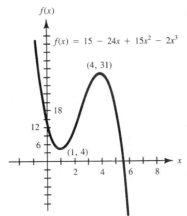

$f(x) = 15 - 24x + 15x^2 - 2x^3$

(4, 31)

18

12

6

(1, 4)

11.

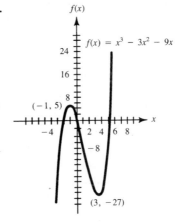

$f(x) = x^3 - 3x^2 - 9x$

24

16

8

(−1, 5)

−8

(3, −27)

13.

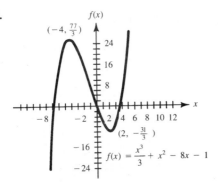

$(-4, \frac{77}{3})$

24

16

8

$(2, -\frac{31}{3})$

−16

$f(x) = \dfrac{x^3}{3} + x^2 - 8x - 1$

−24

15.

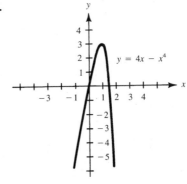

$y = 4x - x^4$

17.

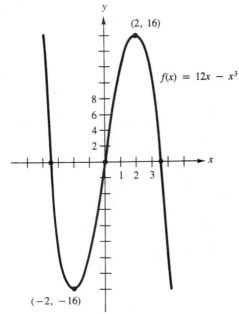

(2, 16)

$f(x) = 12x - x^3$

(−2, −16)

19.

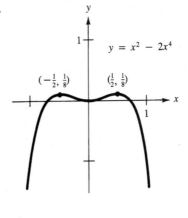

$y = x^2 - 2x^4$

$(-\frac{1}{2}, \frac{1}{8})$ $(\frac{1}{2}, \frac{1}{8})$

21. No   **23.** None   **25.** $f'(x) > 0$ for $x > 3.5$ and $-2 < x < 0$
$f'(x) < 0$ for $x < -2$ and $0 < x < 3.5$

27. Absolute minimum: -55; no absolute maximum

29. Symmetric with respect to y axis; intercept: $(0, 1)$; asymptotes: $x = 1, x = -1, y = 0$

31. No absolute extrema

33.

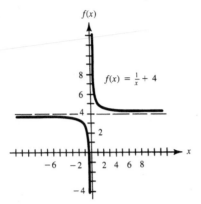

35.

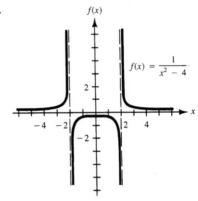

37.

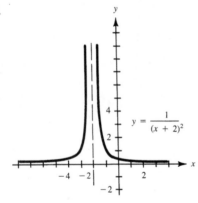

39.

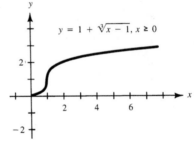

41.

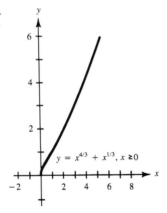

43. Maximum profit: $560

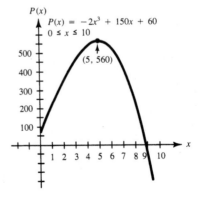

45. a. $P(x) = R(x) - C(x)$;
$P(x) = -8x^3 - 6x^2 + 36x$

b. Marginal cost = $C'(x) = 4x + 16$; marginal revenue = $R'(x) = -24x^2 - 8x + 52$

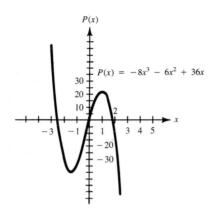

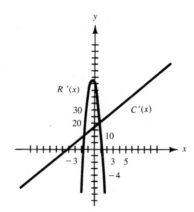

c. At $x = 1$, $R'(x) = C'(x)$. In general, if x^* is a relative maximum, then $P'(x^*) = 0 = R'(x^*) - C'(x^*)$. Thus, $R'(x^*) = C'(x^*)$.

4.6   EXERCISES   (page 336)

1. 80 and 80   **3.** -4 and -4 or 4 and 4   **5.** 4 and 12   **7.** 8 cm by 8 cm   **9.** $6\sqrt{2}$ in. by $6\sqrt{2}$ in.
11. 75 in.^{3} **13.** 400   **15.** $13\frac{2}{3}$ thousand $\approx 13{,}667$ **17.** 200   **19.** 250   **21.** $t = \sqrt{3} \approx 1.7$ hr
23. 190   **25.** Radius: $\sqrt[3]{32} \approx 3.17$ cm
Height: $2\sqrt[3]{32} \approx 6.35$ cm
27. Base: 5.04 ft by 5.04 ft; height: 6.3 ft $(4\sqrt[3]{2}$ by $4\sqrt[3]{2}$ by $5\sqrt[3]{2})$
29. Width of rectangle: $\dfrac{500}{\pi + 2}$ ft ≈ 97.25 ft; length of rectangle: 125 ft
31. Width: $\dfrac{4\sqrt{3}}{3}$ ft ≈ 2.31 ft; depth: $\dfrac{4\sqrt{6}}{3}$ ft ≈ 3.27 ft   **33.** $\dfrac{20\sqrt[3]{2}}{\sqrt[3]{2} + 1} \approx 11.15$ mi from A (or 8.85 mi from B)   **35.** 4°C

CHAPTER 4 REVIEW EXERCISES   (page 341)

1. $(c, f(c))$ **2.** $(d, f(d))$ and $(h, f(h))$ **3.** $a < x < d, h < x < b$ **4.** $d < x < h$ **5.** $f(a)$ **6.** $(e, f(e))$
7. a. Increasing: $\dfrac{-2\sqrt{3}}{3} < x < \dfrac{2\sqrt{3}}{3}$; decreasing: $x < \dfrac{-2\sqrt{3}}{3}$ and $x > \dfrac{2\sqrt{3}}{3}$
b. Concave up: $x < 0$; concave down: $x > 0$; point of inflection: $(0, 0)$
c. Local minimum: $\left(\dfrac{-2\sqrt{3}}{3}, \dfrac{-16\sqrt{3}}{9}\right)$; local maximum: $\left(\dfrac{2\sqrt{3}}{3}, \dfrac{16\sqrt{3}}{9}\right)$ **d.** Absolute extrema: none
e. Intercepts: $(-2, 0), (0, 0), (2, 0)$; symmetry: with respect to origin; asymptotes: none

f.

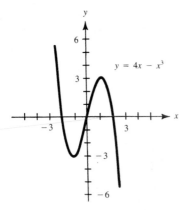

$y = 4x - x^3$

8. a. Increasing: $0 < x < \frac{8}{3}$; decreasing: $x < 0$ and $x > \frac{8}{3}$
 b. Concave up: $x < \frac{4}{3}$; concave down: $x > \frac{4}{3}$; point of inflection: $(\frac{4}{3}, \frac{128}{27})$
 c. Local minimum: $(0, 0)$; local maximum: $(\frac{8}{3}, \frac{256}{27})$ **d.** Absolute extrema: none
 e. Intercepts: $(0, 0)$, $(4, 0)$; symmetry: none; asymptotes: none   **f.**

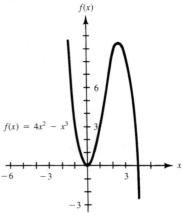

$f(x) = 4x^2 - x^3$

9. a. Increasing: $-\sqrt{5} < x < 0$ and $x > \sqrt{5}$; decreasing: $x < -\sqrt{5}$ and $0 < x < \sqrt{5}$
 b. Concave up: $x < \dfrac{-\sqrt{15}}{3}$ and $x > \dfrac{\sqrt{15}}{3}$; concave down: $\dfrac{-\sqrt{15}}{3} < x < \dfrac{\sqrt{15}}{3}$;

  points of inflection: $\left(\dfrac{-\sqrt{15}}{3}, \dfrac{-44}{9}\right)$, $\left(\dfrac{\sqrt{15}}{3}, \dfrac{-44}{9}\right)$

 c. Local maximum: $(0, 9)$; local minimum: $(-\sqrt{5}, -16)$, $(\sqrt{5}, -16)$ **d.** Absolute minimum: -16
 e. Intercepts: $(-3, 0)$, $(-1, 0)$, $(0, 9)$, $(1, 0)$, $(3, 0)$; symmetry: y axis; asymptotes: none

f.

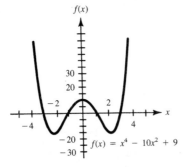

$f(x) = x^4 - 10x^2 + 9$

10. **a.** Increasing: $-\dfrac{\sqrt{10}}{2} < x < 0$ and $\dfrac{\sqrt{10}}{2} < x < 3$; decreasing: $-2 < x < \dfrac{-\sqrt{10}}{2}$ and $0 < x < \dfrac{\sqrt{10}}{2}$

   b. Concave up: $-2 < x < \dfrac{-\sqrt{30}}{6}$ and $\dfrac{\sqrt{30}}{6} < x < 3$; concave down: $\dfrac{-\sqrt{30}}{6} < x < \dfrac{\sqrt{30}}{6}$;

   points of inflection: $\left(\dfrac{-\sqrt{30}}{6}, \dfrac{19}{36}\right)$ and $\left(\dfrac{\sqrt{30}}{6}, \dfrac{19}{36}\right)$

   c. Local maximum: $(0, 4)$; local minimum: $\left(\dfrac{-\sqrt{10}}{2}, \dfrac{-9}{4}\right), \left(\dfrac{\sqrt{10}}{2}, \dfrac{-9}{4}\right)$

   d. Absolute maximum value: 40; absolute minimum value: $\dfrac{-9}{4}$

   e. Intercepts: $(-2, 0), (0, 4), (-1, 0), (2, 0), (1, 0)$; symmetry: none (because of the restriction on x); asymptotes: none

   f.

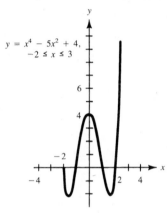

$y = x^4 - 5x^2 + 4,$
$-2 \le x \le 3$

11. **a.** Increasing for all x **b.** Concave up: $x < 1$; concave down: $x > 1$; point of inflection: $(1, 0)$
   c. Local extrema: none   **d.** Absolute extrema: none
   e. Intercepts: $(0, -1), (1, 0)$; symmetry: none; asymptotes: none   **f.**

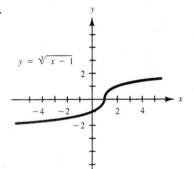

$y = \sqrt[3]{x - 1}$

12. **a.** Increasing: $x > 0$; decreasing: $x < 0$ **b.** Concave up: nowhere; concave down: all x; no points of inflection
   c. Local minimum: $(0, 0)$ **d.** Absolute minimum value: 0
   e. Intercepts: $(0, 0)$; symmetry: y axis; asymptotes: none   **f.**

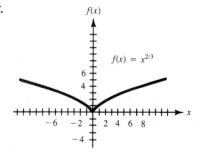

$f(x) = x^{2/3}$

13. **a.** Increasing: $x < \dfrac{-2\sqrt{15}}{5}$ and $x > \dfrac{2\sqrt{15}}{5}$; decreasing: $\dfrac{-2\sqrt{15}}{5} < x < \dfrac{2\sqrt{15}}{5}$

   b. Concave up: $\dfrac{-\sqrt{30}}{5} < x < 0$ and $x > \dfrac{\sqrt{30}}{5}$; concave down: $x < \dfrac{-\sqrt{30}}{5}$ and $0 < x < \dfrac{\sqrt{30}}{5}$;

   points of inflection: $\left(\dfrac{-\sqrt{30}}{5}, \dfrac{84\sqrt{30}}{125}\right)$, $(0, 0)$, $\left(\dfrac{\sqrt{30}}{5}, \dfrac{-84\sqrt{30}}{125}\right)$

   c. Local maximum: $\left(\dfrac{-2\sqrt{15}}{5}, \dfrac{192\sqrt{15}}{125}\right)$; local minimum: $\left(\dfrac{2\sqrt{15}}{5}, \dfrac{-192\sqrt{15}}{125}\right)$

   d. Absolute extrema: none   **e.** Intercepts: $(-2, 0)$, $(0, 0)$, $(2, 0)$; symmetry: origin; asymptotes: none

   f.

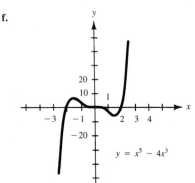

$$y = x^5 - 4x^3$$

14. **a.** Decreasing: for all $x > 0$ **b.** Concave up for all $x > 0$ **c.** Local extrema: none

   d. Absolute extrema: none   **e.** Intercepts: $(\sqrt[3]{4}, 0)$; symmetry: none; asymptotes: $x = 0$

   f.

$$y = \dfrac{4}{x^2} - x, \ x > 0$$

15. **a.** Increasing: $x < -4$, $-4 < x < 4$, $x > 4$; decreasing: nowhere

   b. Concave up: $x < -4$ and $0 < x < 4$; concave down: $-4 < x < 0$ and $x > 4$; point of inflection: $(0, 0)$

   c. Local extrema: none   **d.** Absolute extrema: none

   e. Intercepts: $(0, 0)$; symmetry: origin; asymptotes: $x = 4$, $x = -4$, $y = 0$

f.

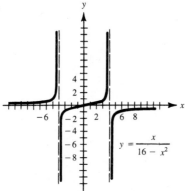

$$y = \frac{x}{16 - x^2}$$

16. **a.** Increasing for all x $(-1 < x < 1)$
  b. Concave up: $0 < x < 1$; concave down: $-1 < x < 0$; point of inflection: $(0, 0)$
  c. Local extrema: none   **d.** Absolute extrema: none
  e. Intercepts: $(0, 0)$; symmetry: origin; asymptotes: $x = \pm 1$ **f.**

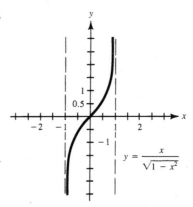

$$y = \frac{x}{\sqrt{1 - x^2}}$$

17. 20 m by 40 m   **18.** 12 and 36   **19.** **a.** $R(x) = 40\sqrt{x}$ **b.** 1600   **c.** \$400   **d.** \$1
20. **a.** Lowest: 2 mi; highest 0.7 mi
  b. Lowest: 2 mi; highest $\dfrac{254 - \sqrt{25,708}}{126} \approx 0.74$ mi

---

CHAPTER 4 TEST (page 343)

1. b   **2.** b   **3.** b   **4.** b   **5.** c   **6.** c   **7.** a   **8.** b   **9.**

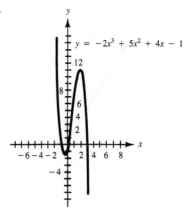

$$y = -2x^3 + 5x^2 + 4x - 1$$

10. **a.** $x < -\frac{2}{3}$ and $x > 4$ **b.** $-\frac{2}{3} < x < 4$ **c.** $x < \frac{5}{3}$ **d.** $x > \frac{5}{3}$ **e.** $(\frac{5}{3}, \frac{-610}{27})$ **11.** $4000

12. **a.** $R(x) = 75x - \dfrac{x^2}{4}$ **b.** $x < 150$ **c.** $x > 150$ **d.** $37.50

13. **a.** Increasing   **b.** $0 < s < 60$ **c.** $60   **14.** $x = 19$ maximizes revenue at $249   **15.** 1600 in.^{2}

5.1 EXERCISES (page 354)

1. Exponential growth

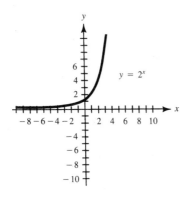

3. Exponential decay

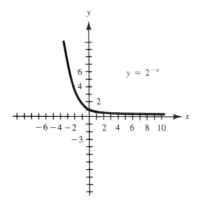

5. Exponential decay

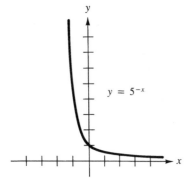

7. Neither growth nor decay

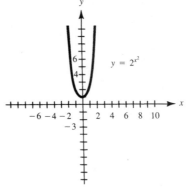

9. Exponential growth

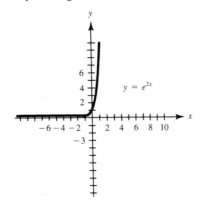

11. Exponential growth

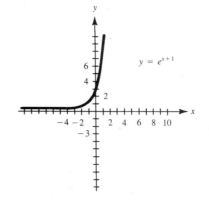

13. Exponential growth

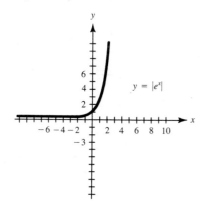

$y = |e^x|$

15. $x = 5$ **17.** $x = -2$ **19.** $x = 1$ **21.** $x = -1$
23. $x = 4$

25. a.

t	$(1.06)^t$
10	1.791
11	1.898
12	2.01

← about 12 years (11.89 years)

b.

t	$(1.12)^t$
5	1.762
6	1.974
7	2.211

← a little more than 6 years (6.12 years)

c.

t	$(1.10)^t$
3	1.331
4	1.464
5	1.610

← between 4 and 5 years (4.25 years)

27. $2442.74   **29. a.** $2442.81   **31. a.** $1.10669 \approx \$1.11$ **b.** Approx. 10.669%   **c.** Approx. 10.515%

33. $125,971   **35. a.**   **b.** $2000   **c.** $843.75

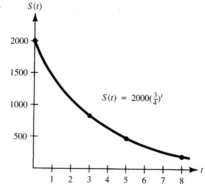

$S(t) = 2000(\frac{3}{4})^t$

37. 15.83 roentgens   **39. a.** 23.2 cm   **b.** 69.5 cm

5.1A   EXERCISES   (page 365)

1.

```
WINDOW FORMAT
Xmin=-2
Xmax=4
Xscl=1
Ymin=-.5
Ymax=16
Yscl=2
```

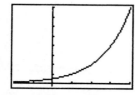

3.

```
WINDOW FORMAT
Xmin=-3
Xmax=3
Xscl=1
Ymin=-.5
Ymax=16
Yscl=2
```

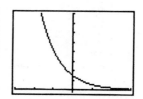

5.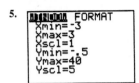

7.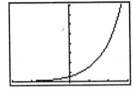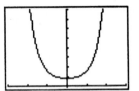

9.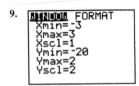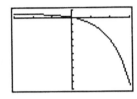

11. $x = 0, x \approx -0.57$ 13. $x = 0, x = -0.5$

15. $x = 0$ 17. $x \approx -1.69, x = 2$ 19. $x = 4$

21.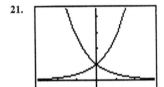

23.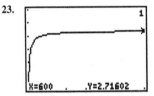

25.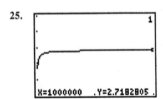

27.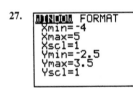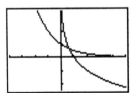

29.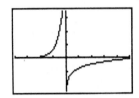

31. $-\infty < x \leq 0$

5.2   **EXERCISES**   (page 374)

1. $\log_4 16 = 2$ **3.** $\log_5 \dfrac{1}{125} = -3$ **5.** $\log 0.01 = -2$ **7.** $3^2 = 9$ **9.** $10^2 = 100$ **11.** $e^{-1} = \dfrac{1}{e}$

13. $x = -15$ **15.** $x = 3$ **17.** $x = 3$

19.

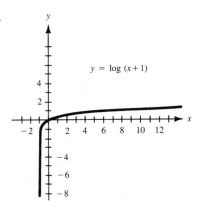

$y = \log_2 x$

21.

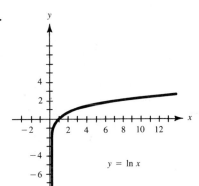

$y = \ln x$

23.

$y = \log (x + 1)$

25. 6.7940   **27.** 5.4600   **29.** $\log_b x + 2 \log_b y$ **31.** $\log_b x - \log_b (y + z)$ **33.** $\frac{1}{3} \log_b (x + y)$

35. -2.7480704 **37.** $x = 0.3203$ **39.** $x = 1.8495$ **41.** $x = \pm 1.5174$ **43.** $x = -1.0292$

45.   **a.** $f[g(x)] = \ln e^x = x$
$\qquad g[f(x)] = e^{\ln x} = x$

b.

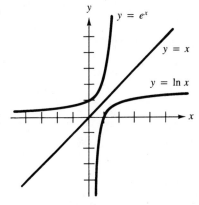

$y = e^x$
$y = x$
$y = \ln x$

47. a. 2500 units   **b.** 3049 units   **c.** 3261 units   **49.** $t \approx 6.8$ yr   **51. a.** \$3000   **b.** 1 month

53. a. $R(x) = \dfrac{300x}{\ln(x+4)}$ **b.** $R(20) \approx \$1887.95$ **55. a.** 70   **b.** 109

57. a. $K \approx 4.9 \times 10^{-6}$ **b.** $N(t) = \dfrac{5}{1 + 4e^{-0.245t}}$

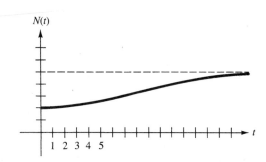

59. a. 6   **b.** $A = (1.5849 \times 10^7)A_0$ **61. a.** 11,140 yr   **b.** 7363 yr

---

5.2A EXERCISES (page 384)

1.

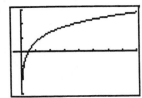

3.

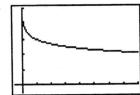

5.

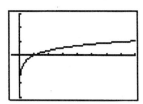

7.

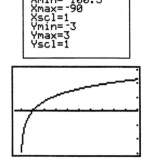

9. 19.08   **11.** 1

13. a.

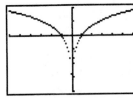

Domain: $x \neq 0$

b.

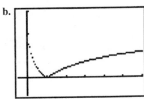

Domain: $x > 0$;   Range: $y \geq 0$

15.

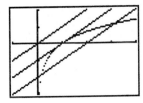

17. a. $x > 0$ and $x \neq 1$ **b.** $y = e$, $x > 0$ and $x \neq 1$
   c. Because the exponent's denominator is 0. The calculator erroneously drew in the point.

5.3   EXERCISES   (page 391)

1. $\dfrac{dy}{dx} = \dfrac{3}{3x - 1}$ **3.** $\dfrac{dy}{dx} = \dfrac{1}{2x}$ **5.** $\dfrac{dy}{dx} = \dfrac{3(\ln x)^2}{x}$ **7.** $\dfrac{dy}{dx} = \dfrac{2}{2x + 1} - \dfrac{1}{x - 1}$ or $\dfrac{dy}{dx} = \dfrac{-3}{(2x + 1)(x - 1)}$

9. $\dfrac{dy}{dx} = x + 2x \ln x$ **11.** $\dfrac{dy}{dx} = \dfrac{-1 + \ln x}{(\ln x)^2}$ **13.** $\dfrac{dy}{dx} = \dfrac{\ln 3x - \ln 2x}{x(\ln 3x)^2}$ **15.** $\dfrac{dy}{dx} = \dfrac{-2}{x(\ln x)^3}$

17. $\dfrac{dy}{dx} = \dfrac{x + 1 - x \ln x}{x(x + 1)^2}$ **19.** $f'(x) = \dfrac{1}{x \ln 10}$ **21.** $f'(x) = \dfrac{1}{2 \ln 10}(1 + \ln x)$ **23.** $\dfrac{d^2y}{dx^2} = \dfrac{-4}{(2x + 1)^2}$

25.

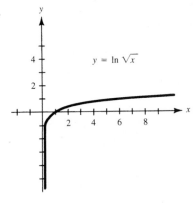

$y = \ln \sqrt{x}$

27.

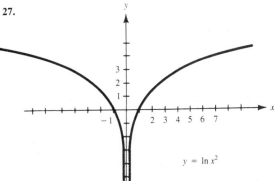

$y = \ln x^2$

29.

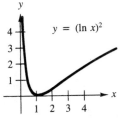

$y = (\ln x)^2$

31.

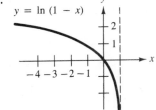

$y = \ln (1 - x)$

33. $\dfrac{dp}{dx} = \dfrac{-10}{x}$ **35. a.** 49 **b.** \$580.40 **37. a.** $S'(2) = 20, S'(4) = 12$ **b.** Decreasing rate

39. $N'(30) = \frac{30}{31} + \ln 31 \approx 4.4017$; therefore, the number of respondents is increasing by approximately 440.

41.

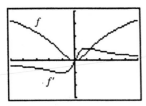

43.

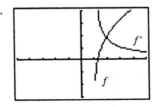

___

5.4 EXERCISES (page 399)

1. $\dfrac{dy}{dx} = 3e^{3x}$ **3.** $\dfrac{dy}{dx} = 72e^{6x}$ **5.** $\dfrac{dy}{dx} = -2xe^{1-x^2}$ **7.** $\dfrac{dy}{dx} = e^x(x+1)$ **9.** $\dfrac{dy}{dx} = \dfrac{e^x}{(e^x+1)^2}$

11. $\dfrac{dy}{dx} = xe^{-x}(2-x)$ **13.** $\dfrac{dy}{dx} = 1$ **15.** $\dfrac{dy}{dx} = 2x$ **17.** $\dfrac{dy}{dx} = 1$ **19.** $\dfrac{d^2y}{dx^2} = 16e^{4x}$

21. $\dfrac{d^2y}{dx^2} = 4e^{2x} + 2e^x$

23.

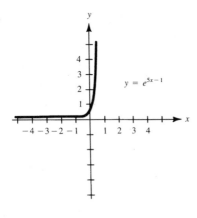

25.

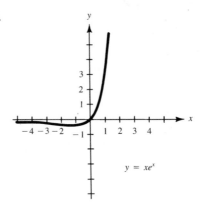

27.

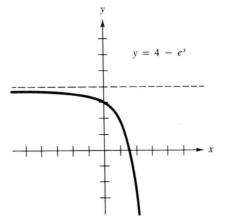

29. $\dfrac{dy}{dx} = \dfrac{3^x(x \ln 3 + \ln 3 - 1)}{(x + 1)^2}$ **31.** $\dfrac{dy}{dx} = 2^{-x^2}(-2x \ln 2)$ **33.** $\dfrac{dA}{dt} = Pre^{rt}$ **35.** **a.** 6.69   **b.** 1.49

37. **a.** 10.79   **b.** Increasing   **c.** Yes   **39.** 0.37   **41.** 2 hr   **43.** **a.** 130   **b.** 67   **c.** −0.25

5.5 EXERCISES (page 407)

1. **a.** $-\frac{1}{3}$; inelastic   **b.** −1; unit elasticity   **c.** −3; elastic

3. **a.** $\frac{-2}{35}$; inelastic   **b.** $\frac{-2}{3}$; inelastic   **c.** −1.6; elastic   **5.** **a.** −2; elastic   **b.** −2; elastic

7. **a.** −1; demand decreases 5%   **b.** $\frac{-1}{3}$; demand decreases $1\frac{2}{3}$%   **c.** −3; demand decreases 15%

9. **a.** $\frac{-8}{21}$; demand decreases approximately 1.9%   **b.** $\frac{-32}{9}$; demand decreases approximately 17.8%

11. **a.** −2; demand decreases 10%   **b.** $\frac{-1}{2}$; demand decreases 2.5%   **13.** Demand will decrease by 4%

15. Demand will decrease by approximately 13.3%.   **17.** **a.** $350 < p < 700$ **b.** $0 < p < 350$

19. **a.** $2700 - 6p$ **b.** $0 < p < \$450$ **c.** $\$450 < p < \$900$ **d.** $\$450 < p < \$900$ **e.** $0 < p < \$450$

21. **a.** $\dfrac{p}{p - \dfrac{a}{b}}$ **b.**

p	$E(p)$
$\dfrac{a}{4b}$	$-\dfrac{1}{3}$
$\dfrac{a}{3b}$	$-\dfrac{1}{2}$
$\dfrac{a}{2b}$	-1
$\dfrac{2a}{3b}$	-2
$\dfrac{10a}{11b}$	-10

23. **a.** $1200 < p < 2400$ **b.** $0 < p < 1200$ **c.** $p = 1200$

CHAPTER 5 REVIEW EXERCISES (page 409)

1.

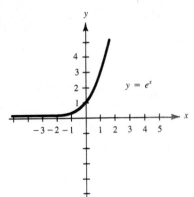

$y = e^x$

2.

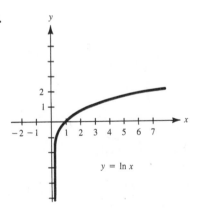

$y = \ln x$

3. $\log_5 25 = 2$ **4.** $\log_{10} 0.01 = -2$ **5.** $\log_6 \frac{1}{36} = -2$ **6.** $\ln 1 = 0$ **7.** $10^2 = 100$ **8.** $e^{-1} = \dfrac{1}{e}$

9. $b^0 = 1$ **10.** $10^{-3} = 0.001$ **11.** **a.** 4   **b.** $\dfrac{120}{e} \approx 44.15$

12. $R(x) = 80xe^{-x/2}$; $x = 2$ maximizes revenue at $\dfrac{160}{e} \approx \$58.86$

13. a.

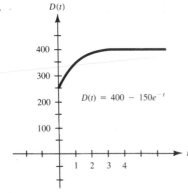

b. $D'(2) = 150e^{-2} \approx 20.3$
$D'(4) = 150e^{-4} \approx 2.75$
$D'(10) = 150e^{-10} \approx 0.007$

14. 11.9 yr **15.** $\dfrac{dy}{dx} = -e^{-x}$ **16.** $\dfrac{dy}{dx} = \dfrac{2}{2x+1}$ **17.** $\dfrac{dy}{dx} = \dfrac{-2x}{4-x^2}$ **18.** $\dfrac{dy}{dx} = -2xe^{-x^2}$

19. $\dfrac{dy}{dx} = x^2e^x(x+3)$ **20.** $\dfrac{dy}{dx} = \dfrac{1 + \dfrac{1}{x} - \ln x}{(x+1)^2} = \dfrac{x+1-x\ln x}{x(x+1)^2}$ **21.** $\dfrac{dy}{dx} = \dfrac{e^x - e^{-x}}{2}$ **22.** $\dfrac{dy}{dx} = \dfrac{x}{x^2+1}$

23. a. $R = 2700p - 6p^2$ **b.** $\dfrac{p}{p-450}$ **c.** $\$225 < p < \$450$ **d.** $0 < p < \$225$

24. a. $\dfrac{2p^2}{p^2 - 300}$ **b.** $-\dfrac{2}{11}$ **c.** -1

CHAPTER 5 TEST (page 411)

1. c **2.** d **3.** c **4.** a **5.** b **6.** b **7.** d **8.** a **9.** c **10.** b **11.** 0.0623

12.

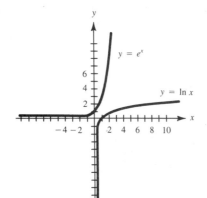

13. $2(xe^{x^2-1} + 5)$ **14.** $\dfrac{2x}{x^2+1}$ **15.** $\dfrac{1 - \ln x}{x^2}$ **16.** -1 **17.** 2 **18.** \$66,119.76

19. a. 11.04   **b.** 4.06   **20. a.** $R(x) = 1000xe^{-0.01x}$ **b.** $R'(x) = -10e^{-0.01x}(x - 100)$

6.1   EXERCISES   (page 422)

1. $5x + C$ **3.** $x^4 + C$ **5.** $\frac{2}{3}x^6 + C$ **7.** $x^3 - x^2 + 5x + C$ **9.** $y^6 - 10y^3 + \frac{7}{2}y^2 - 4y + C$

11. $\frac{3}{7}x^7 - 2x^2 + 7x + C$ **13.** $\ln |x| + C$ **15.** $5x - 3 \ln |x| + C$ **17.** $\frac{5}{4}x^4 - \frac{3}{4}x^2 + \frac{1}{4}x + C$

19. $\dfrac{-15}{x^2} + \dfrac{2}{x} - \ln |x| + C$ **21.** $\dfrac{x^2}{2} - e^x + C$ **23.** $\dfrac{3\sqrt[3]{x^4}}{4} - \dfrac{8\sqrt{x^3}}{3} + C$ **25.** $80\sqrt[4]{x} + C$ **27.** $\dfrac{3\sqrt[3]{x^4}}{4} + C$

29. $\frac{4}{3}x^3 + 2x^2 + x + C$ **31.** $f(x) = x^3 - 3x^2 + x - 10$ **33.** $f(x) = 2x^4 - \dfrac{x^3}{3} + \dfrac{x^2}{2} + 10$

35. $f(x) = 3 \ln |x| - 5x + 5$ **37. a.** $f'(2) = 2$ for all three curves (slope of tangent line is 2)

b.

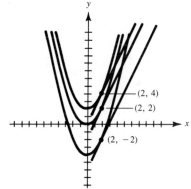

39. a. $C(x) = x^2 - 6x + 400$ **b.** $C(50) = \$2600$ **41.** $\$160,000$ **43.** $S(t) = 0.4t^2$

45. $P(x) = x^3 - 180x^2 + 10,800x - 215,200$

47. a. $N(t) = -500t^2 + 80,000$ **b.** Approximately 12.6 months

49. a. $P(t) = 2t + \frac{15}{4}t^{4/3} + C$ **b.** Approximately 3601   **51.** $s(t) = \frac{1}{6}t^4 + \frac{1}{3}t^3 + 5t$

53. a. $s(t) = -16t^2 + 160; v(t) = -32t$ **b.** $t = \sqrt{10} \approx 3.16$ sec   **c.** $-32\sqrt{10} \approx -101.2$ ft/sec

6.2   EXERCISES   (page 433)

1. $\frac{1}{6}(x + 3)^6 + C$ **3.** $\frac{1}{16}(4x - 7)^4 + C$ **5.** $\frac{1}{6}(x^3 + 1)^6 + C$ **7.** $\frac{2}{3}(x^5 - 1)^{3/2} + C$ **9.** $\frac{3}{4}(y^3 - 1)^{4/3} + C$

11. $\frac{2}{3}(x^3 - 1)^{1/2} + C$ **13.** $\frac{1}{3} \ln |x^3 - 1| + C$ **15.** $\frac{1}{3}e^{x^3} + C$ **17.** $\frac{1}{6} \ln |3t^2 - 1| + C$

19. $\frac{1}{2} \ln |x^2 - 4| + C$ **21.** $-\frac{1}{2}e^{-x^2} + C$ **23.** $\frac{5}{3}(2x - 1)^{3/2} + C$ **25.** $\ln |x^3 - 3x^2 + 5| + C$

27. $\frac{1}{3} \ln |x^3 - 5x^2 + 7x - 1| + C$ **29.** $-2.5e^{-0.2x^2} + C$ **31.** $0.4(1 + 0.5x)^5 + C$ **33.** $2e^{\sqrt{x}} + C$

35. $\frac{3}{5}(x^3 - 5x)^{5/3} + C$ **37.** $3 \ln |x^2 + 7x + 12| + C$ **39.** $\dfrac{1}{(1 - \sqrt{x})^2} + C$ **41.** $\dfrac{(\ln x)^2}{2} + C$

43. $\dfrac{(\ln 2x)^2}{2} + C$ **45.** $4x(2x^2 - 1)^7$ **47.** $x^2(x^3 + 1)^{1/2}$ **49.** $\dfrac{x}{1 - x^2}$

51. a. $\frac{2}{5}(x - 2)^{5/2} + \frac{4}{3}(x - 2)^{3/2} + C$ **b.** $\frac{2}{3}(x + 2)^{3/2} - 4(x + 2)^{1/2} + C$

   c. $x - 2 \ln |x + 2| + C$ **d.** $\dfrac{-1}{2(x - 2)^2} - \dfrac{2}{3(x - 2)^3} + C$ or $\dfrac{2 - 3x}{6(x - 2)^3} + C$

53. $\$3700$ **55.** $\$3031.42$ **57.** 772.9 g

6.3 EXERCISES (page 446)

1. 8.25 square units **3.** 21.5 square units **5.** 24 **7.** 64 **9.** 8 **11.** 12 **13.** 7 **15.** $\dfrac{e-1}{2}$

17. ln 3 **19.** 4 **21.** $\frac{1}{2}$ **23.** 2 **25.** $\frac{1}{3}$ **27.** $\frac{1}{2}(\ln 15 - \ln 3)$ or $\frac{1}{2}\ln 5$ **29.** $\frac{13}{3}$ **31.** $\frac{122}{5}$

33. 10 **35.** $\frac{1}{2}e^2 + e - \frac{3}{2}$ **37.** $\frac{-1}{12}$ **39.** $\frac{-8}{5}$

41. $\frac{11}{3}$

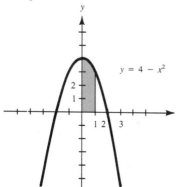

43. $\frac{1}{2}$

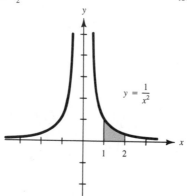

45. $\frac{125}{6}$

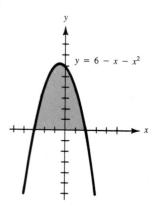

47. $\frac{2}{3}$

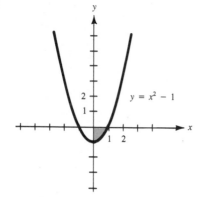

49. 20

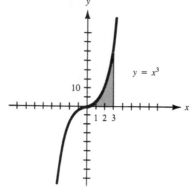

51. $\frac{1}{2}$

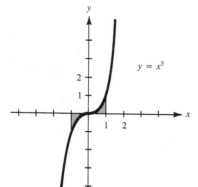

53. 3.19

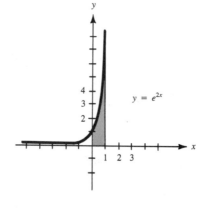

55. 4.61   **57.** 2

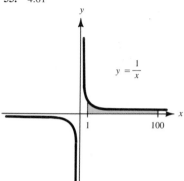

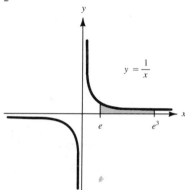

59. $\frac{14}{3}$ **61.** 4.433044   **63.** \$3816   **65.** \$235,000   **67.** Approximately 172   **69.** 140 ft

---

6.4   EXERCISES   (page 458)

1. 16   **3.** $\frac{20}{3}$ **5.** $\frac{32}{3}$ **7.** $\frac{4}{3}$

9. $\frac{9}{2}$

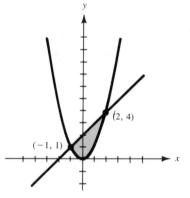

11. 36

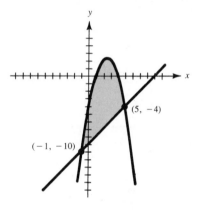

13. 9

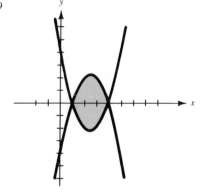

15. 0.49

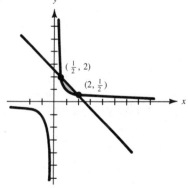

17. $\frac{15}{4}$ **19.** 4 **21.** $\frac{71}{6}$ **23.** 11.05

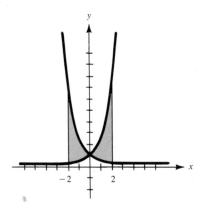

25. a. 6 **b.** $\frac{8}{3}$ **c.** $\frac{26}{3}$ **d.** Same
27. Equilibrium point: $(20, 30)$; consumer surplus: \$200; producer surplus: \$266.67
29. Equilibrium point: $(10, 480)$; consumer surplus: \$133.33; producer surplus: \$2400 **31.** 50 **33.** 40
35. 48 **37.** 9 **39.** \$438,844

6.5 EXERCISES (page 468)

1. 2 **3.** $\frac{5}{4}$ **5.** $\frac{2}{3}$ **7.** $\frac{4}{3}$ **9.** $\frac{37\pi}{3}$ **11.** $\frac{16\pi}{15}$ **13.** $\frac{\pi}{7}$ **15.** $\frac{15\pi}{2}$ **17.** 10.036 **19.** 2.178

21. 16π **23.** $\frac{\pi}{5}$ **25.** \$50,500 **27.** 410 **29.** 5.233

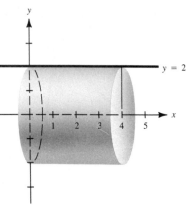

31. 123.2 lb

CHAPTER 6 REVIEW EXERCISES (page 471)

1. $x^3 - \frac{1}{2}x^2 + x + C$ **2.** $x^4 - 2x^2 + C$ **3.** $\frac{-2}{9}(1 - x^3)^{3/2} + C$ **4.** $\frac{-2}{3}(1 - x^3)^{1/2} + C$

5. $\frac{1}{12}(2x - 1)^6 + C$ **6.** $\frac{5}{12}(2x - 1)^{6/5} + C$ **7.** $\frac{1}{2}\ln|2x + 5| + C$ **8.** $\frac{1}{4}\ln(2x^2 + 5) + C$

9. $-\dfrac{2}{\sqrt{x}}+C$ **10.** $\dfrac{2}{5}x^{5/2}-2x^{1/2}+C$ **11.** $\dfrac{17}{6}$ **12.** 9   **13.** $\frac{32}{5}$ **14.** $\frac{32}{5}$ **15.** 13,400   **16.** 1

17. $4-2\sqrt{2}$ **18.** $\frac{1}{2}$ **19.** $\frac{8}{5}$ **20.** 4   **21.** 12   **22.** $\dfrac{1}{2}$ **23.** $\dfrac{1}{12}$ **24.** $\dfrac{1}{6}$ **25.** $\dfrac{4}{3}$ **26.** $\dfrac{1}{4}$

27. 40π **28.** $\dfrac{3\pi}{7}$ **29.** $f(x)=\dfrac{x^3}{3}+\dfrac{1}{x}$ **30.** $f(x)=-\dfrac{1}{3}(1-2x)^{3/2}+6$ **31.** $P(x)=600\sqrt[3]{x}-5$

32. \$4000   **33.** 3.6 parts per gallon   **34.** \$2666.67   **35.** \$513

CHAPTER 6 TEST (page 473)

1. b   **2.** c   **3.** d   **4.** b   **5.** a   **6.** d   **7.** c   **8.** a   **9.** d   **10.** b   **11.** $\dfrac{1}{18}(x^3+1)^6+C$

12. $f(x)=x^2+\dfrac{1}{x}$ **13.** Consumer surplus: \$450; producer surplus: \$1800   **14.** \$10,507.50

15. $3x+\frac{4}{3}x^3-\frac{1}{2}e^{-x}+\frac{11}{2}$ **16.** 1255

7.1 EXERCISES (page 481)

1. $\dfrac{x^3\ln x}{3}-\dfrac{x^3}{9}+C$ **3.** $\dfrac{1}{2}e^{x^2}+C$ **5.** $\dfrac{-xe^{-3x}}{3}-\dfrac{e^{-3x}}{9}+C$ **7.** $x^2e^x-2xe^x+2e^x+C$

9. $x\ln 2x-x+C$ **11.** $\dfrac{1}{5}xe^{5x}-\dfrac{1}{25}e^{5x}+C$ **13.** $-\dfrac{1}{2}xe^{-2x}-\dfrac{1}{4}e^{-2x}+C$ **15.** $\dfrac{1}{2}(\ln x)^2+C$

17. $\dfrac{1}{2}x\ln x-\dfrac{1}{2}x+C$ **19.** $2x\ln x-2x+C$ **21.** $\dfrac{x^2}{2}(\ln x^2-1)+C$ **23.** 0.39   **25.** 0.64   **27.** $\frac{326}{15}$

29. 1   **31.** 0.36   **33.** 0.39   **35.** 1   **37.** 0.74   **39.** $2x\ln(x+1)-2x+2\ln(x+1)+10$

41. $P(x)=500[\ln(x+1)]^2-100$ **43. a.** $S(t)=50t(\ln 2t-1)+465$ **b.** 910

7.2 EXERCISES (page 486)

1. $\dfrac{1}{3}\ln\left|\dfrac{x}{x+3}\right|+C$ **3.** $\dfrac{1}{6}\ln\left|\dfrac{x-3}{x+3}\right|+C$ **5.** $\dfrac{1}{2}\ln(x^2+1)+C$

7. $\dfrac{1}{2}(x\sqrt{x^2+9}+9\ln|x+\sqrt{x^2+9}|)+C$ **9.** $\sqrt{x^2+9}-3\ln\left|\dfrac{3+\sqrt{x^2+9}}{x}\right|+C$ **11.** $\dfrac{1}{6}\ln(2+3x^2)+C$

13. $\dfrac{1}{24}\ln\left|\dfrac{3+4x}{3-4x}\right|+C$ **15.** $\dfrac{-x}{2}-\dfrac{5}{4}\ln|5-2x|+C$ **17.** $\dfrac{1}{4}\left[\dfrac{-1}{5-2x}+\dfrac{5}{2(5-2x)^2}\right]+C$

19. $\dfrac{x^4\ln x}{4}-\dfrac{x^4}{16}+C$ **21.** $x^2e^x-2xe^x+2e^x+C$ **23.** $x-\dfrac{1}{2}\ln|1+e^{2x}|+C$

25. $\dfrac{-\sqrt{x+4}}{x}+\dfrac{1}{4}\ln\left|\dfrac{\sqrt{x+4}-2}{\sqrt{x+4}+2}\right|+C$ **27.** 0.718   **29.** 0.193   **31.** 1.099   **33.** 0.152   **35.** 0.718

37. $\dfrac{1}{6}\ln\left|\dfrac{1+x^3}{1-x^3}\right|+C$ **39.** $\dfrac{1}{2}x^2-2\ln|4+x^2|+C$ **41.** $\dfrac{1}{2}x^2-\dfrac{1}{2}\ln(1+e^{x^2})+C$ **43.** 5.056   **45.** 5.018

47. \$541,137   **49.** $R(x)=xe^{-x}$ **51.** 4.99 billion

7.3 EXERCISES (page 496)

1. a. 22   **b.** $\frac{64}{3}$ **c.** $\frac{64}{3}$ **3. a.** $\frac{21}{32}$ **b.** $\frac{2}{3}$ **c.** $\frac{2}{3}$ **5. a.** $\frac{85}{128}$ **b.** $\frac{2}{3}$ **c.** $\frac{2}{3}$

7. **a.** 1.7223 **b.** 1.7182889 **c.** 1.7182818 **9.** **a.** 1.4054 **b.** 1.3877 **c.** ln 4 ≈ 1.3863
11. **a.** 5.1463 **b.** 5.2522 **c.** $\frac{16}{3}$ ≈ 5.3333 **13.** **a.** 1.4907 **b.** 1.4637 **15.** **a.** 1.4820 **b.** 1.4987
17. **a.** 0.9068 **b.** 0.9097 **19.** 22 **21.** $176,400 **23.** 125 **25.** 12.02 **27.** Approximately 2
29. 108.83 amp. minutes **31.** $n = 5$ **33.** $n = 6$

7.3A EXERCISES (page 506)

1. ln 3 ≈ 1.098661 **3.** ln 3 ≈ 1.098612 **5.** $\frac{3}{4}$ ≈ 0.7668301 **7.** $\frac{16}{3}$ ≈ 5.25221 **9.** 1457.768
11. 11.82057 **13.** 2.541117 **15.** 15.78667 **17.** $y = -728x^2 + 1456x + 700$

7.4 EXERCISES (page 513)

1. 4 **3.** Diverges **5.** Diverges **7.** 1 **9.** Diverges **11.** Diverges **13.** 5 **15.** $\frac{1}{4}$
17. Diverges **19.** Diverges **21.** $\frac{1}{3}$ **23.** 1 **25.** Diverges **27.** Diverges
29. 2 **31.** Area ≈ 0.736 square unit

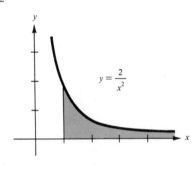

35. $187,500 **37.** Because the integral diverges, the number of pollutants increases without bound.

CHAPTER 7 REVIEW EXERCISES (page 516)

1. $x \ln 4x - x + C$ **2.** $\frac{x}{2}(\ln x - 1) + C$ **3.** $\frac{3}{4}\left[\frac{1}{1+2x} + \ln |1+2x|\right] + C$ **4.** $\frac{1}{24}\ln\left|\frac{3x-4}{3x+4}\right| + C$
5. $\frac{e^{4x}}{16}(4x-1) + C$ **6.** $-2e^{-x}(x+1) + C$ **7.** $x_i - \frac{1}{2}\ln|1 + e^{2x}| + C$
8. $\sqrt{x^2+9} - 3\ln\left|\frac{3+\sqrt{x^2+9}}{x}\right| + C$ **9.** 2.098 **10.** 2.944 **11.** 0.116 **12.** 7.733
13. **a.** 1.1032 **b.** 1.0987 **14.** **a.** 0.2207 **b.** 0.2005 **15.** **a.** 3.1383 **b.** 3.1908
16. **a.** 0.5191 **b.** 0.5177 **17.** 3 **18.** Diverges **19.** $\frac{1}{4}$ **20.** $\frac{1}{2}$ **21.** Diverges **22.** −1
23. 13.045 **24.** $R(x) = 1000xe^{-0.2x} + 4975e^{-0.2x} - 4975$

CHAPTER 7 TEST (page 518)

1. b **2.** a **3.** d **4.** b **5.** a **6.** c **7.** a **8.** $\frac{x^3}{3}\ln 3x - \frac{x^3}{9} + C$ **9.** 6.25
10. 4 **11.** $C(x) = 5e^x + xe^{-x} + e^{-x} + 794$

8.1 EXERCISES (page 526)

13. $y = 3x + C$ **15.** $y = x^2 + x + C$ **17.** $y = \frac{3}{2}x^4 + \frac{1}{2}x^2 - 2x + C$ **19.** $y = \frac{2}{3}\sqrt{x^3} + 2x + C$

21. $y = 2x^2 + C$ **23.** $y = \frac{1}{2}x^3 + C$ **25.** $y = \frac{4}{x} + C$ **27.** $y = x^3 + C_1 x + C_2$

29. $y = 2e^{2x} + C_1 x + C_2$ **31.** $y = 6x^2 + 2$ **33.** $y = 3 + 7x - 2x^2$ **35.** $y = \frac{1}{2}x^2 - \frac{1}{3}x^3 + \frac{5}{6}$

37. $y = \frac{2}{3}x^2 - 2$ **39.** $y = x + \frac{1}{x} - 1$ **41.** $y = -3 \ln |x| + 4$ **43.** $y = \frac{3}{4}\sqrt[3]{x^4}$ **45.** $y = x^2 + 3x + 5$

47. $C = 6x - 0.02x^2$ **49.** **a.** $y = 200e^{3t} + C$ **b.** $y = 200e^{3t}$ **c.** $6.538 \cdot 10^8$

8.2 EXERCISES (page 530)

1. $y^2 = 2x^2 + C$ **3.** $\frac{1}{y} = -x^2 + C$ **5.** $\frac{1}{y} = \frac{1}{x} + C$ **7.** $\ln |y| = 2x + C$ **9.** $3e^{-y} = -x^3 + C$

11. $\ln |y - 1| = x^2 + C$ **13.** $\frac{1}{y} = -x^2 + x + C$ **15.** $y^3 = 3x^2 + 27$ **17.** $\frac{1}{y} = \frac{-1}{x} + 2$

19. $y^2 = x^2 - 8x + 4$ **21.** $4 \ln |y| = x^2 - 4$ **23.** $y^2 = 4x^2 + C$ **25.** $2y^3 = -3e^{2x} + C$

27. $2e^{-2y} = -x^4 + C$ **29.** $2 \ln |y| = e^{x^2} + C$ **31.** $3y^2 = 2x^3 - 24x + 48$ **33.** $\ln |y + 3| = \frac{x^2 - 1}{2x^2}$

35. $2 \ln |y| = e^{x^2} + 1$ **37.** $37,829$ **39.** $y = a - \frac{1}{kt + C}$

8.3 EXERCISES (page 537)

1. $\frac{dA}{dt} = kA; A = Ce^{kt}$ **3.** $\$7288.48$ **5.** $\$33,333$ **7.** $y = \frac{20,000}{1 + 19e^{-1.27t}}$ **9.** $\frac{dx}{dt} = k(4000 - x)$

11. $28,051$ **13.** $\frac{dS}{dt} = kS^2$ **15.** $Q = 200e^{0.37t}$; approximately 1272 **17.** $\frac{dR}{dt} = kR; R = Ce^{kt}$

19. $T = 70 + 330e^{-0.18t}$ **21.** $\frac{dT}{dt} = k(70° - T); T = 70° + Ce^{-kt}$, where C represents the initial temperature of the object

CHAPTER 8 REVIEW EXERCISES (page 540)

1. $y = 4x + C$ **2.** $y = 3x - 2x^2 + C$ **3.** $y = \frac{5}{3}x^3 + \frac{3}{2}x^2 - 2x + C$ **4.** $y = \frac{8}{5}x^5 - \frac{1}{4}x^4 + 5x + C$

5. $y = \frac{6}{x} + C$ **6.** $y = 2x^2 + e^x + C$ **7.** $y^2 = x^2 + C$ **8.** $\frac{1}{y^2} = -x^2 + C$ **9.** $2e^{-y} = -x^2 + C$

10. $\frac{1}{y} = -e^x + C$ **11.** $y = 5x - \frac{3}{2}x^2 - 4$ **12.** $y = \frac{3}{4}x^2 + 4$ **13.** $y = x + \frac{2}{x} + 1$ **14.** $y^3 = 3x^2 + 729$

15. $\frac{1}{y} = \frac{3}{x} + 3$ **16.** $y^3 = x^2 + 3x + 512$ **17.** $\ln |y + 1| = \frac{-1}{x} - 2$ **18.** $2 \ln |y| = e^{x^2} - 1$

19. $A = 2000e^{0.1t}$ **20.** **a.** $\frac{dy}{dt} = k(50 - y)$ **b.** $y = 50 - 50e^{-kt}$ **c.** 45

CHAPTER 8 TEST (page 542)

1. b   **2.** d   **3.** d   **4.** b   **5.** b   **6.** $y = x^2 + e^{-x} + C$ **7.** $2y^3 = 3x^2 + 12x + C$
8. $y = \sqrt{x + 1} + 5$ **9.** $\$8580.03$ **10.** $Q = 4000e^{0.2t}$

9.1 EXERCISES (page 547)

1. 3 **3.** 16 **5.** 0 **7.** 50 **9.** 100 **11** $\sqrt{75} = 5\sqrt{3}$ **13.** 0 **15.** $\sqrt{96} = 4\sqrt{6}$ **17.** 54
19. 12 **21.** 0 **23.** 3 **25.** 25 **27.** -203 **29.** 88 **31. a.** \$40,200 **b.** \$40,800
33. a. 10 **b.** 80 **35.** 400,000 foot-pounds

9.2 EXERCISES (page 555)

1, 3, 5.

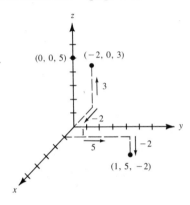

7.

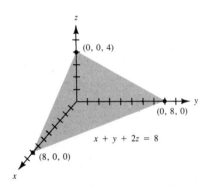

9.

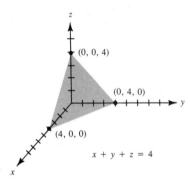

11.

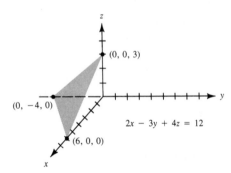

13.

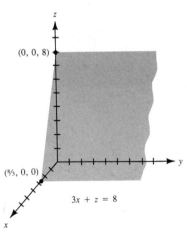

15.

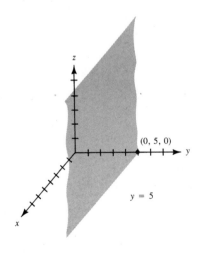

17. $\sqrt{13}$ **19.** 5   **21.**

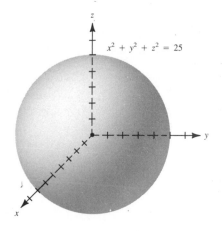

$x^2 + y^2 + z^2 = 25$

23.

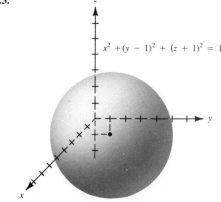

$x^2 + (y - 1)^2 + (z + 1)^2 = 16$

25.

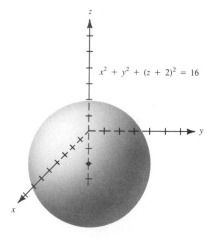

$x^2 + y^2 + (z + 2)^2 = 16$

27. A circle of radius 4 centered at $(2, 0, 0)$; $(x - 2)^2 + y^2 = 16$ **29.** A hyperbola whose center is $(0, 0, 0)$; $5x^2 - y^2 = 1$

9.3   **EXERCISES**   (page 561)

1. a. 2   **b.** 3   **3. a.** $1 + 12xy$ **b.** $-1 + 6x^2$ **5. a.** $2e^{(2x + 3y)}$ **b.** $3e^{(2x + 3y)}$

7. a. $2x\sqrt{y}$ **b.** $\dfrac{x^2}{2y^{1/2}}$ **9. a.** $\dfrac{4}{4x + 3y}$ **b.** $\dfrac{3}{4x + 3y}$ **11. a.** $\dfrac{2x}{x^2 + 2y^2}$ **b.** $\dfrac{4y}{x^2 + 2y^2}$ **13.** 16

15. $\tfrac{3}{5}$ **17.** $\tfrac{1}{7}$ **19.** 2   **21.** $\tfrac{3}{2}$ **23.** $10y^3 - 12$ **25.** $30xy^2$ **27.** $6e^{3x^2 + 2y}(6x^2 + 1)$ **29.** $12xe^{3x^2 + 2y}$

31. -32 **33.** -34 **35.** $f_{xx} = 0$, $f_{xy} = f_{yx} = 12ye^{y^2}$, $f_{yy} = 12xe^{y^2}(2y^2 + 1)$

37. $\dfrac{\partial^2 z}{\partial x^2} = 10y - ye^x$, $\dfrac{\partial^2 z}{\partial y\, \partial x} = \dfrac{\partial^2 z}{\partial x\, \partial y} = 10x - e^x$, $\dfrac{\partial^2 z}{\partial y^2} = 0$

39. a. \$59,000   **b.** $\dfrac{\partial p}{\partial x} = 6x$

   c. $\left.\dfrac{\partial p}{\partial x}\right|_{(3, 4)} = 18$. The profit increases by approximately \$18,000 if the number of gas stoves is held constant at 400 and the number of electric stoves is increased by 100 ($x = 1$).

41. a. $\dfrac{500y^{0.5}}{x^{0.5}}$ **b.** $\dfrac{500x^{0.5}}{y^{0.5}}$ **43. a.** $\dfrac{-375y^{0.75}}{x^{1.75}}$ **b.** $\dfrac{375}{x^{0.75}y^{0.25}}$

45. a. $\dfrac{\partial x}{\partial q} = 5$. The demand for the ranch model increases by approximately 5 units per dollar increase in selling price.

b. $\dfrac{\partial y}{\partial p} = 3$. The demand for colonial models increases by approximately 3 units per dollar increase in selling price.

47. a. 16   **b.** No; $\left.\dfrac{\partial s}{\partial x}\right|_{(3,\,5)} = 14$, so the rate of change in sales with respect to magazine advertising would be less.

49. a. $-2 - x$ **b.** $-3 - d$ **51.** $0.5a + 0.03$

9.4   EXERCISES (page 571)

1. Critical point: $(0, 2, -4)$; -4 is a relative minimum   **3.** Critical point: $(4, 2, 4)$; 4 is a relative minimum
5. Critical point: $(0, 2, -88)$; -88 is a relative maximum   **7.** Critical point: $(\frac{1}{2}, -1, 68)$; 68 is a relative maximum
9. Critical points: $(5, 3, -62)$ and $(5, -1, -30)$; $(5, 3, -62)$ is a saddle point; -30 is a relative maximum
11. Critical points: $(0, 0, 0)$ and $(\frac{2}{3}, \frac{1}{9}, \frac{4}{27})$; $(0, 0, 0)$ is a saddle point; $\frac{4}{27}$ is a relative maximum
13. a. $x = \frac{7}{2}, y = 2$ **b.** \$127,250   **17.** 150 mixers, 100 blenders; \$4500
19. $w = \sqrt[3]{150} \approx 5.31$ in.
  $l = \sqrt[3]{150} \approx 5.31$ in.
  $h = \dfrac{200}{\sqrt[3]{150^2}} \approx 7.08$ in.

9.5   EXERCISES (page 577)

1. $x = 1$ and $y = 4$ give a maximum of 8 for f **3.** $x = 2$ and $y = \frac{2}{3}$ give a minimum of 20 for f
5. $x = 2$ and $y = 18$ give a minimum of 20 for f **7.** $x = \frac{7}{8}$ and $y = \frac{49}{12}$ give a maximum of $2\frac{1}{24}$ for f
9. $x = 4$ and $y = 1$ give a maximum of 13 for f **11.** 62 and 62 yield a product of 3844   **13.** 400 cm^2
15. Maximum volume is 6000 in.^{3} with dimensions 10 in. by 20 in. by 30 in.   **17.** 4 in. by 4 in. by 4 in.
19. The radius should be 3.8 cm and the height should be 7.6 cm.   **21.** 50 ft by 100 ft
23. Base: 4 in. by 4 in.; height: 6 in.

9.6   EXERCISES (page 586)

1. $y = 2x + \frac{26}{3}$ **3.** $y = -\frac{12}{7}x + \frac{69}{7}$ **5.** $y = \frac{17}{8}x - \frac{8}{3}$ **7.** $y = \frac{6}{5}x + \frac{24}{5}$ **9.** $\frac{44}{3}$ **11.** 3   **13.** $\frac{121}{12}$
15. $\frac{84}{5}$ **17.** $y = \frac{9}{4}x + \frac{34}{3}$ **19.** $y = -\frac{9}{5}x + 10$ **21.** $y = 2.7x + 0.1$
23. a. $y = 6.25x + 16.5$ **b.** 41.5 (tens of billions of dollars)
25. a. $y = 0.12927x + 0.45244$ (where 1948 is $x = 0$)   **b.** About \$6.27   **c.** No
27. $y = -0.11439x + 10.6471$ **29.** $m = 1.4, b = -\frac{2}{15}$

CHAPTER 9 REVIEW EXERCISES (page 591)

1. 2   **2.**

3.

4. $\sqrt{13} \approx 3.6056$

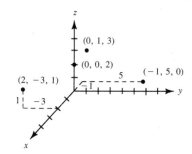

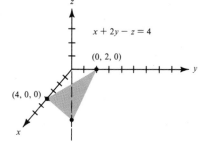

5.

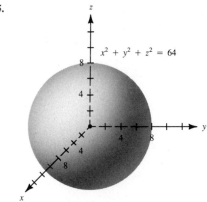

6. $ye^{xy} + 12x^2y^2$ **7.** $xe^{xy} + 8x^3y - 12y^3$ **8.** $y^2e^{xy} + 24xy^2$ **9.** $xye^{xy} + e^{xy} + 24x^2y$
10. $xye^{xy} + e^{xy} + 24x^2y$ **11.** $x^2e^{xy} + 8x^3 - 36y^2$ **12.** $(\frac{16}{5}, \frac{-14}{5}, \frac{-82}{5})$ is a relative minimum for f.
13. **a.** $x = 2, y = 3.5$ **b.** 127.25 (thousand dollars)   **14.** $x = 2$ and $y = 18$ give a maximum of 20 for f.
15. **a, e.**

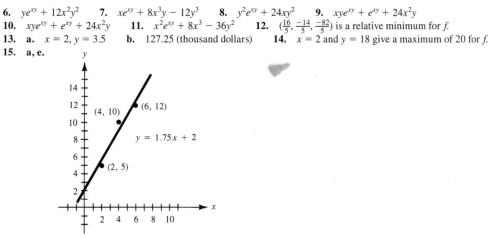

b. $(5 - 2m - b)^2 + (10 - 4m - b)^2 + (12 - 6m - b)^2$
c. $f_m = -244 + 112m + 24b, f_b = -54 + 24m + 6b$ **d.** $m = 1.75, b = 2$

CHAPTER 9 TEST   (page 592)

1. b   **2.** d   **3.** d   **4.** a   **5.** c   **6.** a

7.

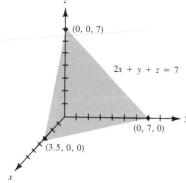

8.

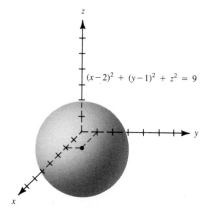

9. 3   **10.** $8x^3 - 10xy^3$ **11.** $-15x^2y^2 + 6y$ **12.** 0   **13.** $-30xy^2$ **14.** $-30x^2y + 6$ **15.** 10

16. $(-1, -1, 7)$ is a saddle point of f; $(3, 3, -25)$ is a relative minimum   **17.** 13

18. **a.** Approximately 826 units of the first (x) item and 195 units of the other (y) item
    b. Approximately \$2596.51

19. **a.** $f(m, b) = (-3 + 2m - b)^2 + (-1 + m - b)^2 + (2 - b)^2 + (-m - b)^2 + (2 - 2m - b)^2$
    b. $f_m(m, b) = -22 + 20m$ **c.** $f_b(m, b) = 10b$ **20.** $y = 1.1x$

GLOSSARY

Acceleration Given the equation of motion, $s = f(t)$, the acceleration is given by the second derivative, $\dfrac{d^2s}{dt^2} = f''(t)$.

Antiderivative A function F, whose derivative is f, is called an antiderivative of f provided $F'(x) = f(x)$ for all x in the domain of f.

Average value of a function $f_{av} = \dfrac{1}{b-a}\displaystyle\int_a^b f(x)dx$ is the average value for a function f continuous over the interval $[a, b]$.

Break-even point The point where revenue equals cost or profit is zero.

Chain Rule Given $y = f(u)$ and $u = g(x)$, then $\dfrac{dy}{dx} = \dfrac{dy}{du} \cdot \dfrac{du}{dx}$ provided f is differentiable at u and u is differentiable at x.

Circle Equation, with center (h, k) and radius r, is given by $(x - h)^2 + (y - k)^2 = r^2$. Area, A, is given by $A = \pi r^2$. Circumference, C, is given by $C = 2\pi r$.

Concavity A function f is concave up on an interval whenever $f''(x) > 0$ and concave down whenever $f''(x) < 0$.

Consumer surplus The amount saved by consumers who were willing to pay a price higher than the equilibrium price (point where supply and demand curves intersect).

Continuity A function f is continuous at $x = c$ if and only if the function is defined at c, $\lim\limits_{x \to c} f(x)$ exists, and $\lim\limits_{x \to c} f(x) = f(c)$.

Cost function $C(x)$ is the cost of producing x units. *See also* Marginal cost.

Critical values For a function f, $x = a$ is called a critical value if $f'(a) = 0$ or $f'(a)$ is undefined but $f(a)$ is defined.

Demand equation An equation that gives a relationship between the demand for an item and the price for the item.

Derivative The derivative of $y = f(x)$ at the point $(a, f(a))$ is given by $f'(a) = \lim\limits_{x \to a} \dfrac{f(x) - f(a)}{x - a}$ provided the limit exists. *See also* Tangent line.

Differential For a function $y = f(x)$, the differential of y is given by $dy = f'(x)dx$.

Distance formula The distance, d, between the points (x_1, y_1) and (x_2, y_2) is represented by $d = \sqrt{(x_2 - x_1)^2 + (y_2 - y_1)^2}$.

Elasticity of demand For the demand equation, $x = f(p)$, the elasticity of demand is given by $E(p) = \dfrac{p \cdot f'(p)}{f(p)}$. Inelastic state: $E(p) > -1$ Elastic state: $E(p) < -1$ Unit elasticity: $E(p) = -1$

Equilibrium point The point where the supply and demand curves intersect.

Exponential growth/decay $y = b^x$ exhibits exponential growth if $b > 1$; it exhibits exponential decay if $0 < b < 1$.

Function Of a single independent variable: "y is a function of x" means that for each

value of x there is associated a unique y value. This can be written as $y = f(x)$. Of two independent variables: "z is a function of x and y" means that for each ordered pair (x, y) there is associated a unique z value. This can be written as $z = f(x, y)$.
See also Inverse function.

Inflection point	Point where the concavity changes.
Integral	Indefinite: $\int f(x)dx = F(x) + C$, where F is an antiderivative of f. Definite: $\int_a^b f(x)dx = F(b) - F(a)$, where f is continuous over the closed interval $[a, b]$ and F is an antiderivative of f.
Inverse function	For the function $y = f(x)$, the inverse function $f^{-1}(x)$ has the property that $(f \circ f^{-1})(x) = (f^{-1} \circ f)(x) = x$. The domain and range of f^{-1} are, respectively, the range and domain of the function f.
Limit of a function	The limit, L, of the function f as x approaches c is written symbolically as $\lim_{x \to c} f(x) = L$. Left-hand limit: "x approaches c from below (or from the left)" is written $\lim_{x \to c^-}$. Right-hand limit: "x approaches c from above (or from the right)" is written $\lim_{x \to c^+}$.
Line	Point-slope form: $y - y_1 = m(x - x_1)$ Slope-intercept form: $y = mx + b$ See also Slope of a line.
Marginal cost	The derivative of the cost function.
Marginal profit	The derivative of the profit function.
Marginal revenue	The derivative of the revenue function.
Maximum value	Local maximum: $f(c)$ is a maximum value if there exists an interval around c such that $f(x) \le f(c)$ for all x in the interval. Absolute maximum: $f(c)$ is a maximum value provided $f(x) \le f(c)$ for all x in the domain.
Minimum value	Local minimum: $f(c)$ is a minimum value if there exists an interval around c such that $f(x) \ge f(c)$ for all x in the interval. Absolute minimum: $f(c)$ is a minimum value provided $f(x) \ge f(c)$ for all x in the domain.
Partial derivative	For $z = f(x, y)$ the first partial derivative of f with respect to x is the derivative of f considering y as a constant and x as a variable, denoted by $\dfrac{\partial f}{\partial x}$, $f_x(x, y)$, or f_x. The first partial derivative of f with respect to y is the derivative of f considering x as a constant and y as a variable, denoted by $\dfrac{\partial f}{\partial y}$, $f_y(x, y)$, or f_y.
Power Rule	If $f(x) = x^n$, then $f'(x) = nx^{n-1}$.
Producer surplus	Increased profit realized by those producers who would have been willing to sell the item at a price lower than the equilibrium price (the point where the supply and demand curves intersect).
Product Rule	If $f(x) = u(x)v(x)$ where u and v are differentiable at x, then $f'(x) = u(x)v'(x) + u'(x)v(x)$.

Profit function	$P(x) = R(x) - C(x)$; that is, profit is revenue less cost. *See also* Marginal profit.
Pythagorean Theorem	$a^2 + b^2 = c^2$

Quotient Rule	If $f(x) = \dfrac{u(x)}{v(x)}$, where u and v are differentiable, then $f'(x) = \dfrac{v(x)u'(x) - u(x)v'(x)}{[v(x)]^2}$.
Rate of change	Average rate of change over the interval $[a, b]$: $\dfrac{f(b) - f(a)}{b - a}$
	Instantaneous rate of change at $x = a$: $\lim\limits_{x \to a} \dfrac{f(x) - f(a)}{x - a}$
Revenue function	$R(x)$ is the total revenue (or income) for producing and selling x units. Revenue equals the number of units times the price per unit. *See also* Marginal revenue.
Simpson's Rule	The area under a continuous function f over an interval $[a, b]$ can be approximated by
	$$\frac{b - a}{3n}[f(x_0) + 4f(x_1) + 2f(x_2) + 4f(x_3) + \cdots + 4f(x_{n-1}) + f(x_n)]$$
Slope of a line	The slope, m, of the line joining the points (x_1, y_1) and (x_2, y_2) is given by
	$$m = \frac{y_2 - y_1}{x_2 - x_1}.$$
Supply equation	An equation that gives a relationship between the number of items supplied and the price for the item.
Tangent line	A line that touches a curve at a point $(a, f(a))$. The slope of the tangent line is given by
	$$m_{\text{tan}} = \lim\limits_{x \to a} \frac{f(x) - f(a)}{x - a} = f'(a)$$
Trapezoidal Rule	The area under a continuous function f over an interval $[a, b]$ can be approximated by
	$$\frac{b - a}{2n}[f(x_0) + 2f(x_1) + 2f(x_2) + \cdots + 2f(x_{n-1}) + f(x_n)]$$
Velocity	Given the equation of motion, $s = f(t)$, the (instantaneous) velocity is given by the first derivative, $\dfrac{ds}{dt} = f'(t)$.

INDEX

TO THE OWNER OF THIS BOOK:

We hope that you have found *Applied Calculus*, Fourth Edition, useful. So that this book can be improved in a future edition, would you take the time to complete this sheet and return it? Thank you.

School and address: _____

Department: _____

Instructor's name: _____

1. What I like most about this book is: _____

_____

_____

2. What I like least about this book is: _____

_____

_____

3. My general reaction to this book is: _____

_____

4. The name of the course in which I used this book is: _____

_____

5. Were all of the chapters of the book assigned for you to read? _____

   If not, which ones weren't? _____

6. In the space below, or on a separate sheet of paper, please write specific suggestions for improving this book and anything else you'd care to share about your experience in using the book.

_____

_____

_____

_____

_____

Optional:

Your name: _____ Date: _____

May Brooks/Cole quote you, either in promotion for *Applied Calculus*, Fourth Edition, or in future publishing ventures?

Yes: _____ No: _____

Sincerely,

Claudia Dunham Taylor
Lawrence Gilligan

FOLD HERE

BUSINESS REPLY MAIL
FIRST CLASS   PERMIT NO. 358   PACIFIC GROVE, CA

POSTAGE WILL BE PAID BY ADDRESSEE

ATT: *Claudia Dunham Taylor & Lawrence Gilligan*

Brooks/Cole Publishing Company
511 Forest Lodge Road
Pacific Grove, California 93950-9968

FOLD HERE

LIST OF APPLICATIONS

BUSINESS/ECONOMICS

Accounting
   appreciation
   budgeting
   depreciation
   income tax
Actuarial science
Advertising
Agribusiness
Assembly line
   learning curves
   production
Average cost
Average variable cost

Banking
   Banker's Rule
   continuous interest
   effective yields
   interest
Break-even analysis
Budgets

Car rental industry
Catering industry
Cobb-Douglas production formula
Commission sales
Construction industry
Consumer demand
Consumer price index
Consumer surplus
Container/package manufacturing
Cost functions

Data processing
Demand curve
Demand functions
Depreciation

Earnings
Economics
   consumer price index
   inflation
   mortgage rates
Elasticity of demand
Electronics industry
Equilibrium point

Factory production
Farm prices
Finance
   investment growth
   portfolio management
Fixed costs

Housing starts

Income
   from sales
   tax
Inflation
Insurance industry
Investment strategies

Labor
   employee earnings
   marginal productivity
   productivity
Learning curves

Management
Manufacturing
   package design
   production tolerances
Marginal analysis
Marginal average cost
Marginal cost
Marginal productivity
   of capital
   of labor
Marginal profit
Marginal revenue
Marketing/market research
Mortgage rates

Packaging
Partial marginal demand
Pricing
Producer surplus
Production
   average cost
   costs
   learning curves
   levels
   rates
Product life

Profit
Publishing industry

Quality control

Real estate
   mortgage rates
Revenue

Sales
   absolute ceiling on
   commission on
Shipping and receiving
Stock market
Supply and demand

Taxes
Total cost
Total profit
Training programs
Travel industry
Trust funds

Unemployment

Variable cost

Warehouse design

BIOLOGY/MEDICINE

Air pollution
Animal populations
Athletic performance

Bacteria growth/bacteriology
   unbounded growth model
Body fat
Birthrates
Blood sugar levels
Botany

Cancer death rates
Cell reproduction

DDT levels
Drug concentration in the blood
Drug effectiveness
Drug research
Drug therapy